VITICULTURE AND ENVIRONMENT

I0065368

Awarded

SPECIAL DISTINCTION IN VITICULTURE

by the Office International de la Vigne et du Vin

Paris 1994

Also by John Gladstones

Wine, Terroir and Climate Change

(Wakefield Press, 2011)

VITICULTURE
and
ENVIRONMENT

A study of the effects of environment on grapegrowing and wine qualities, with emphasis on present and future areas for growing winegrapes in Australia

JOHN GLADSTONES

TRIVINUM PRESS

First published in 1992 by Winetitles, Adelaide SA

Reprinted 1994, 1997, 1999

This revised edition published in 2016 by

Trivinum Press Pty Ltd
PO Box 7
Tanunda
South Australia 5352

www.winebookcellar.com.au

Copyright © John Gladstones, 1992 & 2016

All rights reserved. This book is copyright. Apart from any fair dealing for the
purposes of private study, research, criticism or review, as permitted under the
Copyright Act, no part may be reproduced without written permission.
Enquiries should be addressed to the publisher.

Front cover designed by Stacey Zass
Text typeset by Michael Deves, Trivinum
Printed by IngramSpark

ISBN 978 0 9945016 0 8 (paperback)

ISBN 978 0 9945016 1 5 (ebook)

National Library of Australia Cataloguing-in-Publication entry: (paperback)

Creator:	Gladstones, J.S. (John Sylvester), 1932– , author.
Title:	Viticulture and environment: a study of the effects of environment on grapegrowing and wine qualities, with emphasis on present and future areas for growing winegrapes in Australia / John Gladstones.
Edition:	Second edition.
ISBN:	978 0 9945016 0 8 (paperback)
Notes:	Includes bibliographical references and index.
Subjects:	Viticulture – Australia – Climatic factors.
	Vineyards – Australia – Location.
Dewey Number:	634.8

Contents

Foreword Bryan Coombe		viii
Preface to the First Edition		ix
Preface to the Second Edition		xi
1	**Introduction**	1
2	**History of Climate Selection for Australian Viticulture**	4
3	**Individual Climatic Factors**	8
	Ripening period mean temperature	8
	Mean temperatures and vegetative growth	11
	Mean temperatures and vine fruitfulness	12
	Soil temperature, and the roles and interactions of cytokinins	12
	Continentality	15
	Temperature variability	18
	Sunshine hours	20
	Light quality	26
	Rainfall and its seasonal distribution	27
	Relative humidity and saturation deficit	29
	Windiness	30
	Optimum combinations of climatic factors	31
	General discussion and conclusions	32
4	**Soil Type and Topography**	33
	Soil type	33
	Soil physical properties	34
	Soil inorganic chemistry	37
	The roles of soil organic matter	38
	Soil type: summary	39
	Topography and aspect	41
	Altitude	45
5	**Management Practices and their Interactions with Environment**	46
	Vine spacing, training, and management of the canopy – general considerations	46
	Height of heading	46
	Row orientation	47
	Vine and row spacing	48
	Trellising systems	48
	Canopy light relations, crop load and fruit quality	49
	Interaction between canopy light relations and temperature	51
	Supplementary watering	52
	Fertilizing and green manuring	53
	Summer mulch or bare surface?	58
	Other management practices	59
	The concept of vine balance	60
	Management versus environment	61

6   **Predicting Maturity Date**   62

     1. Adjustment for the vine temperature response curve   62

     2. Adjustment for latitude/day-length   64

     3. Adjustment for diurnal temperature range   64

7   **Maturity Groupings of Winegrape Varieties**   66

8   **Construction and Use of the Climate Tables**   69

     Notes on the individual columns of the climate tables   69

     Practical use of the tables   71

9   **Viticultural Environments of Western Australia**   72

10   **Viticultural Environments of South Australia**   104

11   **Viticultural Environments of Victoria**   124

12   **Viticultural Environments of New South Wales**   151

13   **Viticultural Environments of Queensland**   179

14   **Viticultural Environments of Tasmania**   183

15   **Viticultural Environments of New Zealand**   192

16   **Viticultural Environments of France**   197

17   **Viticultural Environments of Germany**   216

18   **Viticultural Environments of Central and Eastern Europe**   220

19   **Viticultural Environments of Italy**   229

20   **Viticultural Environments of Portugal, Madeira and Spain**   233

21   **Viticultural Environments of England**   242

22   **Viticultural Environments of the United States of America**   246

23   **Changes in Climate**   255

     Past climatic fluctuations, and their effects on viticulture   255

     Causes of natural decadal temperature fluctuations,
and their projection to future decades   260

     The greenhouse effect   260

     Have temperatures risen already?   262

     Grapevines as a thermometer   264

     Temperature forecast for the 21st century   266

     Direct effects of carbon dioxide concentration on vine growth,
grape quality and optimum temperatures   267

     Viticulture in foreseeable future climates   269

24   **Summary and Practical Recommendations for Australia**   272

     New South Wales and South-East Queensland (Figure 19)   272

     Victoria (Figure 20)   273

     Tasmania (Figure 21)   274

     South Australia (Figure 22)   275

     Western Australia (Figure 23)   275

     Research needs for Australia   276

     Vine breeding   277

APPENDIX 1

Plant Hormones: Their Functions and Interactions with Environment 279
    1. Auxins 279
    2. Gibberellins 279
    3. Cytokinins 280
    4. Abscisic acid 280
    Summary 280

APPENDIX 2

Sources and Treatment of the Climate Data 281
    Sources of climate data 281
    Periods of climatic records, and period adjustments 281

APPENDIX 3

Critique of the Papers of Jackson and Cherry (1988) and Moncur et al. (1989) 284
    Jackson and Cherry (1988) 284
    Moncur et al. (1989) 285

References 286

Indexes
    Authors 299
    Localities and appellations 301
    Grape varieties 305
    General index 306

FOREWORD
to the First Edition

The appearance of this book will be a surprise to many viticulturists. Attuned though they may be to the powerful 'grapevine' that dispenses viticultural gossip, few will have known that it was being written – indeed not many know John Gladstones. This might at first seem unusual for the writer of such an authoritative book. His preface shows that he is a West Australian agronomist whose major contribution has been the breeding of lupins. He is, of course, much more than this, as is attested by the numerous awards that have been bestowed upon him. His role in the establishment of viticulture in the Margaret River and Mount Barker regions of Western Australia is better known, but perhaps still not well enough appreciated. For our purposes, the most important considerations are his university lecturing – which has given him time and encouragement to think and ponder – and his early exposure to the mysteries of wine quality. From the interaction of these has evolved this book, largely written as a hobby!

I hope readers will reflect on the circumstances of the writing of this work and their significance for the future of viticultural science in Australia. Here is a scholar who has delved into the literature on an impressively broad range of botanical and agricultural science, who has been able to pursue his interests without having to fight for funds or give undue attention to short-term problems. Present trends in viticultural research, particularly within the universities, are creating conditions that run counter to those that John Gladstones enjoyed when compiling these pages.

No-one else could write such a volume. Although I have only seen the manuscript briefly, it was sufficient to remind me of the feelings I had, many decades ago, when I read *The Origin of Species* – the similar attention to detail, the precise but mesmerizing prose, and the gradual building towards broad conclusions new to the world. It is a work of significant scholarship that presents a new set of methods for considering the effects of climate on plant performance, in this case winegrape ripening, wine style and wine quality. Because the assessment of wine quality is subjective, and difficult to quantify, the author has reported anecdotes and opinions together with formal scientific data. I found myself agreeing with the flavour of most – but not all – of these views. And I wonder how we are going to be able to test them scientifically. But the suggestions that have been made will foster discussion, argument and experiments from which more knowledge will emerge.

This is a seminal work. I congratulate and thank John Gladstones.

Bryan Coombe
Reader in Horticultural Science The University of Adelaide

PREFACE
to the First Edition

The origins of this book go back many years. In 1955 Professor Harold Olmo, of the University of California, spent eight months in Western Australia studying problems of the Swan Valley vine industry. He was based at the University of Western Australia, Perth, where I was doing postgraduate work on lupins at the time. I came to know him a little, and to admire his infectious and kindly enthusiasm.

Olmo's report, published in 1956, suggested the cool southern areas of Western Australia as a place to produce light table wines of western European style. On the basis mainly of temperature analogies to California and Europe, he pointed especially to the Mount Barker-Rocky Gully area, just inland from Albany, as having the best prospects. This led some ten years later to the Western Australian Department of Agriculture establishing experimental vines at Forest Hill, under the supervision of the then Government Viticulturist Bill Jamieson, and to subsequent commercial plantings in the area.

Meanwhile a budding undergraduate interest in Swan Valley wines had been given further scope and encouragement through the kindness of the late Jack Mann, winemaker at Houghton Vineyard, who in the mid to late 1950s allowed me the use of about a hectare of vacant land on the vineyard for my lupin experiments. Happily the plot was convenient to the century-old cellars, with their haunting smell and magnetic invitation to learn more about the wines and their making.

In 1960 I joined the Faculty of Agriculture in the University of Western Australia, with responsibility for teaching horticulture and climatology, and later, crop physiology and crop ecology. It was thus a natural progression to take up the vine ecological studies where Olmo had left off.

Thinking about it, I came to suspect that the possibilities for cool-climate viticulture in southern Australia were much wider than Olmo's rather brief reference had suggested. In particular it seemed that Margaret River, on the southern west coast of Western Australia, had unusually close climatic analogies during the vine growing season with Bordeaux, but with less spring frost, more reliable summer sunshine, and less risk of hail or excessive rain during ripening. There were similar apparent advantages over Mount Barker, combined with more reliable winter-spring rainfall. These conclusions were published in 1965 and 1966.

Few could have foreseen the speed with which they were taken up in commercial practice. Main credit for the viticultural pioneering of Margaret River (as in many other Australian wine-producing regions) belongs to its early medical vignerons: most notably Tom Cullity, who planted the Vasse Felix vineyard in 1967. He was closely followed by Bill Pannell at Moss Wood in 1969, and Kevin and Di Cullen at Cullens in 1971.

The success of these pioneers, and equally of those in the Mount Baker region – Tony and Betty Pearse at Forest Hill, Merv and Judy Lange (Alkoomi) and the Roche family at Frankland, Tony Smith (Plantagenet Wines) at Denbarker, and Mike and Alison Goundrey on the Hay River – inspired me back into the field of vine ecology in about 1973. The fact that the wine styles from each region seemed to conform well to what might be expected from the earlier crude comparisons encouraged me in the thought that the environmental requirements for different types of viticulture might be capable of still closer definition. In particular I became interested in learning how to predict, as accurately as possible, the environments best suited to individual grape varieties, and to more finely distinguished wine styles. This book is the ultimate result.

I owe thanks to many who have helped in various ways. The vignerons and winemakers mentioned above, together with many other friends in the Western Australian wine industry, have been consistent sources of stimulus (both mental and physical!) without which what has been essentially a hobby interest could not have flourished.

To my wife Pat, and to Robert and Helen, I owe a particular debt of gratitude for their patience and forbearance over many years, not least through interminable inspections of Western Australian countryside, and later, visits to vineyards. Without Pat's encouragement and support the work could not possibly have been brought to completion.

Scientific and wine industry colleagues have given technical and practical help in a number of ways. The list cannot be complete, but I would like specifically to thank Bill Jamieson, Tony Devitt, John Elliott, Barry Goldspink, Dorham Mann, Bryan Coombe, Brian Croser, Andrew Pirie, Joe Gentilli, John Kosovich, Bernard Abbott and Mike Peterkin. Erland Happ made an especially valuable contribution through discussions and criticisms of the manuscript, and in locating several useful references I had been unaware of.

I am indebted to the Library staff of the Western Australian Department of Agriculture for their ready assistance, and above all to the staff of the Department's Word Processing Centre, for carrying out a big job with unfailing cheerfulness and efficiency. The Australian Bureau of Meteorology, Melbourne, provided a great deal of climatic information, most of it unpublished; while the detailed treatment of present and potential vine-growing areas throughout Australia would not have been possible without the excellent topographical maps published by the Australian Surveying and Land Information Group, Canberra.

Special thanks to my former assistant Colin Smith for his work in setting up the climatic tables on computer, and for a number of useful comments on early drafts of the manuscript; to the staff of the Western Australian Government Printing Office, who put the climatic tables into their final form; to my neighbour Hugh Manners, who kindly drew Figures 3 to 7; and to Craig Paton, of Fairweather Cross Drafting, Perth, for his work on Figures 1–2 and 8–18.

Finally, I acknowledge with gratitude the tolerance shown by the Administration of the Western Australian Department of Agriculture in allowing me freedom and facilities to carry out the studies described in this book, despite the fact that they were not part of my official duties.

J.S. Gladstones
Perth, September 1992

PREFACE
to the Second Edition

Viticulture and Environment has been out of print for some years now, leaving a gap in the literature that has been too long unfilled. My second book *Wine, Terroir and Climate Change* (Wakefield Press, Adelaide, 2011) updated and extended its underlying science, but largely did not deal with the practical applications of most direct interest to many readers, which have changed little. I therefore welcome Trivinum's decision to issue a new edition, which I trust will fill the gap for a new generation of students, grapegrowers and winemakers, and those in the wine trade or who are simply interested.

In preparing it I have not introduced any new reference material, preferring to leave the book essentially as a statement of its time. However I have sought to clarify some issues, most notably the relationship of vine phenology to temperature (Chapter 6). Some of the discussion on the viticultural merits of the various Australian regions, particularly in New South Wales and Tasmania, has also been modified in the light of longer-term experience there.

Readers should note three significant changes between this book, first published in 1992, and the later *Wine, Terroir and Climate Change*. The most important is that I now treat irrigation, especially trickle irrigation, with greater reservation. With growing experience of terroir and the pursuit of quality, we in the New World are starting to realize that Old World attitudes to it have some valid reasons. The same applies to vine vigour and grape yields versus wine quality. Secondly, the later book improves the calculation of phenologically effective degree days and provides a more complete account of how it is done. Thirdly, *Wine, Terroir and Climate Change* expands and slightly modifies the listing of grape variety maturity groupings and their associated degree day requirements. Although the difference is not great, I suggest that the new listing be used in preference.

A further point needs emphasizing. The climate data used for the tables of this book were those accessible in the 1980s, and came from varying periods going back to the early 20th century. These tables will in time be progressively replaced and extended, using more recent and better controlled data. However, for reasons fully explained in both books, I believe they still provide a fair interim basis for assessing and comparing viticultural climates.

It remains to acknowledge my gratitude to several colleagues and friends. Firstly, to Trivinum Press for undertaking the new edition, and again to Michael Deves for his considerable input into resurrecting the manuscript and seeing it back into print.

James Halliday and Brian Croser contributed immensely to the success of the first edition through their tireless advocacy, both in Australia and overseas, where I believe it helped influence current viticultural moves towards cooler and more equable coastal areas of California, Chile and possibly South Africa.

Peter Dry and the late Bryan Coombe were also leading advocates within Australia. Bryan's foreword for the first edition is reproduced here.

Particularly I wish to pay tribute to Bryan Coombe as one of the true gentlemen of viticultural science, whose work is, I think, still under-appreciated. His meticulous research and teaching over a long career provided a basis for much of our present understanding of vine physiology, and certainly to my own thinking.

Finally I wish to thank the many grapegrowers and winemakers who expressed appreciation of the first edition. I trust their successors will find this edition equally helpful.

John Gladstones, Perth, August 2015

PREFACE

to the Second Edition

1

Introduction

The modern scientific literature, until very recently, said surprisingly little about the effects of natural environment on winegrapes and the wines made from them. Traditional European beliefs, by contrast, always placed great emphasis on environment. Consistent differences in grape and wine qualities among regions, among districts within regions, among vineyards within districts, and even among parts of a vineyard, were widely accepted as existing more or less regardless of management, and were ascribed to local differences in climate and to subtleties of topography and soil.

Twentieth century researchers tended to be sceptical of such ideas, if only because they were unmeasurable and therefore unprovable. Moreover, this was a time of genuine achievement in winemaking technology. Advances in the winery held out the hope that top quality could be achieved whatever the source and nature of the grapes. Especially was this so in the New World, where social and economic factors favoured production from highly mechanized, high-yielding vineyards in irrigated and usually hot areas. The success achieved in making sound, if undistinguished, wine at low cost from such grapes stands as a tribute to the technology developed.

But it was not enough. Over the last two or three decades consumers throughout the world have come to demand higher quality in their wines. Technically sound, but mediocre, wines are in world-wide glut, whereas fine wines (principally, at this stage, dry table wines) are in strong and steadily growing demand everywhere. Price differentials in the New World between the better bottled wines and ordinary 'cask' or 'jug' wines now approach those that have long existed in Western Europe.

At the same time the realization has grown that winemaking technology may be reaching the limit of its potential to improve wine quality.[1] Technology, by

its very success, has again brought to the fore the limiting importance of wine's basic ingredient: the grape.

This revised view is now accepted generally among progressive American and Australian winemakers and researchers (for example see Smart 1988; Rankine 1989). The Australian winemaker Brian Croser gave it perhaps ultimate expression when he wrote in 1978:

> Grape quality is the most significant factor affecting the quality of table wine. The winemaking process is to a large extent an exercise in quality control which is aimed at achieving the highest possible grape quality and then protecting those qualities to the bottle.

Others might allow the winemaker a more creative role. Nor should the task of 'protecting' those qualities to the bottle be under-estimated; much wine, in Australia as elsewhere, continues to show the effects of imperfect winemaking. Nevertheless most agree that, with winemaking technology now available to conserve most of the grape's virtues intact, the remaining frontier is the quality of the grapes.

Ultimate wine quality can in fact only be assured when both grape *and* winemaking are at their best. And the viticultural industry can be assured that the wine-consuming public, having enjoyed superior wines at reasonable prices, will no longer tolerate inferior or mediocre wines which they might once have happily accepted.

The tide has turned in other ways, too. From World War II to the 1970s, the Western World suffered chronically from shortage of labour. All industries sought to replace labour. In viticulture, mechanical mass production methods were especially applicable to the inland irrigation areas of the New World: partly because of their flat terrain and straight rows, and partly because many of these areas were isolated from ready sources of labour. (This was more of a problem in Australia than in California, which has traditional sources of migratory labour.)

Now, however, changes have taken place in both the labour market and the wine market, neither of which seems likely to be fully reversed in the future. Wineries of the New World have been forced at last to offer prices for top quality grapes that justify their greater cost of production. At the same time labour has become more readily available; it may become still

---

1 I shall not attempt to define wine quality in any strict sense, because it varies with markets, fashions and individual palates. Readers of a book such as this will have their own subjective concepts of quality, which will probably include an absence of 'off' smells and flavours, good and distinctive ripe fruit characters, palate balance, and in most young bottled wines, a perceived ability to improve with age. Note that I use the plural term 'qualities' simply to denote wine, grape or environmental characteristics, without implications as to superiority or otherwise. 'Style' is treated as being fully distinct from quality.

more so in the future. The combined circumstances are now moderating many of the New World's attitudes to grape growing, formerly dominated by big technology, to something closer to those traditionally held in the better wine-producing areas of Europe. Or perhaps more accurately it might be said that a rapprochement is developing, in which the wine industries of both old and new worlds take best advantage of the technology available, at the same time recognizing that the traditional skills of the vigneron are needed as much as ever.

Two main aspects of vine growing influence grape quality. The first is the choice of vineyard location and natural environment. That is the primary focus of this book. The second is management within the vineyard, including choice of grape varieties, trellising, training and pruning methods, soil husbandry and fertilizer treatments, irrigation, disease and pest control, and so on. I discuss management, but only insofar as it interacts with environment.

In concentrating on the natural environment of vineyards, I do not wish in any way to minimize the contribution that management can make to grape quality. It is great, and can be overriding. Recent research has produced big advances in the scientific understanding of vine management, for instance the outstanding work on trellising and training by Shaulis and his students in the United States of America, by Carbonneau and colleagues in France, and by Smart and colleagues in Australia and New Zealand.

With careful management it is often possible to produce good winegrapes and wines in seemingly unfavourable environments. A hot area like the Swan Valley in Western Australia can, and does, produce some excellent table wines, given skill and dedication. Such production under often less-than-ideal conditions is by no means doomed, especially where it enjoys advantages of market access. Indeed, the traditional pattern of decentralized production for local markets, across a wide range of environments, may well re-emerge in the decades to come as a result of social and economic changes. Its success hinges on maintaining a small, self-contained scale and a high degree of personal technical and marketing skills. Among the technical skills must be a deep understanding of the limitations of environment, and of ways to circumvent them. It is a case of knowing one's enemy.

At the other end of the spectrum are the large-scale commercial winemakers, producing for mass metropolitan markets and export. Economies of scale, or alternatively outstanding quality which commands universal prestige and high prices, must here make up for the high costs of commercial distribution. Because the markets are far-flung, the location of the vineyards and/or wineries plays only a small part in determining total distribution costs.

Finding the most suitable natural environments for intended wine styles is therefore crucial to the commercial success of such enterprises.

Wine-making technology can be changed. Within limits, vineyard management can be changed. Vine varieties can be grafted over, if desired, with minimal loss of time and production. But the climate and other external elements of a vineyard's natural environment cannot be changed or controlled, other than by site selection in the first place. It follows that the choice of vineyard site or sites is the most important decision to be made in any commercial winemaking venture, irrevocably committing as it does a large investment over many decades.

It is, of course, not always possible to find ideal, available *and* accessible environments for particular wine styles to which producers may aspire. Burgundy styles from Pinot Noir are a frequent case in point in Australia. But the methods described here will at least help to identify the environments which approach an ideal most closely; or alternatively, they will point to the wine styles and perhaps grape varieties to which any environment is best suited. That, anyway, is what I have attempted.

The study focuses on Australia, but necessarily entailed comparisons with Northern Hemisphere viticultural areas. Indeed, the classic vineyards of Western Europe remain the benchmark for all such comparisons, although that does not rule out the possibility of finding still better environments elsewhere.

In making the comparisons, one thing stood out. The geography and topography, and often the soils, of the world's most famous vineyards are generally well documented, most notably in Hugh Johnson's magnificent *Wine Atlas of the World* (1971, and subsequent editions). By contrast little has been published on their climates, either over the whole season or, more critically, during grape ripening. Such figures as have been published are usually from near or distant sites which are often unrepresentative of the vineyards, and most are confined to crude monthly, quarterly or annual averages of temperature and rainfall.

One of the first steps, therefore, was to find or estimate relevant climatic figures for representative existing vineyards. The estimates entailed making adjustments to available figures to allow for actual vineyard latitudes, altitudes, and certain topographic features, as described in Chapter 4.

These estimates gave a basis for extrapolating to new viticultural environments, given standard climatic data and information on their topographies and soils. The following chapters tell how this was done, and use the methods developed to construct detailed profiles of viticultural climate for old, new and potential viticultural environments throughout Australia. Climatic tables with more limited comment are also given for the main quality wine-producing areas of Europe, North America and New Zealand. These could be of wider interest; to the best of my knowledge, no comparable documentation has been previously published or attempted.

A final, personal thought. In an increasingly standardized, mechanized and computerized world, fewer and fewer things remain that are truly individual. Wine is one of them. Despite the high technology of its production, wine remains essentially a natural product, with infinite variation from area to area, maker to maker, grape variety to grape variety, and even bottle to bottle. It has subtleties and a sensual fascination which take it far beyond being a mere alcoholic beverage. It has a rich social and medical history, as Becker (1988) so eloquently recounts. The moderate consumption of table wines, traditionally with meals within the family or in good company, is a pleasant and entirely healthy custom. These are the things to which post-industrial man is instinctively returning, as a remaining link with the natural world, and as an antidote to the barbarity of his mechanistic surroundings.

Quality in wine is an artistic goal in its own right. Like other artistic goals to which humans aspire, it is a civilizing influence. The world needs such influences.

2

History of Climate Selection
for Australian Viticulture

Vine plantings followed quickly after European settlement in Australia. The aim was to provide fresh and dried fruit, and wine for mainly local consumption. Wine style, and even quality, were secondary considerations at first, so long as the vines succeeded and the products were acceptable. True commercial developments followed as the population rose, and the pioneer viticulturists found, by trial and error, which were the best areas and soil types for vines and wine production. The story has been well told by James Halliday in his excellent *Australian Wine Compendium* (1985) and *Wine Atlas of Australia and New Zealand* (1991); and by Gregory (1988) and others.

Although the early vignerons quickly identified the soil types on which vines thrived, any serious study of viticultural climates – other than in the broadest sense – had to wait until much later. That is hardly surprising, because at the time there were few climatic records.

James Busby, in his *Treatise on the Culture of the Vine* (1825), had only a little to say about climate, and none of it specifically about Australia. He observed that:

Warm climates, in favouring the production of saccharine matter, generally produce strong spirituous wines, sugar being necessary for the formation of alcohol; while the wines produced in cooler climates, though sometimes agreeably perfumed, are characterized by their want of strength, and tendency to degenerate into acetous fermentation.

A little later, in a passage which shows a clear understanding of the problems in hot climates, he wrote:

… in Burgundy, where the sun's rays do not act so powerfully in the production of saccharine matter, the wines are distinguished by a richness and delicacy of taste and flavour, while those produced under the burning sun of Languedoc and Provence, possessing no virtue but spirituosity, are generally employed in distillation. Not that the existence of a large portion of saccharine matter is incompatible with that of flavour and perfume, but it generally happens, that the volatile principles on which these depend, are dissipated during the lengthened fermentation necessary to convert into alcohol, the excess of saccharine matter.

Dr A.C. Kelly's *Wine-Growing in Australia* (1867), published in Adelaide, likewise said nothing about the local climate other than to assert its high suitability for vine growing, as did Despeissis (1895, 1902, 1921) in Western Australia. The Victorian Royal Commission on Vegetable Products, in its *Handbook on Viticulture* (1891) was more specific. It divided Victoria into three regions. Region 1, south of the Great Dividing Range, was considered very suitable for the production of light wines. The Commission expressed surprise at its neglect by viticulturists. Region 2, largely along the northern slopes of the range, was warmer and had more sunshine, and produced good full-bodied table wines. The hot, northern Region 3 was best for fortified wines. The distinctions remain valid. The problem in Region 1, we now know, is its liability to spring frosts in all but a few favoured locations, such as around Port Phillip Bay (Chapter 11). The present work shows that this is related largely to temperature variability.

Australian vineyard developments in the first 60 years of the 20th century paid scant regard to climatic suitability for winemaking, consisting as they did mainly of plantings in hot irrigation areas for dried fruit or brandy production. Only after the visit in 1955 of Professor Harold Olmo, Professor of Viticulture in the University of California, did climate again come to be accepted widely in Australia as a major criterion for growing winegrapes.

Olmo spent eight months in Western Australia at the invitation of the Western Australian Vine Fruits Research Trust, primarily to study and report on problems of vineyard decline in the Swan Valley. These proved largely to be related to nematodes and poor drainage. Olmo also pointed out, however, that as a wine-producing area the Swan Valley has climatic limitations. Its high summer temperatures and sunshine hours, combined with low summer and early autumn rainfall, make it very suitable for dried vine fruits and full-bodied fortified wines. On the other hand he considered it too hot for making table wines, for which the grapes need cooler ripening conditions to achieve the desired natural balance of sugar, acid and flavour. He applied the temperature criteria developed by Amerine and Winkler (1944) in California, and concluded (Olmo 1956) that the south-coastal area around Mount Barker and Rocky Gully-Frankland had the best prospects in the State for table wine production: a view later endorsed by the Western Australian Grape Industry Committee (1964).

Amerine and Winkler based their temperature summation system on that developed in France by A.P. de Candolle (1855), which relied on the observation that vines there start active growth in the spring when mean air temperature reaches about 10°C (50°F). Amerine and Winkler's system measures the 'available' heat for growth, over a period, in terms of the number of degrees Fahrenheit by which the average mean temperature[1] for the period exceeds 50, multiplied by the number of days in the period. In practice this usually entails using monthly average means to calculate monthly temperature summations. Monthly totals are in turn summated over an assumed vine growing season of April to October inclusive (Northern Hemisphere), or October to April inclusive (Southern Hemisphere).

On this basis Amerine and Winkler defined five regions of California for grape growing:

Region I (the coolest), having less than 2500 day° Fahrenheit;

Region II, 2500–3000 day°F;

Region III, 3000–3500 day°F;

Region IV, 3500–4000 day°F; and

Region V, over 4000 day°F.

In California, Regions I and II are said to produce the best table wines, with light to medium body and good balance. Region III produces full-bodied dry and sweet table wines and lighter-bodied dessert wines (including some of the best port styles). Region IV is best for dessert wines and generally inferior for table wines. Region V (which in California is more or less entirely in the irrigated inland valleys) is regarded as suitable mainly for fresh and drying grapes. It produces only bulk table wines of inferior quality, together with some fortified wines.

By the Californian criteria, Mount Barker and Rocky Gully-Frankland in Western Australia are borderline between Regions I and II. The Swan Valley is in Region V, although it differs from the Californian Region V in being more coastal and in relying mainly on natural rainfall. Other representative Australian locations by

this classification are: Region I: Hobart, Launceston, Yarra Valley, Drumborg, Coonawarra and Eden Valley; Region II: Geelong, Barossa Valley, Keppoch Valley, Margaret River; Region III: Seymour, Bendigo, Clare; Region IV: Mudgee, Rutherglen, Southern Vales-Adelaide, Berri; and Region V: Griffith, Hunter Valley.

Many viticulturists (especially from the Hunter Valley!) have criticized the classification as being too simplistic, which undoubtedly it is. Nevertheless Amerine and Winkler's temperature summations have given mostly useful first approximations of viticultural climate, in Australia as in California. They have been the basis for much of the New World's viticultural expansion into cooler climates over the last three decades.

My own early studies (Gladstones 1965, 1966) grew out of Olmo's. Olmo had based his conclusions on seasonal temperature summations, together with some consideration of rainfall and its distribution. It seemed logical to think that, in any detailed search for new viticultural areas, other climatic factors of known importance to vinegrowing should also be considered, such as sunshine hours and the risk of spring frosts. More sunshine hours, together with a lesser risk of spring frosts, pointed to a more consistently favourable viticultural climate along Western Australia's lower west coast than at Mount Barker, which is inland from the south coast.

A comparison of known average dates for the beginning of vintage from given grape varieties in Australia and Europe also suggested that more day degrees (as measured by Amerine and Winkler) are needed to achieve ripeness in Australia than in Europe. I tentatively ascribed this (Gladstones 1965) to Australia's lower latitude, and therefore shorter summer days, and suggested that in Australia the optimum climates for table wines might be found more in Regions II and III than in Regions I and II.

An essential climatic difference between the south and west coasts of south-western Australia appeared to lie in their relative variabilities of temperature. Whereas Busselton on the west coast is considerably warmer than Mount Barker and Rocky Gully-Frankland, and lies in Region III of Amerine and Winkler, it gets less days per year with temperatures over 90°F (Gladstones 1966). Margaret River, in high Region II, gets a lot less. At the same time the west coast is less subject to spells of very cool, cloudy weather in summer, such as occur quite often on the south coast in association with persistent on-shore winds. In short, the west coast is both more sunny in summer and less variable (i.e. more equable) in its day-to-day or week-to-week temperatures within seasons, which seemed more likely to promote quality in winegrapes than to detract from it. On these bases I proposed the Margaret River-Busselton area as having some potential advantages over Mount Barker-Frankland, although the latter still compared well with established vinegrowing areas elsewhere.

---

1 Strictly speaking, a mean is the mid-point between two values, and the daily mean temperature is correctly calculated as (maximum + minimum)/2. This is not the same as the true average of the temperatures as measured continuously over the 24 hours; in fact the mean and the true average can differ substantially (McIntyre et al. 1987). Nevertheless for most regions the monthly averages of the means are the only temperature data available for use in ecological comparisons. Preserving the conceptual distinction between 'mean' and 'average' is essential for clear thinking in a study such as this.

The publication of Hugh Johnson's *A World Atlas of Wine* (1971) led me to examine the subject of temperature variability further. Looking at his topographical maps in conjunction with the available regional climatic data, it seemed that not only are the world's greatest wines made in regions of low short-term (day to day, and within-day) temperature variability; but also that, within a region, the local features associated with the best vineyards are precisely those which restrict such temperature variation still further. I suggested (Gladstones 1976, 1977) that short-term temperature variability should be considered as a climatic factor in its own right, and proposed an index of variability on which to base comparisons. The argument is developed further in Chapters 3 and 4 of this book.

Shortcomings nevertheless remained in the fit of climatic data to vine behaviour and wine qualities. In particular, none of the existing climatic measures accurately predicted maturity date of the grapes, so as to give a basis for estimating average weather conditions during ripening in projected new viticultural areas. Arguably that is the most important period for determining grape and potential wine qualities. Any detailed comparison of viticultural environments needs to include direct comparisons for it. The development of a method to achieve that end using standard climatic data, for vines under 'normal' management (that is, management which does not result in delayed grape maturity), is described in Chapter 6.

But average date of ripeness in any environment depends also on grape variety. Varieties differ greatly in their phenology.[2] A classification of grape varieties into eight maturity groupings (Gladstones 1976, 1977) is here revised and expanded (Chapter 7).

The end point is a set of climatic profiles, showing regional climatic conditions through the growing season and, adjusted for representative vineyard sites (see Chapter 4), for the final ripening 30 days of each of the eight vine maturity groups.[3]

2 Phenology refers to the timing of the successive growth stages, for instance of budburst, flowering and (most importantly) grape maturity for winemaking. For a detailed discussion of grape vine phenology, see Coombe (1988). In this work I arbitrarily define grape maturity for calculation purposes as that most suitable for making dry or semi-sweet table wines, and the ripening period as the 30 days immediately preceding maturity so defined.

3 I follow largely the terminology of Smart (1977) in describing scales of climate. A regional climate (= macroclimate) is that of a general region, as exampled by the data from one or more well-established recording sites, normally over a period of 30 years or more. A mesoclimate is that of a more limited area, distinguished by its altitude and particular topographic characteristics which modify the climate locally (Chapter 4). 'Site climate' can be synonymous with mesoclimate, but is usefully kept distinct for application to individual vineyards or parts of vineyards. The term microclimate is properly confined to that within a particular vine canopy or part of the canopy, or at some point relative to the vine or the soil.

Several other systems for classifying or describing the climates of Australian vinegrowing areas have been proposed in recent years.

Prescott (1965, 1969a, 1969b) defined climates in terms of annual average mean temperature, amplitude between the average mean temperatures of the hottest and coldest months, the existence of a six-month period with average mean temperatures over 10°C, the average mean temperature of the warmest month, and the lag of the warmest period behind the summer solstice. He concluded (1969b) that the average mean temperature of the warmest month is as good an indicator as any, both of the limit of vine cultivation and of the limits for individual grape varieties. This was a useful simplification compared with the existing temperature summations, and may well be as accurate. But like all crude measures it can give at best only a very general guide to viticulture. Too many factors of great practical importance are overlooked for it to be a basis for detailed prediction. The same applies, only more so, to the simple use of latitude as a guide to grape ripening capacity (Jackson and Cherry 1988; Jackson and Spurling 1988).

Smart (1977), and later Dry and Smart (1988b) used a homoclime approach. They sought to get as close as possible a match in Australia to European viticultural climates across a range of climatic measures. The studies laid special emphasis on matching April with October (vintage) and October with April (budburst) in maximum and minimum temperatures, together with solar radiation, rainfall and relative humidity throughout the growing season. But despite their detail, these studies clearly illustrate the limitations of using raw climatic averages on any basis. Seeking to duplicate the climate of Mâcon (broadly representing Burgundy and Beaujolais), Smart found the nearest Australian match to be on the tablelands of the Great Dividing Range in New South Wales and Victoria. Oberon, near Lithgow (west of Sydney) was taken as representative. But as the comparable data for Bombala and Lithgow (Chapter 12, Tables 80 and 96 of the present work) show, these environments in Australia are ruled out by their excessive temperature variability in spring, leading to an extreme risk of killing frost. Even without this factor, the combination of lower latitude and wider diurnal temperature range so reduces the *biological effectiveness* of the temperature summation (see Chapter 6), that only the very earliest-maturing grape varieties could ripen if they survived.

Smart and Dry (1980), and Dry and Smart (1988a) later proposed a system which classified Australian viticultural climates in terms of five measures: average mean temperature of the hottest month (five categories); continentality, or the difference between the January and July average mean temperatures (five categories); total sunshine hours for October-March (four categories); aridity, as measured by the difference between total rainfall and 0.5 of the evaporation from an A-class pan for October-March (four categories); and average 900 hours relative humidity over the same period (four categories). They associated high yields with high sunshine hours and with irrigation in warm to hot areas; and good grape quality for winemaking with cool to warm temperatures, high relative humidity, and strong continentality. It is notable that the best quality combination, according to this classification, does not exist in Australia or New Zealand.

Smart and Dry defined temperatures in joint terms of the January (July in the Northern Hemisphere) average mean temperature and the index of continentality. To the extent that continentality is meaningful, this might represent an advance over January average mean temperature alone, and probably over crude temperature summation. However the implications of continentality are complex (see Chapter 3). I question whether it finds a place as part of an overall index of climatic suitability for vine growing, although it does add a descriptive dimension.

Using all five dimensions of Smart and Dry's classification gives a fuller description of viticultural climates than the previous methods, while remaining tolerably concise. Nevertheless it still gives only a broad delineation. It does not give a basis for detailed or localized predictions, nor does it encompass the vital concept of temperature variability.

Pirie (1978a, 1978b) took a different approach, arguing that wine quality and style depend most on vine water relations. He pointed out that the classic table wine producing areas of Europe have moderately high rainfall and relative humidity in summer, so that the vines suffer little moisture stress. Such climates support perennial pastures. Most of the long-established Australian vineyard areas have little summer rain, and in general do not support perennial pastures. Their solar radiation and sunshine hours are high, and relative humidities low to very low. Even with irrigation, the vines suffer from moisture stress (see Chapter 3). Pirie suggested that this leads to coarseness and lack of subtlety in the wines, regardless of temperature. He did not construct a new climatic classification for viticulture, but pointed out that the desired climate type for table wines corresponded with the 'cf' climatic class of Köppen (1931): that is, medium-temperature climates with an average of more than 30 mm of rainfall in the driest summer month. In Australasia, such climates are confined to the east and south-east coastal regions of the Australian mainland, together with the whole of Tasmania and New Zealand. Pirie found his ideal climate for European-style wines in north-eastern Tasmania.

As will be seen from later discussion, I find much common ground with Pirie, though not complete agreement.

Several other proposals for classifying viticultural climates have consisted merely of refining or rearranging Amerine and Winkler's temperature summation system, and need not be discussed here.

The New Zealand researchers Jackson and Cherry (1988) recently proposed a new, statistically-based temperature-latitude index of viticultural climates. Moncur et al. (1989) have also given results from controlled-temperature experiments which caused them to question the use of 10°C as a base temperature for viticultural summations. I consider Jackson and Cherry's study to have been seriously flawed, however, while neither study offers any improvement over traditional temperature summations for practical purposes. Appendix 3 examines them in greater detail.

3

Individual Climatic Factors

Ripening period mean temperature

Comparisons among the world's wine-producing areas show a broad association between average mean temperature during the final ripening 30 days and the styles of wine produced (Gladstones 1976, 1977).

Grapes with ripening-month average mean temperatures below about 15°C reach ripeness, if they do so at all, with high acid levels. Some unfermented sugar is usually retained (or added) in winemaking to balance the acidity and the frequent thinness of body in the wines. German still wines and many sparkling wines fall into this category.

Ripening-month average mean temperatures between about 15 and 21°C, with sound vine management and moderate yield levels, provide mostly well-balanced musts for dry table wines; or if the grapes are allowed to become very ripe, for sweet table wines. Wines from the cooler end of the range tend towards lightness of body, freshness and delicacy; and those from the warmer end towards fullness of body and flavour, provided that there is also enough sunshine (see below).

Grapes lose malic acid at greater rates as ripening mean temperatures are higher. Above about 21°C in the final ripening 30 days they tend to reach too low an acid level for making naturally balanced table wines. If sunshine hours and exposure, and resulting sugar levels, are high enough, such grapes at full ripeness usually make good fortified wines. From 21 to about 24°C is an ideal ripening mean temperature for ports, muscats and similar styles. Table wines mostly need acid addition, which is less satisfactory than having naturally balanced musts; but with skilled vine management and winemaking they can still be very good. Indeed, the capacity in warm climates to make wines of full flavour and relatively low acid, which drink well when young, can be a commercial advantage.

Picking very early in such environments, when acid and sugar are theoretically better balanced, produces wines lacking in fruit flavour. Red wines from early-picked grapes tend also to be harshly astringent, a result of the tannins having not yet condensed to their fully mature polymeric forms (Peynaud and Ribéreau-Gayon 1971). This is one of the reasons why white table wines are commonly better than the reds in hot climates, and why red grapes are often more

successfully used there for making rosé styles, in which there is little extraction of the tannins.

Above about 24°C average mean temperature in the final ripening period, the combined effects of heat and usually moisture stress nearly always reduce grape quality for all table wines. Conditions are perhaps also starting to get too hot for the classic styles of sweet fortified wines. Some very good full-bodied, 'liqueur'-style wines may still be made from partially raisined grapes of exceptionally high sugar content. Being mainly in regions of very low summer rainfall and relative humidity, this is also the natural zone for drying grapes.

But while the relative rates of acid loss and sugar accumulation during ripening determine the potential style and balance of a wine, it is a third component of ripening which most governs potential quality. That is the process known as 'physiological', or 'flavour' ripening. It includes physical softening of the fruits and, most importantly, the accumulation of various pigments together with flavours and aromas (between them known as flavourants), which are the products of ripening metabolism. The synthesis of these products from sugar takes place mainly in the berry skin, and depends chemically on the activity of various intermediary enzyme systems, driven by the energy available from the respiration (combustion) of sugar back to carbon dioxide and water.

Enzyme activity in general increases with temperature in the low and intermediate temperature ranges, but falls again at high temperatures. Very high temperatures, such as in sun-exposed berries in hot climates, can inactivate or completely destroy the enzymes (see, for instance, Kliewer 1977). Maximum rate of pigment and flavourant synthesis thus depends on there being a good supply of sugar as the basic chemical substrate and energy source, together with favourable intermediate temperatures for best activity of the enzymes.

Flavour ripening is not necessarily synchronous with the other ripening processes. At temperatures above the optimum for enzyme activity, full flavour is often not reached – if it is reached at all – until acid is too low and pH too high for naturally balanced musts. At the other extreme very low ripening temperatures (or excessively variable temperatures, as discussed below) may so retard enzyme metabolism that flavour ripening can remain incomplete even when sugar is ample for winemaking. Such grapes, and the wines made from them, lack colour, flavour and aroma intensity.

A precise optimum mean temperature for flavour and aroma formation in grapes has not been defined, but experimental and general field evidence seems to indicate an optimum mean for pigment formation in red grapes in the vicinity of 20 to 22°C, provided that light and other environmental factors besides temperature are not limiting (Kobayashi et al. 1967; Kliewer and Torres 1972; Hale and Buttrose 1973; Pirie 1978a, 1979). Colour intensity and hue are traditionally held to be indicators also of flavour intensity and overall quality in red wines (Somers and Evans 1974; Somers 1975; Jackson et al. 1978; Pirie 1978b; Leforestier 1987). In wines the final colour depends on additional chemical factors besides anthocyanin content. Nevertheless the original anthocyanin content of the grapes contributes, and is reckoned to be a measure of the quality of the grapes themselves. In the absence of better information, it seems reasonable to conjecture that whatever affects the synthesis of anthocyanins and other pigments will affect the synthesis of flavour and aroma compounds in a broadly similar way.

Certain other considerations point to the likelihood of an optimum ripening mean temperature, for grapes, of about 20°C or slightly higher. It is a little below the mean temperature of around 23–25°C at which grape vines reach their greatest rates both of photosynthesis (Kriedemann 1968) and of potential growth (see following section, and Chapter 6). Below that temperature, provided that there is ample light intensity and leaf exposure, growth rate is limited more by the rates of biochemical and physiological processes than by photosynthesis; therefore a surplus of sugar can accumulate which is available to be stored or transported to the ripening fruit, or to form leaf pigments and other 'secondary', or 'luxury' products not essential to plant growth. At the temperature giving greatest growth rate, net sugar production and the capacity to use it for growth are in balance. But as temperature rises above that optimum, the demand for sugar by rapidly accelerating respiration, combined with no further increase in photosynthetic rate, leaves a smaller amount available for growth, which therefore itself diminishes. Sugar availability is now the limiting factor, so less goes to the fruits and little to forming secondary compounds, or for storage as a reserve in the stems or roots. That is why anthocyanin colours in plants disappear at high growing temperatures, and why grapes and other temperate fruits have poor colour and flavour when grown (if they will grow) in sub-tropical and tropical climates.[1]

The best temperatures for sugar flux into ripening grapes, then, should be below the optimum for growth rate of the vines. Experimental results (Buttrose et al.

1971; Hale and Buttrose 1973; Alleweldt et al. 1984) confirm that even quite low ripening temperatures do not necessarily reduce berry sugar content at ripeness.

Maximum development of colour and flavour, however, requires also that temperatures during ripening be high enough for optimal activity of the enzyme systems in the berries. Only then will the greatest possible proportion of the incoming sugar be converted into other desired end products. Rate of enzyme activity increases right up to the optimum temperature for growth rate, and possibly beyond. Balancing the two relationships leads logically to the expectation that the greatest production of pigments and flavourants will be with growing and ripening temperatures just a little below the optimum for vine growth.

Indirect supporting evidence concerning the purely biochemical processes of ripening comes from studies of certain fruits that are normally ripened in temperature-controlled storage. Pears, tomatoes, avocados and even bananas have shown a remarkably uniform optimum ripening temperature, for quality, of about 20–21°C (Spencer 1965; Truscott and Warner 1970; Westwood 1978; Kadar 1985; Burt and Parr 1988). In other interesting parallels, Ough and Amerine (1966), Daudt and Ough (1973) and Killian and Ough (1979) found approximately the same temperature to be optimal for ester production in fermenting wine musts; and Nordstrom (1965), for ester formation in brewing beer.

These parallels suggest, among other things, that the enzymic biochemistry of ripening has much in common across many fruits; and more speculatively, with that of fermentation. Daudt and Ough (1973) pointed out that ester formation more or less ceases with the end of fermentation, showing that (like ripening) it is an energy-requiring process. Might the biochemical changes during fermentation represent, in some respects, a further progression of ripening? One is tempted to think that the ethereal aromas and bouquet which can develop in good wine perhaps parallel the similar aromas occasionally – but rarely – seen in the very best fruits, when they have reached their fullest possible ripeness in still-perfect condition. On the other hand the parallels may simply reflect a common temperature optimum for the whole range of enzyme reactions in temperate plants, fruits and their immediate products. Either way, they provide some indirect support for the idea that about 20–22°C is optimal for physiological ripening in grapes, and for the formation of desired colour, flavour and aroma compounds in them.

There is, however, a reservation concerning grapes ripening on the vine, which parallels what can happen during fermentation. The anthocyanins, and other condensed phenolics such as the tannins, are non-volatile, fairly stable compounds. The total present at ripeness is therefore essentially the total formed during ripening. The aromas and associated fruit flavours, on the other hand, are due to compounds which are to

---

1 The effects of sugar supply to the fruit on fruit quality are discussed more fully in Chapter 5, in the context of vine canopy management.

varying degrees volatile; they may also be chemically unstable. They include monoterpene alcohols, phenols, and various esters. Boiling points at normal atmospheric pressure range from 200–250°C for some of the higher molecular weight compounds (e.g. geraniol 230°C, d-linalool 198°C) down to around that of water for some of the more volatile esters (Weast and Astle 1982; Windholz et al. 1983).

Direct loss of volatile compounds from the berries during ripening, unlike that from fermenting musts, has received little attention. Any discussion is therefore highly speculative. The assumption that it occurs seems subjectively logical, however, if only because ripe fruit can be smelt. A possibly overriding factor is the relationship to temperature of cell wall permeability to the responsible compounds, since they may have to diffuse through several to many cells to reach sites of direct evaporation into the atmosphere. According to Collander (1959), cell membrane permeability increases very steeply with temperature: two-fold to five-fold, or even more, per 10°C rise in temperature. Moreover, compounds of lower molecular weight (and therefore, generally, lower boiling point) permeate more readily than those with high molecular weights. Differences in volatility among compounds, and between temperature levels, may therefore be greatly magnified in determining relative rates of evaporation of the different compounds from fruit.

That substantial losses of aroma compounds probably do occur is suggested directly by the progressive loss of aroma and flavour that is commonly perceived in grapes after picking (and similarly in other fruits which ripen only on the plant: see Westwood 1978). It is also well known that the flavours, and especially aromas, of fruits are best preserved in low-temperature storage. Reynolds and Wardle (1988) similarly found a large difference between sun-exposed and shaded berries of Traminer grapes in their balance of free, volatile monoterpenes to fixed terpene glycosides, showing a disproportionate loss of the volatile form from the exposed berries.

According to this concept, whether or not individual flavourants accumulate in the ripening grapes, and if so how fast, would depend on their relative rates of formation and loss. Lower ripening temperatures should make possible the accumulation of compounds with lower molecular weights and boiling points, down to the temperature for each compound where the slower evaporative loss is just balanced by its slower synthesis. Went (1953) cited tea and strawberries as products which become more aromatic with lower growing/ripening temperatures. I suggest that the same principle applies to grapes, with potential aroma tending to increase and become more complex – because of the greater range of volatile compounds retained – down to the point where low ripening temperature excessively slows the synthesis of all compounds responsible for aroma and flavour.

Perhaps as important is the recent finding that the monoterpenols, together with (probably) the aromatic phenols, are in a constant cycle of formation, limited accumulation, and finally conjugation with sugar into their odourless, flavourless glycoside forms (Williams et al. 1987, 1988; Strauss et al. 1987). The free compounds contribute the characteristic aromas of a number of grape varieties, including most notably the Muscat family (Marais 1983); but once conjugated into glycosides they are ineffective as flavourants.[2] The temperature dynamics of the system are at present unstudied, but it may be guessed that the size maintained of the pool of free, 'active' flavourants would vary directly with their rates of formation.

The free terpene compounds have fairly high boiling points, so their optimum temperatures for accumulation might be expected to be higher than for the more volatile esters. Very probably they approximate to those for maximum enzyme activity, and for the accumulation of anthocyanins and tannins. The remarkably uniform ripening mean temperature of about 22°C in areas producing the world's greatest Muscat wines – see the discussion on Rutherglen, Chapter 11 – seems consistent with this idea.

All these relationships could help to explain why delicate, fruit-estery aromas are most characteristic of wines produced in cool to mild climates, employing cool and/or closed fermentation. Warm-climate and most red wines have robust, less fugitive flavours and aromas, which depend less on cool or closed fermentation for their retention.[3]

---

2 The extent, if any, to which the glycosides might later hydrolyse during fermentation and wine maturation, to release again the free monoterpenols or phenols, remains uncertain. Recent thinking (Abbott et al. 1990a, 1990b) accepts it as probable. In any case the issue does not greatly affect the argument relating rate of accumulation to temperature.

3 One other factor possibly contributes. In vegetable oils it is known that the degree of unsaturation (presence of double or triple carbon-to-carbon bonds in the fatty acid chains) increases as ripening temperature falls. According to Creveling and Jennings (1970) and Mazliak (1970), many of the volatiles of apples, pears and a number of other fruits are formed during ripening as, or from, degradation products of fatty acids derived from plant oils and waxes; so their chemical constitutions, and perhaps average molecular weights, would presumably reflect the temperature regimes under which the oils and waxes were formed. 'Leaf aldehyde' (*trans*-2-hexenal) and 'leaf alcohol' (*cis*-3-hexenol) have been thought to be associated with 'grassy-leafy' odours in grapes and wine, and are themselves breakdown products of the triple-unsaturated fatty acid linolenic acid (Stevens et al. 1967; Joslin and Ough 1978). Recently the characteristic 'herbaceous' aroma of many Sauvignon Blanc and some Cabernet Sauvignon wines has been linked to their content of methoxypyrazines, which are highly unsaturated monocyclic compounds with two nitrogen atoms in the ring structure, and occur in greatest concentrations in wines from grapes incompletely ripened in low temperatures and/or shaded canopies (Lacey et al. 1988; Allen et al. 1988, 1989, 1990a, 1990b). The temperature regulation of their formation (or persistence) is suggestively analogous to that of the polyunsaturated fatty acids, while the substitution of

Partly for the same reason, cool-climate wines are, in general, best drunk chilled or at cool room temperature; whereas warm-climate wines need higher drinking temperatures to bring out their aromas. (Another factor is that the high acids of cool-climate wines can become harsh to the palate at high temperatures.) The combined result is a set of broad parallels among the ripening temperatures of grapes, the optimum temperatures for their fermentation, and the best drinking temperatures for the wines made from them.

To summarize, a ripening average mean temperature of about 20–22°C appears to be optimal for pigment intensity in grapes, and probably for intensity of flavour and at least some aromas. To this however must be added that lower ripening mean temperatures could be desirable to retain the more highly volatile aromas, such as contribute complexity and floral character (especially in wines to be drunk chilled); and also, to retain high acid levels for some wine styles, and low pH. The optimum ripening mean temperature therefore hinges on an appropriate balance among all these factors for each individual wine style. Intensity of colour and flavour is sacrificed at lower ripening temperatures in favour of potential delicacy, freshness and certain kinds of aromatic character in the wine.

A point to be noted in conclusion is that all discussion of optimum temperature hinges on light supply to the leaves not being seriously limiting to photosynthesis. Limitation by insufficient light exposure or intensity will tend to reduce the growing and ripening temperatures giving best yield and fruit quality. Reasons for this are explained later in this chapter and in Chapter 5.

Mean temperatures and vegetative growth
The control of temperature over vine growth is not easy to define in practice, for two reasons.

1. At most places, temperature and sunshine hours are closely correlated through the season. To sort out their individual contributions is difficult or impossible, unless one or both of them can be manipulated artificially.
2) In most controlled-environment experiments the tops and roots of the vines have a common temperature environment; whereas in the field, disparities between air and soil temperatures are normal and probably important. I attempt to deal with these problems in the next two sections.

Despite the uncertainties, trial results from controlled environments with common top and root temperatures do point to some conclusions that are consistent with field experience. Buttrose (1968, 1969) grew five grape varieties at temperatures ranging from 15°C day/10°C

night up to 35°C day/30°C night. The small vines that grew at 15/10°C had stems with only half the dry weight of the leaves. Those at 20/15°C were much more robust and had nearly as much stem as leaf dry weight. These vines were (relatively) the stemmiest of all. Fastest total dry weight production was at 25/20°C (or 30/25°C in the case of Ohanez), consistent with an optimum mean temperature for vine growth of about 23–25°C. These vines again had much more leaf than stem dry weight, and the proportion was maintained when total dry matter production fell away with above-optimum temperatures. Thus, the greatest relative investment of assimilates in stem growth was at temperatures moderately below the optimum for total growth.

Growth rate as measured by main stem length behaved differently, increasing with temperature to 30/25°C; while the production of new main stem nodes was fastest at either 30/25°C or 35/30°C. It can thus be deduced that both stem thickness and internode length declined progressively above 20/15°C. This agrees with the field-observed tendency to rank vine growth in cool to mild climates, as described by Due (1988).

Downes and Gladstones (1984) got similar results under controlled temperatures with the narrow-leafed lupin, *Lupinus angustifolius*, except that the apparent optimum temperature for growth of this species was lower (21/16°C). With reduced temperatures below that the stems were progressively thicker and the internodes longer. The ultimate heights and weights of the plants became much greater with lower temperature (if they were allowed to grow for long enough), right down to 15/10°C. However the large plants at 15/10°C set relatively few pods and seeds, investing only a quarter of their post-flowering dry matter gain in seed growth as compared with approximately half at 21/16°C and above. This corresponds exactly with the coarse growth but poor fruitfulness of vines growing under consistently cool conditions.

In both cases it seems clear that any surplus sugar available in the plant at temperatures moderately below the optimum for growth rate is directed preferentially to stem growth, at least during the main vegetative growth period. This corresponds with a strong action of auxins and, especially, of gibberellins in controlling growth at that time (see Appendix 1).[4] It is a logical enough strategy for plants which have evolved in temperate climates, where early growth is normally at sub-optimal temperatures. It is not in a species' survival interest to switch its main efforts to reproduction until the plants are large enough, and until temperatures and day-length are approaching or passing optimum later in the season.

N for C atoms of the benzene ring might, very speculatively, be related to the low ratio of available carbon to nitrogen in non-photosynthesizing shaded leaves, and in the berries themselves before full ripening.

4 Appendix 1 outlines a general scheme for the functions of the main hormones (growth substances) in plants, and how I see them interacting with environment in controlling the development of grape vines and berries. Familiarity with this is desirable for understanding the discussion that follows.

Mean temperatures and vine fruitfulness

The relationship between vine fruitfulness and temperature is mediated initially through the fruitfulness of individual buds. These form during spring and summer in the leaf axils of the current season's growing shoots, and grow into new shoots the following season if they are retained at pruning and burst satisfactorily in the spring. The embryonic bunches are already formed within the parent buds, and can be seen in them by mid-summer if dissected under a microscope. The differentiation of fruitful as opposed to purely vegetative buds (which produce only tendrils on the new shoots instead of bunches), and of multiple bunches per bud and shoot, are favoured by high temperatures during early bud development in late spring (Baldwin 1965; Buttrose 1970, 1974; Possingham 1970). Adequate warmth and sunshine are then needed in the following spring and summer for early bunch growth, flowering and setting, and the final development and ripening of the berries.

The need in practical management is, of course, to achieve a suitable and sustainable balance between vigour and fruitfulness. This will vary in part with the natural habits of individual grape varieties, as discussed a little later in this section. Other considerations include the commercial aim of the grape production, quantity versus quality. Fruitfulness which regularly leads to cropping beyond a vine's capacity is always undesirable, because it both reduces wine quality and weakens the vine. Varieties giving the best wine are mostly fairly vigorous but not over-fruitful, although the same result can be attained in fruitful vines by bunch thinning.

Relatively unfruitful grape varieties in any environment tend to develop fruitful buds only well out along the canes, these being the ones which start forming after the weather has become warm and sunny enough. Not surprisingly, the least inherently fruitful varieties tend to be those naturally adapted to hot climates, such as Sultana and Ohanez. By contrast Riesling, a variety well adapted to cool climates, forms fruitful buds at quite low temperatures (Buttrose 1970).

In the light of these differences it is clearly not possible to define a universally optimum temperature regime for vine growth, bud formation, flowering and early fruit development. As a broad guide, however, practical experience seems again to point to an optimum somewhere in the vicinity of 20–22°C for most winegrapes, i.e. just a little below the range of 23–25°C which gives potentially the fastest growth.

The work of Buttrose (1970, 1974) showed that quite short high-temperature pulses are effective in promoting the differentiation of fruitful buds, suggesting that the influence of temperature on fruitfulness differs in kind from that over growth rate. Reductions in the sensitivity of vines to their own gibberellins seems a likely cause (see later discussion of cytokinins, and their interactions with gibberellins). Together with abundant sunshine hours, additional factors favouring fruitfulness include good leaf and direct bud exposure to sunlight, as will also be discussed in later sections of this chapter.

In summary, two complementary temperature responses appear largely to govern the temperature adaptations of grape varieties: 1) their natural tendencies towards rank, stemmy growth under low to moderate temperatures (assuming that moisture, nutrition, etc. allow); and 2) their needs for heat during bud formation, to promote fruitfulness for the following year. The two interact. Sparse bunch formation and set will directly exacerbate any tendency to rank vegetative growth, by reducing competition for assimilates and growth substances in mid and late season. Vegetative vigour in turn tends to reduce fruitfulness, by shading and reducing the temperature of the differentiating buds (and perhaps also their immediately associated leaves). The combination can set up a developing vicious circle of unfruitfulness and rank growth in susceptible varieties in consistently cool to mild climates. Conversely, weak vegetative growth combined with high bud fruitfulness leads readily to over-cropping, poor fruit quality and vine exhaustion of other varieties in warm to hot climates.

A general principle follows. Where growing conditions are conducive to vegetative growth, such as in most cool southern Australian coastal areas, the best-adapted varieties will in general be those with only moderate to average vigour, combined with naturally good bud fruitfulness. Hot areas require more vigorous but less fruitful varieties to achieve the right balance; or if varieties are highly fruitful, such as Muscat Gordo Blanco or Zinfandel, they need strong boosting of vegetative growth and often bunch thinning.

Obviously other management practices and features of the environment make their own contributions. These will be discussed further in Chapter 5, under the heading 'The concept of vine balance'.

Soil temperature, and the roles and interactions of cytokinins

Implicit in the older and general vine literature is a belief in the importance of soil temperature. Readily warmed soils, and a resulting early and quick growth start in spring, have always been recognized as essential to ripening at the cool limit of viticulture. The traditional, more or less universal, viticultural belief in the value of stones and rocks in the soil (Chapter 4), and similarly of slopes facing the sun, stems at least in part from this need.

The 19th century French literature nevertheless emphasized the value of readily warmed soils even in quite hot areas. For instance the famous 'pudding stones' of the best Châteauneuf du Pape vineyards in southern France were (and still are: see Johnson 1971) believed to play an essential role in ripening, through heat absorption and its re-radiation to the bunches at night. I cite other examples from warm to hot climates in Chapter 4, and suggest there that such re-radiation has a qualitative, as well as quantitative, effect on ripening.

Although such effects on above-ground microclimate are undoubtedly important, recent experimental evidence has also underlined the central role of soil temperature in its own right, in controlling whole-plant growth and physiological development. Moreover, a hormonal basis for this control has been substantially elucidated.

Woodham and Alexander (1966) first showed that when sultana grapevines were held at a common air temperature, but with root temperatures ranging from 11 to 30°C, their growth varied enormously depending on root temperature.

These authors suggested that production by the vine roots of a plant hormone of the cytokinin group might be a basis for the difference. Skene and Kerridge (1967), and later Zelleke and Kliewer (1981), confirmed the specific effect of root temperature on growth, and demonstrated changes in the amount and form of cytokinins in the sap rising from the roots which could account for it.

Kliewer (1975) and Zelleke and Kliewer (1979) further showed a substantial rise in the proportion of Cabernet Sauvignon buds bursting with higher root temperatures: a finding consistent with the known function of cytokinins in promoting cell division and bud activation in plants generally, and with the observed retarding effect of cold soils on vine budburst in the field. The results paralleled those of Buttrose (1968, 1969), cited earlier, in which the vine tops and roots were under common temperature control. They suggest that an important part of the temperature response of grape vines is associated specifically with soil and root temperatures: at least in spring and early summer, when these are low enough compared with air temperatures to be the more limiting.

Weaver (1965, 1966) showed that direct treatment of flowering clusters with cytokinins promoted the setting and development of Black Corinth (Currant) grapes, an effect similar to that of certain auxins when applied to the clusters (Coombe 1950, 1965; Weaver and Williams 1951); and further that, like auxins, the cytokinins powerfully attract nutrients to the parts of the plant where they accumulate in highest concentration. Mullins (1967, 1968) was able to demonstrate with detached vine cuttings that the growth of the young fruit clusters pre-flowering depended on cytokinins, and on the presence of growing roots to supply them. Consistent with this, Mullins and Rajasekaran (1981) showed that a warm soil and vine root system, relative to the temperature of the aerial parts, prevented coulure (failure of berry setting). Early and vigorous root growth ensured a more than sufficient cytokinin supply to meet all needs of the cooler top parts.

Control by cytokinins, and therefore implicitly by root temperatures and other soil conditions controlling root growth, has also been shown for the differentiation of fruitful buds. Srinivasan and Mullins (1976) described the development of the cluster rudiment, or anlage (plural anlagen) within the still-minute bud. They showed, consistent with the earlier work of Baldwin (1965) and Buttrose (1970), that high temperatures in late spring and early summer favoured its differentiation into an embryonic cluster, whereas with low temperatures it developed only into an embryonic tendril (a tendril being the anatomical equivalent of the bunch stalk). Later work by Srinivasan and Mullins (1978, 1980, 1981) showed that artificially supplied cytokinin could substitute for high temperature in promoting cluster differentiation. Gibberellic acid substituted for low temperature in promoting only tendril development. The decisive factor appeared to be the balance of cytokinin versus gibberellin, with high temperature favouring the production of cytokinin and fruitful buds, and low temperatures, gibberellin and tendrils.

Srinivasan and Mullins found that gibberellins are in fact necessary to form the anlage in the first place; but there was no evidence that in practice this limits fruitful bunch formation in the grape vine. Therefore, one of the main factors governing potential bud fruitfulness in the grape vine appears to be soil temperature through the spring and early summer, via cytokinin production in the roots and its export to the developing buds.

Evidence for an effect of soil conditions and root-produced cytokinins on current-season berry growth and ripening in grapes is less firmly established, but is nevertheless suggestive.

Winkler (1962), Coombe and Hale (1973), Coombe (1976) and others have described three fairly distinct stages or periods in the growth of the grape berry following fertilization. Period I is a stage of initial rapid fruit growth after fertilization, during which the seeds (if present) remain largely undeveloped. Other than in the seeds, all cell multiplication in the berry is completed before or during this period. Period II is one mainly of seed development, with little increase in flesh or in the overall size of the berry. It is foreshortened in seedless berries, and those with early-aborted seeds, such as Sultana. Period III begins at véraison, which marks the start of ripening. Sugar begins to flow in, the cells of the flesh expand dramatically, and the berries become translucent. Pigmented varieties start to colour. Sugar content increases continuously up to near maturity, but aroma and flavour do not, in general, develop until the latter half of ripening.

Pre-flowering development of the clusters depends directly on cytokinin supply, as described above. Setting of the fertilized flowers appears then to be related to the supply of sugar and other nutrients in their immediate vicinity (Coombe 1962, 1965). That in itself probably depends on total cytokinin supply being ample for all needs, since young vegetative growth competes more strongly than the clusters for cyto-kinin, as well as for nutrients (Coombe 1962; Mullins 1968; Mullins and Osborne 1970). What appears to count most in determining the relative allocation of available nutrients to the setting fruits is the balance between the cytokinin locally available to them on the one hand, and the auxins and gibberellins produced by the competing vegetative growth on the other.

Promotion of post-fertilization berry growth through Period I and Period II seems to be adequately accounted for by nutrient attraction by the auxins produced in the young seed embryos; or in seedless varieties, by gibberellins probably imported from outside the berry (Coombe 1960).

The end Period II and initiation of ripening (Period III), at véraison, appears normally to have to await maturation of the seeds in seeded varieties (Cawthon and Morris 1982). This, together with the known decline in natural berry auxin content at véraison (Coombe 1960), and the finding that applied auxin delays ripening (Weaver 1962; Hale 1968), suggests that the main barrier to ripening during Period II is auxin produced by the growing seeds. The normally early ripening of seedless grape varieties, and of individual seedless berries of seeded varieties, supports such a conclusion.

Much evidence now also points to the 'stress' hormone abscisic acid (ABA) as having a complementary role in ripening initiation (Coombe and Hale 1973; Coombe 1973, 1976). As noted in Appendix 1, abscisic acid forms in plants under a variety of stress conditions, including those of heat and moisture lack, together with declining day-length in late summer and autumn. It is antagonistic alike to cytokinins, auxins and gibberellins, and when dominant causes a slowing or cessation of vegetative growth and a diversion of assimilates, together with other nutrients and growth substances, to the fruits and seeds.

In summary, it can be hypothesized that rising abscisic acid levels in the vine, combined with declining production of auxins and gibberellins by the reduced vegetative growth, and of auxins by the maturing seeds, culminates in the dramatic 'switching point' of véraison, when the bunches suddenly become established as the dominant sink for assimilates and other nutrients within the vine.[5]

5 This phenomenon has its parallel in the well-known 'self-destruct' mechanism of annual crop plants, in which a snowballing mobilization and near-complete withdrawal of nutrients from the leaves, stems and roots, and their transfer to the ripening seeds, causes the mother plant to mature and die. The perennial plant retains a functioning permanent structure, although the drain of over-cropping can progressively weaken it. Equally, under-cropping leaves a surplus of assimilates which encourages continued vegetative growth during fruiting. In the latter situation, and particularly under low temperature and other conditions conducive to vegetative growth, the normal dominance of the vegetative sinks for cytokinins and assimilates can be re-asserted over the needs of the ripening fruit, with results potentially as harmful to fruit quality as those of over-cropping.

Abscisic acid probably has a continuing role in grape ripening biochemistry (Pirie and Mullins 1976; Kataoka et al. 1982; Palejwala et al. 1985). But whether lack of it is ever a major limiting factor in practice, once ripening is in full process, seems less certain. Certainly in grape vines an excess of stress during ripening is adverse to ripening, as will be emphasized later in this chapter and in Chapter 5.

On the other hand there is a good deal of indirect evidence that cytokinins again play an important part during ripening, perhaps complementary to, and in balance with, that of abscisic acid. Specific primary roles might include attracting sugar to the berry, in that respect replacing the auxin produced earlier by the developing seed embryos, and secondly maintaining the berry in sound metabolic condition and non-senescent through to full maturity. Van Overbeek (1962, cited by Skoog and Schmitz 1972) long ago suggested such a function for grapes, and Westwood (1978) for fruits generally.

Consistent with this idea, Lilov and Temenuschka (1976) and Zelleke and Kliewer (1981) both found a secondary peak in the cytokinin concentration in grapevine sap during ripening (the first having been in the earliest stage of berry growth). Niimi and Torikata (1978) likewise found that sap cytokinin in bound form remained high throughout berry ripening, although 'free' cytokinin fell away rapidly once vegetative growth had slowed down. It may be relevant, in this context, that direct xylem connections to the berries appear to break down at véraison (Coombe 1990). This suggests that during ripening the bound form of cytokinin goes first to the leaves with the xylem sap, and thence to the berries via the phloem system. Assuming only the free form to be active, the necessary hydrolysis must take place either in the leaves or in the berries themselves.

Lilov and Temenuschka (1976) showed that cytokinin accumulates in ripening grapes, and that up to a certain saturation level it does so preferentially in the skin. This is where sugar first accumulates at véraison (Coombe 1973), and is thought to be the main site of pigment, flavour and aroma biosynthesis. Whether cytokinin acts only to attract sugar to the berries, or whether it has additional metabolic roles there, is unclear. However, the equivalence of cytokonins to red light exposure (see discussion later in this chapter on light quality, and in Appendix 1) points to at least some direct biochemical role.

A plausible overall picture of ripening, then, is that abscisic acid (combined with auxin disappearance) promotes cytokinin supply to the berries rather than to the leaves and growing points, and also directly initiates the main biochemical processes of ripening; cytokinins then maintain sugar flux into the berries, protect them against senescence, and probably participate in some of the desirable metabolic processes of ripening, including anthocyanin formation. By implication, satisfactory full ripening hinges on suitable, probably changing, abscisin/cytokinin balances in the berries through the various stages of ripening.

The question remains as to whether, in practice, soil temperature is ever limiting to root activity and cytokinin production during ripening. Traditional wisdom in cool viticultural climates holds that soil temperature does influence ripening. A well-warmed soil probably allows it to continue longer into a cold autumn than would otherwise be possible. Several mechanisms for this suggest themselves, not all of them necessarily involving cytokinins and none of them mutually exclusive. They include direct microclimatic warming of the lower vine canopy and bunches by the soil, prolonged root activity and cytokinin supply to the bunches, and prolonged leaf retention and photosynthesis for the same reason. Probably all are involved, in complementary ways.

Lack of root warmth is unlikely to limit ripening in hot climates, where the grapes ripen early and at a time when upper soil temperatures are close to their peak. In Australian mainland environments it is more likely to be lack of moisture (discussed later in this chapter), and perhaps too high a soil temperature, that limits root activity and cytokinin production during ripening. Fear and Nonnecke (1989) found in Iowa, USA, that mid-summer straw mulching of strawberries cooled the soil at 10 cm depth by some 5°C, to an average of about 22°C, and at the same time promoted fruiting (which I interpret as a cytokinin-associated process) and inhibited runner development (a probably gibberellin-associated process). This suggests the existence of a distinct optimum soil temperature for root function and cytokinin export, very likely paralleling that for above-ground growth. Some implications are developed further under the heading 'Summer mulch or bare surface?' in Chapter 5.

Whatever the environmental controls, the general principle remains that the bunches need a continuous supply of sugar, and almost certainly of cytokinin and possibly abscisic acid, since physiological ripening of grapes ceases as soon as the bunches are picked. The temperature and other physical conditions of the soil during ripening can be expected to have a significant influence on these processes, and therefore on the potential qualities of the grapes and wine.

Continentality

As noted in Chapter 2, this is the range of temperature between winter and summer, normally measured as the difference between the average mean temperatures of the coldest and hottest months. The difference is large in continental climates and small in maritime climates, due to the much slower heating and cooling of the oceans as compared with the land.

Some authors (e.g. Smart and Dry 1980) have used the simple difference between January and July average mean temperatures, but that can be misleading. It gives unduly small figures for some of the more maritime climates, in which the ocean lag in heating and cooling means that these are not the hottest and coldest months. True continentality values for a world range of viticultural climates are given in Table 1. Figure 1 shows the annual march of monthly average mean temperatures for representative continental, intermediate and maritime viticultural climates.

Smart and Dry (1980) suggested that superior wine quality is associated with greater continentality. I can find little evidence to support this as a generalization. Some of the points covered below indeed suggest a greater quality potential in continental-type climates, but at the same time perhaps a lesser chance of the potential being achieved. Continentality nevertheless has interesting implications for wine style, and quite importantly, for vine agronomy.

It can be argued, for instance, that continentality underlies the particular wine styles of Germany and similar cool areas of Central Europe. Ripening there is under conditions of rapidly falling mean temperatures (Figure 1), and also of diminishing sunshine and rainfall. Favourable temperatures and sunshine early in ripening set in train a vigorous ripening process, with

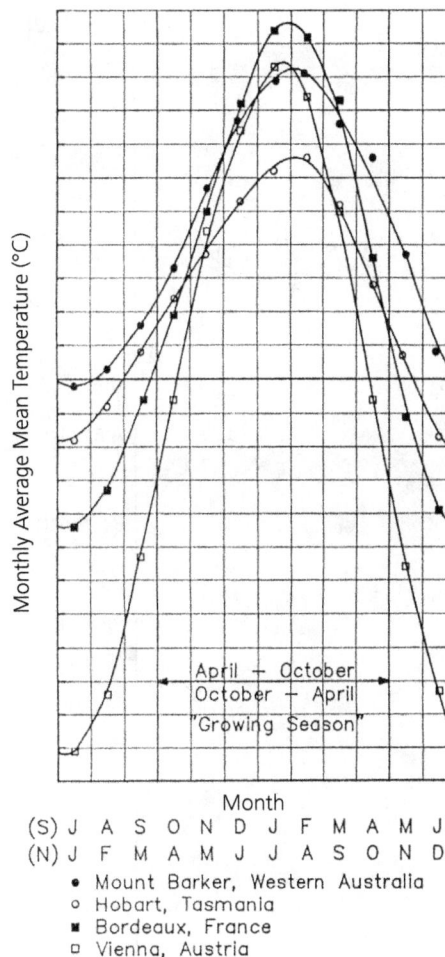

Figure 1. Annual and growing-season temperature curves, showing differences in continentality: Vienna, Austria (continental); Bordeaux, France (intermediate); Hobart, Tasmania, and Mount Barker, Western Australia (maritime).

Table 1. Continentality of selected vinegrowing areas. Measured as average mean temperature of hottest month minus average mean temperature of coldest month. Raw climatic data (not adjusted for vine sites).

France		USA		South Australia	
Bordeaux	15.3	Fresno, California	20.0	Adelaide/Southern Vales	10.8
Dijon (Burgundy)	18.1	San Francisco, California	6.7	Berri	12.6
Reims (Champagne)	17.2	San Jose, California	10.6	Clare	13.3
Toulon	15.0	Sacramento, California	15.8	Kybybolite	10.1
		Portland, Oregon	15.3	Mount Gambier	8.9
Spain		Walla Walla, Washington	23.6	Strathalbyn	9.6
Ciudad Real	19.2			Port Lincoln	8.4
Malaga	13.6	**Canada**			
		London, Ontario	26.4	**Western Australia**	
Portugal				Albany (Airport)	8.0
Braganca	16.1	**Queensland**		Bakers Hill	12.6
Lisbon	11.7	Killarney	13.2	Bridgetown	10.8
Funchal (Madeira)	6.1	Stanthorpe	13.8	Bunbury	8.8
				Busselton	9.3
Germany		**New South Wales**		Denmark	7.4
Bad Durkheim	18.8	Bega	11.6	Donnybrook	11.0
Frankfurt/Main	18.1	Cessnock (Hunter Valley)	13.2	Guildford (Swan Valley)	11.5
		Cowra	16.3	Karridale	7.1
Austria		Griffith	15.3	Katanning	11.0
Vienna	20.6	Mudgee	15.6	Mandurah	10.2
		Parramatta	11.7	Manjimup	9.7
Italy				Margaret River	8.9
Perugia (Chianti)	18.3			Mount Barker	9.2
Turin	23.1	**Victoria**		Perth	10.4
Verona	18.1	Ararat (Great Western)	11.7	Pemberton	8.6
		Avoca	12.9	Wokalup	11.0
Croatia		Bairnsdale	9.9		
Zagreb	22.0	Geelong	9.9		
		Merbein	13.7	**Tasmania**	
Hungary		Mornington	9.8	Hobart	8.7
Eger	22.8	Seymour	13.3	Launceston	10.8
		Wangaratta	14.8	St Helens	8.8
				Swansea	8.6
Bulgaria					
Plovdiv	23.9			**New Zealand**	
				Blenheim	10.4
				Napier	10.1
				Nelson	9.7

persisting warm soil and a postulated good supply of root-produced cytokinin as we saw in the last section. The atmospherically cool, equable late and post-ripening period then allows retention of acid and of delicate, volatile aromas during juice concentration under the influence of late infection by the noble rot (*Botrytis cinerea*).

The work of Hale and Buttrose (1974) provides evidence of other potential quality benefits from high – even very high – temperatures in mid-summer, as a prelude to cooler ripening. Maintaining Cabernet Sauvignon vines at 35°C day/30°C night through Period I of berry growth resulted in irreversibly smaller berries, compared with 25/20 or 18/13°C. These berries ripened to the greatest sugar concentrations whatever the Period III temperature. Again, for any Period I temperature the berries were smaller, and their sugar concentrations greater, as Period III temperature was lower. The smallest berries and greatest sugar concentrations of all were when Period I was at the highest and Period III at the lowest temperatures. These berries would also have had the greatest ratio of skin to pulp. On both counts they should have more colour, flavour and aroma per unit of juice, must and wine. The greater quality would, however, be at the expense of grape yield (although as Hardie and Martin (1990) stress, this might be much less so when measured cumulatively over successive seasons, since low yields in one year will usually be compensated for by higher following yields, depending on the overall capacity of the vine).

The study of Burgundy vintages by Gadille (1967, cited by Johnson 1971) points to the same conclusion. Apart from the benefits of a warm, sunny spring in promoting early development, the seasonal characteristics which most clearly distinguished the best quality vintages were high mid-summer temperatures and sunshine hours. The next-most important factor was sunshine in August, and to a lesser extent thereafter. Temperatures during the ripening period (August-September) on average bore little relationship to vintage quality. Considering that the coolest ripening seasons would also have tended to be unfavourably wet and/or sunless, this implies that the very best ripening periods, though sunny, were probably at most only of average temperature. A factor analysis based on seven years of field results in Germany (Alleweldt et al. 1984) drew essentially parallel conclusions.

Perfect ripening and balance are, however, precarious in a continental climate. Normal seasonal variations can equally result in early ripening under high temperatures, or in not ripening fully at all. Continental vintages are therefore notoriously variable.

A slow decline in temperatures and sunshine through the ripening and post-ripening period, such as occurs in coastal California and in all Australian viticultural regions, may seldom capture the delicacy, but at the same time fullness, of the greatest German and other European vintages; but it allows more reliable ripening, to more consistent balances for winemaking, across the range of seasons. Moreover, a greater range of grape maturity types can be grown at one site and still ripen under close to optimum conditions. Besides increasing the range of wine types that can be made, there is commercial advantage in the longer vintage that becomes possible.

Nor is the continental advantage of relatively high mid-summer temperature in reducing berry size and concentrating sugar and flavour necessarily lost. Many of the more maritime viticultural climates have summers that are hot and sunny enough to achieve this. In any case, control over water supply can achieve the same result wherever summer rainfall is limited. I take this point up again in later sections of this chapter, and in Chapter 5.

Therefore, even if maritime environments might not normally aspire to the occasional superb vintage which an anomalous continental season can produce, their potential average wine quality may well be higher. Certainly it will be more consistent.

Insufficient continentality can nevertheless create viticultural problems. A slow rise of temperature after budburst prolongs the period of frost risk. Where the winters are very mild, some varieties are not winter-dormant enough. Budburst can then be very early and erratic, and there is a consequent strong tendency to dominance by a few terminal shoots and substantial failure of budburst lower on the canes.[6] According to Hamilton (1986) and Dr Michael Peterkin (personal communication), the problem is worst in young vines with not yet fully developed root systems, which have suffered from moisture stress late in the previous summer. It is further exacerbated if the young vines have been cropped beyond their capacity (Erland Happ, personal communication).

The combination of early terminal budburst with often prolonged cool, wet and windy weather into late spring makes the young growth both unusually lanky and vulnerable to storm damage. Cool, cloudy or rainy weather at flowering will often cause poor berry set. Lack of high temperatures in late spring/early summer can also reduce the differentiation of fruitful buds for the following season. Such fertile buds as are differentiated will tend to be further out along the cane than in continental climates. Finally, a mild, maritime summer may have temperatures low enough to encourage a prolongation of active, stemmy growth through the fruiting and even ripening periods.

---

6 Any shoot which is already growing tends to suppress the bursting and development of all buds below it on the cane. The suppression is principally mediated by auxins, which are produced in the young growing leaves and travel only downwards along the stem. The resulting monopoly enjoyed by higher shoot(s) over both nutrients and probably cytokinins gives rise to the phenomenon of 'apical dominance', which is more or less universal in plants. See Appendix 1 for a fuller account.

The continental pattern of temperature provides a much more definite control over budburst. Very low temperatures early in spring hold the buds firmly in the temperature-controlled state of 'imposed', or 'enforced' dormancy (Antcliff and May 1961; Baldwin 1966). The release from dormancy is clear-cut because of the steep rise in temperature; the result is an even emergence of shoots along the whole cane, with minimal opportunity for any to attain strong dominance over the others. The shoots then grow under rapidly increasing temperatures. By the time of flowering, fruit set and fruit development, temperatures in a moderate continental climate are usually close to optimal for fruiting and beyond those conducive to stemminess. The dominance of the fruit over vegetative growth, once established in such a climate, normally persists through later fruit development and ripening. This is especially so at high latitudes, where rapidly shortening day-length in late summer probably encourages abscisin production. Even ample rain at this time does not greatly tempt the continental vine into renewed vegetative growth.

The result of all these factors is that vine management is more difficult in cool to mild maritime climates than in continental climates, and needs greater viticultural skills to achieve the best fruit quality.

There are also clear implications for the adaptation of individual grape varieties. The most important for Australia is a general upward displacement of the optimum temperature summation for particular grape varieties, as compared with the much more continental climates of Europe. This can, of course, result in them ripening under higher temperatures. Be that as it may, varieties adapted to cool, but continental, climates need overall warmer maritime climates if they are to experience optimum temperatures during the critical flowering, setting and early fruit growth periods. This is especially so if they are naturally prone to shy setting (coulure) and strong vegetative vigour – as is the case with many of the classic wine varieties. The varieties best able to fruit reliably in cool or mild maritime climates are in general those with high natural fruitfulness and only moderate vegetative vigour.

A corollary is that the need for easily warmed, well drained soils may be just as great in cool maritime climates as in cool continental climates. In the latter a quickly warmed soil is needed for ripening to occur at all in the short growing season available. In the maritime climates it is needed to offset the slow rise of temperature in spring, and thereby to attain adequate cytokinin production and fruitfulness.

Viticulture in the mild, very maritime climate of Margaret River, Western Australia, reflects all these relationships. Varieties most affected by irregular setting or rankness have in general been those which originated in cool and very continental climates (Riesling, Traminer, Chardonnay), whereas Bordeaux varieties are in the main well adapted. The 'problem' varieties have in Australia tended to perform best in warmer and/or more inland and drier environments, such as the Barossa Valley, the Barossa Hills, and Clare and even the Hunter Valley. Riesling and Chardonnay can still be good varieties in the hot Swan Valley. The Malbec, which suffers from coulure and excessive vigour in Bordeaux, does so even more at Margaret River; however it has performed quite well in the Swan Valley, while in the southern region of Western Australia it does best on the dry inland fringe at Frankland. Conversely the fruitful and high-yielding varieties Sémillon, Chenin Blanc and Zinfandel have proved well adapted to Margaret River, and give wines of – for them – surprising intensity and excellence of character.

Temperature variability

Discussion so far has referred mainly to the monthly averages of mean temperature, and their march through the year (continentality). However variability of temperature is a further factor in its own right.

Temperature variability among seasons is, of course, a well-recognized factor in viticulture. At the cold limit and in cool, continental viticultural climates it is responsible for much of the variation among vintages, with more or less complete failure of grape ripening in the coolest seasons and incomplete ripening in many others.[7] The opposite applies at the hot limit of viticulture, with relative (though seldom complete) failure in hot years and the best vintages in cool years. In optimal climates, wine styles fluctuate within generic limits from year to year, but nearly all are potentially good. Obviously this has great importance for commercial viticulture.

But beyond such seasonal fluctuations, there exist forms of short-term temperature variability which greatly affect viticulture and, I argue, grape and wine quality. They include the normal fluctuation between day maxima and night minima (diurnal range), and the variability of the maxima and minima from day to day or week to week, within a given part of the season. As an overall index of such variability I have taken, for a given month, the sum of the range between its average maximum and minimum daily temperatures, and that between its average highest maximum and average lowest minimum temperatures for the month. These data are available from many of the standard climatic records.[8]

---

7 On the other hand, individual warm to hot seasons towards the cold limit of viticulture have historically produced many of the world's acknowledged greatest table wines. Why this should be, rather than that they be regularly produced in comparable normal seasons in warmer climates, is a paradox which arguably holds the climatic key to winegrape quality. I suggest that temperature variability, or more strictly its converse, equability, is an important part of that key.

8 The temperature variability index (TVI) for the full growing season then becomes:

$$\Sigma[(TD_{max} - TD_{min}) + (TM_{max} - TM_{min})],$$

As an example of how this type of temperature variability affects vines, killing frosts after budburst in spring are associated not so much with low average temperatures, or even with low average minimum temperatures, as with too great a variability of the minimum. Low average daily means, and, perhaps almost equally, low average minima, result in delayed budburst and consequent frost escape. It is when average means and minima have been high enough to initiate spring growth, but sporadic frosts still occur, that damage results.

Similarly, heat damage in summer to vines and fruit results more from occasional very marked extremes (usually with strong winds and very low relative humidities) than from high average temperatures. Exposed bunches are especially liable to heat damage, and are at their most sensitive around véraison, so that constitutes a critical period. Again it is red wines which are the most affected by heat, because they depend more than white or rosé wines on skin-derived components, and the skin is the first part of the berry to suffer heat damage.[9]

Australian inland, and some south coastal, environments have very marked temperature variability: to the extent that, because of spring frosts, viticulture is safe only in areas with warm to hot ripening periods. This is especially so in New South Wales. True cool-climate wines as a rule cannot be made in such regions, and the frequent collapse of heat-exposed berries results in both loss of yield and a prevalence of 'cooked' characters in the wines. Temperature variability indices are typically 45 or more, for both growing season and final ripening 30 days, compared with only 30 to 35 in most European vinegrowing areas.

That minimal temperature variability is essential to avoid killing frosts at the cold limit of viticulture is self-evident enough. Nevertheless the same features of topography and soil (as will be discussed in Chapter 4) that locally raise night temperatures appear also to be associated regularly with top vineyard and wine quality in warm, and even hot, regions. This is clear from the maps of Johnson (1971), and suggests that more is involved than simple frost avoidance. Avoidance of heat damage cannot be a full explanation, either. Even in hot climates the best vineyards typically have sunny, relatively warm, aspects. Fairly clearly the association

between temperature equability and wine quality must have a further basis besides the simple avoidance of frost and heat injury.

My interpretation is that relatively constant, intermediate temperatures during ripening specifically favour the biochemical processes of colour/flavour/aroma development in the berries. These processes are capable of continuing 24 hours a day, provided that sugar substrate is available and temperatures are favourable for enzyme activity. Excessive day temperatures may inhibit the enzyme systems even in the absence of visible berry injury, and also accelerate the evaporative loss of volatile aroma compounds from the berries. On the other hand low night temperatures inhibit all metabolism during that period, including respiration. Cold nights can therefore be expected to retard physiological ripening and quality development, and instead promote the accumulation of unchanged sugar in the berries. Went (1953) has described this type of relationship for a number of plant species.

Few directly relevant experiments have been reported for grape vines. One was that of Kliewer and Torres (1972) in California, who studied the effects of controlled day and night temperatures on ripening and colouring of Pinot Noir and Cabernet Sauvignon grapes. Moderate (15–20°C) as opposed to high (30–35°C) mean temperatures increased colouring, as expected. However additionally there was an unexplained strong, and to some extent overriding, effect of day/night temperature range. Best colouring was where the day and night temperatures were the same. Where the range exceeded 10°C, colouring was greatly inhibited. Other experimental results of Kobayoshi et al. (1967), Kliewer (1970) and Buttrose et al. (1971), while less specific, were all consistent with this finding.

A general conclusion can be drawn. The narrower the range of variation about a given mean or average ripening temperature, the greater the grape flavour, aroma and pigmentation will be at a given time of ripening and sugar level. In cool climates the grapes will achieve full flavour ripeness even though sugar may remain low and acid high. In warm, sunny climates they will attain full flavour ripeness before sugar level and pH have become too high, and acid too low; or before breakdown processes have started to predominate in the berries. In all cases the colour, flavour and desirable aroma qualities in the wine that can only come from fully ripe grapes are enhanced relative to potential alcoholic strength.

The implications do not end there. As well as reducing the risk of 'off' aromas and flavours due to berry scorching, equable temperatures can be expected to enhance in particular the retention of those compounds that are most subject to loss at high, but not damaging, temperatures: whether through their low boiling points or through chemical instability. Minimal temperature variability during ripening should thus increase especially the more volatile and unstable

where TD_{max} and TD_{min} are the average daily maximum and minimum and TM_{max} and TM_{min} the average highest maximum and average lowest minimum temperatures, for each of the seven months of the season.

9 Swan Valley winemakers say that excessive heat at, or very soon after, véraison can prejudice the berries' ability to develop good colour and flavour in later ripening, even when no visible damage has occurred; whereas similar heat once ripening is well under way may have little effect on final fruit quality (John Kosovich, personal communication). This suggests either that sugar in the near-ripe berry (or its skin) is protective, or that véraison is a critical time for establishing the enzymic apparatus of ripening. Perhaps both mechanisms operate and are complementary.

grape components, and therefore add to the life and complexity, as well as to the fullness, of wine aroma and flavour.

The best wine-producing regions of Europe are all characterized by narrow ranges of day-night and day to day temperature variation through the growing and ripening seasons, as are, even more, the best vineyard sites within them (Chapter 4). A pattern of regional differences consistent with this appears now to be emerging also in Australia and west coastal USA, although it is not yet fully clear and there are many confounding factors. Some of these are touched on in discussions of the individual countries and regions (Chapters 9–22).

The temperature variability thesis in some ways contradicts the widely-held idea that a 'long, cool ripening period' is needed for table wine quality. Not in dispute is the fact that moderately low ripening temperatures (and therefore perhaps slower ripening) are needed for certain highly-regarded wine styles. But slow ripening is not, in itself, a virtue. In fact physiological ripening should be both as rapid and as complete as possible, for any given mean or average temperature. This not only gives the best composition balance, with the desirable flavour characteristics of full ripeness and least loss of aromatics from the grapes; rapid ripening also reduces the time available for loss of quality due to environmental insults such as heatwaves, hail, or heavy rain.

Very good grape quality for winemaking is in fact fully compatible with warm, and even quite hot, ripening conditions, provided that they are equable. But picking in hot climates must be carefully timed, because the optimum harvest stage can be brief. That is one of the main advantages of mechanical harvesting in hot climates. At the same time, restricted temperature variability during ripening can be expected to minimize the problem. Because the different ripening processes then proceed in balance with each other, largely regardless of mean temperature, picking can be tolerated over a longer period. Differences will be in wine style rather than in quality. Early picking produces wines that are light and fresh; late picking, wines that are full-flavoured and soft; but each will be in harmonious balance for its style. With great temperature variability during ripening, such harmony may be approached only at a fleeting optimum maturity, if at all.

A final point about ripening period temperature variability is that there is a trade-off against average temperature. Minimal variability makes viticulture and good grape quality possible across the greatest range of average temperatures: with least damage from extremes in either direction, and with greatest tolerance of temperature differences among seasons. Conversely, an optimum average temperature permits the greatest tolerance of variability. Ideally, there should be an optimum average temperature during ripening (for a given wine style), combined with the least possible day-night and day to day variability of temperature throughout both growth and ripening.

Sunshine hours

These have their own effects, both directly and by interaction with temperature. They can be considered in relation to specific times of season and vine growth stages, or over the whole season as a general climatic feature.

Universal experience shows that ample sunshine at flowering promotes fruit set. This may be partly because of good conditions for pollination, and partly through enhanced current photosynthesis and assimilate availability. Indirect mechanisms are probably also involved. For instance a strong supply of assimilates to the roots because of abundant sunshine, combined with warming of the soil, results in a good returning supply of cytokinins.

In combination with high enough temperature, ample sunshine and/or high light intensities in late spring and early summer favour fruitfulness of the new buds then being formed (May and Antcliff 1963; Baldwin 1965; Buttrose 1970, 1974; Srinivasan and Mullins 1981). This has again been generally interpreted as resulting from improved assimilate supply. However May (1965) found that direct sun-illumination of the individual differentiating buds on Sultana vines enhanced their fruitfulness regardless of sunshine received by the surrounding leaves, suggesting that assimilate supply is not the only factor. A likely explanation, in the light of subsequent research, is that local heating and the spectral quality of direct sunlight both help by reducing the bud's sensitivity to gibberellins.

For currently fruit-bearing canes, much sunshine during ripening predictably leads to full and rapid sugar accumulation in the berries. Plentiful sunshine prior to ripening can also contribute, via sugar or starch reserves built up within the vine. As much as 40 per cent of the sugars moving into ripening grapes can come from storages in the stems and roots (Kliewer and Antcliff 1970; May 1987).

Gadille (1967), cited by Johnson (1971), showed that the great Burgundy years have above-average sunshine hours throughout the growing season; but the periods when sunshine appears to be most influential are the spring, and again just before and at véraison in early August. As already noted, temperatures in great years also tend to be above-average in early spring (resulting in early flowering and usually early ripening), and again in July; but they are not obviously so during August-September. All these relationships support the experimental indications from the work of Pirie and Mullins (1977), and of Carbonneau and Huglin (1982), that a rapid flux of sugar into the berries *at and soon after* véraison is especially significant for quality development in winegrapes.

Everything discussed so far seems to suggest that

unlimited sunshine hours are desirable for viticulture. In practice, the issue is less simple. Very sunny climates mostly have marked temperature variability. They usually have low relative humidities, which are also detrimental, as described later in this chapter. A reasonable conclusion is that sunshine hours are positively related to both yield and quality over a fairly wide range, *but only if temperature variability and relative humidity remain favourable.* That understood, sunshine hours can now be examined in relation to temperature, and in terms of minimum seasonal requirements for particular wine styles.

Figure 2 graphs whole-season sunshine hours against biologically effective day degrees (explained in Chapter 6), for representative climates of Europe, Australasia and western USA. Keeping in mind the limitations of whole-season summations of any kind as predictors of vine performance, some broad conclusions follow.

The first is that vines in warm climates need more sunshine hours to achieve a given fullness of wine body and style than those in cool climates. And indeed this is a logical expectation, since plant respiration consumes progressively more assimilate as temperature increases, whereas assimilate production by photosynthesis reaches a plateau at intermediate temperatures. Thus the ratio of growing season sunshine hours to temperature summation is a meaningful measure.

Temperature, not sunshine, appears to be the primary limiting factor and determinant of potential wine style in most cool vinegrowing climates: that is, up to about 1450 biologically effective day degrees. Temperature summation governs whether particular vine varieties will ripen at all; and if so, when, and on average under what temperatures and sunshine hours.

Differences in sunshine hours over the growing season, with temperature summation constant, still probably influence sugar levels at given physiological ripening stages in cool climates. There will also be some indirect effect of late-season sunshine through soil warming. The best seasons are always sunny throughout. Nevertheless at the cool margin it remains true that flavour development is limited primarily by the degree of physiological ripeness achieved, and that is controlled directly by temperature. To the extent that sunshine hours contribute to physiological ripeness, it is through their effects on temperature. Just as importantly, flavour is limited by the very early maturity and inferior winemaking quality of the grape varieties that must be grown to ensure ripening in such climates (Chapter 7).

Becker (1977) suggested a minimum of 1250 sunshine hours from April to October for Riesling to succeed in northern Europe; but he noted that all European environments that are warm enough normally meet that requirement. Figure 2 supports this as a generalization for virtually all viticulture. If 1,200 growing-season sunshine hours is a generous estimate of the requirement for the earliest-maturing grape varieties, of which Riesling is far from being one, it can be seen that all existing cool viticultural environments amply exceed the minimum hours in average seasons. Added evidence that temperature, not sunshine, is the main limitation is the fact that no truly cool environment produces naturally full-bodied wines (other than occasionally with noble rotting), no matter how much sunshine it gets.

Between 1450 and 1700 biologically effective day degrees, temperature and sunshine hours are to varying degrees co-limiting. All grape varieties ripen; but effective temperature summation, together with grape maturity type, still has a strong influence on ripening date and therefore on ripening-period temperature and sunshine hours.

From 1450 to perhaps 1550 effective day degrees and with less than 1600 sunshine hours, wine styles are confined strictly to table wines, mainly medium-bodied. This is the typical environmental range for Bordeaux-style wines. Of particular interest is the fact that the cooler southern areas of Western Australia, such as Albany-Mount Barker and Manjimup-Pemberton, together with the warmest areas around Port Phillip Bay in Victoria, form a close cluster in Figure 2 around Bordeaux and Saint Emilion (points 94 and 103). Viticultural regions in this temperature range with more than 1600 sunshine hours still make almost exclusively table wines, but with generally more alcohol and body. Examples include Valence (Hermitage) in France; most wine-producing areas of Bulgaria; Seymour, Avoca and the Strathbogie Range in Victoria; and Padthaway and the Barossa Valley in South Australia.

From 1550 to 1700 effective day degrees is a truly transitional zone, making mostly full-bodied table wines and some fortified wines from early to midseason grape varieties which achieve a high sugar content. Sunshine hours are now at least as influential as temperature. The cooler and/or less sunny parts still make predominantly table wines, from a full range of grape varieties; but their production also includes light, fruity fortified wines made from mainly early-maturing white varieties such as Frontignac, Muscadelle and Verdelho (see Chapter 7, Table 5). Representative areas include much of the south of France, as typified by Montpellier and Orange; Bendigo in Victoria; Langhorne Creek in South Australia; and Yallingup-Busselton in Western Australia. The warmer, sunnier parts of this transitional zone include the Douro Valley of Portugal, Inverell, Mudgee and Cowra in New South Wales, Wangaratta-Rutherglen in Victoria, Clare in South Australia, and Bunbury-Donnybrook in Western Australia. Such areas are noted equally for their mainly medium-bodied fortified wines, from both white and red grape varieties of early to midseason maturity, and for their full-bodied table wines. They also produce tablegrapes of excellent quality.

Legend to Figure 2 on page 24

Location	Point number on graph	Corresp. table number	Location	Point number on graph	Corresp. table number
Western Australia			**Victoria**		
Albany (airport)	1	6	Ararat (Great Western)	39	51
Bakers Hill	2	7	Avoca	40	52
Bridgetown	3	8	Bairnsdale	41	53
Bunbury	4	9	Ballarat	42	54
Busselton	5	10	Bendigo	43	56
Collie	6	11	Euroa (Strathbogie Range)	44	59
Denmark	7	12	Geelong	45	60
Donnybrook	8	13	Healesville (Yarra Valley)	46	62
Esperance	9	14	Heywood (Drumborg)	47	63
Guildford (Swan V.)	10	15	Leongatha	48	64
Kalamunda	11	16	Maffra	49	65
Karridale	12	17	Melbourne	50	66
Kojonup	13	19	Merbein	51	67
Mandurah	14	20	Mornington	52	68
Manjimup	15	21	Orbost	53	69
Margaret River	16	22	Portland	54	70
Mount Barker	17	23	Seymour	55	72
Pemberton	18	25	Swan Hill	56	73
Perth	19	26	Wangaratta	57	75
Porongurup	20	27	Werribee	58	76
Rocky Gully-Frankland	21	28			
Wandering	22	29	**New South Wales**		
Wilyabrup	23	30	Armidale	59	77
Wokalup	24	31	Bega	60	79
Yallingup	25	32	Canberra	61	82
			Cessnock (Hunter Valley)	62	83
South Australia			Coonabarabran	63	84
Adelaide (city)	26	34	Corowa	64	86
Adelaide, Waite Institute	27	35	Cowra	65	87
Barossa Valley	28	36	Forbes	66	88
Belair	29	37	Glen Innes	67	89
Berri	30	38	Goulburn	68	90
Clare	31	39	Griffith	69	91
Coonawarra	32	40	Inverell	70	92
Eden Valley – Springton	33	41	Jerry's Plains (Hunter Valley)	71	93
Nuriootpa	34	43	Leeton	72	95
Padthaway (Keppoch Valley)	35	44	Mudgee	73	98
Port Lincoln	36	45	Murrurundi	74	99
Stirling	37	47	Parramatta	75	101
Strathalbyn	38	48	Picton	76	102
			Richmond	77	103
			Scone (Upper Hunter Valley)	78	104
			Tenterfield	79	106
			Young	80	107

Location	Point number on graph	Corresp. table number	Location	Point number on graph	Corresp. table number
Queensland			**Austria**		
Stanthorpe	81	109	Vienna	113	149
Tasmania			**Hungary**		
Bushy Park	82	112	Eger (Tokay etc.)	114	151
Hobart	83	113	Keszthely (Lake Balaton)	115	152
Launceston	84	114	Mosonmagyarovar	116	153
Risdon	85	115			
			Bulgaria		
New Zealand			Pleven	117	154
Auckland	86	117	Plovdiv	118	155
Blenheim	87	118	Sliven	119	156
Napier	88	119			
			Italy		
France			Bolzano (Alto Adige)	120	157
Angers (Anjou etc.)	89	120	Siena (Chianti)	121	158
Angoulême (Cognac)	90	121			
Auxerre (Chablis)	91	122	**Portugal**		
Beaujolais	92	123	Braga (Minho etc.)	122	161
Bergerac	93	124	Caldas da Rainha	123	163
Bordeaux	94	125/6	Dão	124	163
Cahors	95	127	Lisbon	125	164
Colmar (Alsace)	96	128	Pinhão (Middle Douro)	126	165
Cosne (Upper Loire)	97	129	Régua (Lower Douro)	127	166
Dijon (Burgundy)	98	130	Vila Real (Mateus etc.)	128	167
Montpellier, Orange	99	132/4			
Nantes (Muscadet)	100	133	**England**		
Orléans (Loiret)	101	135	Greenwich	129	171
Reims (Champagne)	102	137	Southampton	130	173
Saint Emilion	103	138			
Serrières (Mâconnais)	104	139	**California, USA**		
Toulon	105	140	Fresno, (San Joaquin Valley)	131	174
Tours (Vouvray)	106	141	Napa (Napa Valley)	132	175
Valence (Hermitage etc.)	107	142	San Jose (Santa Clara Valley)	133	177
Germany			**Oregon, USA**		
Bad Durkheim (Rheinpfalz)	108	143	Portland	134	178
Freiburg (South Baden)	109	144	Roseburg	135	179
Geisenheim (Rheingau)	110	145			
			Washington, USA		
Switzerland			Prosser	136	180
Geneva	111	147	Seattle	137	181
Lugarno (Ticino)	112	148	Walla Walla	138	182

Figure 2. Total sunshine hours vs summation of biologically effective day degrees C for October-April (Southern Hemisphere) or April-October (Northern Hemisphere). For definition of biologically effective day degrees, see Chapter 6.

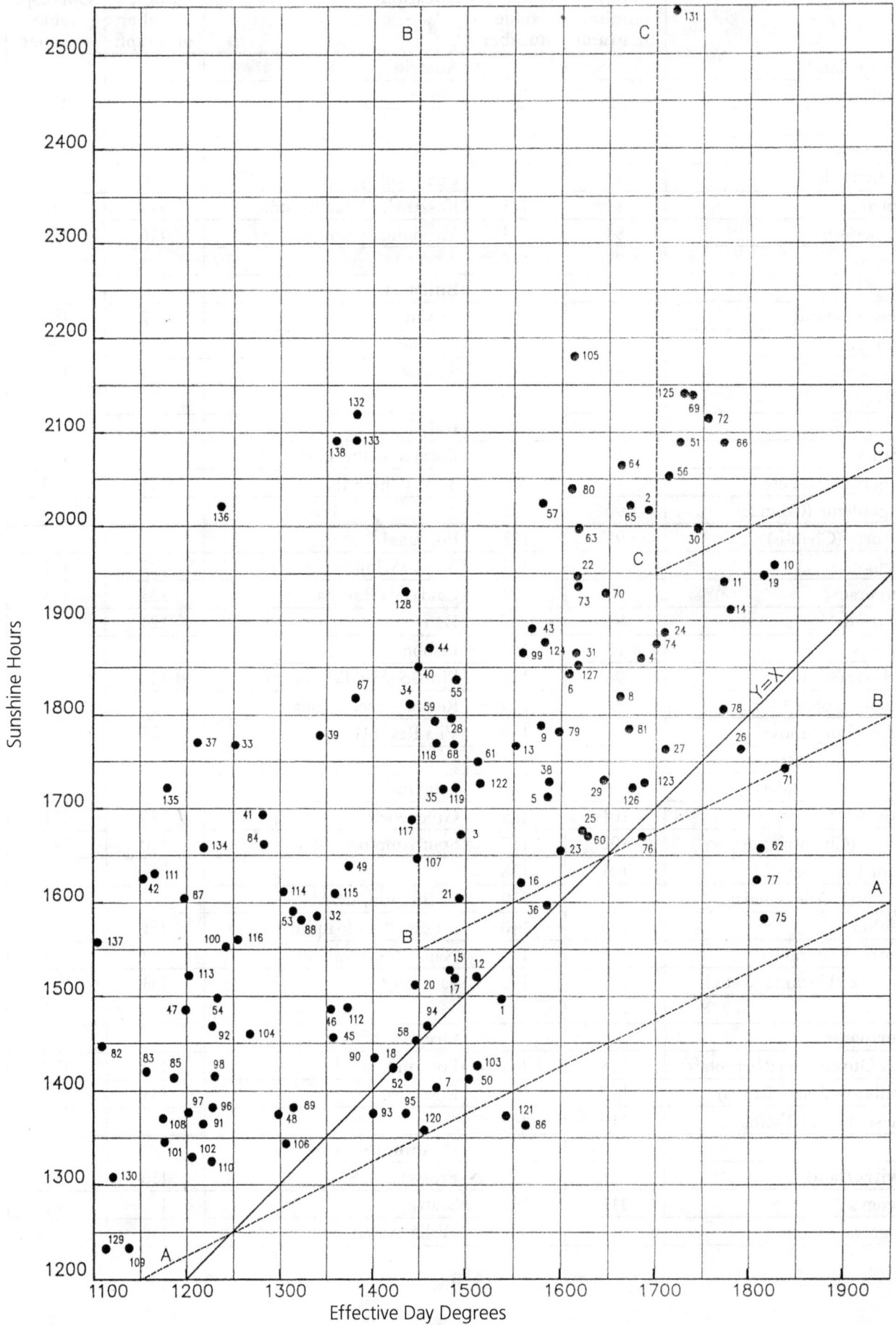

Above 1700 biologically effective day degrees, temperature is no longer a significant limiting factor and all grape varieties ripen under warm to hot conditions in average seasons. With a high rate of assimilate loss through vine respiration, it is now primarily sunshine hours (together with canopy light relations and crop load) which limit berry sugar levels and determine potential wine style. The lower limit of growing-season sunshine hours at these temperatures for acceptable winegrapes is uncertain, but probably lies between 1500 and 1600. With less than about 1750 hours, sugar accumulation is only enough to make table wines (for instance see points 62 and 71 in Figure 2, for the Hunter Valley in New South Wales). Such hot-climate table wines tend naturally towards low acidity and softness.

About 1750–2000 sunshine hours in moderately hot areas give potentially good fortified wines and full-bodied, naturally low-acid table wines. For wine quality these areas usually depend on high enough relative humidities or regular afternoon sea breezes to moderate stresses on the vines, as discussed later in this chapter. The hotter west coastal vinegrowing areas of Western Australia fall into this category, including the Swan Valley and Gingin-Bindoon, Kalamunda, Perth and Mandurah (points 10, 11, 19 and 14 on Figure 2); also the Adelaide-Southern Vales area in South Australia (points 26 and 27) which is a little cooler and less sunny, but has lower afternoon relative humidities. These also are good environments for high-quality tablegrapes.

Hot environments with more than 2000 whole-season sunshine hours nearly always have very low relative humidities as well, leading readily to raisining of the grapes. Most suffer from periodic extreme high temperatures. Such environments are beyond the limit for making regularly good table wines, and perhaps also for the best fortified wines, other than those of very high sugar content which depend on natural raisining. They are best suited for early-market tablegrapes and for the production of dried vine fruits. Examples include Merbein in Victoria, Griffith in New South Wales, and Fresno in California.

The lines AA, BBB and CCC in Figure 2 give a broad delineation of the natural wine styles with differing combinations of growing season sunshine hours and effective temperature summations. Environments between AA and BBB are more or less restricted to table wines; those from BBB to CCC can produce both table and fortified wines; while those above and to the right of CCC are beyond the limit for good winemaking, and best confined to early table and drying grapes. Obviously there is a continuous gradation. Scope also exists at intermediate temperatures for manipulation via grape varieties of differing maturities, and through vineyard management. Technical adjustments can be made in winemaking, e.g. deacidification and/or addition of sugar or concentrated must in cold or sunless

environments and seasons. In hot climates, careful timing of picking and acid/pH adjustment can equally result in greatly improved wines.

Nevertheless the best wines, other things being equal, nearly always come from environments giving an appropriate natural balance in the grape for their particular style. Graphing of the individual data for new locations or sites against Figure 2 should give a broad guide to the styles (though not fully to the quality) of the wines they will most easily produce.

Figure 2 also allows a rough estimate of the quantitative relationship between effective temperature summation and sunshine requirement.

The line AA defines the approximate minimum sunshine hours needed over the seven-month season to produce satisfactory table wines from *vinifera* grapes. Among the existing viticultural regions covered in this study, only Auckland, New Zealand, falls markedly below the line; and it is notable that viticulture there has in the past depended largely on species hybrid varieties. The angled arms of BBB and CCC define respectively the approximate transitions to full-bodied table wines plus fortified wines, and to mainly drying grapes, as governed by sunshine hours. The slopes of all three lines suggest that for every increase of two effective day degrees for the season, approximately one extra hour of sunshine is needed to produce wines of equivalent body or sweetness.

The paucity of hot climates that are naturally suited to table wine production reflects other important climatic limitations besides temperature and sunshine hours. Most such climates are excessively humid, leading (at those temperatures) to fungal diseases in *vinifera* vines. Only a few have predominantly low enough relative humidities and disease risk. They include notably the Hunter Valley of New South Wales, which escapes the worst disease risk through its frequent dry morning land breezes. (See further discussion under relative humidity, below.)

The simple ratio of sunshine hours to effective temperature summation has some predictive value in its own right in warm to hot climates, assuming that other climatic elements are appropriate for viticulture. As a general rule, ratios less than 1 are associated with only low to moderate sugar contents in the berries at full physiological ripeness. Colour, flavour and aroma can still be good, because such ratios tend to be correlated with minimal temperature variability. Softness follows from the low acid content and fully ripe tannin structure of the grapes by the time sugar reaches the necessary level.

With increases in the ratio beyond 1 the wines tend progressively to become more tannic, astringent and alcoholic. However any adverse changes are probably due less to the ratios themselves than to the other factors with which high ratios are correlated. These include excessive temperature variability, as already discussed, and atmospheric aridity and moisture stresses on the vines, as discussed later in this chapter.[10]

Figure 2 raises three final points of interest. First, the north American west coast stands out in its very high ratios of sunshine hours to effective temperature summation. The more inland sites also have wide diurnal temperature ranges and low relative humidities. The strongly tannic nature of many Californian red wines, as compared with those of Europe and even Australia, is thus perhaps as much a reflection of natural environment as of winemaking methods.

Toulon in France (point 105 in Figure 2, and Chapter 16, Table 140), and to a lesser extent Lisbon in Portugal (point 125 in Figure 2, and Chapter 20, Table 164), also have very wide ratios of sunshine hours to temperature summation; but they are anomalous in having at the same time very moderate temperature variability and quite high afternoon relative humidities. The combination is typical of south-coastal locations in the Northern Hemisphere, and equally of north-coastal locations in the Southern Hemisphere: see Yallingup (point 25 in Figure 2, and Chapter 9, Table 32) in Western Australia; Launceston (point 84 in Figure 2, and Chapter 14, Table 114) in Tasmania; and Blenheim (point 87 in Figure 2, and Chapter 15, Table 118) in New Zealand. Such environments have unique advantages for viticulture.

The last point concerns the two central/northern Italian sites for which I have sunshine data: Bolzano (point 120 in Figure 2, and Chapter 19, Table 157) and Siena (point 121 in Figure 2, and Chapter 19, Table 158). Their climates are surprisingly continental and cloudy considering their Mediterranean location, and they have apparently minimal sunshine hours for their temperatures to make good wines. Much rain falls during ripening. It may therefore be no accident that the traditional northern Italian grape varieties, while they presumably give well-balanced wines in their home environments, often give acid, astringent wines elsewhere. Might a systematic trade-off exist between sunshine hours to temperature summation ratio and the natural winemaking adaptations of individual grape varieties? If so, it could be very useful in predicting new areas of adaptation. For example, central and northern Italian varieties should be well adapted to coastal environments of New South Wales; to New England in northern New South Wales and the Granite Belt of south-east Queensland; and to the Auckland region in New Zealand.

Light quality

Only recently have researchers come to recognize the viticultural importance of light spectral quality: that is, the composition of the light in terms of its component wavelengths and colours; although with

hindsight, some aspects of vine lore could readily be explained in terms of it. The evidence now indicates that gradations of spectral quality within the canopy control very important aspects of vine growth physiology, and might also directly influence fruit qualities (Smart et al. 1982a; Smart 1985a, 1987b; Morgan et al. 1985). Morgan and Smith (1981) and Smith (1982) give general accounts of the effects of light quality on plant growth.

Red wavelengths, and full sunlight of normal spectral quality, are known universally to elicit leafy, stocky growth, vigorous branching, fruitfulness, dark green leaf colour, and often a generous production of pigments. These responses are broadly analogous to those of an ample cytokinin supply from the roots, as already described under root temperature and in Appendix 1.

A very different quality of light prevails in the shaded internal parts of vine and other plant canopies. Most of the light that is useful for photosynthesis, including the red wavelengths, has been absorbed by the outer leaves which first intercepted the sunlight. Wavelengths which are reflected or filter through to the shaded canopy interior consist largely of the non-absorbed green part of the spectrum, together with some in the near-invisible 'far red', or 'near infra-red', zone which are likewise poorly absorbed by leaves. The ratio of red (wavelengths about 660 nm) to far red (about 730 nm) light is reduced in full canopy shade to only about a tenth of that in direct sunlight. The plant senses this ratio via a light-sensitive pigment known as phytochrome: see Siegelman (1969). Phytochrome can reportedly discriminate among ratios of red to far red wavelengths at light intensities as weak as bright moonlight.

The effects on plant growth of a low red to far red wavelength ratio are the opposite of those associated with direct sunlight and red light. They agree instead with those associated with dominance by the hormone gibberellin, as described in Appendix 1: that is, greater elongation of stem internodes, less lateral branching, reduced intensity of green colour, and nonproduction of anthocyanin pigments. The mechanism for this appears to lie in an enhanced sensitivity, mediated by phytochrome, to the plant's own gibberellins. Kasperbauer et al. (1984) and Kasperbauer (1987) also showed for soybeans that there is a greater allocation of assimilates to the plant tops and less to the roots, which are the site of cytokinin synthesis, so an indirectly-caused reduction in cytokinin production may contribute.

As we have already seen, the differentiation of unfruitful buds in grapevines is likewise associated with a dominance of gibberellins over cytokinins, in a balance which can be greatly influenced by variation in cytokinin production due to root environment. A low red to far red wavelength ratio within the canopy thus appears to work in conjunction with low root

10 Dr Bill Pannell, founder of the Moss Wood vineyard in Western Australia encapsulated the concept nicely in a conversation with the author: 'Soft climate, soft wines; hard climate, hard wines'.

temperature and poor soil aeration in encouraging gibberellin dominance, and thus lanky, non-fruitful growth: the typical scenario for coulure in susceptible vine varieties.

Evidence that the red to far red ratio of the light illuminating the bunches and berries themselves can influence their qualities for winemaking is still only indirect. Several recent studies (Kliewer and Bledsoe 1987; Koblet 1987; Smart 1987b; Morrison 1988) have confirmed that the winemaking composition of grapes growing in dense canopies can be improved by removing leaves from immediately around the bunches. Such results could hardly have been due to increased photosynthesis. One possible mechanism is that berry exposure to direct light favours cytokinin dominance, and therefore sugar accumulation, in the exposed skins. A possible complementary explanation is that the removal of photosynthetically mostly useless leaves surrounding the bunches eliminates a sink locally competing with them for cytokinins moving up from the root system. Additionally there is evidence, reviewed by Smart (1987b), of direct links between light quality and the biochemistry of ripening in the exposed berries, via the phytochrome system. The evidence is strong for anthocyanin synthesis, but one may reasonably suspect that the implications for ripening biochemistry are much wider.[11]

It is in any case well known from practical field observation that well exposed bunches (provided that they are not heat-damaged) generally have the best colour and flavour. If the quality of the light received by the shaded bunches and shaded lower leaves is also important, some other interesting implications follow as well. For instance it could help to explain the traditional value placed on red soils for viticulture, especially for red wines, because a considerable part of

the light illuminating the bunches and under-leaves is reflected from the soil. It could also have implications for soil management. These questions are taken up again in Chapters 4 and 5.

Rainfall and its seasonal distribution

The adequacy of total or whole-season rainfall for vine growth is an obvious climatic criterion for purely rain-fed vines, or for where good-quality water for supplementary watering is limited. It needs no further discussion here.

Moisture availability at particular growth stages, on the other hand, has complex and important implications. Van Zyl and van Huyssteen (1984) and Ludvigsen (1987a) have discussed some of these in detail.

Flowering and berry setting are both sensitive to moisture stress, with marked effects on potential yield. The differentiation of fruitful buds on the new shoots in late spring-early summer is also somewhat susceptible to moisture stress.

Equally, however, both setting and fruitful bud differentiation can be upset by heavy rain: the former by direct interference with pollination, and both by the often accompanying lack of sunshine and photosynthesis. Vigorous vegetative growth encouraged by heavy spring rains may further affect setting and fruitful bud differentiation, by shading the lower leaves and new buds, and by monopolizing assimilates and producing plentiful auxins and gibberellins.

From bud differentiation to just before véraison, the vines appear to tolerate moisture stress well. Hardie and Considine (1976) found that the berries, from being very sensitive to moisture stress in their initial growth stages, gain resistance to it once they reach about 4 mm diameter. Some stress then becomes desirable in cool-summer environments to encourage a natural cessation of vegetative growth before véraison: both to divert a maximum of assimilates and cytokinins to the fruit, and perhaps to supply enough abscisin to help initiate ripening. It is also desirable that the existing vegetative growth be 'hardened', so that it will best tolerate any stresses during the ripening period to follow. This stage corresponds well with the July-August period when high temperatures and sunshine hours were clearly associated with the best grape quality in Gadille's study of Burgundy vintages.

The week or two up to and around véraison appears again to be a period of some sensitivity to moisture stress, as well as to heat. Adequate moisture then is needed for the full development of berry size and realization of yield (if that is what is wanted). A continued mild moisture stress through this period is said to increase colour and flavour in the wine by limiting berry size (Ludvigsen 1987a; Matthews and Anderson 1988), but at the expense of current season's yield. Any stress extending to véraison should nevertheless probably be no more than enough to give moderate control over berry size, and to hold back vegetative growth and ensure full diversion of

---

11 It is also of interest to speculate on the ecological significance of anthocyanin formation in plants, which is near-universal. One function is very probably that of a natural, in-built fungicide. However at least one other function can reasonably be hypothesized on the grounds of influence on light relations within the canopy. Except in fruits which are strong sugar accumulators, and in senescent leaves where starches are being converted to sugar for export prior to leaf-drop, anthocyanin normally forms only in surface tissues with good illumination and a consequent surplus of assimilates. Its presence shows that these tissues have more light and/or sugar than they immediately need. Anthocyanin formation, in proportion to the surplus of sugar, might then be seen as constituting a self-regulating mechanism whereby excess red wavelengths are reflected away, to be received by less well illuminated leaf and green stem tissues which can use them to advantage. In other words, anthocyanins provide a plausible mechanism for rationing photosynthetically useful wavelengths among plant tissues, so as to achieve their most efficient overall use. This seems consistent with the author's observation over many years of plant breeding, that anthocyanin-free genotypes of all plants species are in general less vigorous than their normal, flexibly anthocyanin-producing relatives; also, that they are less fruitful, consistent with a light quality within the canopy which is less conducive to cytokinin effects and more to gibberellin dominance.

assimilates to the fruit. Carbonneau and Huglin (1982) considered that a rapid flux of sugar to the berries in the immediate post-véraison period is especially important for later flavour development, so severe moisture stress at véraison is likely to be counter-productive.

From then until harvest it is universally essential that there be continuously enough moisture to keep the vines healthy and actively nurturing the crop to full maturity. Root growth must continue so as to supply cytokinins, and thereby maintain the resistance of vines and fruit against stress and senescence, including the integrity of the ripening biochemistry. The prime importance of avoiding moisture stress during ripening is illustrated by the fact that even in a cool, summer-rainfall climate such as Germany's, supplementary water at this time can improve wine quality (Ruhl 1988). Long (1987) reported that Cabernet Sauvignon vines in California, subjected to moisture deficit up to véraison but not thereafter, produced wine that was 'more intensely coloured and flavoured, with a 'bigger' tannin structure and more 'ripe, fruity, berry' characters'. Normally the need for moisture stress to discourage continued vegetative growth is removed in fully-cropping vines from véraison onwards, because by that stage the bunches are sufficiently established as the vine's dominant sink for assimilates and growth substances. Nevertheless a danger remains of strong continued growth in under-cropping vines and in cool, maritime climates. The nutrient and hormone needs of vigorous growth can then still prevail over those of the fruit, to the detriment of fruit quality as we saw earlier under 'Continentality'.

Stress leading to leaf drop is especially injurious in hot climates if it exposes the bunches fully to direct overhead sun. Stress without leaf drop can still markedly reduce photosynthesis, and hence the passage of sugar (but not necessarily of potassium) to the ripening berries. Freeman et al. (1982), and Iland (1988) have suggested that the export of potassium from stressed leaves is actually enhanced, with a resulting much higher ratio of potassium to sugar moving through the vine's phloem system to the berries. This can have undesirable effects on must pH and wine quality, as explained below in the context of relative humidity and saturation deficit..

Happ (1987) has reported experience at Yallingup, Western Australia, which corroborates both the above points.

> Avoiding pre-vintage stress and premature leaf drop on an unacceptable scale involves keeping the water up to the vines after véraison and particularly in the last five weeks of ripening. We have observed that stressed vines produce grapes with less acid and a higher pH for a given sugar level. The berries are definitely less palatable.

A continued adequate water supply from harvest until normal leaf-fall enables the canes to mature fully and the vines to build up assimilate reserves, as a springboard for vigorous, normally-timed budburst and early growth in the spring. Any over-supply of water that might encourage a resumption of vegetative growth is to be avoided, however.

The effects of excess winter or early spring rainfall depend mainly on soil drainage. Waterlogging can be very injurious if it persists until budburst and the resumption of active root growth. Young growing vine roots are very sensitive to the lack of soil aeration that results from waterlogging, with the deeper roots normally the most affected. On the other hand the soil moisture reserves need to be re-charged.

Good spring rain is essential in Mediterranean-climate vineyards which rely partly or wholly on natural rainfall, together with enough depth and water-holding capacity of the soil to carry the vines through summer largely on soil reserves (see further discussion of soil types in Chapter 4). Preferably it should not come during flowering. The balance between too little and too much rain in spring, and the optimum timing of spring rains, can be delicate in any environment.

Rain during ripening is another case where requirements can conflict. On the one hand enough moisture is needed during this period to avoid undue vine stress, as noted above. On the other, much rain during ripening can be harmful for many reasons – especially if it is accompanied by hail, which can be devastating. The best combination appears to be that of a largely rain-free (but non-stressful) ripening period, together with a capacity for controlled supplementary watering if needed (Chapter 5).

In hot, atmospherically arid climates, even slight rainfall during ripening can cause serious berry splitting in susceptible varieties such as Chenin Blanc and Zinfandel. Berry splitting occurs due to 'a mismatch between the expansion growth of the flesh and the extensibility of the skin' (Lang and Thorpe 1988). Compared with those growing in a continuously moist, mild climate, grapes which have developed under hot, arid conditions experience a proportionately greater flux of water into the berries when moisture suddenly becomes available; their possibly stress-toughened skins may also be less extensible. Lang and Thorpe additionally emphasize the importance of direct water intake through the skins of wet berries, and therefore of ventilation to promote quick berry drying as a way to reduce splitting.

In summary, no one environment can be expected to provide regularly the perfect rainfall regime. The best, in practice, appear to be either those with moderate temperatures and rainfall distributed fairly evenly through the year, but still having enough sunshine hours – such as the cooler vine regions of Europe and the warmer parts of Tasmania and the South Island of New Zealand; or alternatively, Mediterranean-type climates with high and reliable autumn-winter-spring rainfall but a sunny, mostly rain-free summer. The latter climatic type must be combined with a deep, retentive but well-drained soil, or if that is sub-optimal, the capacity for controlled supplementary watering in summer.

Relative humidity and saturation deficit[12]

These play a subtle but major role in their own right. As well as contributing directly to vine and fruit injury under extreme temperatures, saturation deficit is the overriding determinant of normal moisture responses in an arid climate. Freeman et al. (1980) found that even fully irrigated vines were unable to take up water fast enough to meet transpirational (leaf evaporative) demands under hot, dry atmospheric conditions. Photosynthesis ceased in the mid to late morning, almost as early as on non-irrigated vines, because the leaf pores (stomata) closed in order to conserve moisture.

The critical influence of atmospheric saturation deficit extends further. Rawson et al. (1977) and others have shown that a greater saturation deficit results in more moisture being transpired by the leaves per unit of carbon dioxide intake through the stomata. Growth and yield per unit of water transpired is therefore reduced as saturation deficit increases. And because potassium uptake into plant tops via the xylem (sap-conducting) system is closely related on any given soil to the amount of water absorbed from it (Russell and Barber 1960; Barber 1985; J.F. Loneragan, A.D. Robson, personal communications), this means that the amount of potassium accumulating in the leaves and stems per unit of photosynthetically-produced sugar increases with saturation deficit. Since much of both the potassium and the sugar ultimately migrates to the ripening berries, the result is increased juice, must and wine potassium content. This is so regardless of irrigation or lack of it, and of soil potassium supply.

A high must potassium content is now firmly established as one of the major factors that has limited table wine quality in Australia (Somers 1975). Excess potassium leads to the loss of tartaric acid through its precipitation as insoluble potassium bitartrate (Somers 1977; Boulton 1980; Iland 1987). While this may be advantageous in some cold-climate wines, it is clearly not so in warm and hot climates, where natural grape acid levels at maturity are low already. Results include not only a loss of flavour freshness, but also much greater dangers of wine oxidation and microbial spoilage (Rankine 1989). Practical remedies in winemaking include the

addition of extra tartaric (or other) acid, and/or the use of ion exchange methods. However, such measures are regarded generally as inferior to having naturally well-balanced musts to start with. Any environmental or management factor which increases berry and must potassium content is therefore normally regarded in Australia (and similar environments) as being detrimental to wine quality.

Irrigation is one factor which increases total potassium uptake; but judiciously used, it normally also increases both growth and yield to a similar extent, so that any direct effect on potassium concentration in the vine and fruit is more or less cancelled out. Indeed, irrigation during ripening can probably help by preventing the undue mobilization of potassium out of stressed leaves. The commonly observed adverse effects of irrigation on must and wine pH are due, in fact, primarily to the impaired light relations that a denser leaf canopy can bring about, as will be discussed under canopy management in Chapter 5. The fault lies with vine training that is inadequate for the extra growth achieved, not with watering as such.

Similarly, the use of potassium fertilizers is often wrongly blamed for high must and wine pH. Potassium-deficient soils may indeed help to ensure low wine pH, but this will be at least partly because of the improved canopy light relations caused by poor vine growth. It will be at the expense of yield and the general welfare of the vine, including, very probably, its resistance to diseases (see Chapter 4). Provided that the resulting extra growth is not excessive for the available trellis, the reasonable use of potassium fertilizers to overcome deficiency is unlikely to increase must and wine pH to any appreciable degree, as Goldspink et al. (1989) showed clearly. In cool and humid climates with tolerably adequate trellising there should be little problem of excess potassium uptake even at quite high soil potassium levels, because vines in those climates use water efficiently for photosynthesis and sugar production.

High natural must and wine pH, then, is in part a direct and inescapable result of hot, atmospherically arid climates. The relationship of pH to atmospheric saturation deficit is almost certainly one of the reasons why most of the world's best table wines come from areas, not only of equable ripening temperatures, but also of fairly high daytime relative humidities.

Very high relative humidities can of course promote fungal diseases, especially at higher temperatures, and are usually associated with too much rain and/or not enough sunshine. Early afternoon relative humidities of about 55 to 65 per cent through summer and ripening would appear to be ideal for table wine quality in cool to mild climates, or perhaps 50 to 60 per cent as a best compromise between quality and potential disease problems. About 40 to 50 per cent in the early afternoon seems appropriate for sweet fortified wines in warm climates; or possibly a little lower for very sweet wines, for which partial berry raising may contribute

---

12 Saturation deficit is the difference between the actual water vapour content of a given air body and what it could contain if vapour-saturated (100 per cent relative humidity) at the same temperature. Because the saturation water vapour content of air rises greatly with temperature, heating of air containing a given amount of water vapour simultaneously reduces its relative humidity and increases its saturation deficit. As a direct indicator of the 'evaporative power' of air, saturation deficit is the climatic measure most nearly related to potential moisture stress in vines and other plants. It can be calculated from relative humidity and temperature via standard tables.

to the style. Still lower relative humidities post-ripening may be needed for drying grapes.

All of these considerations point to advantages for winemaking, on a dry continent such as Australia, of a coastal or near-coastal location. The normally regular humid afternoon sea breezes that occur in summer should allow an earlier afternoon resumption of photosynthesis following any late morning or noon closure of the leaf stomata due to stress. Much of the active photosynthesizing period is then under quite moderate saturation deficits. And because the effects on transpiration, photosynthesis and potassium uptake reflect primarily what happens while the stomata are open, this should favourably affect both the vines' water use efficiency and must and wine pH.

Wine style in areas with regular summer afternoon sea breezes can therefore be expected to shift towards that of areas which lack the benefit of sea breezes but which are, by standard climatic statistics, cooler and more humid.

A near-coastal location gives the vine potentially the best of both worlds: predominantly dry land winds at night and in the morning, to minimize the risk of fungal diseases; and mild, moderately humid, sea breezes in the afternoon when they can do the most good. Additionally, the evening fall in temperature is usually slow. A typical summer or autumn day therefore has prolonged periods through the afternoon and evening, and again in the morning, when conditions are close to optimal for photosynthesis and physiological ripening.

Inland areas which have high enough afternoon relative humidities in summer to avoid severe stress normally have too high relative humidities at other times of the day, and often excessive rainfall as well during critical periods: as in northern New South Wales and Queensland. Conversely, inland localities with low enough overall relative humidities and summer rain to avoid serious disease problems mostly have extreme saturation deficits through the middle of the day. Even if the stomata re-open and photosynthesis resumes in the late afternoon, it is still usually under conditions of marked saturation deficit. Must and wine pH therefore always tend to be high. This is additional to the fact that inland areas of Australia tend to experience a sharp alternation between often above-optimum day and often below-optimum night temperatures, leaving only short periods of optimum temperatures in the morning and evening.

Severe continuous stresses occur occasionally in coastal vines when the sea breeze fails, and hot, dry land winds persist throughout the day. The risk of this happening increases with distance inland, and behind hill or mountain barriers which can block the free convectional circulation of air between land and sea. Valleys which penetrate inland have a funnelling effect on the afternoon ingress of cooler, and therefore heavier, marine air: a factor of particular importance in the Hunter Valley on the east coast of Australia, and in the Swan Valley on the west coast. On the generally humid east coast the major valleys probably help at least equally as conduits of drier inland air at night.

I have chosen to use 1500 hours relative humidities to describe viticultural climates, rather than the 900 hours figures used by Smart and Dry (1980), because in most Australian vineyards it is saturation deficit in the middle of the day and afternoon which most limits vine productivity and wine quality. Also, morning figures vary less, and discriminate less among places. While they are directly significant in humid climates as a partial measure of disease risk, in Australia generally that is not the main problem.

It should be noted that the published Northern Hemisphere figures for afternoon relative humidities are mostly from times earlier than 1500 hours, and are therefore not fully comparable with those available for Australia. The viticultural climates of cool-continental Europe also differ from those of Australia and west coastal USA in that cloudiness and rainfall in summer, if anything, increase inland. The differences between European coastal and inland relative humidities are therefore minimal, and the time at which afternoon relative humidities are measured has little effect on the values obtained. In Australia and west coastal USA, by contrast, both the timing of relative humidity measurements and the distance inland are critical.

Windiness

Windiness, as such, has both positive and negative effects. The positive effects are seldom remarked on. As well as being related to the incidence of summer land and sea breezes, they include the maintenance of air circulation and prevention of excessive humidity build-up within the vine canopy. Moderate atmospheric turbulence also helps to maintain equable temperatures in and around the vines, and is especially helpful in preventing frosts. Just as importantly, the movement of leaves allows the intermittent illumination of internal leaves by shafts of sunlight – which is then used with greater photosynthetic efficiency than by leaves which are illuminated continuously (Kriedemann et al. 1973).

Strong winds, on the other hand, are nearly always harmful. Obvious wind damage is of two main types. The first is by strong winds in spring or early summer which injure the young, tender growth and bunches up to and including the time of setting, as described by Hamilton (1988). The problem is greatest in cool and maritime climates, especially in west coastal areas, and on the exposed parts of hills and mountains. The southern coastal plain of France is much subject to such winds.

The second kind of damage is by hot, dry winds in summer. In Australia these are most prevalent inland and in south coastal areas, bringing heat waves from the north and north-west in Victoria and South

Australia, and from the north-east and north along the south coast of Western Australia. Heat waves also occur on the west and to a lesser extent east coasts, but are generally less prolonged. However their timing in these generally warm climates often coincides with the vulnerable periods of véraison and (to a lesser extent) ripening. Ripening on the cooler south coast is later and escapes more often. The greater concern there is is damage to the vines, although heat not infrequently still extends to the time of véraison, and can cause collapse of exposed berries and the imperfect ripening of some which survive. During a severe heat wave on the south coast the dominant land winds may suppress the normal coastal sea breezes for as much as a week.

There is growing evidence in California and Australia (Freeman and Kliewer 1984; Kobriger et al. 1984: Kliewer and Gates 1987; Ewart et al. 1987; Hamilton 1988) that quite moderate winds can cause closure of stomata in the leaves and thus inhibit photosynthesis, despite ample moisture availability to the roots and even reasonable atmospheric humidity. This can reduce yield in the absence of any visible injury. However the little evidence available tends to confirm the theoretical expectation that, because the primary effect of wind is temporarily to close stomata, it does not result in raised must and wine pH in the same way as constant atmospheric aridity (above).

The greater windiness of coastal, as compared with inland, areas thus has both virtues and drawbacks. Protection from strong winds remains important everywhere. This is one reason why east-facing slopes are preferred for viticulture almost universally (Chapter 4). Many viticulturists are now also realizing the value of natural and artificial windbreaks (Chapter 5). But given adequate protection to the vines, the relatively windy coastal and near-coastal areas still offer the best climatic conditions for producing high-quality winegrapes in predominantly arid regions such as southern Australia and west coastal USA.

Optimum combinations of climatic factors

The climatic factors discussed in this chapter are mostly interrelated. For instance high relative humidities are correlated with restricted temperature variability and mild night temperatures, because water vapour in the air acts as a blanket slowing the radiative loss of heat at night. Its condensation into mist or fog at night also releases latent heat of vaporization, which further prevents or delays temperature fall.

However not all the correlations among climatic elements are positive for viticulture. Desirably high relative humidity is often associated with lack of sunshine, which is undoubtedly adverse to some viticulture; and with high rainfall, which may be adverse or otherwise. To the extent that high relative humidity is a feature of coastal regions, it is often associated with strong winds as noted above; nevertheless some exposed inland sites can equally be subject to damaging winds (Hamilton 1988).

Such conflicting correlations mean that natural environments which combine all the best climatic attributes for viticulture are rare and anomalous. Taking all the evidence into account, I conclude that the best environments combine, to the greatest possible degree: ample sunshine hours; moderately high relative humidities; and minimal temperature variability around optimum means for particular wine styles, both through the vine growing season and particularly during ripening. The best vineyard sites within regions accentuate these characteristics, and are sheltered from damaging winds. Enough continentality to ensure even budburst and early growth in spring is helpful, but its lack can be offset to some degree by grape variety selection and management practices, as discussed in Chapter 5.

Table 2 summarizes suggested best combinations of

Table 2. Ideal averages for ripening month,[1] for different wine styles

	Table wines				Fortified wines and drying grapes		
	Delicate, white, sweet[2]	Light, dry or sweet	Medium, dry or sweet	Full, dry, soft	Light aromatic styles	Medium, port styles	Liqueur styles; drying
Mean temp., °C	12–15	15–18	18–21	21–24	20–22	22–24	24–26
Sunshine hours	> 150	> 180	> 210	> 240	> 240	> 270	> 300
Rainfall, mm	< 50	< 75	< 75	< 75	< 50	< 25	< 10
Early p.m. R.H.%	60–65	55–60	50–55	50–55	45–50	40–45	35–40
T.V.I.[3]	< 30	< 34	< 38	< 38	< 42	< 42	< 42
Highest max., °C	< 27	< 30	< 33	< 36	< 36	< 38	< 40

1 That is, the final 30 days to ripeness appropriate for making dry or semi-sweet table wines. Grapes for full-bodied sweet table wines and fortified wines will normally be left on the vine beyond this period to gain sugar concentration.

2 With at least some residual sugar, to balance acidity; including sparkling wines made by the champagne method.

3 Temperature Variability Index: see text. Figures apply equally to the full growing season and to the ripening month.

the main climatic elements during the final 30 days to grape maturity, for the different wine styles. It provides a quick reference against which to compare the estimates of actual ripening-month conditions in different winemaking regions, as developed in Chapters 6–8 and tabulated in Chapters 9–22.

General discussion and conclusions

The conclusions from the present study agree in many respects with those of Pirie (1978a, 1982, 1988). Pirie sought environments with moderately high summer rainfall and relative humidity (or low evaporation), so as to avoid moisture stress. He maintained that vines stressed for moisture produce coarse wines, irrespective of growing season and ripening temperatures, and of the eventual sugar/acid balance in the grapes. He therefore confined his attention to the temperate, uniform-rainfall Cfa and especially Cfb climatic classes of Köppen (1931), as opposed to the summer-drought, or Mediterranean, classes Csa and Csb. I find that the uniform-rainfall climates on average also have fairly moderate temperature variability through much of the growing season and during ripening.

An equable ripening climate, in terms of temperature variability, humidity, sunshine and evaporative demand, undoubtedly explains the ability of the Hunter Valley in New South Wales to produce good table wines despite its high average temperatures: as viticulturists there have long claimed. Nor can there be any doubt that north-eastern Tasmania, which was Pirie's eventual choice, has excellent climatic conditions for producing European-style, cool-climate wines of the highest quality.

The possibilities for consistent production of premium-quality wines in Australia are nevertheless much wider. To the extent that the east-coast climates, especially that of Tasmania, provide a close match to some of Europe's premium wine-producing areas, they share their weaknesses as well as their strengths. Like much of Europe they are liable to excessive rainfall and humidity at critical periods, including ripening. Many also have a serious spring frost risk.

Certain other regions of Australia have broadly the same strengths, but fewer of the weaknesses. The south-west coastal areas of Western Australia have equally high relative humidities and moderate temperature variability during summer, despite their summer drought. They have more reliable sunshine, mostly less spring frosts, and less hazardous ripening. Some coastal areas, such as Margaret River, have possibly greater late-spring wind problems in exposed areas, while a lack of continentality can adversely affect some grape varieties. However these factors can largely be countered by careful selection of sites and grape varieties, and by appropriate management. With supplementary watering in summer and shelter from winds where needed, such areas provide a combination of climatic elements very closely approaching the theoretical ideal for medium to full-bodied table wines, and for fortified wines in the warmer parts. Similarly, the cooler areas of South Australia such as Coonawarra and the Adelaide/Barossa Hills, even though their early spring and summer conditions may be stressful, can still have mild, equable conditions conducive to grape and wine quality during the critical final ripening period. Several other parts of South Australia, not yet exploited, have at least equally good climatic conditions for table wine production.

A question still to be faced is that of possible future climatic change. Dry (1988) and Smart (1989) have argued that optimum conditions for viticulture will migrate to cooler regions, such as the Australian southern highlands and southern New Zealand, as a result of general climatic warming due to the 'greenhouse effect'. I question this and argue a contrary case in Chapter 23. For now it is enough to state my belief that foreseeable atmospheric changes will, if anything, greatly benefit Australian viticulture. The benefit should be greatest in existing intermediate and moderately warm viticultural regions, rather than at the cold limit. There will be no major viticultural migrations, except from the hot, arid inland to more humid and equable coastal climates. Past and present experience will continue to be a proper basis for viticultural planning.

4

Soil Type and Topography

Soil type

The importance of soil type to vinegrowing is well recognized. On the other hand its relationship to wine quality, or qualities, remains controversial. Nineteenth century French viticulturists saw definite associations between soil type and wine qualities, seemingly beyond those resulting simply from effects on vine growth. Busby (1825) reported that in France: 'a sandy soil will, in general, produce a delicate wine, the calcareous soil a spirituous wine, the decomposed granite a brisk wine'. Guyot (1865) claimed that: 'a calcareous and chalky soil always gives the purest wine and the freest from earthy taste'. Perhaps consistent with this, Quintana Gana and Gomez Pinol (1989) found an inverse relationship between soil sodium content and wine pH, and with the content of undesirable phenolic compounds in several white grape varieties. Portes and Ruyssen (1886) held that wines from sandy soils are light, delicate, lacking strength and colour, but perfumed and lively; that limestone and chalk increase alcoholic strength; and that soils high in iron and clay give depth and richness of colour to red wines. Similarly Despeissis (1902) stated that in Western Australia the ironstone gravelly loams and sandy loams associated with the marri (redgum) tree (*Eucalyptus calophylla*) are: '*par excellence* the best suited for making of high-class wine; clean to the taste, rich in colour, and of pleasant bouquet'. Peynaud and Ribereau-Gayon (1971) considered that: 'clay soils produce more acidic, less delicate grapes, rich in pigments and tannins. Limestone soils give a particularly high content of odoriferous constituents to some varieties of grapes.'

Carbonneau and Casteran (1987a, 1987b), in Bordeaux, more recently demonstrated experimentally a difference in wine style which could reasonably be ascribed to soil type. Despite matching trellising and other management practices, Cabernet Sauvignon vines grown on a moist sand were 'significantly more balanced and soft or velvety', whereas those grown on a dry gravelly soil 'significantly possess better aromatic quality, fruit, colour quality and intensity, structure, but higher astringency, bitterness and acidity'. These differences seem broadly consistent with those described by the earlier French writers.

Portes and Ruyssen (1886) cited a general belief among scientists and practical vine-growers of the time, that reddish soils favour the making of red wines, whereas grey or yellow soils are more suitable for white wines. They ascribed this at least partly to red soils absorbing and re-radiating more heat: which, in cool climates, is needed to bring grapes to the full ripeness essential for making red wines. On the other hand Gregory (1963) and Lake (1964) noted a similar relationship in the Hunter Valley of New South Wales, which could hardly be described as cool. Champagnol (1984) related red colour of the soil to its general suitability for viticulture in the south of France, and ascribed this primarily to good drainage.

Recent research on the spectral quality of light (Chapter 3) provides another explanation for these widely observed relationships. Reflection of white, yellow, or especially orange or reddish light into the lower vine canopy and bunch area could contribute by raising the ratio of red to far red wavelengths there, and so tip the balance in all grape varieties towards cytokinin dominance and fruitfulness. Direct links between the wavelength-controlled phytochrome system and anthocyanin synthesis (Smith 1972; Smart 1987b) point to a particular significance of orange to red wavelengths for ripening red grapes. This in turn may help to explain the very specific environmental and management requirements for making good red wines from certain grape varieties with marginal anthocyanin production, such as Pinot Noir.

All nineteenth century writers agreed that rocky, stony or chalky soils give the best wines.

Most modern scientific writers have minimized the influence of soil type on wine qualities, other than as exerted through vine vigour and moisture relations: for example see Saayman and Kleynhans (1978), and Seguin (1983, 1986). Champagnol (1984) presents a more traditional view, acknowledging that moisture relations are important, but maintaining that soil mineral characteristics can still influence the subtler qualities of wines very significantly, particularly at the higher quality levels. Likewise I find it hard to dismiss entirely the long-held beliefs of so many practical growers, and of earlier generations of researchers.

The next two sections cover mechanisms by which soil composition might exert an influence on winegrape qualities: firstly in terms of soil physical properties, and secondly in terms of soil inorganic chemistry. I then discuss the various roles of soil organic matter.

Soil physical properties

Soil depth, drainage and moisture retentiveness have already been mentioned in the context of rainfall distribution: with special relevance to regions of mediterranean-type rainfall distribution, where capacity to store moisture without waterlogging, and to supply it to the vine in summer, are crucial in the absence of supplementary watering.

Seguin (1983, 1986; see also Johnson 1971) carried this argument further. He postulated that in all climates a steady, moderate availability of moisture is essential for production of the highest quality grapes. Excess moisture leads to high juice acidity and the dilution of phenolic constituents. A sudden increase in available soil moisture during ripening can also result in berry splitting and bunch rot. A consistent lack of enough moisture is damaging in a number of ways (Chapter 3).

In a shallow soil, with limited potential rooting depth, heavy rain can quickly raise the root zone to beyond its moisture storage capacity. Such soils can readily alternate between waterlogging and drought, following only average fluctuations in rainfall. By contrast well-drained, deep soils permit deep and more dispersed rooting; which, provided that the soil has a reasonable water-holding capacity per unit volume, buffers the vine against most fluctuations in moisture supply. Vine performance and wine qualities are then much more consistent from year to year. The best vineyards are characterized by their ability to produce good wines even in poor vintages, whereas on marginal sites the effects of poor vintage years are magnified.

According to Seguin the best vineyards of Bordeaux, acknowledged as such over the centuries, are nearly always those closest to natural drainage channels. This ensures that both internal and surface drainage are good, so that the vine roots can grow to great depth and function well and continuously there.

Seguin pointed out that soils of diverse geological origins can fulfil the requirements. They include gravelly alluvials, such as at Bordeaux; many limestone-based soils; and even heavy clays, as in Pomerol. Provided that clays are well enough structured and drained to allow successful root development and function, they can serve well because of their very high water-holding capacity. Under dry conditions the stored water is released to the roots slowly and progressively. Saayman and Kleynhans (1978), in South Africa, attributed the same advantage to a duplex soil with good subsoil moisture-holding capacity.

Limestone or chalk subsoils and rubbles, even if they cannot themselves be penetrated by roots, soak up water and act as a reservoir which slowly releases its moisture to the overlying or adjacent soil by capillary action. The fine-soil fraction of limestone soils, being calcium-saturated, is also typically well structured and well drained. But sands have only very limited water-holding capacity, and almost all the stored moisture is readily available to the roots to be quickly dissipated at luxury consumption rates. Great rooting depth, with sparse root densities, may offset this to some extent.

All these considerations have less force in climates where water is supplied partly or mainly by irrigation. The capacity to control water supply, and thereby to avoid problems of both too little and too much soil moisture during the growing season, results in a wider tolerance of soil physical types than where the vines depend entirely on uncontrolled rainfall and soil-stored moisture.

The distinction has become especially important in mediterranean climates such as those of west coastal North America and much of southern Australia, where supplementary drip irrigation in summer and autumn has become near-universal for the production of high-quality winegrapes. Previously, successful viticulture in these wet-winter, summer-drought climates was confined more or less completely to deep, water-retentive soils of the valley floors and flat land. Vines had to be spaced widely, and at the same time hard-pruned to low yields, in order to conserve enough of the winter-spring moisture for the critical ripening period. The pattern followed was that of traditional dryland viticulture in the Mediterranean Basin. At the same time the usual fertility of these soils, combined with still-ample moisture early in the season, often led to over-vegetative early growth, and to either or both of unfruitfulness (especially in shy-bearing 'noble' grape varieties) and poor fruit quality due to excessive shading within the canopy. The problems, which are common also on fertile soils in temperate uniform-rainfall regions, and with full irrigation in arid climates, have already been touched on in Chapter 3. Their management is discussed in Chapter 5.

One need which remains absolute in all regions is for good soil drainage. In mediterranean climates this is especially so where the winter-spring rainfall is very high, resulting in a potential prolongation of waterlogging into the spring growing period.

Much lore and all early writers, e.g. Rendu (1857) and Portes and Ruyssen (1886), lay great emphasis on the proportion of rock or gravel in soils for winegrape growing, drawing parallels across a wide range of climatic types between the percentage of rock or gravel in the soil and wine quality.

Many reasons have been advanced for this, and for the traditional preference for such soils in viticulture. One is that they are so used merely because they are too poor for other uses. This contains some truth, but is hardly enough. A second, and more convincing, reason is that the usual infertility of stony soils is needed to maintain vine balance and fruitfulness in cool, summer-rainfall climates. Stony soils have the

Figure 3. Heat absorption and re-radiation by soils of high and low thermal conductivity.

a. Heat-absorbent soil

b. Non heat-absorbent soil

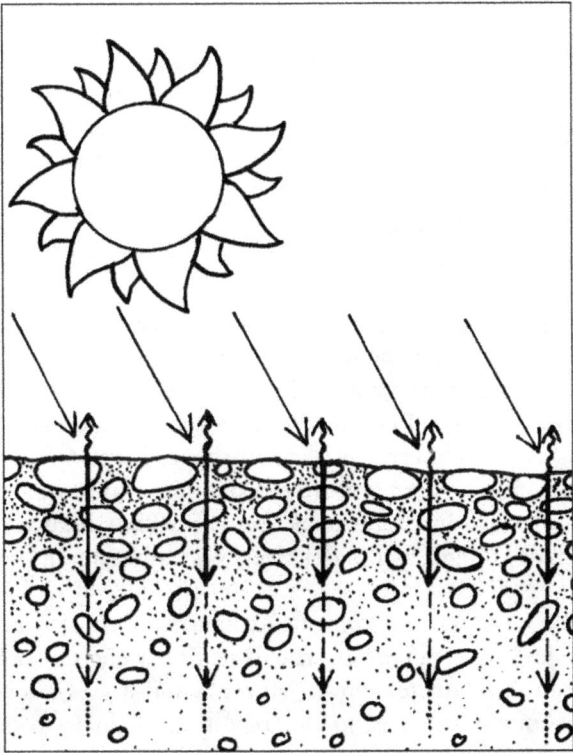

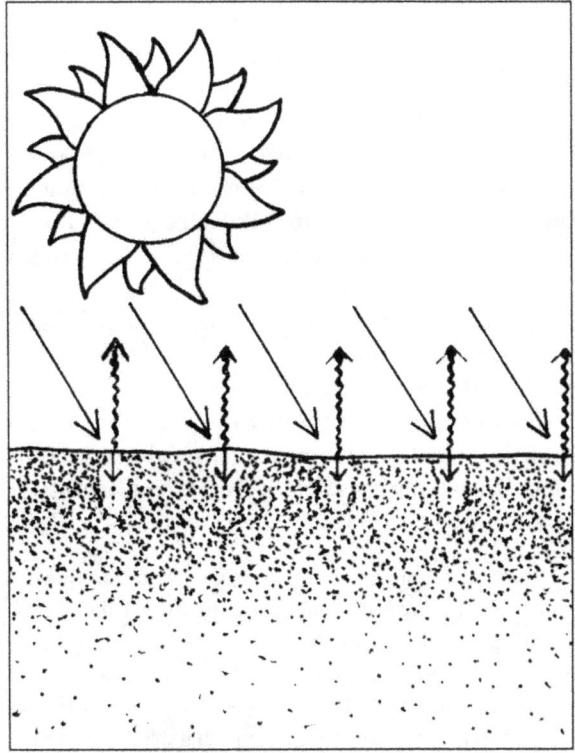

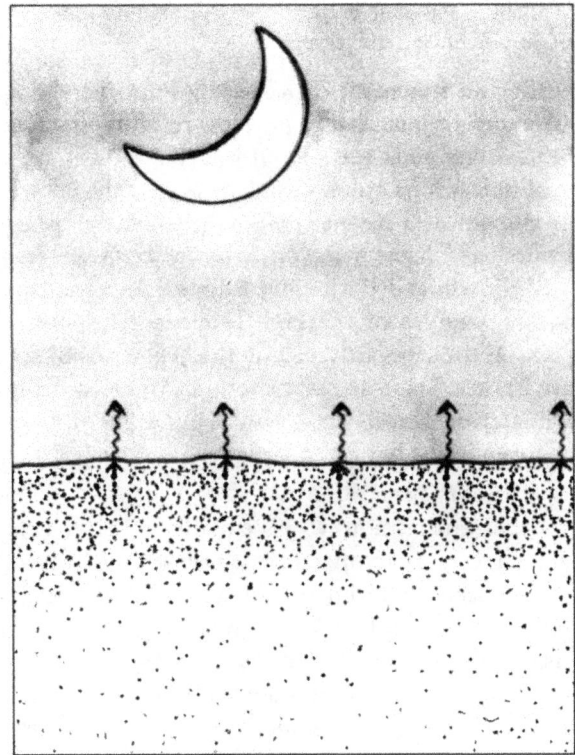

additional advantage of being usually well drained.

A stony or rocky surface also gives valuable resistance to soil erosion. This is especially important on slopes – which happen, in most cases, to provide the best mesoclimates for viticulture, as discussed in the second half of this chapter.

Traditional wisdom says, however, that stony or rocky soil surfaces bestow an additional benefit for viticulture. By efficiently absorbing and storing more heat than normal soils during the day, and re-radiating it to the vines through the night and during cloudy periods, they are held to reduce frost and allow fuller ripening. This is illustrated in Figure 3. Among other things, it makes viticulture possible in colder climates than would be so otherwise. Present knowledge of the thermal properties of soils, reviewed by Geiger (1965) and van Eimern (1966), confirms this.

But the old beliefs seem to have gone further. Not only is ripening on stony or rocky soils more complete; it is also qualitatively better. Portes and Ruyssen (1886) stated:

> That principle which give wines their bouquet, and which the great Bordeaux chemist has called 'oenanthine', or flower of wine, above all needs complete maturation of the grapes for its production. It is always much greater on dry, stony or pebbly soils than in the same grape varieties planted on soils that are fat, strong and clayey.

Such a statement could hardly refer merely to an assurance of quantitative ripeness resulting from extra heat. Stony soils feature as much, or almost so, as a quality factor in quite hot climates as at the cool limit of viticulture. I have already mentioned the 'pudding stones' of Châteauneuf du Pape in southern France. Both Rendu and Portes and Ruyssen document many other examples of the high proportion of stones or rocks in the vineyards making the best wines of southern France, Spain and elsewhere. In Western Australia the lateritic gravels are highly reputed for wine quality, even in the hot Swan Valley.

Several mechanisms seem possible for such a relationship. Undoubtedly one is generally low soil fertility, and its effect on vine vegetativeness and canopy light relations, as discussed in Chapter 3 and further in Chapter 5.

At least part, however, is equally or better explainable in terms of temperature variability. Heat absorbed during the day, and re-radiated at night or under cloud cover, can be seen as enhancing berry metabolism and completeness of flavour ripening relative to sugar accumulation and acid loss (Chapter 3). Even in hot ripening environments, the air temperature at night commonly falls below optimum for the enzyme processes involved. Indeed, physiological ripening at night could conceivably assume particular importance when temperatures through much of the day are above optimum for it.

The direct effects of stony soils on canopy microclimate during the day are more speculative. The better heat conductivity of rock, compared with fine (and especially dry) soil, means that more heat is absorbed and transmitted to a greater soil depth, so that the surface does not reach such high temperatures. Presumably less heat is immediately re-radiated to contribute to the heat load on the vines and bunches during the day. Thus at least there is no ground for thinking that a stony or rocky soil surface will impair vine temperature relations during the day. More probably it is beneficial.

A stony surface has the additional benefit that it acts to capture moisture because of its unevenness, but later forms a shield against surface evaporation. To the extent that soil beneath the stones remain moister, both its heat storage capacity per unit mass (specific heat) and thermal conductivity are greatly increased. This further increases the depth of warming and minimizes soil temperature rise per unit of heat input. Re-radiation in the heat of the day is thus reduced, and that at night and under cloud is steadier and more prolonged.

The same factors might be expected to benefit vine root development between the rows. Experience in cool viticultural climates shows that rocks and stones accelerate soil warming in spring, resulting in earlier root growth, earlier and more even budburst, more rapid early top growth, and greater fruitfulness. In summer there is earlier and greater subsoil warming, combined with a moderation of high temperature extremes and a reduced diurnal amplitude of temperature in the surface layers. The likely result across all climates is a greater capacity for root growth and feeding in both the deeper and the near-surface layers, starting earlier in spring and continuing with less interruption through the season. Consequences later in the season should theoretically include greater cytokinin supply during ripening, and probably better vine balance.

A light to reddish-coloured rocky soil surface will increase the direct reflection of useful light wavelengths back into the vine canopy, as already noted. This is instead of the re-radiation being more or less exclusively as heat wavelengths, as takes place from dark-coloured soils. If at the same time the surface is rocky and heat-absorbent, it achieves the best possible combination of directly reflecting useful light wavelengths, while absorbing a maximum of the heat wavelengths and holding the heat long enough to be re-radiated to the vines when it is most useful.

What, then, was the 'oenanthine' of the 19th century researchers? Clearly it was something which develops only in ultimate ripeness, as they themselves believed. But more specifically, that surely means ultimate *physiological* ripeness.[1] To reach that, the berries must

---

1 The 19th century French speculations find a parallel in the contemporary Australian and New Zealand debate on wines made from grapes picked at differing stages of maturity, and

maintain their full health and metabolic integrity to the end. This requires in turn continuous sufficient cytokinin supply to both berries and leaves, from a strongly functioning root system throughout ripening. Ample sugar flux to the berries from well-illuminated leaves, and from various buffering storages within the vine, must continuously supply the surplus needed for greatest colour and flavourant biosynthesis in the berry skins. Finally, equable optimum berry temperatures are needed to allow maximum rates of biosynthetic and physiological ripening, relative to acid loss and pH rise. Full physiological ripeness can then be attained while natural balance is still optimal for winemaking, and ahead of any onset of senescence in the berries.

None of this detracts from the major contribution of soil moisture relations to vine growth and wine qualities, as expounded by Seguin (1983, 1986) and others. It is fully consistent with a prime control by soil moisture. Nevertheless the circumstantial and theoretical evidence suggests that other soil physical qualities can also contribute significantly. They should not be overlooked.

Soil inorganic chemistry

Little *objective* evidence is available that soil chemical attributes directly influence wine qualities, provided that there are no gross nutritional deficiencies or imbalances.

A luxury supply of nitrogen can over-stimulate vegetative growth. This may reduce both yield and quality, especially if not adequately accommodated by trellising. Excess nitrogen exacerbates disease susceptibility (Bavaresco 1989). It can also precipitate deficiencies of other elements, both through increased growth and nutrient demand, and, in some cases, probably by complexing them into inactive forms within the plant.

On the other hand there is no evidence at all that a deficiency of nitrogen is beneficial, unless it is the only way to contain growth which is over-vigorous for other reasons. Given proper trellising to accommodate

the growth, responses to nitrogen at least up to maximum fruit yield seem unlikely to threaten fruit and wine quality. Indeed, there is strong evidence that wine quality can be reduced by *lack* of sufficient nitrogen in the grapes and musts. The question of vine and yeast nitrogen nutrition is discussed fully in Chapter 5, under 'Fertilizing and green manuring'.

Unnecessarily high potassium in the soil is undesirable in hot, arid climates, because of its potential effect on wine pH. But as we have already seen in Chapter 3, there is much less risk of this in cool and humid climates. Deficiency is more to be feared in such climates. Besides reducing growth and yield, potassium deficiency is well known to increase vine susceptibility to a number of fungal and bacterial diseases (Huber and Arny 1985; Bavaresco 1989). According to Bavaresco this is associated with elevated contents of cell constituents having low molecular weights, and a reduction in those with high molecular weights such as proteins and cellulose, which are responsible for the organizational and structural integrity of the cells. Such a change might arguably affect the ability of grapes to reach the desired state of full maturity before senescence and/or the onset of Botrytis infection. More speculatively, weaker or leaky cell membranes in potassium-deficient berries might allow the faster passage and evaporative loss of aroma compounds. Loss of leaves due to potassium deficiency in any case reduces photosynthesis, and exposes the bunches to greater aroma loss and sunburn.

Greater bunch exposure may, of course, be beneficial in cool climates; but any such benefit is likely to be more than offset by adverse factors if caused by nutrient deficiencies other than of nitrogen. Grapes from potassium-deficient vines typically lack colour and sugar content (Champagnol 1984). At the same time neither the vine literature (Winkler 1962) nor, by and large, the literature on fruits generally (Usherwood 1985) contains any strong suggestion that fruit quality continues to improve with potassium supply beyond that needed for fully normal growth and yield.

The effects of luxury potassium availability in the soil are likely to be minimal in cool, humid climates, as already explained. But in hot, atmospherically arid climates, the postulated greater potassium uptake per unit of vine growth and yield (Chapter 3) will cause soil differences to be more strongly reflected in the vine and fruit. Stress, and the often unfavourable canopy light relations with irrigation in these environments, also lead to the danger that this potassium will be unduly mobilized from the leaves etc. to the fruit (Chapters 3 and 5). Large increases in must and wine pH can result.

The general conclusion is that adequate potassium for the vine is essential, both for yield and for fruit quality. Potassium availability greatly above that can, however, contribute seriously to reduced wine quality in hot, arid climates.

Calcium (as in limestone) and magnesium (present

---

from ripening conditions of differing temperatures and canopy shading. In particular the debate touches on the incidence of 'herbaceous' and 'vegetable' characters, such as are associated with methoxypyrazines, as opposed to the ripe berry and other fruit characters of wines made from fully ripe and well illuminated bunches. See, for instance, Jordan and Croser (1984); Champagnol (1984, p.314); Smith et al. (1988); Duval (1988); Brakjovich (1988); Allen et al. (1990a, 1990b); and Judd (1990). I suggest that the initial welcome afforded to herbaceousness in Australia and New Zealand arose largely from a misinterpretation of cool-climate wine style. It is a legitimate enough subsidiary component lending freshness and contributing to complexity in some white wines, for instance those made from Sauvignon Blanc and Semillon. A small element from the use of a controlled range of grape maturities can also lend complexity to red wines. But in general, I believe that obvious herbaceousness and vegetable characters will come in future, as they are in Europe, to be regarded as a fault, symptomatic of incomplete or imperfect ripening. Great wines can only be made when most of the grapes are fully and perfectly ripened.

with calcium in dolomitic limestone) are both antagonistic to potassium in plant nutrition, and may limit its uptake from the soil. At first sight one might expect that to be an advantage in hot, dry climates and a disadvantage in cool, moist climates. However, the reverse, if anything, seems to apply, with calcareous soils valued most towards the cool limit of viticulture (Winkler 1962) but not especially so in hotter regions (see below).

The overriding reason for the value of calcareous soils in cool regions is almost certainly physical. Calcareous soils are nearly always well structured and drained; they provide a steady water supply to the vine; and, most importantly in cool areas, they warm easily. Grapes can therefore ripen earlier and more fully, and later-ripening varieties can be grown, than on most other soils in the same climate. Equally, the relative warmth of the soil in autumn probably allows ripening to continue longer into late autumn than would otherwise be so. This should be especially significant for varieties which are very sensitive to heat during ripening, such as Pinot Noir, by allowing them to ripen fully under atmospherically cool and non-damaging temperatures. Another mechanism might be the roles of must calcium and magnesium in protecting against the inhibition of fermentation by high alcohol contents (Kunkee 1991): thereby perhaps allowing fermentation to continue to the desired extent under low winter temperatures, or under other conditions where this might not otherwise occur.

According to Winkler, limestone-based soils are not specially valued in Italy. Similarly Champagnol (1984) notes that the reputed superiority of calcareous soils for wine quality does not extend to the Mediterranean region of southern France. Indeed, non-calcareous soils derived from schist, softly cemented conglomerates, sandstone and ancient alluvium are reckoned there to give wines that are finer, rounder and more harmonious than those from calcareous soils, despite the usually better moisture relations of the latter. Champagnol acknowledges, however, that important interactions exist between soil type and vine variety, with Shiraz, Gamay, Cabernet Sauvignon, Carignan and Chenin Blanc giving their best qualities on non-calcareous soils, and Pinot Noir, Cabernet Franc, Chardonnay and Sauvignon Blanc their best on calcareous soils. Champagnol is emphatic that at least some of the so far undefined differences in wine qualities associated with soil result directly from mineral nutrition, and that the differences become more apparent and consistent, the higher the quality levels of the wines.

A role of trace elements such as copper, zinc, manganese and boron could be readily explained, because these are typically components of co-enzymes, which participate chemically in enzyme reactions. Balances among available trace elements could therefore conceivably influence the balances among enzyme reactions responsible for flavourant formation in the grapes and fermenting musts. Soil pH, and the presence or absence of limestone, could have indirect effects through their profound influence in the availability of the individual trace elements to plants. This varies greatly from element to element.

Across all soils a red, brown or yellow colour in the subsoil is an extremely useful indicator of good drainage, being due to oxidized iron compounds. Bleached subsoils contain their colourless reduced equivalents, indicating waterlogging. Mottled subsoils are found where there is intermittent waterlogging.

The most important overall chemical attribute of any soil, however, is undoubtedly its capacity to provide steady, balanced nutrition for growth of the vine. Fully adequate vine nutrition cannot, in itself, have deleterious effects on grape quality. Indeed, the reverse is almost certainly the case for most nutrients (Bavaresco 1989). The only potential ill-effects from more-than-adequate nutrition (short of toxicity) appears to be in the case of nitrogen, together with potassium in some circumstances.

Soils with moderate to high clay contents have the best ability to supply steady nutrition to the vine. This is partly because they have the greatest capacity to store nutrients within their finely-divided clay colloid structure, in physically or chemically bound forms which to varying degrees are only slowly soluble. The same applies whether the nutrients are native or whether they are residual from earlier fertilizer applications. Natural rock materials in the soil also break down to soluble, and therefore plant-available, forms, though mostly at an extremely slow rate. If finely enough divided, rock material can be a significant – though seldom large – source of mineral nutrients to the vines and to associated green manure crops.

In very sandy soils there is little fine material for nutrients to adhere to or react with. Nutrients applied in fully soluble form can therefore readily produce an immediate surfeit, followed by rapid leaching and ultimate disappearance from the soil profile. This is especially so with nitrogen. The result for the vine is an alternation of feast and famine, unless fertilization can be in very frequent small doses. A practical approach for sandy soils in dry climates is to supply nutrients continuously, at very low concentration, through irrigation water ('fertigation'). An alternative for intensive systems is to incorporate nutrient-rich clay or rock in finely divided form, or special fertilizers having very low solubility, deep into the soil before planting.

The roles of soil organic matter
Organic matter plays several extremely important roles in the soil. It acts as a store of nutrients which slowly become available to plants as it breaks down by oxidation. In very sandy soils it may constitute the only major store and steady source of the more soluble nutrients. In all soils it is a major contributor to

maintaining a friable, absorptive and erosion-resistant physical structure. It also contributes usefully to the ability of the soil to store moisture.

A very high surface organic matter content may to some degree be counter-productive when fully dry, by insulating against heat transmission in and out of the soil. The surface darkening of red or yellow soils by organic matter will also reduce the reflection of red and other photosynthetic wavelengths back to the fruit and canopy. However these are unlikely to be significant problems on most soils that are otherwise well suited to viticulture.

A unique advantage of organic matter as a source of mineral nutrients is that, being formed originally from plant material, its composition and that of its breakdown products bear an approximate relationship to the requirements for new plant growth. Older and more weathered forms of organic matter, or humus, are lower in nitrogen than freshly decayed material, and therefore less stimulating to vegetative growth. The original nature of the decaying material must also influence composition. Within broad limits, however, there is a reasonable supposition that the nutrients becoming available from plant-derived soil organic matter will be tolerably balanced and complete for new plant growth.

Mineral nutrients and nitrogen are not the only significant products of organic matter breakdown. Emitted carbon dioxide, mainly from microbial respiration, may also contribute to new above-ground growth and to yield.

The soil atmosphere typically contains 10 times as much carbon dioxide as that above ground (Milthorpe and Moorby 1979). In well aerated soils this constantly diffuses into the outside atmosphere, resulting in slightly enhanced concentrations in the air just above the soil surface (provided that atmospheric turbulence there is not too great). The amount entering the atmosphere will clearly depend on the amount of organic matter in the soil, and the vigour of the microbial activity breaking it down.

Milthorpe and Moorby found some 6 to 8 per cent of the carbon dioxide taken up by barley in an Australian environment to come from this source. Saugier (1977) states that the soil contribution to carbon dioxide uptake by plants, which would include that re-cycled from the respiring roots of the plants themselves, may amount to between 10 and 50 per cent for healthy vegetation.

The effects of enhanced carbon dioxide availability are detailed in Chapter 23, in the context of rising atmospheric concentrations from other sources. Here, it may merely be noted that any such contribution will nearly always be extremely beneficial, both for plant yield and fruit quality, and probably also for plant resistance to adverse environments and disease.

A further advantage of organic sources of nutrients and carbon dioxide is that their rate of breakdown to forms available to the plant depends on temperature

(the rate increasing as temperature rises), combined with the presence of enough air and moisture in the soil to support microbial activity. Precisely the same factors govern, in a parallel way, the potential for root development and plant growth. Nutrient and carbon dioxide availability from breaking down organic matter therefore parallels quite closely the plant's needs through the season, leading to a consistent plane of nutrition. This contrasts with the situation where highly soluble inorganic fertilizers are used, particularly on sandy soils. Not only does it benefit the growing plants. It also minimizes the loss of soluble nutrients from the soil profile by leaching, and their undesired removal into the groundwater and river systems.

A continuous, sufficient supply of nutrients closely matched to the vine's needs helps to promote balanced growth and fruiting. There is also an intimate relationship between mineral nutrition and resistance to diseases. I have already referred to the specific case of potassium, but the relationship is almost certainly more general (Bavaresco 1989). Plants that are weakened by marginal or clinical deficiencies, or by nutritional imbalances, succumb more readily to fungal and bacterial diseases, and probably to at least some viruses. An excess of nitrogen is likewise widely accepted as accentuating plant susceptibility to many diseases and pests.

A further conclusion follows logically: that the vine's need for balanced nutrition must be met continuously, if it is to maintain fully its natural defences against diseases and pests. Transient deficiencies or nutritional imbalances could provide a 'window of opportunity' for disease or pest organisms to become established and build up, potentially to the point where even a totally restored level of plant resistance is unlikely to prevail. By contrast, a continuous and fully expressed, but quite moderate level of resistance can usually prevent the normally small initial population of a pathogenic organism from gaining a foothold. The nearest analogy to plant disease resistance over time is that of a chain, of which the strength is that of its weakest link. And at least in lighter-textured soils the continuous, balanced supply of nutrients needed to optimize a plant's natural defences is most likely to be supplied by predominantly organic means. The question is taken up again in Chapter 5, under 'Fertilizing and green manuring'.

Soil type: summary

Much historical and practical evidence points to a major role of soil physical conditions in governing potential wine quality and qualities. The evidence on soil chemical attributes is less compelling, but there are several probable avenues of influence. Despite a lack of rigorous scientific proof, and an acknowledged dominant control over wine-grape qualities by climate, soil type remains very important in the choice of vine-growing sites within a region.

The drainage and thermal properties of soil, and

Figure 4. Air drainage from slopes versus hollows or flat land.

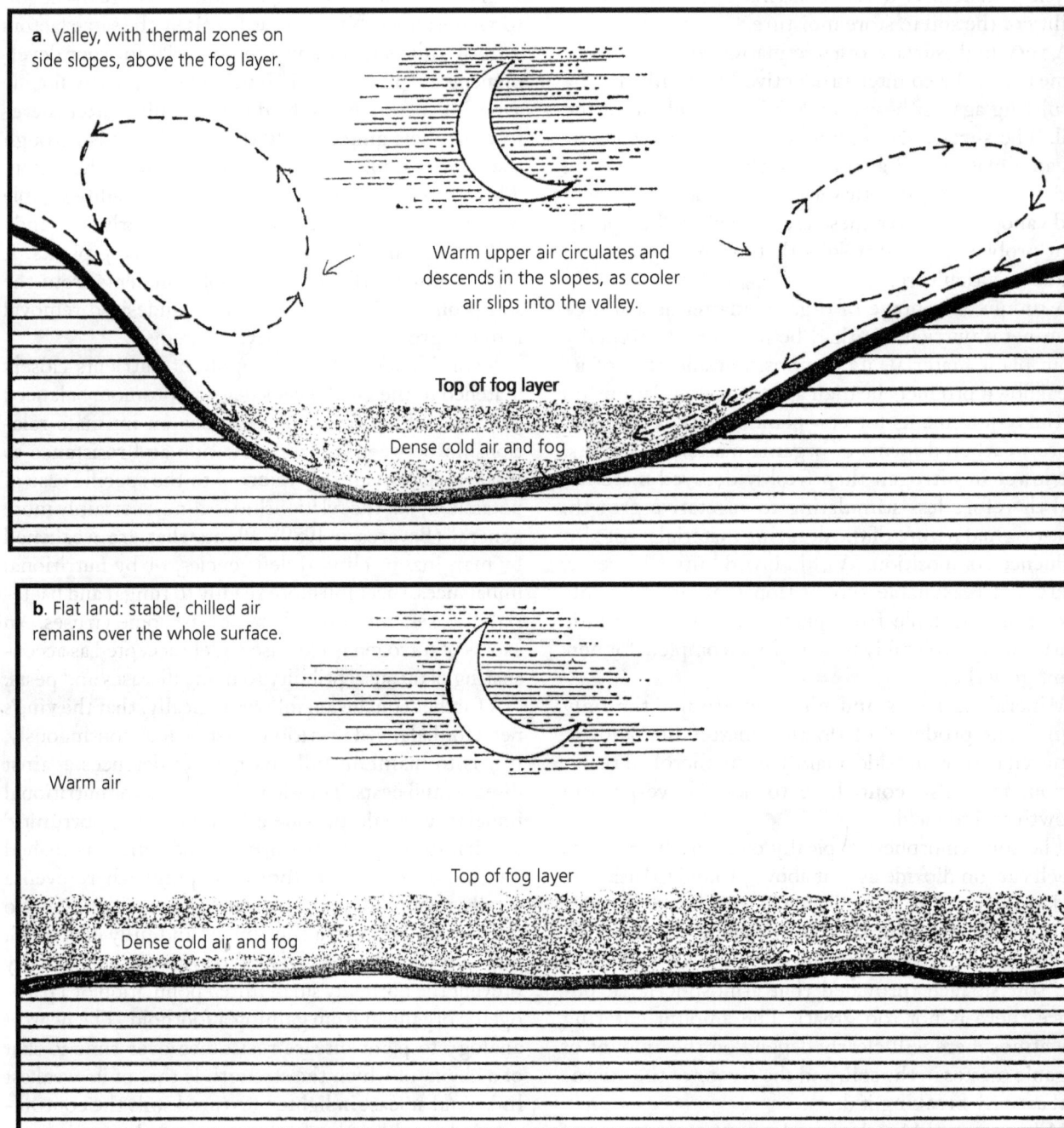

a. Valley, with thermal zones on side slopes, above the fog layer.

Warm upper air circulates and descends in the slopes, as cooler air slips into the valley.

Top of fog layer

Dense cold air and fog

b. Flat land: stable, chilled air remains over the whole surface.

Warm air

Top of fog layer

Dense cold air and fog

Table 3. Temperature adjustments (°C) for topographic and soil factors[1]

	Minimum temperature	Maximum temperature	Variability index	Full-season day°
Moderate slopes, undulating terrain	+1.0	−0.5	−3.0	+50
Steep slopes, free-standing hills	+1.5	−0.5	−4.0	+100
Slopes facing sun	+0.75	+0.25	−1.0	+100
Stony, rocky or calcareous soil	+1.0	−0.5	−3.0	+50

1 Relative to an assumed flat terrain and normal loamy soil for the recording station. Partial presence of a factor attracts a proportionately reduced adjustment. Where the recording station is itself known or thought to be on sloping terrain etc., all adjustments to be estimated relative to it. Altitude differences to be allowed for at 0.6° per 100 m. All adjustments are additive, and are assumed to apply uniformly through the growing season.

in particular the advantages of stony-surfaced soils, have special importance in cool viticultural climates. Nevertheless they remain significant in warm to hot climates, particularly for wine quality.

The properties of good vineyard soils are partly explainable in terms of heat absorption and its re-radiation to the vines at night and during cloudy periods. Soil depth and drainage also govern vine root development; deep, vigorous roots result in a steady supply of moisture and nutrients, together with root-produced growth substances which promote disease and stress resistance and balanced, fruitful growth. Mineral nutrition needs to be balanced, with a continuously adequate supply of all minerals including potassium, and enough, but not too much, nitrogen. Organic sources of nutrients are desirable, to help ensure that the supply remains balanced and closely matched to the vine's needs and potential growth rate through the season.

Perhaps, at least in Australia, the pioneer viticulturists can rightly claim the last word. They quickly found a good relationship between the suitability of soils for vine growing and particular kinds of natural vegetation. I note several well-established relationships in Western Australia (Chapter 9), and they doubtless have their equivalents elsewhere. Such relationships represent an integrated biological outcome of many factors of soil physics, soil chemistry and climate, well beyond what objective science can yet tell us. They should not be under-estimated, and in many situations may still be the best practical guide.

Topography and aspect

The role of topography in frost avoidance and surface water drainage is universally appreciated. There is often also a link between topography and suitability of soil type for viticulture, as discussed in the earlier part of this chapter. Well-drained, stony-surfaced soils of good depth are commonly found on the lower slopes of hills and valley sides. These tend at the same time to be the warmest and most frost-free sites locally available, and are so selected of necessity towards the cold limit of viticulture.

Certain very small patches or strips of land in Europe have, over the years, nevertheless consistently produced wines generally accepted as being superior to those from surrounding land, despite apparently similar climates, soils and vineyard management.[2] Why this should be has never been fully resolved, although Seguin (1983, 1986) has advanced evidence to implicate local differences in soil drainage and vine moisture relations; while a relationship of wine quality to soil rockiness or stoniness is also widely accepted. Like

drainage, this can vary greatly over small distances.

But if their topographies are carefully examined (see, for instance, Johnson 1971), the very best vineyards will usually be found as well to have two or more of the following features (Gladstones 1976, 1977).[3] All of these features are associated with reduced temperature variability, through reduced diurnal range and, especially, higher evening, night and early morning temperatures.

1. They are on slopes with excellent air drainage, and are situated above the fog level (Figure 4).
2. The very best are usually on the slopes of *projecting or isolated hills*. These have outstanding air drainage characteristics (Figure 5).
3. Even in hot areas they directly face the sun during at least some part of the day (Figure 6). Part-easterly aspects are common.
4. If inland, they tend to be close to substantial rivers or lakes (Figure 7).

The effects of topography on grape and wine quality can thus be interpreted in terms of the temperature variability thesis of Chapter 3. Because each acts through a comparable temperature mechanism, the individual topographic features can to a large degree reinforce or substitute for each other. Soil physical characteristics contribute in an exactly analogous way, to the extent that they act by modifying above-ground microclimate.

An estimate of the magnitudes of individual topographic and soil effects on effective temperature, as perceived by the vine, can be gained by comparing the ripening dates of individual grape varieties within and among nearby localities. Altitude differences must be allowed for ($0.6°C$ per 100 m of altitude), and vine management must be comparable. I have attempted to do this on the basis of various published information on vineyards and their environments, in the light of existing knowledge of mesoclimates generally (Geiger 1965).

Table 3 lists the resulting maximum adjustments that I use. All are as compared with an assumed flat recording site and 'normal' loamy soil. Proportionately reduced adjustments apply to part-effects, such as shallow slopes or only slightly stony soils, or where there is reason to think that the source recording site is other than flat and loamy. It has to be acknowledged that the adjustments can at best only be approximate. Nevertheless I find that they greatly improve

---

2 It is, of course, undeniable that the more highly valued sites will tend to receive closer management attention, as is economically warranted by the higher wine prices received. The question remains as to how they came to be so in the beginning, and why site reputations have remained so constant over the centuries.

3 Gerald Asher appears independently to have arrived at some of the same conclusions and explanations as I do, based on his experience in Europe and California: see pages 168, 179 and 180 of his delightful book *On Wine* (1986). Interestingly, he cites what I take to have been the Californian research of Kliewer and Torres (1972), which also played a key role in my thinking on temperature variability. Their paper did not greatly emphasize the role of diurnal temperature range, however, nor has the significance of their results been brought out in later viticultural research literature.

Figure 5. Cross section of a valley showing the air drainage of projecting and isolated hills compared with the valley floor and normal valley sides.

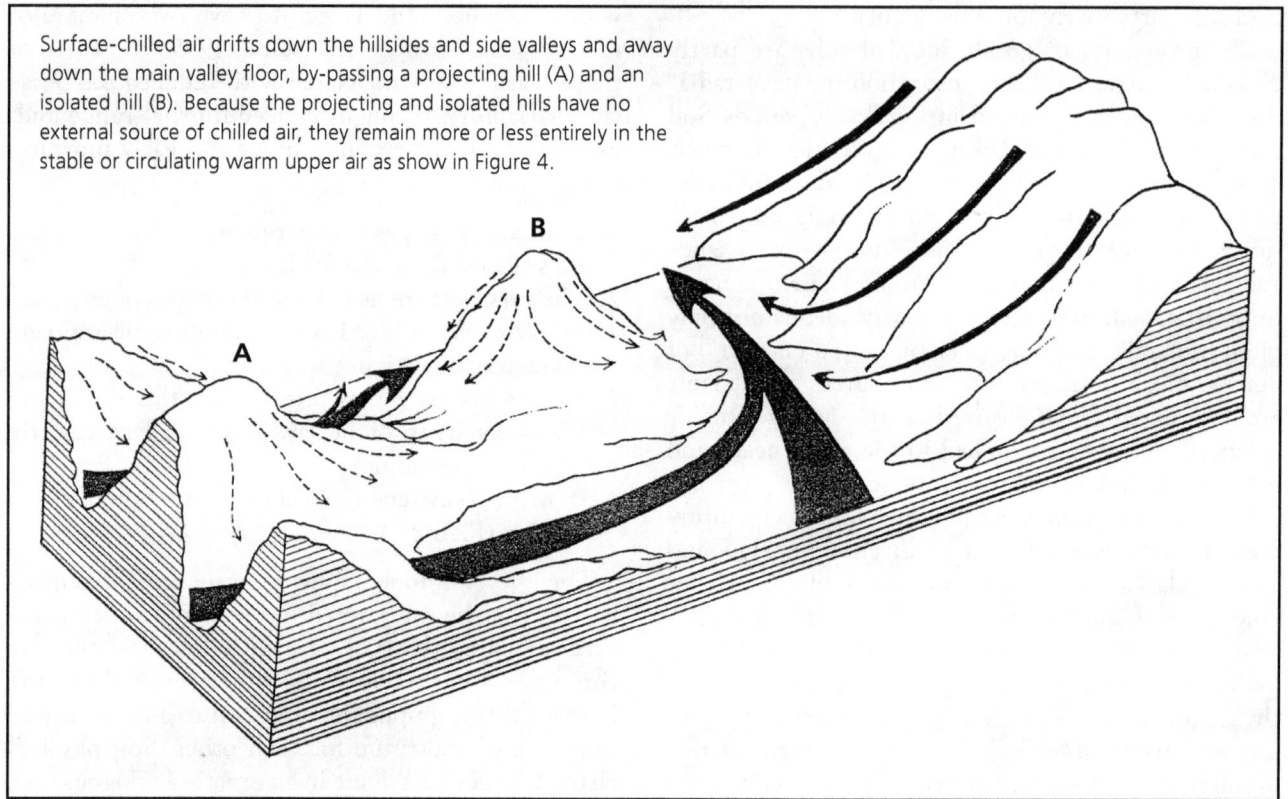

Surface-chilled air drifts down the hillsides and side valleys and away down the main valley floor, by-passing a projecting hill (A) and an isolated hill (B). Because the projecting and isolated hills have no external source of chilled air, they remain more or less entirely in the stable or circulating warm upper air as show in Figure 4.

B

A

Figure 6. Effect of aspect on soil heating and diurnal temperatures.

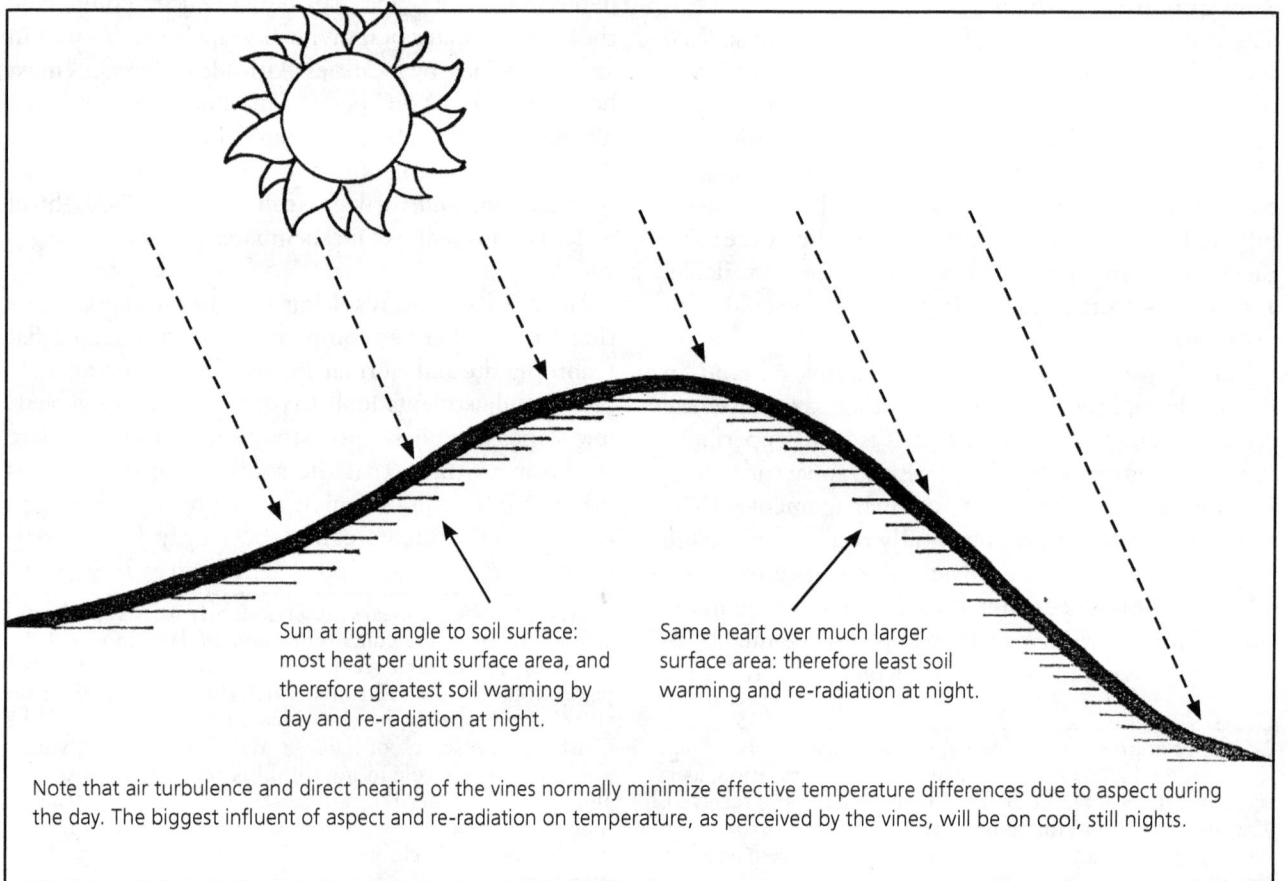

Sun at right angle to soil surface: most heat per unit surface area, and therefore greatest soil warming by day and re-radiation at night.

Same heart over much larger surface area: therefore least soil warming and re-radiation at night.

Note that air turbulence and direct heating of the vines normally minimize effective temperature differences due to aspect during the day. The biggest influent of aspect and re-radiation on temperature, as perceived by the vines, will be on cool, still nights.

Figure 7. Influence of water bodies on diurnal circulation and land temperatures.

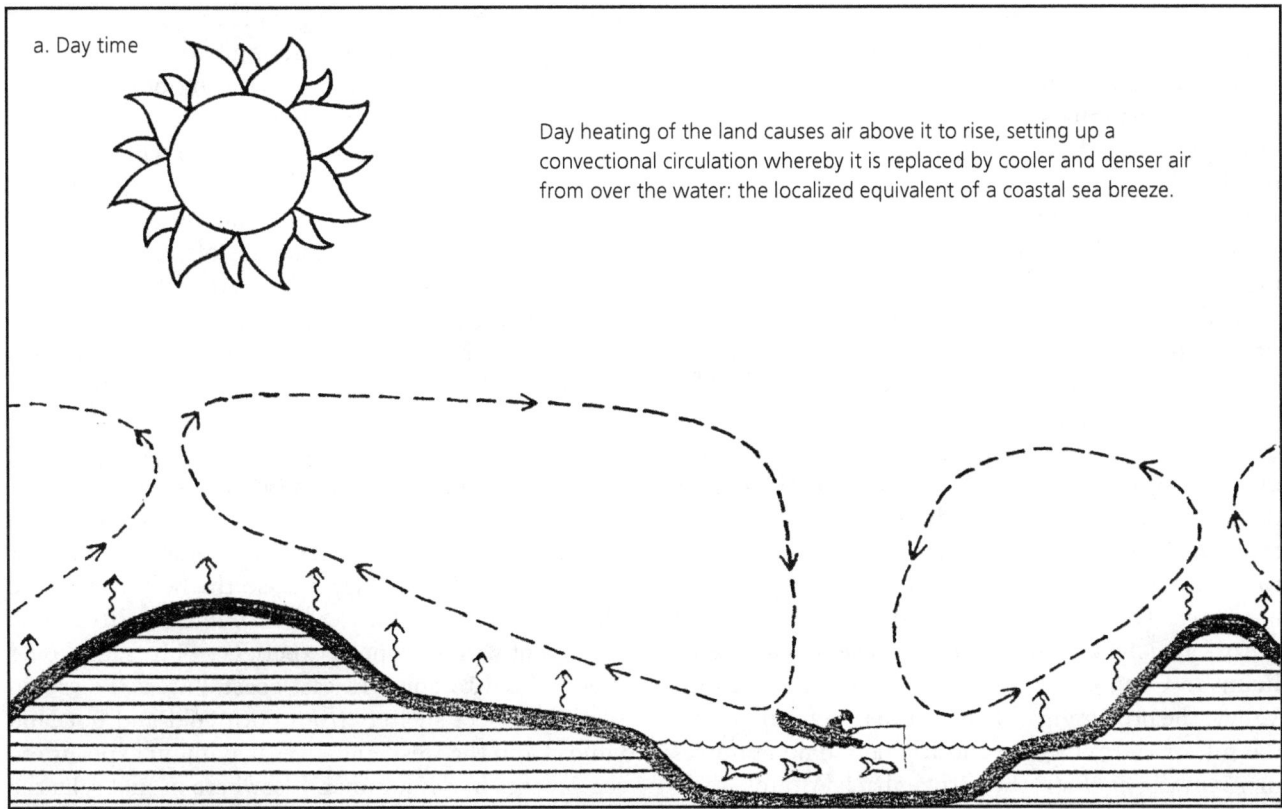

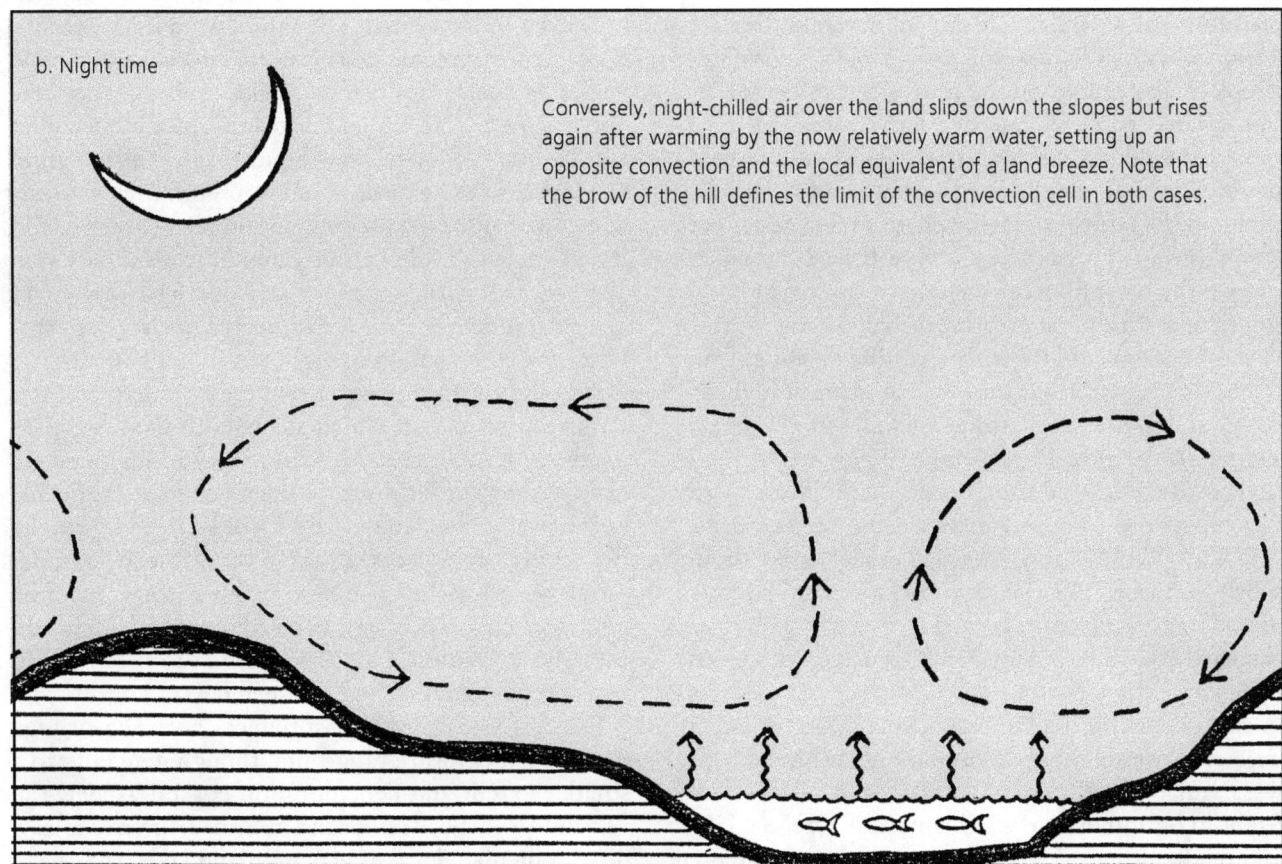

a. Day time

Day heating of the land causes air above it to rise, setting up a convectional circulation whereby it is replaced by cooler and denser air from over the water: the localized equivalent of a coastal sea breeze.

b. Night time

Conversely, night-chilled air over the land slips down the slopes but rises again after warming by the now relatively warm water, setting up an opposite convection and the local equivalent of a land breeze. Note that the brow of the hill defines the limit of the convection cell in both cases.

predictions of vine phenology, compared with those based on the raw published figures for climate. Their further refinement is a task for future research.

The climatic characteristics of isolated or projecting hills, or of isolated hill ranges, are of special interest for inland Australia because it has so few temperature-modifying water bodies. Together with soil type, they provide the main chance to counteract the prevailing temperature variability there.

The lower and middle slopes of isolated and projecting hills develop an especially pronounced 'thermal zone' (Geiger 1965): that is, a layer of warm night air sitting above the 'inversion', or fog boundary which caps the dense, cold air settled on the flats or valley bottom below. Because there is no source of new chilled air from above (Figure 5), the thermal zones of such slopes are remarkably stable. Up-slope from them the mean temperature diminishes with altitude at about the normal lapse rate (0.6°C per 100 m), but the diurnal range remains narrow or becomes even narrower. Examples of where such thermal zones greatly benefit viticulture include the Montagne de Reims in Champagne; the Kaiser Stuhl in the Rhine Valley; the projecting hill of Corton in Burgundy; and the hill of Hermitage on the Rhône. Some examples in Australia include the Brokenback/Mount View Range in the Hunter Valley of New South Wales; the Warby Range in north-east Victoria; the Pyrenees and Strathbogie ranges in Victoria; and the Porongurup and Bennett ranges in the south of Western Australia. Smaller topographic features in Australia which very probably play a similar role include several hills in the Yarra Valley of Victoria; the projecting ridge at Pipers Brook in Tasmania; and the north-south ridge of hills immediately east of Clare, in South Australia.

Aspect is also traditionally regarded as important in viticulture. The soils of east, south and west-facing slopes in the Northern Hemisphere, and of east, north and west-facing slopes in the Southern Hemisphere receive the most direct insolation, resulting in higher soil temperatures and greatest re-radiation of warmth to the vines either in the early morning or at night, and under cloud cover. This is most significant at high latitudes, and particularly around budburst in spring and again during ripening in autumn. The low sun angles at those latitudes and times mean that the soil energy differences due to slope and aspect can be very large.

In Europe the highly regarded aspects are east and south (Johnson 1971), i.e. equivalent to east and north in the Southern Hemisphere. The easterly component can partly be explained by shelter from stormy westerly winds, and often by a 'rain shadow' effect of mountains to the west on rainfall and sunshine hours, in otherwise cool, wet climates. Examples of this include the west bank of the Rhine in Germany and Alsace, of the Saône in Burgundy, of the Willamette in Oregon, USA, and of the Tamar, Derwent and Huon in Tasmania. Additionally, slopes facing east are warmed by the sun as soon as it rises, when temperatures are lowest and therefore most limiting. Quick warming of a stony soil surface in the early morning could make an important contribution to effective canopy and ripening fruit temperatures. Conversely, the heating of such slopes is relatively reduced at the hottest time of day. West-facing slopes are worse placed for both wind and day-time sun exposure; but they do receive the greatest late afternoon soil warming, which should enhance night warming of the vines and perhaps offset the worst of the disadvantages. Slopes which never directly face the sun get none of these benefits.

Energy differences due to aspect are relatively smaller in Australia than in Europe, because of the lower latitude and more vertical sun angle. They would be most significant in Tasmania. Probably more important here is the effect on exposure to damaging winds. Westerly winds and storms up to flowering are often a serious limitation in Australia's cooler and more maritime vineyards, so the best aspects are almost certainly south-east through east to north, consistent with European experience. The same could be expected to apply in east coastal areas, where an easterly aspect shelters from both cold and hot land winds from the west, and gives maximum exposure to sea-breezes in the summer. Elsewhere in Australia the most desirable aspects are harder to define, other than that steep slopes facing directly south are probably better avoided.

One of the initially more surprising conclusions from the study is that aspects facing the sun are favourable even in hot climates. It can be seen in the Douro Valley of Portugal, in the not-quite-so-hot Rhône Valley of southern France (consider, for instance, the aptly named Côte Rôtie, Château Grillet etc.), and in the Hunter Valley of New South Wales, although in the last case the exposure of the premium north-east slopes to cooling sea breezes is a complementary factor. The benefits of soil warming and heat re-radiation in the early morning, at night and during cloud cover appear to outweigh any drawbacks of greater day-time heat on slopes facing the sun, even in hot climates.

The significance of water bodies lies chiefly in their temperature inertia, which stems both from a capacity to store more heat per unit mass and temperature change than the land (due to the much greater specific heat of water compared with rock or soil), and from the deeper distribution of the heat in water due to its transparency and turbulence (most active in rivers). The land surface, by contrast, heats and cools quickly and to wider extremes. This is so from season to season, hence continentality, and between day and night.

Proximity to the sea coast reduces both the average diurnal temperature range and, though generally to a smaller extent and depending on coastal aspect, the variability of the maxima and minima from day to day. The immediate local effects of inland lakes and rivers are equally important. Because the surrounding land heats to higher temperatures by day, and then cools to lower temperatures by night, alternating cycles of air convection are set up by day and night (Figure 7) – provided that they are not prevailed over by other winds. The convection cells are the direct equivalents of the land/sea breeze systems of coastal regions, but are normally confined to the immediate valleys or basins in which the water bodies lie: usually to within a few kilometres, or even a few hundred metres for smaller rivers or lakes.

Frosts are greatly reduced in all directions within the reach of these convection cells. Beneficial effects in reducing afternoon temperatures, and in raising daytime relative humidities, are greatest on the lee bank relative to the prevailing or hottest summer winds.

Examples of European vine industries which depend strongly on the temperature-modifying influence of large inland lakes are those around the lakes of Geneva and Neuchâtel in Switzerland, the Neusiedler See in Austria, and Lake Balaton in Hungary. Most of the best vineyards in Germany immediately overlook the Rhine and Mosel rivers. The Médoc of Bordeaux, as well as being close to the Atlantic on the west, faces east across the broad Gironde estuary. The fact that land winds have to cross some 5 km of estuarine water before reaching the Médoc vineyards undoubtedly moderates their climate significantly, and helps to explain why their wines in general are the most highly esteemed of Bordeaux. In Bordeaux it is said that 'the vines should look at the river' (Johnson 1971).

Australia has few lakes or rivers that are permanently, or even predominantly, water-filled. Lake Mokoan at Glenrowan, in north-east Victoria, is one case bringing apparent benefit. The Glenrowan vineyards are on the lower slopes separating Lake Mokoan from the isolated Warby Range on the east side. The combination of lake and isolated hill influences is almost unique among Australian vineyards, and helps to explain this particular strip's long-established fame both for its dry red wines and for its sweet muscats and muscadelles.

Vineyards on the immediate banks of the Murray river probably benefit to some extent. As an example, the effects are marked enough upstream from Wahgunyah, in north-eastern Victoria, for the cool afternoon breeze that blows up the river to be known locally as the 'Wahgunyah Doctor'. Similarly in Western Australia, the sea breeze in the Swan Valley, known as the 'Fremantle Doctor', reputedly follows a preferential course up the river and gives earliest and greatest benefit to the vineyards nearest the water.

Altitude

There is little in the literature to suggest that altitude in its own right has ever been reckoned a significant factor for vine performance or for grape and wine quality. Nevertheless, theoretical reasons can be advanced for thinking that it may be so. They stem from the relationship between atmospheric carbon dioxide concentration and potential photosynthetic rate, which I discuss in some detail in Chapter 23 in the context of the 'greenhouse effect'.

Atmospheric pressure decreases by approximately 1 per cent for every 100 m of altitude, and the partial pressure of carbon dioxide presumably does so in proportion. Studies cited by Woodward (1987) have shown that plants growing at high altitude compensate for this by developing more stomata (breathing pores) per unit of leaf area. This enables them to maintain a normal rate of carbon dioxide intake, and hence photosynthesis and potential growth rate at any temperature. The price, however, is greater evaporative loss of water from the leaves for every unit of carbon dioxide taken in, in much the same way as happens under low relative humidity (Chapter 3).

Woodward grew plants in artificial atmospheres containing 225 parts per million of carbon dioxide, equivalent to an altitude of about 4000 m, and compared them with plants grown at a 'normal' content of 340 parts per million. Water use at the lower carbon dioxide concentration (imitating high altitude) was 2.2 times greater per unit of dry matter produced than at the higher concentration. Such a difference in grape vines would almost certainly result at high altitude in a much higher ratio of potassium accumulation to sugar production in the leaves, and thence movement to the ripening berries and higher must pH. Therefore high altitudes (all other things being equal) can theoretically be expected to reduce grape, must and wine quality in the same way as low relative humidity. Moreover, the two effects should be additive.

One can only speculate how important smaller differences in altitude might be. It is notable, however, that all the world's acknowledged great table wines come from areas of low altitude, as well as fairly high relative humidity. Most come from altitudes below 300 m, and virtually all from less than 500 m. To the best of my knowledge no reputedly great wines have been made, so far, in temperate climates of the high-altitude tropics or sub-tropics.

The implication for Australia is that the same grape and wine qualities might not be attainable from high-altitude inland sites as near the coast, despite similar temperatures and even average relative humidities. The disadvantage inland is made greater by the predominant absence there of the land/sea breeze pattern.

These considerations further underline the case, in Australia, for a predominantly coastal wine industry. Even near the coast, altitude should perhaps preferably not exceed 500 m. Effective differences due to altitude may, of course, diminish as atmospheric carbon dioxide concentrations rise through the coming decades (Chapter 23). It would nevertheless probably be wise to assume, for the time being, that the relative advantage will remain with low-altitude sites.

5

Management Practices and their Interactions with Environment

Vine spacing, training, and management of the canopy – general considerations

Natural environment and management meet in the vine canopy. The last few years have seen a renewed realization of this, and have brought a revolution in scientific thinking on how vines should be trellised and trained.

Raised relative humidity within dense, closed canopies has of course always been recognized as a hazard in wet climates, where it greatly increases the risk of bunch rot and other diseases. Trimming or hedging of the vine in summer is a traditional response to this problem. It is effective in reducing disease incidence, and allows more effective control by spraying or dusting. Several reports (Koblet 1987; Kliewer and Bledsoe 1987; Bledsoe et al. 1988; Freese 1988; Kliewer et al. 1988; Smith et al. 1988) have highlighted additional advantages for fruit quality of properly timed leaf plucking from immediately around the bunches in dense canopies, apparently through better direct illumination and ventilation of the fruit.

Beyond this, researchers have come to realize the importance of exposing as much as possible of the vine canopy to direct sunlight, and particularly the lower leaves on the canes. These most immediately feed the developing and ripening bunches, so their full exposure to sunlight contributes to rapid, complete ripening, and hence to quality. Analogy can be seen here with tree fruits, in which the best-coloured and best-flavoured fruits on a tree are nearly always those which ripen earliest, and are close to fully exposed leaves on the sunny sides of the tree; whereas fruits which struggle to ripen in the shaded interior are poorly coloured and weak in flavour. The writings of Busby (1825) show that this was well known in traditional European viticulture.

Further, good direct sunshine exposure of the lower buds forming on the new canes, and of the leaves closest to them, helps to promote their fruitfulness for the following season, as discussed in Chapter 3 under sunshine hours and light quality. A good pruner recognizes and selects 'sun' canes for retention.

How much of the beneficial effect of direct light comes from greater photosynthesis and sugar availability, and how much from the physiological effects of wavelength differences between direct and leaf-transmitted or leaf-reflected light, is hard to define. In any case the distinction is largely academic in practice. The important thing is that leaves and fruit be exposed to as much direct and soil-reflected light as possible – with the sole proviso of the fruit's need to be largely protected from early afternoon direct sun in other than cool viticultural climates.

Traditional viticulture in many cool climates ensures maximum illumination of the leaves and canes by vertical training to individual stakes, combined as necessary with summer trimming of excess foliage. Elsewhere, fruitfulness and grape quality have in the past largely depended on the combination of low vine vigour (for various reasons) and summer trimming. More recent developments in vine training and management have aimed at achieving the same objective, but with fewer restrictions on vigour and yield.

Height of heading

Heading the vines low, as distinct from low vigour, has long been associated in Europe with superior wine-grape quality. Busby (1825) wrote that while in cold climates the vines must be kept low to ripen their fruit at all, the same practice was followed by the best vineyards of Bordeaux, and even in as hot an area as Sicily. Guyot (1865) described the mechanism:

> ... the rays of sun warm the ground between the rows ... the earth returning to the vine, when the sun ceases to shine, the heat which it had received during the insolation. ... When the rows of the vines are low, and directed from north to south, the sun, from its rising to its setting, strikes constantly the vines, it heats equally all the surface of the ground, which then fulfils the place of a wall, and accelerates vegetation in its several stages. It is the reverse when the vines are trained high ...

Carbonneau and Casteran (1987a) likewise recommend north-south row orientation and low to half-high heading of the vines. Drouhin (1988) notes that in Burgundy: 'Vines are trained low, the first wire being 40 cm from the soil. High trained vines are only allowed in lesser districts.'

Against these views, Champagnol (1984) pointed

out that the low heading practised in cool-climate traditional viticulture is necessitated by the very high vine densities used. Both are needed to ensure that grapes will ripen fully at the cool limit of viticulture. He considered that a heading height of 60–70 cm is more appropriate with contemporary vine spacing in warm climates such as that of Montpellier. Up to 90 cm may be needed for varieties with trailing habits. Such heights are in any case now needed for mechanical harvesting, and for improved training systems such as the Scott Henry (see later discussion), in which some of the shoots are trained downwards. Facility of hand picking also remains an important practical consideration favouring at least half-high heading in many vineyards.

Practicalities of these types tend to be over-riding in deciding optimum heading heights in commercial viticulture. Nevertheless it is worth considering further the microclimatic benefits and costs of the differing heights, and how they interact with environments.

The studies of Chaptal (1943) at Montpellier showed clearly that the temperature summation for the growing season at 20 cm from the soil surface can be as much as 150 day°C more than at 1 metre. Maximum temperatures at 20 cm were substantially higher, but minimum temperatures only slightly lower. Such an increase in summation is critical in cool climates for full ripening, and for what varieties will ripen (Chapter 7). The fact that it comes mainly from higher day temperatures is least detrimental in the coolest climates. In any case the rigorous selection of sites needed there to avoid killing frosts dictates that they will have excellent air drainage, so that the differences in temperature between day and night, and with height above ground, will be minimized in the vineyards.

The situation in warm to hot climates is more complex. Extra heat during the day is only valuable there in the coolest parts of the season, while during summer, including ripening, direct re-radiation from the soil in the heat of the day may be positively injurious. This is perhaps not critical for low bush-trained vines, such as are traditional in many of the hot viticultural climates around the Mediterranean, because foliage shades both the bunches and the ground under and around them. Extra re-radiated warmth to low bunches at night may, in this situation, still be a more-than-offsetting advantage for wine quality. Moreover such bunches and their associated leaves, perhaps even more than with low training in cool climates, stand to benefit most from soil-emitted carbon dioxide (Chapters 4, 23). On general principles, therefore, a good case can be made for low heading, but only when combined with other traditional vineyard technology, and with the types of site, soil and possibly climate that are most conducive to grape quality.

It is different where vertical systems of training are practised in warm to hot climates, especially if combined with plucking of the external lower leaves to increase fruit exposure, or to facilitate disease control and mechanical harvesting. The same applies generally to flat sites with poor air drainage and non heat-absorptive soils, as is the case in many inland irrigation areas. Here, spring frosts can be a hazard to low growth even in hot climates. Heat reflection or direct reradiation during the day can only serve to damage low-hanging fruit, which is also prone to bunch rot if the usually loamy, and sometimes organic, soils remain damp for prolonged periods. In these situations the fruit enjoys more equable temperatures and humidities if well clear of the ground.

Optimum heading height thus varies according to climate, site topography, grape variety, soil type, and management. There is no universal guideline.

Row orientation
North-south orientation of rows is generally thought to improve the light-use efficiency of erect-growing crops. Smart (1973) calculated that the most complete sunlight interception by grape vines is by north-south rows that are tall and fairly closely spaced. North-south row orientation also has the theoretical advantage that it allows full illumination of both faces of the canopy, each for a part of the day. Because leaves and adjacent storage sinks soon become photosynthetically saturated, this should exploit sunlight more efficiently than continuous illumination of one face and continuous shading of the other.

However this reasoning loses force in training systems, such as are discussed below, which achieve desirably dispersed canopies of uncrowded shoots, so that light penetration to most leaves is good regardless of row orientation. North-south rows also have the disadvantage, in a hot climate, that the fruit on the west side of the row is directly exposed to the sun in the early to mid-afternoon, when temperatures are at their peak. Berries undergoing the sudden transition to this from full shade protection through the rest of the day appear to be especially liable to heat damage (Erland Happ, personal communication 1990).

By contrast east-west rows – in a low-latitude environment like most of Australia, where the summer sun is nearly overhead – offer good protection to most fruit through the danger period. Any bunches on the north sides of the rows that do receive considerable exposure have maximum opportunity to acclimatize, since the exposure is relatively continuous. Usually there is also some direct fruit exposure in the early morning and late afternoon to sun shining obliquely along the rows at a low angle, but this is more likely to be beneficial than harmful. Indeed, the fact that east-west rows receive light of higher red to far red ratio at the critical times of early morning and evening than north-south rows (Kasperbauer et al. 1984; Kasperbauer 1987; Hunt et al. 1990a) could have substantial significance for growth and ripening processes. In cool and mild maritime climates, it should theoretically have the advantage of discouraging rank, vegetative growth. The soil between such rows should also enjoy the greatest sun-warming in the early morning and again in the evening.

I conclude tentatively that, in Australia, the orientation of vine rows to optimize light relations has only secondary importance. In hot areas there is probably some advantage to east-west rows. However other factors could have a higher priority in planning decisions, including planting for best erosion control; wind control; and convenience for mechanical harvesting and other vineyard operations.

Vine and row spacing

Traditional European cool-climate viticulture employs close spacing both between and within rows, often with vertical training of the canes on single stakes. It is argued in part that this: (1) makes maximum use of sunlight, and of space on limited premium sites; and (2) helps to prevent unwanted vegetative vigour by creating competition for nutrients and especially moisture, which in these climates is often more than adequate.

Hot, and especially arid, climate viticulture has tended by contrast to employ much wider spacing, usually with the expressed aim of providing a greater soil volume for each vine to exploit for moisture. Trials such as those of Winkler (1959) and Hedberg and Raison (1982) have failed to find any yield or quality advantage in warm climates for close as compared with wide row and vine spacing, provided that trellising is adequate (see below). Large vine size allowed by wide spacing in fact permits a better balance between fruiting and vegetative vigour in environments or cultural systems that are highly conducive to vigour (Smart and Robinson 1991). This is known as the 'big vine effect'. Wide row spacing also serves to allow the use of normal farm tractors and cultivating equipment.

A point which may be decisive is whether or not winter green manuring is employed. I argue later that this is a very desirable practice in mediterranean-type climates, at least on level to moderately sloping land. The rows must be widely enough spaced to grow a worthwhile crop.

But whatever the practical constraints on row and vine spacing, one overriding principle remains. That is, that the basic unit of the vineyard is the individual fruiting shoot.

Getting the best yield and/or fruit quality from each shoot requires that all its leaves enjoy the greatest possible sunshine exposure, while still giving enough protection to the fruit. The most important leaves in both respects are the lower ones on the shoots, which immediately protect and feed the bunches. Getting the best combination of quality and yield per unit vineyard area then entails maximizing the number of fruiting shoots per unit area, while preserving the lower-leaf sunshine exposure of each. Theoretically the best arrangement is a completely uniform distribution of shoots over the vineyard, such as might be attainable with very close vine planting on the square, or alternatively with full overhead trellising. However neither of these seems likely to be practicable in the present physical and economic environment of Australia, except perhaps for very specialized purposes. The aim, therefore, is to combine the greatest possible dispersion and sunlight exposure of the individual fruiting shoots with an economic and manageable arrangement of vines in rows. Attaining this depends in the first instance on the system employed of vine trellising and training, to which we now turn.

Trellising systems

Modern research on trellising began in the USA with the work of Shaulis et al. (1966); in France with that of Carbonneau and associates (Carbonneau et al. 1978; Carbonneau and Huglin 1982; Carbonneau and Casteran 1987a, 1987b); and in Australia and New Zealand mainly with the work of Smart and associates (Smart 1982, 1984, 1985a, 1985b, 1987a; Smart et al. 1985a, 1985b, 1990a, 1990b; Smart and Smith 1988; Cullen 1988). The results of all this research have now been summarized and illustrated in the excellent publication of Smart and Robinson (1991), to which the reader is referred. The following discussion touches only briefly on the main trellising systems as they have been developed historically, or that are now coming into use in Australia.

Shaulis' 'Geneva double curtain' was the first of the new systems to set out deliberately to increase sunlight exposure, both of leaves and of the bunch area. With its divided, hanging canopies, it is perhaps best suited to native American vine species because of their naturally pendulous growth habits, and to very cool climates where direct exposure of the bunches at the cane bases is not undesirable. Even in the mildly warm Bordeaux climate, too much direct bunch exposure to the sun can lead to 'disagreeable aromas' (Carbonneau and Huglin 1982).

For the largely mild to warm conditions of southern Australia's premium viticultural areas, four other new or improved systems are now being widely adopted, in place of the previous standard bilateral cordon without shoot positioning.

The first is simple vertical shoot positioning, as widely employed in northern and western Europe and New Zealand. The shoots, growing from horizontal rods or spur-pruned bilateral cordons, are tucked and held vertical between typically two pairs of high foliage wires. Top and side trimming are carried out through the growing season, and there is commonly some further plucking of lower leaves to expose the ripening fruit. The system is well suited to machine harvesting. Smart and Robinson (1991) maintain, however, that it is appropriate only for vines of moderate or low vigour, being prone to excessive shoot and foliage density where vigour is high. Total exposed canopy area, and therefore yield potential, are limited unless the rows are narrowly spaced.

The Scott Henry system, named after its pioneer in Oregon, USA, differs in that it has parallel pairs of rods or cordons, one above the other. Shoots from the upper rod or cordon are positioned vertically upwards, and those from the lower rod or cordon vertically downwards. The addition of the down-trained shoots greatly increases canopy surface area, and overcomes the yield limitation of simple vertical shoot positioning at workable row spacings. It also reduces the problem of shoot density and vigour in high vigour situations, by giving an effectively bigger vine (see discussion by Smart and Robinson 1991). The system is well adapted to mechanical harvesting. Excessive fruit exposure on the down-trained shoots can be a limitation in hot climates, although less so than with the Geneva double curtain. Henschke et al. (1990) give an example of this in South Australia.

The 'U', or 'lyre' system promoted by Dr Alain Carbonneau, of the INRA Station de Recherche de Viticulture in Bordeaux, France, overcomes the yield limitation and crowding problems of high-vigour vines by providing each vine with parallel, near-vertical canopies, separated by about 90 cm at the bottom and 120 cm at the top. The large vines result in effective divigoration of the individual shoots. Careful trimming of the inside canopy walls is needed where vigour is still high, to ensure their full sun exposure. The slight outward inclination of the canopies, combined with bottom positioning of the fruit, maintains good fruit protection from the vertical sun and makes this a safer system for warm to hot climates than the Scott Henry. Good canopy dispersion can be combined with wide row spacing, allowing room for green manure crops as discussed later in this chapter. The system is not suited to the older types of machine harvesters, but French-designed machines specifically for it are now available.

While both the Scott Henry and Lyre (and similar) systems help to overcome problems of high vine vigour, they may still not fully achieve this where vigour is very high. A suggested remedy (Smart et al. 1990a) is then to combine them with wider vine spacing in the rows. Every second vine might be removed from established vineyards. Light pruning of the resulting very large vines brings vegetative vigour down to a more normal level, and the vines into fruiting balance and satisfactory light relations.

The fourth major system used in Australia is that of 'minimal pruning', amounting at most to a light mechanical trimming in winter of vines which are usually grown as simple bilateral cordons. Developed and strongly advocated by CSIRO researchers over the last two decades (Clingeleffer 1984, 1989; Clingeleffer and Possingham 1987; Possingham et al. 1990), this system remains controversial. Smart and Robinson (1991) conclude that it is probably best adapted to warm

and hot regions, with risks of delayed maturity and bunch rot problems in cool and humid climates. On the other hand, the largely external placement of the fruit carries its own risks in hot climates. Low pruning cost, ease of machine harvesting and high yields are prime attractions for bulk wine production.

Smart and Robinson describe several other systems of vine trellising which are so far little more than experimental in Australia. None at this stage looks obviously better than those described above for producing high-quality grapes under Australian conditions, although further trial is doubtless warranted.

Canopy light relations, crop load and fruit quality
The research by Smart and his colleagues highlighted a relationship of special significance in Australia: that of strong shading within the vine canopy to high content of potassium in the berries and musts. High potassium content is also related to strong atmospheric saturation deficits, as we saw in Chapter 3. That associated with excessive shading within the canopy is almost certainly additional. Moreover, the high potassium content in shaded grapes is typically accompanied by an elevated ratio of malic to tartaric acids, which leads to a pronounced further upward shift in the pH of the new wine if it undergoes malolactic fermentation.

High malic acid contents in musts from heavily shaded berries is presumably related to the effects of ripening temperatures on malic acid loss, as described in Chapter 3. Why potassium should also be high where there are dense canopies has been less clear until very recently, although the concurrence doubtless helps to explain the general correlation between potassium and malic acid contents in grapes (Hale 1977). However the reasons are now fairly well understood.

For a given set of climatic conditions and nutrient supply, total potassium uptake by actively growing vines is closely related to their growth (Freeman and Kliewer 1984). As we have already seen, much of this potassium, and of that taken up later during fruit growth, is remobilized via the vine's phloem system to the developing and ripening fruit, along with sugar (sucrose) from the photosynthesizing leaves and from storage in the stem, roots etc. Shaded leaves manufacture and supply little or no sucrose, but they still contain potassium; and much of this is available for export to the fruit. Indeed, Freeman and Kliewer argue that lack of sucrose in shaded or otherwise stressed leaves results in a compensating increase in potassium loading into the phloem system. Thus, the higher the proportion of shaded leaves, the higher is the ratio of potassium to sucrose moving to the berries.

The problem may be viewed another way. Assuming that crop set is proportional to vine vigour – which will be broadly true in a hot, sunny, inland environment – then the more vigorous and densely canopied a vine, the heavier will be its crop load relative to the area of exterior, and therefore actively photosynthesizing, leaves (Smart 1982, 1985; Smart et al. 1985b, 1990; Smart and Robinson 1991). On top of that, strong atmospheric saturation deficits in mid-summer are likely to reduce the photosynthetic efficiency even of those leaves, through closing of their stomata early in the day. The result is a much reduced availability of sugar per bunch during the critical ripening period, and consequently delayed, uneven, and sometimes incomplete maturity in terms of sugar content. Potassium nevertheless continues to move into the berries from both shaded and unshaded leaves over the whole period, at an unchanged or perhaps enhanced rate, so that berry potassium content is much higher by the time sugar maturity is finally reached. There is also a longer period during which acid levels can fall. The combined result is more potassium and a higher pH in the must and wine.

Slow sugaring can, of course, equally result from simple over-cropping. In that case the net effect on the fruit is one of potassium dilution, with consequently reduced pH: see, for instance, Sinton et al. (1978). The problem of excess potassium and high pH, then, is specifically one of atmospheric saturation deficit and canopy light relations, not of cropping level.

The consequences of slow or retarded sugar flux into the berries, whether from moisture stress, poor canopy light relations, over-cropping, or any combination of the three, nevertheless go further. As noted in Chapter 3, it is generally accepted that colour, flavour, aroma and other 'secondary' compounds in plants and fruits are synthesized from sugars surplus to those required for respiration and basic tissue growth needs. In grapes this synthesis occurs largely in the skin (Winkler 1962).

Pirie and Mullins (1977) measured the contents of anthocyanins and free sugars in the skins of Shiraz and Cabernet Sauvignon grapes, grown at three locations in New South Wales and sampled over the five weeks following véraison. Sugars and anthocyanins in the skins of berries from vigorous, irrigated vines at Griffith were only half those in the skins of berries from low-vigour, dry-land vines growing at Mudgee, despite comparable whole-berry sugar contents at maturity. The correlation between anthocyanins and free sugar content in the skin was almost complete (r = 0.96), and later work showed that the rise in anthocyanin content lagged behind that of sugar by about a week (Pirie and Mullins 1980). The correlation between skin anthocyanin and total berry sugar content, however, was poor. Once sequestered in the storage tissue of the berry pulp, the sugar appears to be metabolically unavailable for conversion into other products.

The existence of a strong correlation between anthocyanins, or anthocyanin colour intensities, and quality in red grapes and wines has long been part of wine lore. Consistent with a direct correlation to anthocyanin, Jackson et al. (1978) found that the flavours and aromas of a range of Beaujolais wines, made from Gamay, were best predicted by their colour intensities when acidified to pH less than one. This procedure converts non-coloured anthocyanin forms in the wine to red-coloured forms, and so gives a measure of total anthocyanins.[1] Similarly Iland and Marquis (1990) showed a very close correlation between grape colour intensity and the overall quality scores of wines made from Pinot Noir grapes grown in South Australia and Tasmania, with different management treatments. In this case quality and colour were both inversely related to yield.

On the evidence it seems reasonable to suggest, as did Pirie (1978b), that the secondary compounds responsible for aroma and flavour are formed in the skins in some kind of proportion to the anthocyanins (in red grapes). That is, maximum production of all compounds contributing to wine quality will depend in similar ways on the rate of sugar flow to the skins at least exceeding a certain threshold. Thus grape quality as a whole can be seen as depending on the rate of sugar inflow during ripening.

The relationship between pigmentation and final flavour and aroma at ripeness may nevertheless be only incomplete, because anthocyanin accumulates in relatively stable forms, whereas the compounds responsible for flavour and aroma are to varying degrees subject to temperature-dependent evaporative losses and chemical degradation or complexing during ripening (Chapter 3). But countering this in the ideal ripening situation is the fact that rapid sugaring and secondary metabolite formation shortens the period available for their loss, as it does also for the loss of acid and accumulation of potassium. Ample sugar, together with a strong supply of cytokinins from the roots that would go with a surplus of sugar throughout the vine, should also help to maintain the berries' defences against disease and senescence, as argued in Chapter 3. The relationship (in red grapes) between rate of sugar influx, anthocyanin intensity and final winemaking quality may thus be very close after all.

In summary, the work of Pirie and Mullins, combined

---

1 On the other hand Somers and Evans (1974) found the colour and quality of young red wines from given grape varieties grown in the Southern Vales, South Australia, to be best predicted by the proportions of their total anthocyanins present in the ionized, red-coloured form, and not at all by the total anthocyanins. In this case the over-riding factors governing wine quality appeared to be the faults of excessive potassium in the wine (leading to tartrate precipitation and high pH) and excessive sulphur dioxide: both of which reduce the proportion of the total anthocyanins present that is in ionized, coloured form. Such a difference in result perhaps partly reflects the fuller maturity of the South Australian as opposed to the Beaujolais grapes, together with differences in the growing and ripening environments. Nevertheless it still does not rule out the possibility of a relationship between wine quality and total anthocyanin, even in a warm climate, once major avoidable viticultural and winemaking faults have been eliminated. Either way, the relationship of wine quality to visible or measurable colour remains valid.

with that done later by Smart and colleagues on canopy light relations, provides a substantial basis for explaining the weak character and high pH of grapes and wines from vigorous, high yielding, but poorly trellised vines grown in arid areas practising full irrigation; also the often disappointing character of wines from vigorous vines in other parts of Australia; and finally, considered together with vine nitrogen nutrition as discussed under 'Fertilizing and green manuring' (below), the more or less general inverse relationship between grape yield and wine colour, flavour and aroma, as demonstrated by Sinton et al. (1978), Freeman et al. (1980) and McCarthy et al. (1987), and widely observed in practice.

The picture is nevertheless still not complete. One further factor remains which contributes to the control of sugar flux to the ripening berries, and which to the best of my knowledge has not previously been canvassed in the context of vine growing. That is the direct interaction between canopy light relations and temperature.

Interaction between canopy light relations and temperature

Figure 2 (Chapter 3) shows that more sunshine hours are needed to attain comparable wine styles as temperature rises. That is because the faster respiration of the vines at high temperatures uses disproportionately more of the assimilates produced by photosynthesis. A parallel relationship can be expected with light distribution within the canopy: greater or more efficient illumination will be needed as temperature rises. The converse result is that the optimum temperature for growth and/or assimilate accumulation in the vine can be expected to fall as light availability falls, or as light relations within the canopy deteriorate with increasing leaf area and mutual shading. This has been clearly demonstrated in other plant species.[2]

Other evidence can be adduced to show that night temperatures are the main factor in this response. Went (1953, 1957) and Went and Sheps (1969) studies the joint effects of diurnal temperature range and light intensity on plant growth. They found that the greater the light intensity and day-time photosynthesis,

the higher was the night temperature needed to give maximum plant growth (day temperatures being kept constant). Similarly, then, the better the light relations within the canopy, and therefore the effective light intensity, the higher the night temperatures that will logically be needed to exploit fully the enhanced products of photosynthesis.

Combined implications for vine canopy light relations and management are as follows:

1 The higher the mean temperatures, the worse will be the relative effects of poor canopy light relations on yield potential and, more markedly still, on the flux of sugar to the ripening berries and therefore fruit and wine quality. The potential for response to improved canopy management will therefore increase with growing and ripening temperatures.

2 In exactly the same way, the potential for response to improved canopy management will be greatest where the diurnal temperature range is narrow and the temperature variability index small.

In brief, the greatest yield and quality response to improved canopy management and light relations can be expected in warm to hot climates, and then, on the sites within them with the lowest temperature variability indices.

Much of the past experimental work on trellising of Vitis vinifera has been in cool to mild regions, such as New Zealand and Bordeaux; while traditional light-efficient training systems have been practised mainly in cool viticultural regions such as Germany. Valuable benefits have been shown in these environments. Our consideration of basic physiology suggests, however, that the benefits should be still greater in warm climates. They might be especially so in warm, but cloudy, regions such as coastal New South Wales and northern New Zealand, where the low ratios of sunshine hours to effective temperature summation (Chapter 3, Figure 2) appear already to limit wine styles.

On the other hand, the advantages of very sunny climates for photosynthesis are often negated in practice by low relative humidities and strong saturation deficits, which can markedly reduce photosynthesis (Chapter 3). Canopy light relations might therefore be as important, perhaps sometimes even more so, in arid inland climates as near the coast.

To the extent that sunlight hours (or intensity) and relative humidity are inversely related and cancel each other in this way, they leave temperature and its variability as the main, consistent environmental factors interacting with canopy light relations across the full range of environments. A logical and universal conclusion to be drawn is that improvements in canopy management and light relations will always result in some upward shift in optimum temperatures for viticulture, especially night temperatures. This will be so whether the criterion is yield, or fruit quality, or any combination of the two.

---

2 The work of Cocks (1973) with the pasture legume subterranean clover illustrates the principle well. Cocks studied response to temperatures ranging from 12°C day/7°C night to 27°C day/22°C night, at a range of foliage densities as measured by leaf area index (i.e. the ratio of leaf area to ground area covered by the plants). 'The response to temperature depended on leaf-area index (LAI). When the LAI was low (0.2), growth rate increased with increasing temperature to a maximum at 22°C day/17°C night. This agreed with the temperature response of single plants of the same species previously reported. However, when the LAI was 3 the growth rate was not influenced by temperature within the range tested. At a still higher value of LAI (5.5), the response in growth became negative with increasing temperature, communities at the highest temperature growing at only half the rate of those at the lowest.'

How far this reasoning can be extended into truly hot regions of Australia remains uncertain. Moderately hot, but equable climates such as coastal New South Wales clearly have much to gain from good canopy light relations. Responses to better canopy management in hot areas inland will also be maximized to the extent that mesoclimates can be selected with reduced temperature variability.

However most hot, inland parts of Australia are subject to extremes of both high temperature and low relative humidity. The amount of bunch protection afforded even by lyre trellising may not always be enough to prevent berry scorching and off-flavours under such conditions: the more so if stress leads to defoliation. So the theoretical possibility of large responses there to improved canopy light relations may not often be attainable in practice.

In such climates it is probably necessary to continue using systems giving greater foliage densities than are desirable for light relations, purely to ensure protection for the fruit. Severe pruning, trimming of unnecessary young foliage to conserve water and minimize stress during ripening, and crop reduction to within the vine's photosynthetic capacity by thinning of excess and unduly exposed bunches, seem likely to remain primary strategies for attaining the best fruit quality. The traditional low bush vines of hot Mediterranean regions represent such an adaptation.

The alternative is to strive for simple productivity at least cost, for instance using minimum mechanical pruning and full irrigation (where water is available). But how far either form of viticulture can remain fully competitive with that in climatically more favoured near-coastal regions, where improving vineyard management can now better combine quality with adequate and reliable yields, is something that looks increasingly doubtful.

Supplementary watering
The moisture needs of the vine through the season are discussed in Chapter 3, where it is shown that the need for ample water, versus some moisture stress, changes through the season according to growth and ripening stage. Many studies and practical experience have shown that continuous full irrigation often results in poor grape and wine quality. But as already noted, these effects are due, not so much to irrigation per se, as to the unfavourable canopy light relations which often result, and the atmospheric aridity that made full irrigation necessary in the first place. The two factors are logically additive in their effects.

It is now abundantly clear the so long as canopy management maintains optimum leaf and fruit exposure for any given external environment, judicious irrigation cannot be harmful to fruit and wine quality. Indeed, quality is more likely to be improved where the main effect is to alleviate undue vine stress during ripening. All evidence shows that stress must be avoided at flowering and fruit set; and probably also for a short time after, to ensure fruitful bud differentiation for the following year. Some mild stress is then useful in early to mid fruit development: firstly, to control potential berry size; secondly, to harden the new growth against later stress; and finally, to stop unwanted further vegetative growth and ensure maximum assimilate diversion to the fruit from the beginning of ripening. But from véraison onwards I believe the principal aim must be to avoid severe stress, and to ensure that the vine can keep on nurturing its crop to full and perfect maturity. The underlying reasons for this are argued in Chapter 3.

Hardie and Martin (1990), in South Australia, adopt a similar view, but with one partial exception. They advocate maintaining a moderate moisture stress, not only through early berry development, but also through the whole ripening period. The aim is to maximize set, but then to restrain berry growth continuously. They report that this increases the ratio of skin area to berry volume, and with it, wine colour and flavour. Any loss of yield in an initial year is claimed to be compensated for in succeeding years, so that long-term yield is maintained, or even (through improved fruitfulness) actually increased.

While the main results of this approach in practice cannot be denied, the question must be asked whether the effects of continuing stress through ripening can be truly beneficial to quality. All the above arguments, together with the studies I have cited in Chapter 3, suggest that this is unlikely. More probably there will be an increase in the coarser, skin-associated components of flavour and aroma, at the expense of the more delicate and aromatic components. This will be especially so if the stress is enough to slow sugar flux into the berries.

In any case it must also be asked how far the maintenance of a mild moisture stress throughout ripening is practicable. Given the natural variability of weather, grapes under mild stress must be vulnerable both to the sudden onset of heat and high saturation deficits on the one hand, and to rain on the other. Keeping them in a state of moderate sufficiency the whole time during ripening should help to buffer against both. Any 'hardening' of the vines that is needed should be complete by the start of ripening if it is to be fully effective.

For several good reasons, drip, or trickle, irrigation is the method mostly used in mediterranean climates. It is feasible over almost any terrain. It is economical in water use, with least loss by evaporation, and so can be combined with seasonal surface catchment of water in areas of moderate or higher total rainfall. It lends itself to use also as a vehicle for nutrient application, as discussed below and in the next section, giving improved control over the supply and timing of both water and nutrients. Finally, because wetting is to depth over only a small area, it does not greatly encourage summer weed growth, or too great a dominance of surface root growth by the vines.

In fully arid climates, wetting of the whole area between the vine rows is needed to allow any consistent root growth and feeding there by the vines; but in mediterranean climates this occurs under natural rainfall during the main growth period of spring and often early summer, and can be extended by the use of mulches between the rows as described later in this chapter. Trickle irrigation is especially well suited to supplementary late-season watering under such conditions.

The impact of trickle irrigation on mediterranean-climate viticulture has been profound in recent years. Dry (1990) has described some of the benefits and wider implications in California. The soil types that are usable are now much more varied and extensive. To the extent that they can be shallower, or have less water-holding capacity than before, the degree of control that is possible over moisture supply to the vine over summer and autumn is greater. Because such soils are usually lower in nitrogen than the loamy soils of the valley floors that were once predominantly used, the potential control over nitrogen supply is also greater. At the same time the topography and meso-climates are usually more favourable for quality than those of the valley floors and flats.

All this means that the mesoclimatic advantages enjoyed by the best cool-temperate vineyards of western Europe are now available for mediterranean-climate viticulture as well. But the latter has the clear further advantages of more sunshine, less danger from summer rains, and with trickle irrigation, potentially a better control over nitrogen as well as water supply during critical periods. If temperatures are favourable and relative humidities high enough, and especially where regular afternoon sea breezes occur in summer, the mediterranean climates can now arguably claim to have the best environmental conditions of all for reliably producing high-quality wines. Largely this has been made possible by trickle irrigation.

Fertilizing and green manuring
Some implications of soil type for vine nutrition were described in Chapter 4. There I emphasized that nutrition needs to be:

(a) balanced, so that all nutrients are supplied in adequate amounts; and

(b) continuous, and matching the vine's needs at its different growth stages.

I suggested that both balance and continuity of nutrition are necessary for maximum natural resistance to diseases and pests, because transient deficiencies or imbalances provide 'windows of opportunity' for undesired organisms to gain a foothold. Finally, I stressed the role of organic nutrient sources, and the desirability of vigorous soil microbial activity to make the nutrients continuously available to the vines.

The inorganic and organic approaches do not necessarily conflict. Many Australian soils have extreme deficiencies of phosphate and other essential plant nutrients in their natural state, and this persists to varying degrees even after years of fertilizer application. Given the relative lack of intensive animal husbandry in most Australian viticultural regions, animal manures are usually unavailable or prohibitively expensive. Where they can be used economically, for instance in small vineyards which are part of a mixed farm, animal manures are an excellent source of both vine nutrients and soil organic matter. But in most Australian commercial vineyards, soil organic matter and fertility must in practice be built up and maintained by growing plants *in situ*; and that is only feasible when they receive the necessary nutrients to make abundant growth in the first place. In addition to nitrogen (for non-legumes) and phosphate, these commonly include the major elements potassium and occasionally magnesium and sulphur, together with the trace elements zinc, copper, manganese and occasionally boron, molybdenum and cobalt.

Further, the fact that some mineral nutrients are removed from the vineyard in the wine, even if the winemaking residues are regularly returned to it, means that the losses must eventually be made good. Some young soils, in the process of active formation from rock, have a slow natural accretion of nutrients that can replace much of those lost; but Australia's soils are predominantly ancient and leached, so that this cannot be guaranteed. Among the mineral elements, potassium in particular is removed with the grapes in large amounts, and on light soils is also fairly readily lost by leaching. Immediate needs for nutrient application can now be assessed with reasonable accuracy by vine petiole analysis (Goldspink et al. 1990).

That said, it remains a worthwhile goal to build up to, and maintain, an optimum state of vineyard nutrition as near as possible to closed-cycle self-sufficiency. This can be done, in climates with enough rainfall, using a system of winter cover or green manuring based on mainly leguminous crops and pasture plants. Legumes can obtain much of their nitrogen from the air, which then becomes available to the vines (mainly in the following year) as the residues break down after turning in, slashing, chemical killing or natural death in the spring.

Deep-rooted green manure plants such as lupins or mustard, grown during the vines' dormant season, have the virtue for light soils that they recapture and recycle other soluble nutrients which might otherwise leach far down the soil profile and be lost. Nutrients taken up by the vines themselves are largely returned to the surface soil by chopping up and mulching of the prunings, apart from the fairly substantial amount of potassium that leaves in the fruit.

Deep-rooted winter green manure crops have the further virtue, pointed out by Happ (1989), that they can cycle a proportion of the *less soluble* nutrients *downwards* from the soil surface in their roots, to where the vine roots can continue to get them after the surface dries in summer. Most of the trace elements, and also phosphate in the iron-rich soils of many Australian vineyards, quickly become chemically bound in the soil and hardly move below the surface if applied there. And because many of the common green manuring crops, including lupins, mustard, cereal rye and triticale, have more-than-usually strong feeding mechanisms for these nutrients from their soil-bound and native mineral forms, they become doubly useful as intermediaries for making nutrients available to the vines. The surface and near-surface residues of all such crops also encourage earthworms, which do a similar job to deep-rooted plants in carrying nutrients (and organic matter) downwards from the surface, and in aerating the soil and improving its drainage.

Any inorganic fertilizers needed to raise initial mineral fertility, or to maintain it at an optimum level, are probably best applied to the green manure crop. Fertilizer nitrogen will be needed as well, for non-leguminous green manure crops such as cereals or mustard. (Rotation of different green manure crops will usually be needed to avoid build-up of diseases in any one of them, and leguminous and non-leguminous crops act as particularly efficient disease-reducing agents for each other when alternated.) The applied nutrients become steadily available to the vines as the organic matter from the green manure crops later breaks down in the soil.

Winter green manuring is particularly suited to mild, winter-wet climates such as characterize nearly all of Australia's vinegrowing regions. The winter crops use excess moisture, and thus help to avoid both nutrient leaching and winter-spring waterlogging. Deeply tap-rooted crops such as lupins, in conjunction with earthworms, have the further vital function of creating channels for later vine roots to follow down, as well as greatly improving water infiltration into the soil and helping to stop plough-pans from forming.

In mediterranean-type climates, green manure crops must normally be slashed, spray-killed or turned in by the time of vine budburst or soon after. This is necessary, firstly, to avoid competition between vine and green manure crop for moisture and nutrients after vine growth has started; and secondly, in most areas, to reduce the risk to the young vine growth of post-budburst frosts. Whether it is better to incorporate green manure crops, or to leave them as a surface mulch, is touched on in the following section.

In either case, a negative aspect of green manuring is that it requires some cultivation and the passage of farm machinery, which can cause compaction and loss of soil structure. Cultivation also reduces the population of earthworms. Addition of organic matter and the 'biological plough' function of green manure tap-roots do compensate for a certain amount of cultivation.

Nevertheless it is important that both cultivation and traffic be kept to a minimum.

An option which in some environments can obviate the need for cultivation over extended periods is to grow self-regenerating pasture legumes. Subterranean clover is well suited to this role, and is shallow-rooted enough not to compete greatly with the vines for moisture. An early-maturing but bulky variety, such as Dalkeith, may set seed early enough to be slashed or sprayed out in spring before starting to compete for water to any significant extent, while still producing enough seed for natural regeneration the following autumn. Alternatively, it and most other present early-maturing sub. clover varieties in Australia, such as Nungarin, Northam and Junee, are hard-seeded enough that one good seed crop will carry over enough seed to allow regeneration for two to three further years without any more being set. Provided that the soil is low in nitrogen to start with (for instance, following a cereal or mustard green manure crop, or a straw mulch) such pasture legumes may persist as a dominant winter stand for two to three years or more, before changing to another type of green manure crop becomes necessary. It is advisable, anyway, that the mainly shallow-rooted pasture plants be rotated periodically with deeper-rooted species. Ludvigsen (1987b) discusses the various types of plant that can be used. Especially among legumes, the most suitable species will depend very largely on soil type.

Much remains to be learned about optimum nitrogen nutrition and supplementary nitrogen fertilizer use for winegrapes. As we saw in Chapter 4, too much exacerbates the natural tendency of many grape varieties to rampant growth under cool to mild temperatures with ample moisture, potentially leading to reductions in both yield and wine quality. Delas et al. (1991) document the typical results of excessive nitrogen application in Bordeaux.

The problems of nitrogen nutrition in Australian and similar environments are perhaps more complex. Most Australian soils are low in nitrogen and many are very low, but in the light of European experience growers have generally been reluctant to apply more than minimal amounts for fear of compromising wine quality. A result has been widespread nitrogen deficiency in vineyards, as described for Western Australia by Cripps and Goldspink (1983), Goldspink and Gordon (1991), Bell (1991) and Dukes et al. (1991). As one result, the nitrogen contents of Australian grapes and musts are generally low (Rankine 1989). On the other hand where nitrogen supply has been greater, the results in Australian vineyards have often duplicated those of cooltemperate Europe. This has been especially so where trellising has been inadequate to accommodate the vines' resulting vigour (Smart 1991). Smart nevertheless remains sceptical as to whether the problems of nitrogen-induced vigour can be fully overcome by improved trellising and canopy management.

The specific further question of nitrogen content in the berries and resulting musts has received little attention, until quite recently. But already there is enough information to suggest the need for a revision of past views on vine nitrogen nutrition and fertilizer use.

In early Californian studies, Kliewer and Ough (1970) found that reduced leaf area per unit of crop weight, whether through over-cropping of normally-leafed vines or by leaf area reduction, resulted in much lower contents of the amino acids arginine and proline in the mature berry juice, together with a less dramatic reduction in total nitrogen. Sinton et al. (1978) later confirmed the existence of a strong inverse relationship between cropping level (as imposed by pruning treatments and bunch thinning) and must nitrogen content in Zinfandel grapes. They showed, further, a strong positive association between the must nitrogen concentrations and the intensities of flavour and aroma in the resulting wines. Wines from low-cropped, and consequently high must nitrogen treatments had the greatest concentrations of esters, but least higher alcohols: both positive factors for wine quality. This was consistent with practical experience of cropping levels, but of course did not prove that high must nitrogen content was the cause.

Subsequently Bell et al. (1979), Ough and Bell (1980) and Ough and Lee (1981) reported similar results where the differences in must nitrogen concentration were due directly to nitrogen fertilizer treatments, and not to cropping levels. Maximum response, with annual spring applications of ammonium nitrate to irrigated vines, was reached at 224 kilograms per hectare with clean cultivation. The higher levels of must nitrogen also substantially speeded fermentation to dryness. Ester formation and/or retention increased, with the greatest increase specifically in the esters of relatively low molecular weight, such as contribute the more volatile and 'fruity' aromas of wine. With increasing ester molecular weights their individual responses to fertilizer and must nitrogen became flatter, and even negative.

Ough et al. (1989) later reported that maximum response in wine tasting scores was reached by 112 kilograms per hectare of applied ammonium nitrate, which was still somewhat above the normal recommended rate for mature, irrigated vines.

Simultaneously with the early Californian work, Vos and Gray (1979) in South Africa showed that the problem of hydrogen sulphide (H_2S) formation during fermentation is associated with low-nitrogen musts. They proposed that hydrogen sulphide is formed largely from the sulphur-containing amino acids released when the yeast has to break down protein as a last-resort source of nitrogen. Jiranek et al. (1990) and Henschke and Jiranek (1991) have since disputed this, and proposed an alternative mechanism for hydrogen sulphide formation. All researchers nevertheless agree that hydrogen sulphide formation during fermentation

signals nitrogen starvation of the yeast.

Other studies reported by van Rooyen and Tromp (1982) and Tromp (1984) in South Africa, Löhnertz (1991) in Germany, and Goldspink et al. (1991) in Western Australia have likewise all showed direct associations between wine quality scores and experimentally manipulated must nitrogen contents, over the normal ranges of must nitrogen encountered. Similarly Bissell et al. (1989) cite the study of Leforestier (1987) in Burgundy, which showed the rated qualities of 150 Pinot Noir wines to be closely correlated with their concentrations of both anthocyanins and nitrogen.[3]

The consensus of results is that the optimum content for rapid and complete fermentation, and for wine quality, is about 500 to 700 milligrams per litre of total elemental nitrogen in the settled juice.

I have cited these findings in some detail because they seem, at first sight, to contradict the hypothesis developed elsewhere in this chapter and in Chapter 3: that the greatest synthesis of wine quality factors depends on a rapid flux of sugar into the berries during ripening. Much evidence suggests that a high plane of vine nitrogen nutrition is normally inimical to this. Like the traditional problem of the best wines coming from cool areas, but from warm seasons in them (Chapter 3), it is a paradox. But like the other for climate, it is one which could hold the key to defining optimum nitrogen nutrition for grapevines, together with methods to attain it. We will return to the practical aspects of the subject a little later.

Meanwhile, a further factor must be considered. Grape ripeness and sugar content are themselves clearly important to fermentation. More sugar requires that there be longer and/or stronger fermentation to reach dryness, especially since high alcohol is inhibitory, and ultimately lethal, to the yeast (Kunkee 1991). Moreover, although ripening is normally accompanied by a rising total amino acid content, much of the increase near and beyond full ripeness is as proline, which is totally unavailable to anaerobically fermenting yeasts; while at the same time, the readily available ammonium content falls away sharply (Kliewer 1968; Ough 1968; Amerine and Joslyn 1970). The resulting problems include not only hydrogen sulphide production, weaker esterification and slow or stuck fermentations, as already described, but also a greater risk of oxidation and other aerobic spoilage during the

---

3 The forms of yeast-available nitrogen in grapes and musts have recently been extensively reviewed (Bisson 1991; Henschke and Jiranek 1991; Kunkee 1991; Rapp and Versini 1991; Sponholz 1991). They comprise mainly the amino acids, which vary amongst themselves in the speed or readiness with which they can be utilized by yeast, together with ammonium ion and small amounts of other soluble nitrogen compounds. The amino acids arginine and glutamine are major, readily used nitrogen sources, whereas proline appears not to be used at all in normal anaerobic fermentation. The hydrolysis of protein to its constituent amino acids for use by the yeast takes place only slowly and with difficulty, after other usable nitrogen sources have been exhausted.

protracted and feeble last stage of fermentation, when protection by carbon dioxide and reducing reactions is greatly diminished (Kunkee 1991). The accumulation of higher alcohols appears likewise to occur at this stage (Bisson 1991).

The findings help to explain why low must nitrogen to sugar ratios and fermentation problems are characteristic, not only of heavily cropping vines with high ratios of crop to leaf area, but also of very sunny climates, which allow ripening to high berry sugar contents. To them can now be added two further management and environmental factors, which can logically be expected to magnify the problem in future. First, improved canopy management and canopy light relations increase sugar production per unit of leaf area, and hence can only increase further the ratio of sugar to nitrogen available to the berries. Second, progressively rising atmospheric carbon dioxide concentrations will with time carry this process still further.

For all these reasons, the problems of must nitrogen and fermentation, and of vine nitrogen nutrition, loom as major areas for research and practical concern in the years ahead. This will be particularly so in warm, sunny environments, such as in much of Australia and California.

A practical remedy for low nitrogen concentrations, now almost universally adopted in Australia and comparable environments, has been to supplement the musts with diammonium phosphate (DAP): a readily utilized nitrogen source in the ammonium form. This has been effective in largely eliminating hydrogen sulphide from Australian wines, together with other sulphides derived from it. There can also be no doubt that fruit flavours and aromas are often greatly improved, and aerobic spoilage problems largely avoided, because of the more vigorous, rapid and (where required) complete fermentations which result.

Important questions nevertheless remain unanswered. Can supplementation with diammonium phosphate (or potentially other synthetic nitrogen sources) fully compensate for a deficiency of the naturally-occurring forms? Even if it can satisfactorily do so now, will it continue to be adequate in future as canopy management and yields improve, and as atmospheric carbon dioxide concentration continues to rise?

Some researchers have reservations even about the present degree of reliance on diammonium phosphate. Jiranek et al. (1990) maintain that it has ill-defined (presumably undesirable) effects on wine aroma profiles. Bisson (1991) stresses the advisability of maintaining a balanced mixture of nitrogen sources for the yeast, having a range of availabilities and perhaps other qualities, rather than relying on a dominant single nitrogen source. Given the complexity of most biological phenomena, this seems reasonable at least as a provisional viewpoint.

The possibility has to be entertained that nitrogen-containing compounds in the grape also contribute directly to wine quality, in their own right or as immediate substrates for other quality factors. The methoxypyrazines (within limits) are one example, as already touched on in Chapters 3 and 4. Positive associations between desired qualities in champagne-style and Chardonnay wines and the amounts of nitrogen excreted into the wine by yeast, or released by yeast autolysis (Feuillat and Charpentier 1982; Kelly-Treadwell 1988; Stuckey et al. 1991a, 1991b) support this idea. Given the desired complexity of wine flavours and aromas, it is hard to imagine that a richness of available nitrogen compounds could not contribute directly in some way.

On the other hand certain nitrogen compounds in wines are generally regarded as detrimented. Dissolved protein can precipitate out and cause protein haze, against which undesirably severe fining may be necessary. Grape protein content increases with maturity, and haze is commonest in wines made from fully ripe grapes in warm climates (Rankine 1989). In addition two other nitrogen compounds, histamine (for those sensitive to it) and ethyl carbamate (urethane) are associated with actual or popularly perceived health problems.

The biogenic amines, of which histamine is one, are widely found in proteinaceous foods that have been subject to the activities of micro-organisms. Levels in wine are usually very low, and more than minute quantities, as elsewhere, usually signify bacterial spoilage (Radler and Fäth 1991). Most is produced in wines of high pH. Normal malolactic fermentation does not appear to be involved.

Modern winemaking largely avoids histamine formation. On the other hand the relationship between must nitrogen content and its *potential* formation remains obscure. Bertrand et al. (1991) reported a doubling of histamine content where vines were very heavily fertilized with nitrogen. Against that, adequate nitrogen in musts might normally be expected to reduce histamine formation, to the extent that it conduces to rapid, sound and complete fermentation, and therefore largely removes the opportunity for bacterial spoilage.

Ethyl carbamate (urethane) is a carcinogen which occurs very widely in fermented foods and beverages. Ough (1991) has reviewed its occurrence and formation in wines, which normally have only minute traces. Most of that in wine forms by the reaction of urea with ethanol, under conditions of abnormal heating and/or long storage. Yeast forms urea mainly from arginine, and under conditions of low nitrogen supply normally re-uses it to synthesize other nitrogen compounds. But when other nitrogen is readily available, including that from diammonium phosphate, surplus urea is excreted into the surrounding medium and remains in the wine. This can then react with ethanol under suitable conditions. At least three studies (Ough et al. 1989; Ough 1991; Bertrand et al. 1991; Kodama et al. 1991) have shown significant increases in wine ethyl carbamate resulting from nitrogen fertilization of the vines.

But it needs to be stressed again that major increases in either histamine or ethyl carbamate have been demonstrated only with nitrogen fertilization well beyond optimum, such as is likely to reduce both yield and fruit quality; moreover, high levels depend as well on bacterial spoilage (histamine) or abnormally high temperature treatments or long storage at above optimum temperatures (ethyl carbamate). Urea excretion, and hence the potential for ethyl carbamate formation, also varies greatly among yeast strains (Ough 1991). Both problems are thus largely avoidable, as is indeed done in modern commercial winemaking. Additionally, Kodama et al. (1991) report that urea can be successfully removed using acid urease, a process now approved in Japan for preventing ethyl carbamate formation in all alcoholic beverages.

We may conclude that nitrogen supply to grapevines can safely be at least enough to give maximum fruit yield. More tentatively, natural berry contents of amino acids and other yeast-available nitrogen forms (as distinct from protein) should desirably be raised from their present generally low levels in Australia, to be as close as possible to the optimum for yeast nutrition. This will in time need greater supplies of nitrogen to the vine as canopy management and yields improve, and as the atmospheric carbon dioxide concentration continues to rise.

The main management problem is clearly one of ensuring that enough nitrogen reaches the fruit, while not stimulating vegetative growth beyond what can be accommodated (for maximum yield and quality) by the best available trellising and canopy managements for any environment. Organic nitrogen sources will continue to meet some of this need, if only as an essential part of sustainable viticultural systems. They also have the advantage that the availability of their nitrogen increases into the fruiting period as the soil warms up. Nevertheless inevitably there will also be a steadily growing need for supplementary fertilizer nitrogen, if both optimum yields and natural fruit qualities are to be maintained and improved. How, when, how much and in what form the nitrogen is best supplied in different environments are thus critical questions needing careful thought and research.

Conradie (1986, 1991) Wermelinger (1991), Williams (1991), Goldspink and Gordon (1991), Dukes et al. (1991) and Peacock et al. (1991) have reviewed various aspects of the dynamics of nitrogen uptake and use by grapevines. Some generalized conclusions can be drawn.

1. Nitrogen applied at budburst tends to be used preferentially for vegetative growth, and potentially for yield. Depending on soil type and spring rainfall, and the chemical form in which it is applied, some or much of it may be lost by leaching before the vines can take it up.

2. The main period for nitrogen uptake is during the latter stages of vegetative growth and during fruiting.

3. With lateness of application up to near fruit ripeness, increasing proportions of that taken up are allocated preferentially to the fruit, with diminishing effects on current season's vegetative growth and yield. However in the absence of soil wetting by summer rainfall or irrigation, both natural and applied nitrogen in the enriched upper soil layer become increasingly unavailable to the vine roots as the surface dries out.

4. Nitrogen taken up after harvest is allocated largely to reserves in the permanent vine structures. Early growth in spring depends more or less entirely on such reserves, and continues to do so to diminishing degrees until about flowering. Virtually all of this nitrogen is used for vegetative growth.

From this it is clear that the balance among vegetation, fruit yield and berry nitrogen responses can be manipulated to a considerable degree by the timing nitrogen fertilization. In this respect low-nitrogen soils have the advantage over rich soils of greater control over nitrogen supply, analogous to that described earlier for moisture supply by trickle irrigation in climates with little summer rainfall. Potential control over nitrogen nutrition, using fertigation, is a further factor favouring commercial viticulture in climates with wet winters and dry (but reasonably humid) summers, in which supplementary watering is needed anyway.

Two last points link vine nutrition and fertilizer use with the environment.

The first is that the arguments concerning nitrogen do not apply to nutrients such as phosphate and the trace elements, which in most soils are highly to extremely insoluble. If applied to the soil in soluble form, they are quickly converted to insoluble forms and are then only very slowly available to the vine or to green manure crops. Moreover, there is no information to suggest that any are needed more in the fruit than for vegetative growth. It is therefore feasible to apply large amounts prior to vine planting, to provide a stock which will last many years. In climates which lead to substantial summer drying of the surface soil, this will need to be deep enough to ensure that vine roots can remain active in their vicinity right through the growth period.

Such considerations apply especially to phosphate. Trace elements can readily be applied later in foliar sprays, if needed. However foliar sprays are more or less ineffectual in supplying any of the major elements, including phosphorus, potassium and nitrogen. The advertisements notwithstanding, the amounts of these elements that can be supplied by foliar spraying are many times less than the vines are likely to need: see Goldspink (1990).

Finally, there may be an immediate or long-term need for liming. Many Australian vineyard soils, particularly in Western Australia and coastal New South Wales, are at least moderately acid to start with and can become more so over the years as a result of fertilizing and green manuring, combined with the residues from sulphur spraying of the vines. Future increases in nitrogen fertilization can only exacerbate the problem.

Grape vines are mostly fairly tolerant of acid, but too low a soil pH increases the risk of nutritional imbalances. These can include a relative deficiency of phosphorus and potential toxicities of manganese (especially in the presence of waterlogging) and aluminium. The microbial life of the soil can become impoverished at low pH, reducing the efficacy of organic matter as a source of nutrients, including nitrogen. Perhaps nearly as importantly, a low soil calcium content reportedly reduces the soil worm population, and therefore the function of these valuable animals in distributing nutrients and organic matter down the soil profile, and in opening up channels which improve soil aeration and water infiltration. Liming to restore the soil to an only moderate degree of acidity can correct these deficiencies or imbalances, and on sandy soils can also help to prevent the loss of potassium by leaching (Darst and Wallingford 1985).

A further aspect is that low soil pH greatly limits the range of legume species which can be grown successfully in a green manuring rotation. Lupins are acid-tolerant, but medics, faba beans, peas and clovers are progressively excluded as pH falls below neutral.

Liming of acid, loamy or clay soils needs to be approached in much the same way as phosphate application, except that it is desirable to incorporate lime throughout the soil as thoroughly as possible, and to as great a depth as is feasible. Liming of light, sandy soils, on the other hand, needs to be approached with caution, because these soils are poorly buffered against changes in pH. Too much lime on a sandy soil can readily induce deficiencies of potassium and magnesium in the vine, and of the trace elements manganese, zinc and boron. Smart et al. (1991) give a more detailed account of the soil acidity problem in vineyards, and of practical remedies.

Summer mulch or bare surface?
Maintaining a green sward between the vines is academic in most Australian environments, because of the water requirement. It may also be questionable because it will result in green, rather than red, light wavelengths being reflected into the lower vine canopy (Chapter 3). The question in practice must therefore be whether it is best to keep the soil surface largely bare through the summer, or whether to cover it with an organic or other mulch.

A bare surface immediately after budburst helps to avoid frost damage to the young vine growth, by promoting day heat absorption and night re-radiation by the soil (Chapter 4). This is further assisted if the soil is firmed down, to eliminate insulating air pockets and improve its heat conductivity.

One can further argue from the temperature variability thesis (Chapter 3) that a bare, heat absorbing and radiating soil through ripening will benefit fruit quality – certainly in cool climates, and probably in warm to hot climates as well. There is some direct evidence to support this (Butcher et al. 1982). Nineteenth century French authors in general advocated clean summer cultivation in the vineyard (Guyot 1865).

Nevertheless such practices exact a price. Bare, exposed soil may be liable to wind and water erosion unless it is stony enough. Heating and drying of the surface – at least in summer-dry climates – is also likely to kill or inactivate many of the feeding vine roots between the rows.

A steady supply of both nutrients and cytokinins is best achieved where roots remain active in the nutrient-enriched surface or near-surface layers through much of the summer. Champagnol (1984) stated that potassium deficiency in southern France is commonly associated with summer drying of the surface soil. The same can be expected in many Australian vineyard areas with mediterranean or sub-mediterranean climates; it will be even more so for the relatively immobile phosphates and trace elements, other than where these have been ripped in deeply. Soil nitrogen is also present mainly near the surface, because most of it is contained in soil organic matter. Anything which reduces surface evaporation therefore helps the summer nutrition of vines in such climates. Release of carbon dioxide may be a further benefit from mulches or organic matter-rich surface soils. Finally, mulches greatly encourage the activities of earthworms. For all these reasons, most contemporary writers agree in advocating a summer mulch between the vine rows in mediterranean climates, such as those of South Africa and southern Australia (e.g. van Zyl and van Huyssteen 1984; Ludvigsen 1987b; Happ 1989).

Not all the effects of mulching are necessarily positive. Surface insulation by dry mulching materials can presumably increase direct heat re-radiation to the vines at the hottest time of day. Certainly there will be less re-radiation of soil warmth to the vines during the evening, though the combination of insulation and the greater and more prolonged heat storage capacity of a moister and cooler soil might reverse the difference by morning. A definite negative factor is that the presence of dry mulching materials heightens fire risk.

The nature of the mulching materials has relevance for vine nutrition, and may also influence the amount and spectral quality of light reflected back into the vine canopy. Sawdust is an effective mulch, but has an extra-wide ratio of carbon to nitrogen, so that microorganisms attempting to break it down monopolize any available soil nitrogen within reach. Vines on already low-nitrogen soils are therefore likely to need much carefully timed extra nitrogen to compensate. Straw has some of the same problems.

An ideal mulching material, apart from its cost and possibly green colour, would be a legume-based hay. As well as being an excellent summer mulch, this would greatly augment the net stock of nutrients available to the vines in the following seasons. Where vines are grown as part of a mixed farm, crops of lupins or beans, or legume-based pastures, cut from other parts of the farm and laid between the vines in late spring, would fill the role admirably.

A stony enough surface, left bare, achieves some of the objectives of organic mulches, while providing good above-ground temperature relations in early spring and in autumn. Soil organic matter levels are maintained better than one might perhaps expect, due to the lack of soil disturbance and the extra vine roots resulting from the moisture that surface stones or rocks save from evaporation. Nevertheless some further mulching cover is probably still desirable in mid summer, to moderate temperatures and evaporation, and to be a later source of nutrients.

The best system in practice will be influenced by how it dovetails with other vineyard operations, and with rainfall and other local conditions. The one principle that spans all systems, however, is that any cultivation should be kept to the absolute minimum that is needed for green manuring (where used). All cultivation and traffic damage the soil to some extent.

A theoretical optimum for Australian environments with enough rain for winter green manure or ancillary hay cropping might be as follows.

- Annual green manuring or pasture cover of the greatest workable area between the vine rows, slashed, sprayed or semi-incorporated to form a summer mulch. Green manure and cover crops to be predominantly leguminous, and rotated as necessary for disease control. Alternatively, complete noncultivation with spraying in early spring for weed control, and legume-based mulches applied between the rows later in spring after any danger of frost is over.

- Any rocks present or readily available to be piled under the vines, to help regulate temperature in the region of the bunches, reduce frost risk, and conserve moisture (either from rainfall or from trickle irrigation). This area to be left bare and undisturbed apart from spraying out of volunteer weed growth in early spring.

The possible efficacy of spray-painting rocks and mulches specifically to reflect the useful red wavelengths into the vine canopy, or of red light reflectors or plastic mulches to achieve the same thing, remains an interesting area for further research. The actual ratios of red to far red wavelengths directed into the canopy would need to be established carefully. For instance some red plastic mulches can actually result in less red, relative to far red, wavelength reflection than a white or black mulch (Hunt et al. 1990b).

Equally worth investigating could be the effects of short exposures to artificial red light during the evening or night. Briefly breaking the dark period has profound physiological effects on many plants, and may do so for vines.

Other management practices
Several other management practices interact with the natural environment in significant ways.

Provision of windbreaks
The often severe effects of wind (Chapters 3 and 4) warrant the expense of windbreaks and loss of planting area in many exposed but otherwise favourable localities. Most types of windbreak are effective, but growing rows of leguminous trees such as tagasaste, or tree lucerne (*Chamaecytisus palmensis*) has practical advantages. The trees do not compete with the vines for nitrogen, and regularly trimmed to hedge form, provide nitrogen-rich mulch for the vines or excellent feed for associated livestock. Appropriate trellising and shoot densities within the vine canopies themselves contribute further to wind control. Dispersed canopies with low shoot densities, such as are desirable for light relations, help greatly by diffusing and breaking the force of air movement. Dense, vertical canopies can form wind tunnels and exacerbate it. Another approach is to plant tall-growing winter cover crops, such as cereal rye, and to delay their mowing or turning in until mid or late spring when the worst wind danger is over. This is most appropriate to sites with little frost risk, and may necessitate compensatory extra water and/or nitrogen to prevent undue competition with the vines. Grape varieties differ in their susceptibility to strong winds – e.g. Chardonnay is very susceptible, and Chenin Blanc, Grenache and Carignan tolerant. For further details see Kobriger et al. (1984); Ewart et al. (1987); Hamilton (1988).

Deep ripping and artificial drainage
The fundamental importance of soil depth and drainage to the vine, and therefore to grape yield and quality, means that deep ripping and artificial deep drainage should always be considered as parts of vineyard site preparation. This is especially so in areas of high winter-spring rainfall, such as the south-west of Western Australia. Researchers in South Africa have reported marked responses to deep soil loosening (70 cm or more) alone, before planting on a soil with moderate subsoil compaction (Saayman 1982; Van Huyssteen 1990). Deep placement of phosphate and trace element fertilizers before planting (and, if necessary, between the rows after planting) is also highly advantageous on many soils, as already discussed.

Rootstocks

Budding or grafting on to genetically different rootstocks is practised in many areas to reduce problems of Phylloxera, nematodes, or soil salinity. Hardie and Cirami (1988) give a comprehensive account of those used in Australia. Rootstocks that exclude chloride were earlier reported to increase potassium uptake into the vines and berries (Downton 1977), thus intensifying problems of high must and wine pH. Some later studies (Whiting 1988; Hardie and Cirami 1988) did not support this finding, suggesting instead that higher berry potassium and must pH from some rootstocks are due to enhanced vine vigour, and thence to impaired canopy light relations in the absence of appropriate vine training. Still more recently, Ruhl and Walker (1990) confirmed that berry potassium contents and must pH can directly reflect differences among rootstock varieties in their potassium uptakes. One can thus reasonably conclude that both potassium uptake by rootstocks and vine vigour differences can contribute. Another important area for future research is the deliberate use of dwarfing rootstocks for vigorous and poorly-setting varieties in cool to mild maritime climates.

Vine age

Long practical experience has shown that mature vines (over about eight years old) produce the best wines. Two main reasons are advanced for this. First, mature vines have established their full rooting depth, thereby stabilizing their nutrition and buffering them (if the soil is deep enough) against short-term fluctuations in moisture availability. The second theoretical advantage is that mature vines have a greater capacity for carbohydrate storage in the trunks, permanent arms and roots. For a given crop size this potentially gives them greater and more assured reserves of assimilates to maintain a constantly strong sugar flux to the ripening berries (May 1987). The advantage is greatest in seasons of difficult or irregular ripening. In high-latitude continental climates, storage of surplus assimilate in mid-summer could have special importance in enabling ripening to continue longer into autumn, under conditions of rapidly diminishing sunshine and temperature; and also to maintain an adequate reserve, despite this, to be a springboard for budburst and early shoot growth in spring. This aspect would be less important in Australia, other than in the coolest climates and where moisture stress causes late-summer defoliation.

Time of pruning

In areas of maritime climate and mild winters, many grape varieties have problems of early and uneven budburst, followed by damage by frost or spring storms as discussed under continentality (Chapter 3). The work of several investigators (Perold 1926 in South Africa; Bernstein 1984 in Israel; Hamilton 1986, 1988 in Western Australia) has shown clearly that late pruning is advisable in these circumstances. Preferably it should be just before budburst, which in the Southern Hemisphere will normally be in August to early October. Budburst is then slightly delayed and more even, and yield and possibly evenness of ripening are improved. Sprays such as hydrogen cyanamide, H_2CN_2 (Lavee 1990) may also assist.

Bunch thinning

This represents a final management tool for ensuring that crop yield is not excessive for the vigour of the vine and/or the fruit quality aspired to. It is normally delayed until véraison or a little later, for two sound reasons. First, this allows the vigneron to assess the final capacity of each vine to ripen its crop, in the light of the season and fruit set as they have already developed, together with the prospects for the remainder of the season and for the provision of water if needed. Second, it helps to ensure that sugar and growth substances thus made available are invested in the remaining bunches, rather than in promoting unwanted late vegetative growth as might happen with earlier thinning. Normally the number of bunches per shoot will be reduced from two to one or from three to two, depending on set and the capacity of the shoot and its exposure to sunlight. Removal of the least developed bunches has the added advantage that it helps to ensure even, synchronous ripening of the rest.

The concept of vine balance

Many of the points developed in this and previous chapters can be integrated in one all-important concept that is familiar to experienced viticulturists: that of vine balance. Balance is when vegetative vigour and fruiting load are in equilibrium, and consistent with high fruit quality. Too heavy fruiting results in slower ripening and markedly reduced quality, together with weakening of the vine. Too light a crop may help encourage excessive vine vigour due to the removal of competition. This is so especially on rich, moist soils and in mild to cool, maritime climates, both of which conduce to vegetative growth right through the season. The result can be poor canopy light relations, slower fruit ripening, and reduced fertility of the buds for the following season. Both over-cropping and under-cropping set up potential vicious circles, which the viticulturist seeks to avoid. Both are inimical to fruit quality.

Much of management, then, is aimed towards vine balance. Trellising and training to give better canopy light relations allow balance and fruit quality to be maintained at higher yield levels and probably higher ripening temperature. Watering and soil management aim to keep the vine healthy, certainly free from any gross nutrient deficiencies or water stresses, but not vegetatively over-stimulated. Timing and amount of nitrogen supply are critical in this respect. Nitrogen needs to be adequate and continuously available (and hence, preferably, from organic sources), but at only moderate levels early in the growing season. Other nutrients should be in balance at optimal levels for vine growth.

All management practices interact with the natural environment in determining balance. For instance stressful environments which subdue vegetative growth may result in naturally good canopy light relations without special trellising. This is one of the main reasons for the traditional belief held in cool temperate climates, which usually have ample moisture, that high quality wines can only be made from grapes grown on infertile soils; and in Australia, why historically some of the best wines have come from inland areas such as Clare and the Barossa Valley, where limited spring rainfall and low relative humidities restrict growth naturally in the absence of irrigation. It also helps to explain why some of the otherwise favourable coastal regions, such as the Mornington Peninsula in Victoria and Margaret River in Western Australia, encountered viticultural difficulties which could not be overcome until the roles of trellising and improved canopy management were clarified.

Important environmental variation also occurs very locally, not only with topography, but also with patchiness of soil fertility and water-holding capacity. The ultimate art of the viticulturist, at pruning and other times, is to recognize this variation, and how it interacts with individual grape varieties, and thence to manage so as to maintain or correct the balance of each vine individually.

Vine balance, then, is the outcome of diverse varietal, environmental and management factors. The important conclusion is that because all these influences are expressed largely through the same key vine attribute, balance, they interact and can be expected to substitute for each other in at least partly predictable ways. For instance, moisture limitation may offset the effects of low temperature on the vegetation/fruiting balance. Favourably high temperatures and/or long days in summer can tip the balance towards fruiting, even though moisture is ample. An only moderate supply of nitrogen has effects analogous to those of moisture limitation. Varietal differences may offset differences in both. Adequate trellising and vine training can to some extent offset the effects of any factor which leads to vigorous growth and resultant poor canopy light relations.

Because of such interactions and the potential input of management, caution is needed when extrapolating the suitability of individual grape varieties from old to new environments on the sole basis of expected climatic conditions during ripening, as estimated in this work. Ripening conditions remain very important, and are a good guide to the style of wine that can be hoped for; but only in part can they predict where a variety will give its best overall performance and quality. That depends as well on where the vines most naturally achieve their best balance.

Management versus environment
If one accepts that vineyard management can greatly improve the quality of winegrapes and wines in Australia (and elsewhere), the next question is whether differences among regions and environments are important at all. Coombe (1987) gave the opinion that:

> When attention is given to vine canopy, crop load, correct picking date and the temperature of the grapes and fermenting must, then a better picture of real regional differences is possible. I believe the recent results of such comparison have been salutary, and I suggest that, in the future, generalizations on wine quality of different regions will be couched in quite different terms ... The evidence that hot regions give grapes that lack compounds contributing to fine flavour is indirect.

Similarly, Croser (1987) wrote:

> Differences in canopy management can produce more significant quality differences than the differences in temperature summation between regions.

These statements are undoubtedly true, not least insofar as they apply to Australia. But while at first sight minimizing the role of natural environment relative to management, they do not in any way rule out important systematic differences in wine qualities among regions or environments, as Coombe acknowledges. Indeed, the logical extension of Coombe's argument, which is also mine, is that it is precisely when both winemaking technology and vineyard management have as nearly as possible been perfected that the natural environment of the vineyard will emerge as the remaining, and primary, discriminating factor in commercial viticulture (as has long been recognized in Europe). And at future higher average commercial quality levels, the remaining differences will probably still be as important to consumers as they are now, and will be equally reflected in prices.

What cannot be disputed is that the commercial cost of attaining a given wine quality level will be least where the vineyard environment is most suitable. An optimum environment gives the greatest possible combination of quality and yield. It also means that normally the grapes and musts will be sound and naturally well balanced for the style of wine in view, resulting in least need for technical intervention in the winemaking process. The wine will be the most natural that is possible, and that in itself will be an incalculable marketing advantage in the decades ahead.

6

Predicting Maturity Date

Two needs must be met before climate during ripening in a new or projected vinegrowing area can be estimated closely enough for practical use.

1. It must be possible to predict grape maturity dates, using standard climatic statistics.
2. Because grape varieties differ in their ripening times in any environment, it is necessary to classify them into groups of varieties having much the same ripening times. This is done in Chapter 7, where the requirements are also specified for ripening of each group in terms of 'effective' temperature summations as defined below.

I arbitrarily take maturity in this work to be that most suitable for making dry or semi-sweet table wines, and the ripening period to be the final 30 days up to the estimated maturity date. Adjustments to maturity date will be needed for other wine styles needing differing degrees of ripeness, but it seems fair to assume that the stages of ripening most critical to flavour qualities will be much the same for all styles.

Management practices can affect date of maturity. Present calculations assume 'normal' management and cropping levels for making quality wines: that is, with maturity not significantly delayed by over-cropping, bad canopy light relations or heavy irrigation.

Differences in altitude, topography and soil type between climatic recording station(s) and actual vineyard sites must also be taken into account in calculating maturity dates, as described in Chapter 4.

Having done that, I found, by a combination of trial, error and basic plant physiology, that it was necessary to make three further adjustments to temperature in order to get a good fit between the regional climatic data (matched for period of records etc.) and observed vine phenology in established viticultural regions. There was no clear evidence that climatic elements other than temperature influence vine phenology. If they appear to, as in the case of solar radiation, their main effect is almost certainly indirect, via temperature, and is therefore already accounted for by the temperature measurements.

The three adjustments to temperature are discussed below in their order of importance, which also

happened to be the order in which they emerged as the calculations evolved.

1. Adjustment for the vine temperature response curve

Plant growth and phenological development respond to temperature in a largely non-linear manner. It is therefore not surprising that simple temperature summations covering the full temperature range above a base temperature, such as used by Amerine and Winkler (1944), Olmo (1956), Gladstones (1965), Kirk (1986) and others, do not accurately predict maturity dates.

It is important here to distinguish between growth as measured by dry weight increase and that measured by rate of phenological development, i.e. of progression through the physiological stages from budburst to final berry maturity. For a given light and atmospheric temperature regime, potential dry matter growth rate for any plant type rises from a base temperature, below which there is no growth, to an optimum range, then falls again to zero. For most well illuminated temperate plants, including vines, the base is around 10°C and the optimum about 23–25°C, falling to zero at around 40°C (Figure 8). Therefore in hot climates a temperature summation that assumes a straight linear response over the entire range greatly overestimates potential dry matter production. Experience shows that it also, to a lesser extent, overestimates rate of phenological development.

On the best evidence I know of, rate of phenological development likewise increases from the base temperature up to a probably comparable optimum. Beyond that, however, instead of falling away it appears to stay relatively constant, as shown in Figure 8. One result of this divergences of responses is that with increasing temperatures, leaves and stem internodes become progressively smaller: as already described for vines and lupins in Chapter 3. This can be seen for vines in the field, where leaves become smaller and internodes shorter and thinner as they are successively formed under increasing temperatures through spring and early summer.

A further difference between the two responses is that whereas dry matter growth in the vineyard

Figure 8. Generalized temperature response curve of vine growth, related to a linear response from 10°C to 19°C and truncation at 19°C.

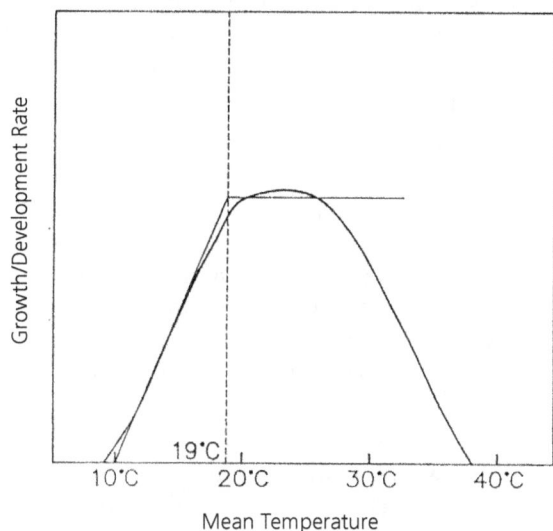

responds to many other factors besides temperature (light, water, nutrition, diseases, etc.), rate of phenological development is largely unaffected by them, and responds more or less exclusively to temperature.

Therefore in practice an appropriate capping of the monthly average means, above which no further increase in summation is recorded, should give a fair first approximation from which to predict vine phenology. That requires finding what cut-off monthly average mean gives best fit to observed phenologies across the full range of grape varieties and environments.

From the beginning (Gladstones 1976, 1977) I approached this by comparing recorded typical vintage dates of given varieties around the world, as compared with the best estimates I could make from available climate statistics (mostly confined to monthly averages of temperature and rainfall) of the growing season temperatures at actual vineyard sites. The reasons for this approach were, firstly, that information on vineyard sites, vintage dates and varieties grown was already available in the scientific and, especially, the popular literature. Secondly, average date of grape maturity is by far the most important statistic in predicting the best adapted varieties for any environment.

Thus by trial and error, using different cut-off temperatures, I found a monthly average mean of 19°C to give the best predictive fit, both to observed phenology in given environments and, more broadly, to the viticultural adaptation of varieties having different phenologies. For warm and hot environments in particular, it gave vastly improved predictions compared with uncapped summations.

A 19°C cut-off goes far towards explaining another common finding, that vine phenological development up to flowering can be broadly predicted from spring mean temperatures, whereas the times from flowering to véraison or from flowering to maturity bear little relationship to temperature (Coombe 1987). Consistent with Coombe's review, Helm and Cambourne (1988) showed that final maturity at Canberra, eastern Australia, can be predicted best from the average mean temperatures of the first three months of the growing season. Pfister (1984) found May-June temperatures to be the best predictors of historical vintage dates in central Europe. In such cool to moderately warm climates the mean temperatures in spring, or up to flowering, are generally below 19°C, and are therefore in the fully temperature-sensitive part of the response curve. After about flowering, this is no longer so. In a

Table 4. Factors for adjusting growth-effective temperature (°C–10) for day length, based on month and latitude[1]

Month/ Lat.	N. April S. October	May November	June December	July January	August February	September March	October April	Average
25°	0.962	0.928	0.912	0.920	0.946	0.987	1.038	0.956
30°	0.972	0.951	0.938	0.942	0.961	0.991	1.027	0.969
35°	0.986	0.974	0.967	0.969	0.979	0.995	1.014	0.983
40°	1.000	1.000	1.000	1.000	1.000	1.000	1.000	1.000
42°	1.007	1.013	1.015	1.014	1.009	1.002	0.995	1.008
44°	1.014	1.026	1.031	1.028	1.019	1.005	0.989	1.016
46°	1.022	1.040	1.049	1.045	1.029	1.008	0.982	1.025
48°	1.030	1.055	1.069	1.062	1.041	1.011	0.974	1.035
50°	1.039	1.072	1.090	1.082	1.053	1.014	0.966	1.045
52°	1.049	1.090	1.113	1.103	1.066	1.018	0.957	1.057

1 Calculated from the Smithsonian Meteorological Tables, Sixth Revised Edition, 1951 (The Smithsonian Institution: Washington).

very warm climate, such as that of the Hunter Valley in New South Wales, the vines are temperature-saturated through most of their growth cycle; therefore little direct relationship of any sort can be expected between temperature and phenology, as McIntyre (1982) found.

2. Adjustment for latitude/day-length

Although truncation of the monthly average mean temperatures at 19°C markedly improved the prediction of grape maturity date, it did not fully correct a tendency for high-latitude grapes to mature with lesser summations than those at low latitudes. I therefore tried an additional adjustment to the temperature excess over 10°C, based on the relative day-lengths at the different latitudes for each month of the growing season. This followed the same principle as the Heliothermic Product of Branas (1946), but was still subject to final truncation at 19°C. I took latitude 40° (about the median latitude of world viticulture) as the neutral, or pivotal point, to allow the most direct comparisons with previous data not adjusted for latitude. That is, temperatures (less 10) at latitudes above 40° were adjusted upwards, and those below 40°, downwards. Monthly adjustment factors based on day-length are shown in Table 4, from which the values for individual site latitudes can be interpolated.

Huglin (1978, 1983) made similar adjustments for European latitudes between 40 and 50°. He maintained that even though total seasonal insolation (sunlight energy) is less at high latitude because of lower sun angle and light intensity, the fact that vine leaves are light-saturated at well below full sunlight intensity means that they use the total more efficiently than when it is concentrated in shorter days. It is not light intensity or total energy that limits, according to this interpretation, but the day-length that is available for photosynthesis. For the same reason Huglin took account only of day-time temperatures and discounted night temperatures.

While the end adjustment is often similar to mine, I believe Huglin's reasoning is wrong. Vine development and ripening at high latitudes are not limited primarily by light or photosynthesis, but by temperature (Chapter 3, and Section 1, above). Rate of phenological development determines when or whether ripening will occur, and its completeness, and is governed directly by temperature. Photosynthetic hours (together with canopy light relations, crop load etc.) influence mainly the sugar level in the berries *at a given physiological stage of ripening*. This is the lesser component governing the ripeness attained in cool climates, and is in any case the component most amenable to correction, by sugar addition to the musts.

Further, the attainment of true physiological ripeness is more likely to be limited by night than by day temperatures (see Chapter 3, and section 3, below). Huglin's exclusive use of day temperature therefore detracts from accuracy.

How, then, is the evident relationship of ripening to latitude, at a given unadjusted temperature summation, to be explained? I suggest that it results from errors in the traditional method of estimating daily temperature in terms of the means, i.e. from half the sum of the maximum and minimum. This does not give a true average, such as might be calculated from measurements every minute or every hour (McIntyre et al. 1987). With long days and short nights, temperatures will tend to be in the upper half of the range for more than half of the 24 hours; and with short days and long nights, for less than half. Therefore the longer the day, the higher will be the true average temperature relative to the mean. Being directly related to day length, such distortions will in turn be systematically related to the combination of latitude and time of season, and can be corrected on that basis.

A second factor leads to under-estimating true temperature summations in some high-latitude, continental climates where the average mean temperatures of the first and last months of the growing season are close to or below 10°C. An actual average mean of 10°C for the month gives no day degrees at all by the normal method of calculation, yet probably half of the month is above 10°C, with active growth or ripening taking place. In such cases I have estimated the summations for those parts of the month that are above 10°C, from areas under the temperature curve as shown in Figure 9. This is a far more logical and accurate method of correction than that of Wendland (1983), as employed by Jackson and Cherry (1988): see Appendix 3.

3. Adjustment for diurnal temperature range

The temperature summations obtained after adjusting for day-length and truncating at 19°C gave mostly good

Figure 9. Estimation of April and October temperature summations above a base of 10°C from the seasonal temperature curve in a cool, continental climate.

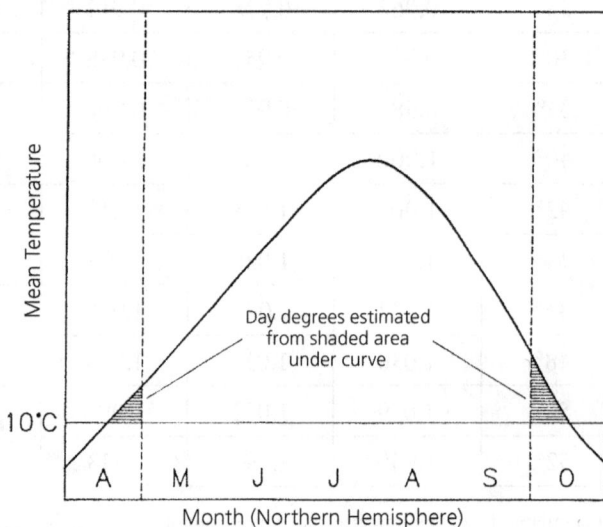

Figure 10. Effect of diurnal temperature range on effective average day and night temperatures (after Went 1957).

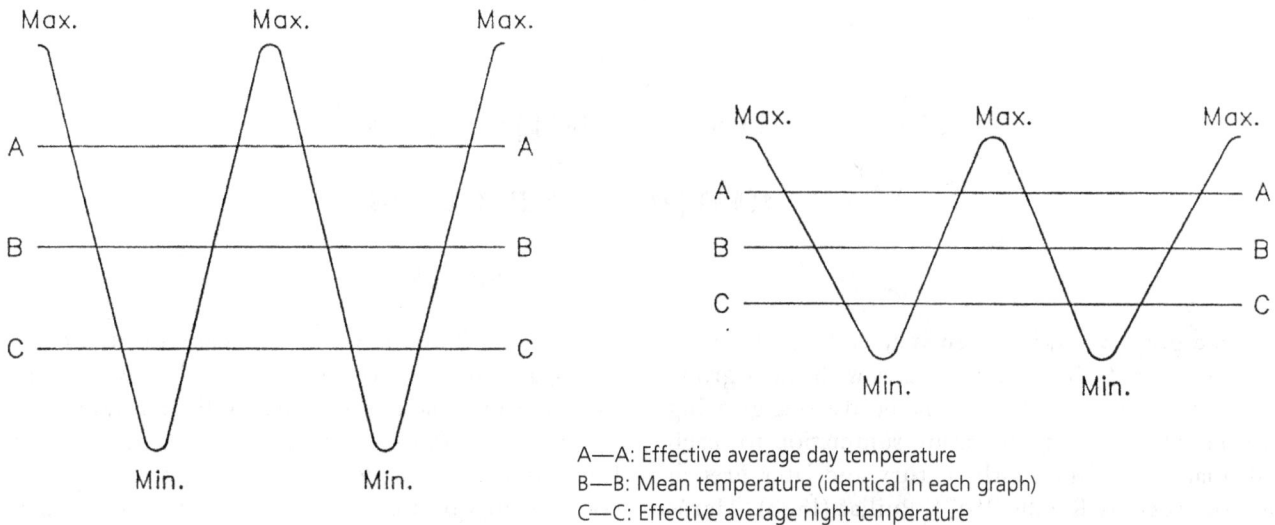

A—A: Effective average day temperature
B—B: Mean temperature (identical in each graph)
C—C: Effective average night temperature

predictions of recorded average vintage dates for given grape varieties throughout the world. A few anomalies remained, however. For instance, the prediction for the elevated inland environment of Mudgee, New South Wales, was that grapes should ripen 10 days or less later than at nearby Cessnock (Hunter Valley) near the coast; yet the generally reported difference is at least three weeks. On the other hand grapes have appeared to ripen in the cool environment of Hobart, Tasmania, earlier and more fully than even the latitude-adjusted data would suggest.

Looking at the climatic data for these two areas, one contrast stood out. The average spring diurnal temperature range at Mudgee, as elsewhere in New South Wales inland from the Great Dividing Range, is unusually large: a result of the mostly dry, clear spring in this environment. That at Hobart is very small. Further inspection of the data for other areas, after allowance for latitude and 19°C cut-off, suggested a general trend towards later-than-expected vintages where the spring diurnal temperature range is wide, and earlier-than-expected vintages where it is narrow.

This suggested that low minimum (or night) temperature might retard budburst and early growth in areas with a wide spring diurnal temperature range. Such a control is in fact reported to be common in temperate plants (Langridge and McWilliam 1967). It constitutes a very logical developmental control where spring frost is a major hazard.

Went (1957) estimated effective average night temperature to be half way between the minimum and the daily mean (Figure 10). An increase in diurnal temperature range of 1.0°C, at a constant mean temperature, thus reduces the effective night temperature by 0.25°C. I have used this principle to adjust the monthly temperature summations, using each month's average diurnal temperature range individually. Because the available field evidence only clearly implicates sites with very wide or very narrow diurnal temperature ranges, I have for the time being chosen conservatively to apply the adjustment only to months in which the range exceeds 13°C (beyond which day degrees are reduced by 0.25°C per 1°C of range over 13°C) or is less than 10°C (day degrees correspondingly increased). That leaves most major vinegrowing areas more or less unchanged. The fact that all monthly average means are ultimately subject to a 19°C cut-off in estimating temperature summations means that the diurnal range adjustment automatically becomes attenuated, and finally disappears, as the months and regions become warmer. Under such conditions night, or minimum, temperatures are presumably less limiting.

The combined adjustments give a basis for calculating 'biologically effective' temperature summations for grape vines. With prior adjustments to the base regional data for representative vineyard location, altitude, slope, aspect and soil type (Chapter 4), it now appears to be possible to predict an average date of maturity for any grape variety, of known maturity type, in all existing viticultural environments to within about a week either way. This seems close enough to permit a worthwhile interpolation of the average climatic conditions during the final 30 days of ripening for each variety, and to extrapolate to new environments.

7

Maturity Groupings of Winegrape Varieties

Some grape varieties ripen very early, and can be grown successfully in cool climates with short growing seasons; others ripen late, and need a long growing season with a high temperature summation to reach full maturity. Nineteenth century and later French authors such as Rendu (1857), Puillat (1888), Marès (1890), Viala and Vermorel (1901–1904) and Galet (1958, 1962) made broad groupings of varietal maturities, mostly into three or four classes with some subdivisions. Allowing for uncertainties of ampelography (identification of grape varieties), and for differences in the climates under which the authors made their observations, their classifications give an approximate guide.

More detailed data on vine phenology are available for some varieties from scientific trials in Australia, the USA and elsewhere.[1] Most importantly, the general literature contains many references to average and relative vintage dates for grape varieties in particular regions.

On the basis of all available information I have classified the main winegrape varieties into eight maturity groups, with specified requirements for 'biologically effective' day degrees (as described in the preceding chapter) to bring them to ripeness. They are listed in Table 5. The groups are at intervals of 50 effective day degrees, which is equivalent to a difference of five or six days in ripening date in warm climates, or a week or more under cooler ripening conditions.

The varieties largely maintain their relative maturities across environments, although individual interactions with environment (especially, perhaps, those stemming from time of budburst) could in practice necessitate minor changes in the rankings according to local experience.

The maturity rankings appear themselves to be related significantly to potential wine style. Wines from the very earliest-maturing groups, like the grapes, tend to be delicate and flavourless. Moderately early grapes (Groups 3 and 4) make aromatic, fruity wines, and include most of the best white wine varieties and those making light, generally non-tannic red wines, such as Pinot Noir and Gamay. The best firm, full-bodied red wines (e.g. Bordeaux and Rhône styles) come from varieties of Maturity Groups 5 and 6. Most late-maturing red varieties make wines in which tannins dominate the fruit characters.

Lateness in grapes appears to be associated with a prolongation of all three periods of berry development (Winkler 1962; Peynaud and Ribéreau-Gayon 1971; Coombe and Iland 1987). A longer Period I probably results in more, but ultimately smaller, cells in the berry, with more cell wall relative to juices. A long Period II is associated with more seeds and greater seed development. Perhaps a long Period III gives more opportunity for tannins to leach from the seeds into the pulp. Slow final ripening (for a given temperature regime) can also be expected to result in a greater relative loss of the volatile aroma and flavour compounds from the surface cells of the fruit, as described in Chapter 3. Tannins and other flavourants of high molecular weight then predominate.

One might also speculate that the ratio of protein to free amino acids will be higher in the late than in quick-maturing varieties. This could be an important factor favouring rapid fermentation and quality in wine from the earlier-maturing varieties, as discussed in Chapter 5 in the context of nitrogen nutrition, under 'Fertilizing and green manuring'.

Such differences broadly parallel those found in other fruits, most notably stone and pome fruits. Early-ripening varieties of all such fruits are usually delicate, soft, easily injured and lacking in flavour – the result of their forming few flesh cells, which then expand to large individual sizes but have only a short time to reach physiological maturity and to accumulate sugar and flavourants. The best eating varieties are generally of early midseason to midseason maturity, and combine to an optimum degree a reasonable firmness without losing juiciness, and maximum flavour and aroma. Late varieties comprise the main export, canning and storage (winter) types. These are hard-fleshed, and lack juiciness and often flavour. Their main virtues are ability to resist injury and decay in transport, handling

---

1 Dry and Gregory (1988) list the relative maturities of the more important grape varieties in three broad classes, on the basis of experimental observations in south-eastern Australia. Their Table 6.3 also summarizes important varietal differences in time of budburst, vine vigour and yield, and bunch and berry characteristics.

Table 5. Winegrape maturity groups and corresponding biologically effective day degrees to ripeness, for making dry or semi-sweet table wines

	Red wines	White or rosé wines
Group 1 1050 day °	—	Madeleine, Madeleine-Sylvaner
Group 2 1100 day °	Blue Portuguese	Chasselas, Müller-Thurgau, Siegerrebe, Bacchus, Pinot Gris, Muscat Ottonel, Red Veltliner, Pinot Noir, Meunier
Group 3 1150 day °	Pinot Noir, Meunier, Gamay, Dolcetto, Bastardo, Tinta Carvalha, Tinta Amarella	Traminer, Sylvaner, Scheurebe, Elbling, Morio-Muskat, Kerner, Green Veltliner, Chardonnay, Aligoté, Melon, Sauvignon Blanc, Frontignac, Pedro Ximenes, Verdelho, Sultana
Group 4 1200 day °	Malbec, Durif, Zinfandel, Schiava (= Trollinger), Tempranillo, Tinta Madeira, Pinotage	Sémillon, Muscadelle,[1] Riesling,[2] Welschriesling, Furmint, Leanyka, Harslevelu, Sercial, Malvasia Bianca, Cabernet Franc
Group 5 1250 day °	Merlot,[3] Cabernet Franc, Shiraz, Cinsaut, Barbera, Sangiovese, Touriga	Chenin Blanc, Folle Blanche, Crouchen, Roussanne, Marsanne, Viognier, Taminga, Cabernet Sauvignon
Group 6 1300 day °	Cabernet Sauvignon, Ruby Cabernet, Mondeuse, Tannat, Kadarka, Corvina, Nebbiolo, Ramisco, Alvarelhão, Mourisco Tinto, Valdiguié	Colombard, Palomino, Dona Branca, Rabigato, Grenache
Group 7 1350 day °	Aramon, Petit Verdot, Mataro, Carignan, Grenache, Freisa, Negrara, Grignolino, Souzão, Graciano, Monastrell	Muscat Gordo Blanco, Trebbiano, Montils
Group 8 1400 day °	Tarrango, Terret Noir	Clairette, Grenache Blanc, Doradillo, Biancone

1 Previously (Gladstones 1991) in Maturity Group 3. However, Dubourdieu (1990) describes Muscadelle as the latest ripening of the Bordeaux white varieties, suggesting that it must be transferred at least to Group 4. The literature variously describes it as 'early' and 'midseason'.

2 The maturity characteristics of Riesling appear to vary significantly with environment. Ripening capacity in Germany fairly clearly indicates Group 4, but in southern Australia it is not infrequently picked after red varieties of Groups 5 and 6, even, apparently, for dry wines. Some of the Australian record may nevertheless be confused by its use for both dry and late-picked, sweet wines.

3 Dubourdieu (1990) describes Merlot as the earliest of the red grape varieties to ripen in Bordeaux, 10–14 days before Cabernet Sauvignon and seven days before Cabernet Franc, suggesting that it might be better placed in Maturity Group 4 along with Malbec. I retain it in Group 5 on the general opinion of the literature, but is appears at least to be among the earliest varieties of the group.

and storage, and to maintain form during cooking.

Early-maturing grapes tend to be easily damaged, for instance by birds, bees and wasps. Many, particularly red varieties such as Pinot Noir, are very sensitive to direct sun exposure; in hot climates their wines readily acquire 'cooked' aromas and flavours. The early red varieties also tend to be deficient in anthocyanins as well as low in tannins, both being groups of compounds with high molecular weights and needing time to accumulate. Their predominant flavourants seem likely to have lower molecular weights and boiling points than those of later varieties. Early-maturing varieties can, therefore, in general be expected to attain and conserve their best qualities under cool ripening.

The berries of late-maturing grape varieties, including the traditional shipping and storage table varieties such as Ohanez (Almeria), are typically less easily injured; but that is at the expense of juiciness and flavour. High ripening temperatures cause less loss of their essential flavourants, such as they have. In fact warmth and ripening as rapidly as possible are needed for them to attain satisfactory flavours at all.

Broadly speaking these differences carry over into winemaking, and into the drinking of the wines. The highly volatile, low molecular weight compounds which I suggest predominate in early-maturing varieties, ripened under cool conditions, need cold and/or closed fermentation for their retention in the wine. Their volatile aromas are fully perceived in wines drunk at low temperatures, such as are also needed to temper the usual acidity, thinness of body, and often retained sweetness of the wines. At high drinking temperatures the acid of such wines becomes sour and dominant, and the remaining flavours merely insipid.

Grapes from midseason and later varieties, ripened under warm conditions, have less to lose with warm fermentation. For red wines generally, a warm fermentation is necessary anyway, for optimal extraction of colour and flavour from the skins. This applies fully as much, if not more, to early-maturing red varieties if enough anthocyanin, tannin and skin-contained flavourants are to be extracted to make true red wines from them. It explains the high fermentation temperatures used in Burgundy, and helps to explain why the grapes for all true red wines are best ripened under reasonably warm conditions. This in turn helps to explain the very restricted climatic adaptation of early varieties such as Pinot Noir for red winemaking. They need warmth during ripening on the one hand, but readily suffer from heat on the other.

All true red wines need warmer drinking temperatures than white wines, to bring out their less volatile aromas and to balance their mostly lower acids and stronger tannins on the palate. White wines made from midseason and late varieties, ripened in warm to hot climates, need higher drinking temperatures than cool-climate white wines for their best appreciation, for parallel reasons.

Red wines made from late-maturing grape varieties grown in hot climates need the highest drinking temperatures of all, in order to bring up what fruit character they have, emphasize what acid they have, and soften their tannin on the palate. Drunk cold, these wines utterly lack fruit and are dominated by their tannin.

In summary, parallels and direct causal relationships can be seen among the maturity rankings of grape varieties, their optimum ripening climates, their optimum fermentation temperatures, and their optimum drinking temperatures. Cold climate grapes are most naturally attuned to cold fermentation, and the wines to being drunk cold. Hot climate grapes are similarly attuned to winemaking and drinking in their own environments. This is indeed a fortunate circumstance, which has meant that wines co-exist naturally with their environments of origin, and with the conditions under which they are normally drunk. Equally it helps to explain deeply ingrained regional differences in wine style and preference, and provides a basis for historically held concepts of wine quality.

Some important exceptions to these broad correlations do exist. For instance the grape variety Verdelho is early maturing (Group 3), yet in Western Australia it produces outstanding dry white wines in the hot Swan Valley and nearby (see Guildford, Chapter 9, Table 15). The moderately hot Douro Valley in Portugal (Chapter 20: Pinhão, Table 165; Régua, Table 166) long numbered the early-maturing Bastardo as one of its chief quality bases, although since Phylloxera the Bastardo has tended to retreat to cooler northern areas because of its susceptibility to sunburn (Viala and Vermorel 1904). Tempranillo remains a major variety for quality throughout Spain, despite its relative earliness.

A pattern seems in fact to exist in Portugal and Spain, and in some other hot environments, of two fairly distinct maturity ranges which are grown in complementary roles. Late midseason and late types are grown largely for their hardiness and yield, and for their ability to ripen after the worst heat of summer has passed. The second group, of early or early midseason maturity, is grown for its quality and more fruity character, despite ripening in the heat and risking heat damage. Indeed, in very hot climates the early varieties can gain some advantage, because their berries often pass through the sensitive véraison stage before the onset of greatest heat. The earlier and later maturity groups in any case lend themselves mutually to blending. They also spread the vintage load.

Such strategies for quality, together with the individual strengths and shortcomings of varieties within the maturity groups, have obvious relevance to grape breeding for hot climates. I take the subject up again in Chapter 24.

8

Construction and Use
of the Climate Tables

The top halves of the tables record the best available published (or in some cases estimated) climatic data for regional weather-recording sites, together with successive adjustments to produce estimates, for typical related vineyard sites, of their biologically-effective temperature summations and temperature variability indices. Commentaries in each chapter explain the individual site adjustments used.

The site-adjusted, biologically-effective temperature summations (column 12 for the Southern Hemisphere, column 10 for the Northern Hemisphere) are then used to estimate average dates of maturity for the individual grape maturity groups, based on the time taken from October 1 in the Southern Hemisphere, or April 1 in the Northern Hemisphere, to reach their respective required numbers of effective day degrees (Table 5, Chapter 7). These dates are shown in the bottom halves of the tables, followed by the average climatic conditions of the final 30 days to maturity for each maturity group, estimated by interpolation.

Appendix 2 gives the literature sources of the climatic data, and the adjustments made to them for differing recording periods etc. where thought necessary.

**Notes on the individual columns
of the climate tables**

Column 1: Lowest recorded minimum temperatures. The lowest-recorded minima are taken direct from the published data for regional recording sites; or from unpublished records of the Australian Bureau of Meteorology in the cases of most Australian sites. Note that the recording periods vary widely. Figures from short periods under-estimate the risk of an exceptional extreme, while those from very long periods tend to over-emphasize it. The figures should therefore be viewed relative to the lengths of recording period.

Column 2: Number of months with less than 0.3°C; and Column 3: Number of months with less than 2.5°C. These data are included for Australia only, and were taken from unpublished records of the Australian Bureau of Meteorology. Most were for the 30-year period 1939–68. Where full records for that period did not exist, I used the available data for

shorter periods as indicated in the column headings. The figures show for each month the actual number of months, among the years specified, which experienced at least one 'heavy frost' (screen minimum less than 0.3°C, or 32.5°F) or 'light frost' (screen minimum less than 2.5°C or 36.5°F). (Use of 32.5 and 36.5°F for frost incidence, instead of the conventional 32.0 and 36.0°F, was necessitated by the fact that many of the records were available only in whole degrees Fahrenheit.) The adjustments to regional minimum temperatures for actual vineyard sites cannot be applied directly to these data; but a fair subjective estimate of vineyard frost incidence can be made by interpolating the net adjustments, as specified in the commentaries, between columns 2 and 3.

Column 4 (*Column 2 for the Northern Hemisphere*): Average monthly lowest minimum temperature. These were taken direct from published data for the Northern Hemisphere and New Zealand, where available. For Australia, they were calculated from unpublished records of the Australian Bureau of Meteorology, normally for the same periods as columns 2 and 3. Net vineyard site adjustments to the average minima can also be applied in parallel to the average monthly lowest minima; and, in Column 1, to the lowest recorded minima.[1] Note that the April and October figures in this column give an exaggerated view of potential frost risk in cool, continental climates, because there the lowest minima for those months normally fall outside the true vinegrowing season.

Column 5(3): Average minimum minus average

---

1 In my original calculations (Gladstones 1976, 1977), I assumed that effects of topography and soil type on average minima should be approximately doubled when applied to the monthly average lowest minima and lowest recorded minima, on the ground that a wide average diurnal temperature range (as shown in column 7) tends to be accompanied by an equally magnified range between each month's average minimum and its average lowest minimum (as shown in column 5). I have since abandoned that assumption in favour of the simpler and more conservative one adopted above. This means that the adjustments to temperature used in this book for topography and soil type will tend, if anything, to under-estimate their effects on temperature variability and frost.

monthly lowest minimum. This column estimates the within-month variability of the minimum, which is a particularly important contributing factor to frost risk after budburst (see 'The temperature variability factor', Chapter 3).[2]

Column 6 (4): Daily mean temperature. The mean is calculated as (maximum + minimum)/2, from the published averages of each for each month.

Column 7 (5): Daily temperature range/2. This is calculated for each month from the difference between its average daily maximum and its average daily minimum. I express it as daily (= diurnal) range/2, so that values can conveniently be subtracted from the daily means (column 6) to give the monthly average minimum temperatures if desired; or alternatively added, to give their average maxima.

Column 8 (6): Average monthly highest maximum temperature. This is an important indicator of the risk of summer stress and heat damage, both to the vines and to the developing and ripening fruit.

Column 9 (7): Average monthly highest maximum minus average maximum. Shows variability of the maximum temperature within a given month, in the same way as column 5 (3) shows variability of the minimum. Least possible variability of the maximum is sought particularly in the hottest part of the year and during ripening.

Column 10 (8): Raw day degree summation above 10°C with 19°C cut-off. Calculated for each month from the raw monthly average mean temperatures in °C, less 10, truncated at 19°C and multiplied by the number of days in the month. Alternatively, for April and October in cool continental climates, by graphic estimation as shown in Figure 9, Chapter 6. Truncation of the average mean at 19°C results in a ceiling of nine day degrees C for any day, and a maximum of 279 day degrees for any month.

Column 11 (9): Day degree summations, adjusted firstly for latitude and daily temperature range. For each month the number of degrees C above 10 is

---

2 Note, however, that in addition to the true day-to-day variability of the minima, the figures in this column have a component resulting from the progressive change in average mean through the month, assuming that average lowest minima progress through the month in parallel with the mean. Clearly it is greatest in spring and autumn, when mean temperatures are changing most rapidly. For the same reason it is greater in continental than in maritime climates. Similar reasoning applies to the range between average maximum and average highest maximum temperatures, in column 8(6). Since both columns contribute to the temperature variability index, it follows that the index gives a relative over-estimate of day-to-day temperature variability in the spring and autumn, and in continental climates.

adjusted: 1) for latitude and day length, in accordance with Table 4, Chapter 6; then 2) for diurnal temperature range by subtracting 0.5° for each degree C by which the daily range/2 (column 7) exceeds 6.5°, or adding 0.5° for each degree by which it is less than 5.0°. The day degree summations are then re-calculated, subject still to an overriding ceiling of 19°C, or nine day degrees per day.

Column 12 (10): Day degree summations, adjusted secondly for representative vine sites. The monthly average maxima and minima from the base regional recording station(s) are adjusted, uniformly across all growing season months, for: vineyard location, in accordance with broad temperature trends with latitude and longitude; altitude difference (0.6° per 100 m); and representative vineyard slope, aspect and soil type, as explained in Chapter 4 and listed in the commentaries to the individual tables. Day degree summations for each month are then re-calculated as in column 11, again subject to an overriding ceiling of 19°C or nine day degrees C per day. Comparison with column 11 thus shows the net vineyard site influence relative to the base recording station. Further broad estimates can finally be made for the ripening potentials and climatic characteristics of other locations throughout a region. Estimates for the Whitfield area south of Wangaratta in Victoria (Chapter 11, Table 50) are an example.

Column 13 (11): Raw temperature variability index. This is calculated from the base regional data for each month, as the average diurnal temperature range (or 2 x daily range/2) plus the range between the average highest maximum and the average lowest minimum temperatures for the month. That is, for the Southern Hemisphere, (2 x column 7) plus column 8 minus column 4; or alternatively, (4 x column 7) plus column 9 plus column 5.

Column 14 (12): Temperature variability index, adjusted for representative vine sites. This can be calculated in the same way as column 13, but incorporating the net site adjustments to maximum and minimum temperatures as used for column 12. Note that in this calculation the change in diurnal temperature range is doubled, because it is included both in its own right and as a component of the range between the average highest maximum and average lowest minimum temperatures.

Column 15 (13): Average sunshine hours. Direct records of monthly total sunshine hours are available for only a few locations, and these are often subject to local factors such as fogs (in valleys) or shadowing by mountains etc. I have therefore preferred, wherever monthly sunshine maps are available, to interpolate from them. Where the resulting values for individual months have deviated markedly from the expected shape of the seasonal curve, based on geographically related sites, I have smoothed them towards the expected curve shape, while maintaining approximately the same total hours for the season.

Columns 16 (14) and 17 (15): Average number of rain days, and **Rainfall.** These are taken directly from the literature. A rain day is defined as one recording 1 mm or more of rainfall.

Column 18 (16): Average per cent relative humidity (early afternoon). Regrettably, the published records of early afternoon relative humidities are not standardized. Most of those for the Northern Hemisphere are for 1330 hours, while all the Australian published records are for 1500 hours. Resulting problems of interpretation are covered in Chapter 3.

All columns: Where the number of years' record is not shown at the head of a column, the data are either derived from other columns or estimated from geographically related stations. Estimates of the latter kind are mostly explained in the individual site commentaries or in Appendix 2.

Practical use of the tables
The bottom half of each table sets out the derived estimates for the predicted final 30 days of ripening, for each of the eight grape maturity groups. The respective ripening period profiles can now be compared, within and across sites and with the postulated 'ideal' profiles for individual wine styles (Table 2, Chapter 3). Combined with the data in the top halves of the tables for risks of spring frost and mid-summer heat or drought stresses, and for the seasonal distributions of rainfall, relative humidity and sunshine hours, these comparisons allow: an assessment of the climatic strengths and weaknesses of each viticultural environment; its suitability for particular wine styles; and, *to some extent*, its suitability for individual grape varieties.

If there proves to be a marked discrepancy between the predicted and the realized vine phenology at any site, its possible causes need to be examined. Assuming that management factors such as heavy irrigation or over-cropping are not responsible, a likely cause is local geographical or site factors. Mostly these will relate to air drainage (or lack of it) at night.

The accuracy and/or relevance of the regional climatic data used for estimation may themselves be questionable. Are they influenced by urban warming? Are they from a short and/or unrepresentative recording period? Have the site characteristics of the recording station(s), if known, been properly allowed for relative to those of the vine sites? Towns and recording stations are often in frosty valleys, but may sometimes be on slopes. Comparison of their raw data with those for comparable stations nearby can give a clue as to whether or not they are anomalous.

Before reaching final conclusions, one more point needs to be remembered. Wide variation exists from season to season, in both climate and vine phenology. Additionally, the phenology of very young vines may not fully predict that of mature vines. Therefore true average vine phenology for a site can only be established over many years. Predictions, provided that they are based on sound climatic data and site assessment, may in fact be a more accurate guide for the long term than actual phenology as recorded in the first few years.

If disparities remain between predicted and actual vine phenology, the tables can still be useful. A small disparity may be reconciled for practical purposes by shifting the whole varietal scale up or down one or two maturity groupings. Anomalous phenological behaviour by individual grape varieties can, of course, be accommodated by shifting them into more appropriate maturity groupings. Some interaction of variety with individual environments is to be expected, although it is seldom likely to exceed one maturity grouping either way.

Tabulated climatic data and ripening period estimates are included for Chapters 9 to 22, taking in both representative existing and (for Australia) potential viticultural sites in each of the regions surveyed. Commentaries on the tables follow the main text of each chapter, and detail the adjustments, and reasons for them, made to the basic regional data used for each table.

For each of the Australian States I also include a map showing the main locations mentioned, together with isotherms (reduced to sea level) of the site-unadjusted average mean temperatures for the full October-April growing season. Approximate actual averages for October-April can be estimated for any location by interpolating from its position between the isotherms, and subtracting 0.6°C per 100 m of altitude above sea level.

9

Viticultural Environments of Western Australia

Tables 6 to 32 give climatic data for all existing Western Australian viticultural regions. Figure 11 shows site locations, together with growing season (October-April) isotherms, reduced to sea level. For good accounts of Western Australia's vinegrowing history, and of its present vineyards, vignerons and wines, see Brady (1982), Halliday (1982b, 1985a, 1991a) and Zekulich (1990).

The main viticultural areas of Western Australia were, until recently, in or near the Swan Valley, close to Perth. Plantings made around the turn of the present century also extended inland to centres such as Bakers Hill and Katanning, but these were later mostly abandoned. Together with Bindoon and Gingin, about 60 km to the north, the Swan produces full-bodied table wines and fortified wines, as well as tablegrapes and currants. It is a hot region in terms of growing season and ripening mean temperatures, but the effects are mitigated by its closeness to the west coast and the afternoon sea breezes that blow fairly regularly in summer.

Vine plantings for table wines in the cooler southern and far south-west coastal regions of Western Australia followed Olmo's (1956) recommendation of the Mount Barker-Frankland area, and my later (1965, 1966) recommendation of Margaret River, as described in Chapter 2. All the data and practical experience point to a close analogy between these regions and Bordeaux. This is so in regard to growing season and ripening mean temperatures, sunshine hours, ratio of growing season sunshine hours to effective temperature summation, and, near the coast, afternoon relative humidities. Raw temperature variability indices, especially inland, are mostly a little higher than optimum by European standards, underlining the importance of careful vineyard site and soil selection to reduce variability and raise effective night temperatures.

Of all the climatic elements in the south of southwestern Australia, only seasonal rainfall distribution differs markedly from that of Bordeaux. South-western Australia gets much less rainfall in summer, but, except on the inland fringe of the Mount Barker-Frankland viticultural area, more in winter. The pattern points to benefits in some, but certainly not all, areas from supplementary watering during ripening. Good soil drainage is essential, to avoid winter-spring waterlogging and allow deep root development. The best soil indicator remains the presence of healthy, vigorous marri (*Eucalyptus calophylla*) trees as a dominant part of the natural vegetation; or in southern high-rainfall areas, of karri, *Eucalyptus diversicolor*. The tuart, *Eucalyptus gomphocephala*, indicates suitable limestone-based sandy soils along the west coast.

Northwards along the west coast the climate is progressively warmer and sunnier, and more suited to fortified wines. The foothills east of Bunbury (see Wokalup, Table 31) and the hills region inland from Perth (Kalamunda, Table 16) have warm to hot conditions very similar to those of the lower and central Douro Valley of Portugal. Nevertheless this area, even as far north as Perth and Gingin, still makes very good, full-bodied table wines from suitable grape varieties. Their quality is almost certainly made possible by the afternoon sea breezes in summer.

Inland areas eastwards from the immediate hinterland of the Darling Scarp have progressively lower relative humidity and autumn-winter-spring rainfall, and greater temperature variability. They are subject to spring frosts. Vines can still be grown where there is enough rainfall, or where fresh water is available for supplementary watering; but the climatic conditions inland are clearly poorer for wine production than those close to the scarp and to the coast.

An important historical factor in Western Australian viticulture has been that many of the world's premium grape varieties were commercially unavailable until very recently. This was due to a ban on the importation of vine materials because of Western Australia's freedom from Phylloxera and (until recently) downy mildew. However in the late 1950s the Western Australian Department of Agriculture began a programme of varietal introduction, through strict quarantine, together with clonal selection within varieties that were already present. Varieties such as Cabernet Franc, Merlot, Pinot Noir, Sauvignon Blanc, Chardonnay, Sémillon and Traminer became available to Western Australian viticulture for the first time during the 1970s – with dramatic consequences for wine quality.

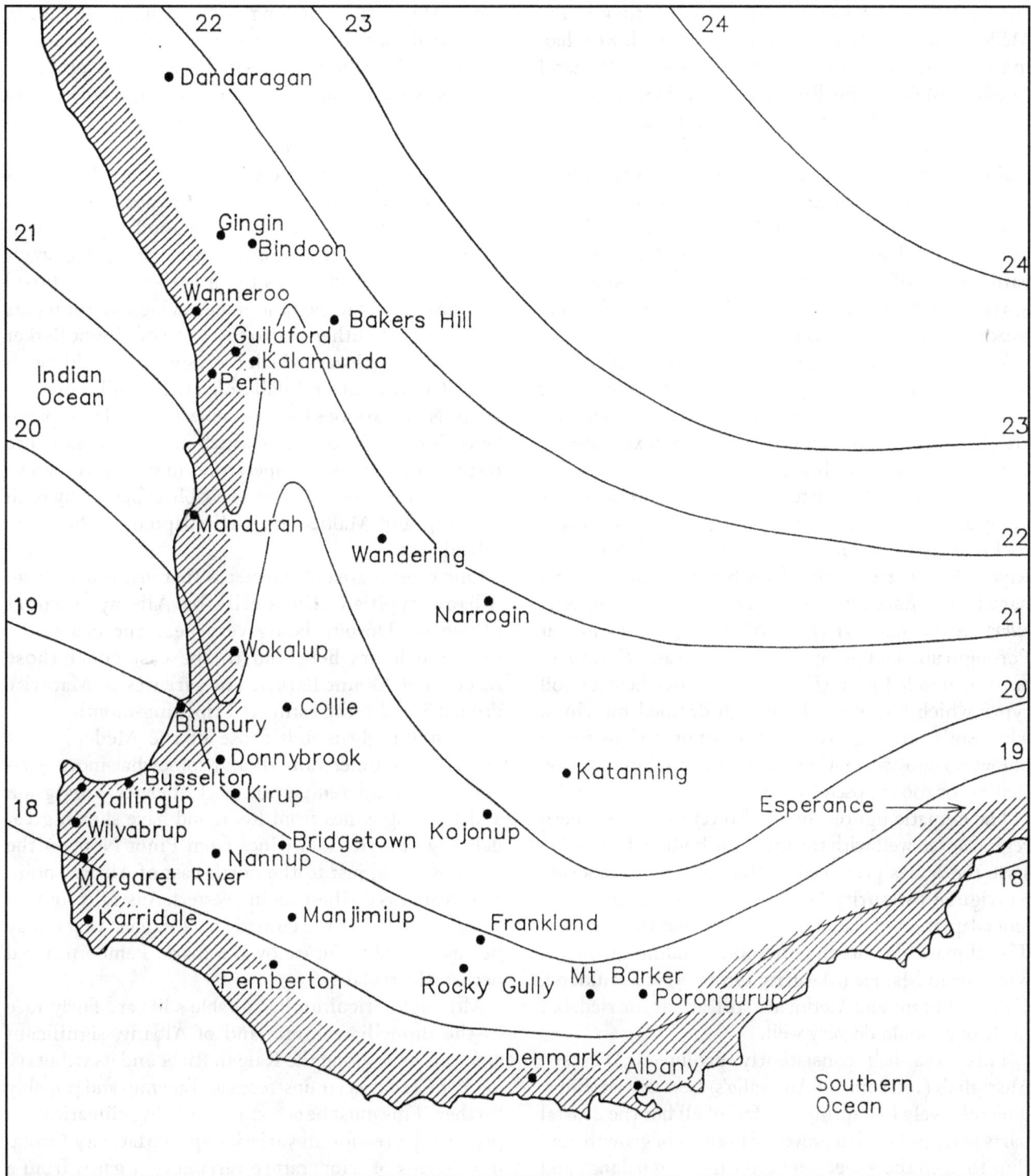

Figure 11. South-western Australia, with isotherms of October-April average mean temperatures, reduced to sea level.

In the absence of Phylloxera, most vines in Western Australia continue to be planted on their own roots. The main exceptions have been in older and warmer areas such as the Swan Valley, where nematode-resistant rootstocks (mostly Schwarzmann) are now generally used. Nematode-resistant rootstocks are also used for some of the new plantings at Manjimup and Pemberton, because of previous potato growing there.

Individual areas of Western Australia

The Lower Great Southern
The 'Lower Great Southern' vinegrowing region, including Albany, Denmark, Frankland, Mount Barker, and the Porongurup foothills, is Western Australia's coolest for vine growing. Ripening-month mean temperatures in the established **Mount Barker**

vineyards (Table 23) are marginally lower than in the Médoc (Chapter 16, Table 126), and clearly lower than in the warmer Bordeaux appellations such as Pomerol (Table 136) and Saint Emilion (Table 138). Ripening temperatures at **Rocky Gully-Frankland** (Table 28) are estimated to be the same as in the Médoc for the main Bordeaux red wine varieties, but a little lower than there for the earlier-maturing white wine varieties. The Mount Barker and Rocky Gully-Frankland figures are calculated for the main established vineyards, most of which are on warm valley slopes at lower altitudes than the Mount Barker township and weather-recording station.

Average ripening-period sunshine hours at Mount Barker, together with the whole-season ratio of sunshine hours to site-adjusted effective day degrees, are almost identical with those of Bordeaux; at Rocky Gully-Frankland, both are a little greater. The viticultural soils are either lateritic gravelly sandy loams or sandy loams derived directly from granite and gneissic rocks, as described in greater detail under Margaret River. The better soils of both types carry dominant marri (*Eucalyptus calophylla*), or karri (*E. diversicolor*) on the loams at Denmark and some of those at Porongurup. The presence of swamp yate (*E. occidentalis*) or flooded gum (*E. rudis*) indicates heavier soil types which are not well enough drained for vines, while soils carrying Banksia species or various forms of low, scrubby vegetation are usually too infertile, too shallow or too exposed for vines.

The climatic figures for the Lower Great Southern region agree well with the medium-bodied, Bordeaux-style red wines produced in the area from Cabernet Sauvignon (Maturity Group 6), and with the excellent adaptation of the slightly earlier-maturing Shiraz. The climate also fits the often outstanding quality of wines from Malbec (Maturity Group 4) at Frankland. Cabernet Franc and Merlot are still largely untried, but in theory should do very well.

This area has consistently produced Western Australia's (and some of Australia's) best dry rieslings. The relatively low spring rainfall of all but the coastal parts perhaps helps to prevent rankness of growth, and thus to keep the variety in better fruiting balance and canopy light relations than, for instance, at Margaret River; while moderate ripening temperatures result in wines of classic Australian dry riesling style with fresh acidity. Chardonnay likewise tends towards the initially lean European style, rather than towards the immediately full-fruited chardonnays of Australia's warmer viticultural regions.

The main climatic weakness of Mount Barker and Rocky Gully-Frankland lies in their rather large temperature variability indices. There is greater variability of the maximum temperature (column 9 in the climatic tables) than on the west coast, with a greater incidence of extreme high temperatures despite lower average mean temperatures. Mostly the extremes occur before the start of ripening, so that fruit quality is not necessarily prejudiced unless there has been damage to the vines. Variability of rainfall is a further problem. This is especially on the dry inland fringe, where even average rainfall is marginal and good water is hard to conserve in many seasons.

The high-temperature extremes diminish towards the coast, while total rainfall and relative humidities increase (as does the availability of good water for irrigation). The risk from spring frost, which is appreciable inland, likewise falls towards the coast. A broad conclusion is that the best vinegrowing environments of the Great Southern lie to the south of Mount Barker and Muirs Highway, which corresponds roughly with the 750 mm annual rainfall isohyet. The only contrary factor is for varieties that are subject to budburst problems close to the coast, or which need some moisture restriction or higher temperatures in spring to achieve vine balance. These factors probably explain why both Riesling and Malbec are well adapted to the more inland parts.

One coastal area of interest is that just inland from Albany, typified climatically by **Albany Airport** (Table 6). Despite being very near the coast, the sunshine hours here and to the east equal those received at Mount Barker. For varieties of Maturity groups 5 and 6 the estimated ripening-month conditions strikingly match those of the Médoc, apart from less summer rainfall and somewhat more variable maximum temperatures. Cabernet Sauvignon and Riesling wines from Redmond have shown great delicacy of style. Red wines from Pinot Noir on the lower Kalgan, just to the north-east of Albany, point to this area as the best yet in Western Australia for that variety. (However the climatic data suggest that it may be challenged in future by Denmark, Pemberton and perhaps Karridale.)

Although viticulturally suitable soils are fairly rare in the immediate hinterland of Albany, significant amounts exist along the Kalgan River and its tributary Napier Creek, as far upstream as Takenup and possibly further. This must be accounted as being climatically a promising area for all varieties up to Maturity Group 6. In terms of temperature variability it gains from a weak east-coast influence. Easterly and north-easterly winds are more likely to be modified by passage over water than they are further inland or to the west.

Another area of great interest is the foothills of the **Porongurup** Range (Table 27), just east of Mount Barker; and especially, the slopes facing north and north-east. The Porongurups are a small, isolated range of intrusive granite, of which the slopes enjoy outstanding air drainage. Good vineyard soils extend along the northern lower slopes to above 350 m, which is 100 m higher than Mount Barker, and higher still compared with the main Mount Barker and Rocky Gully-Frankland vineyards. I estimate that ripening should be about a week later and 1.0°C cooler than in the latter, with significant advantages for temperature variability and extremes during ripening. Cabernet Sauvignon should still ripen fairly safely, to give lighter and more delicate wines than at Mount Barker and Rocky Gully-Frankland. Cabernet Franc, Merlot and perhaps Malbec should be especially well adapted. Gamay and Bastardo (Maturity Group 3) could also be interesting varieties to try on the granite-based soils, although they might suffer from heat injury in some seasons. Such varieties might be better tried in more coastal areas such as Denmark, and possibly Pemberton and Karridale. However Porongurup is strongly indicated for all the high-quality white wine varieties, including especially, perhaps, Riesling and Traminer. Initial plantings of both these varieties have produced some outstanding wines.

The final part of the Lower Great Southern region immediately northwest of **Denmark** (Table 12), on the sheltered, north-facing slopes of the isolated Bennett Range, is perhaps of special interest as one of the few parts of Western Australia capable of producing true coolclimate wine styles. The climate estimates are for about 100–150 m altitude, but should reasonably represent all but the very highest and lowest potential vineyard sites. This area is very near the coast. Estimated ripening conditions are cooler than at Mount Barker and Albany, with mean temperatures similar to those of the Porongurup foothills. Total rainfall and relative humidities are substantially higher, however, and sunshine hours and the ratio of growing season sunshine hours to effective day degrees clearly lower than at all the other Great Southern locations.[1] Despite much more total and winter rainfall than in the rest of the region, the risk of heavy rainfall in March and April appears to be little different. The loamy karri soils, formed directly from gneissic rock, are well drained and relatively fertile.

Wines from grape maturity groups 5 and upwards might be expected on average to be light, because of restricted sunshine. Relative humidities are conducive to fungal diseases throughout the season, and combined with rising rainfall through the autumn, to bunch rot in later-maturing varieties. On the other hand the conditions appear ideal for producing delicate, light to medium-bodied wines from early-maturing grape varieties. These should if possible have good resistance to fungal diseases. On purely climatic grounds the Bennett Range is a logical place for Chardonnay, Pinot Noir and Meunier.

One weakness shown by initial experience has been that the early-maturing grape varieties are locally very subject to bird damage, because of the surrounding forest in which the trees usually do not blossom until after they have ripened. Such varieties necessitate the provision of alternative water and preferred food sources for birds during their ripening. Netting of the vines against birds is an option for potentially high-value grapes; but it requires special care in a climate such as this, to minimize adverse effects on canopy light relations and moisture microclimate. Alternatively, midseason varieties which are resistant enough to rain might be planted, such as Cabernet Sauvignon, Cabernet Franc and Mondeuse, to make mainly light wines. In any case special attention is needed in this environment to vine trellising, training and summer trimming, to maximize leaf exposure and facilitate disease control.

Warren Valley – Nannup

Manjimup (Table 21) has long been thought of as climatically suitable for winegrape growing, and perhaps comparable with Mount Barker-Frankland. It has similar gravelly marri soils, together with probably suitable karri loams in the more southerly parts. Small-scale experimental plantings made in 1977 on karri loam at the Department of Agriculture's old experimental station, about half way between Manjimup and Pemberton, showed that vines grow very well and that they can produce good wines (J.E.L. Cripps 1987, personal communication).

Most of the best potential vineyard sites and soils around Manjimup are in the drainage basins of the Warren and Donnelly rivers, at lower altitudes than the town. They are therefore likely to be warmer, and to have better temperature variability characteristics if well sited. Taking this into account, it now appears that the area immediately around Manjimup should be thought of as intermediate between Mount Barker and Margaret River, though still perhaps closer to Mount Barker in temperature and potential wine qualities.

Manjimup has some marginal climatic advantages over Mount Barker. Winter-spring rainfall is greater and more reliable, but ripening period rainfall is about the same. There is a better availability of good quality surface catchment water. The risk of frost in spring is about the same, and by viticultural standards not serious with adequate site selection. Finally, temperature variability (especially of the maximum) is estimated to be slightly less than at Mount Barker, associated with a greater west coast influence. The only adverse point for wine quality is a lower afternoon relative humidity, but this is not unduly low, and in any case is advantageous for disease control.

Estimated ripening period data for early and midseason grapes suggest that Manjimup has both mean temperatures and sunshine hours very close to those of Bordeaux. Temperature variability is a little greater, and relative humidity a little less, than in Bordeaux. Nevertheless there can be no doubt that, in climatic terms, this is an area of great viticultural potential, fully comparable with the Lower Great Southern and

---

1 The south coastal climate of Western Australia is dominated by on-shore south to south-east trade winds in summer, particularly in January-February between Albany and Cape Leeuwin. The resulting coastal cloudiness typically dissipates about 20 km inland.

Margaret River; and that the wines produced from appropriate grape varieties should be very much in the mainstream of Bordeaux styles.

The main suitable soils around Manjimup are the lateritic gravelly sandy loams and gravelly loams, which in their native state carried dominant marri. Cabernet Sauvignon (Maturity Group 6), as the latest maturing of the varieties likely to be planted, would probably be better confined initially to the warmer sites and soils, to ensure perfect ripening. Cabernet Franc, Merlot and Shiraz (all Maturity Group 5) should be more or less universally adapted, as should the main white wine varieties, to give medium-bodied, well balanced wines.

The Manjimup area can be considered to extend northwards to about Wilgarup. The high point between there and Yornup is the water and air drainage divide between the south-flowing river systems and the northward descent into the Blackwood Valley. It marks a fairly sharp climatic transition from the hotter, more arid, and more temperature-variable and frost-prone climate of the Blackwood Valley to the more maritime-influenced climate to the south: compare Manjimup, Table 21, with Bridgetown, Table 8.

The transition from marri to predominantly karri soils that takes place south of Manjimup, about half way to **Pemberton** (Table 25), marks in turn the southern boundary of the Manjimup area proper, and signals what appears to be a further fairly major environmental change. As well as differing dominant soil types there is a slight southward gradient to lower temperatures (despite Pemberton's lower altitude), together with falling sunshine hours, more rainfall (except in January and February), and greater relative humidity. From the available figures the temperature variability index stays about the same, or more probably falls slightly. The karri loams in their pure state are formed directly from the gneissic country rock, and could reasonably be expected to promote vine vigour. However gravelly soils, similar to those at Manjimup, remain on many of the higher slopes.

The climatic estimates for Pemberton are based on fairly short-term records, which had to be adjusted significantly to reconcile them with the longer-term records of other southern stations. I have assumed, as at Manjimup, that most potential vineyard sites are on slopes lower than the recording station. This may prove not to be entirely correct. The estimates are therefore tentative, and it would be prudent to adopt a conservative attitude to ripening potential until more experience has been gained: especially on soils likely to promote vine vigour.

Certainly, initial main plantings should not be of varieties later than Maturity Group 5, with Cabernet Sauvignon (6) confined to the warmer exposures. Cabernet Franc and Merlot (5) are logical general choices for light to medium-bodied, fruity red wines. Pemberton could also be as suitable an environment as any in Western Australia for Pinot Noir and Meunier

(3 for red wines). The main initial choice of white varieties should probably favour those with good fertility and not prone to excess vigour. Chardonnay (with adequate wind protection) should give excellent wines and is a logical commercial choice. Both red and white wines in this environment are likely to be lighter and perhaps more elegant than at Manjimup and especially Margaret River, and to resemble more closely those from Albany and Denmark. Because of vine vigour, at least on the richer soils, careful trellising and management to improve canopy light relations will be essential.

Should brandy production be contemplated in Western Australia, Pemberton would be a logical site for it, using Trebbiano. Ripening conditions are comparable with those of Cognac, France. Trebbiano is a fairly late variety (Maturity Group 7), but resistant to rain during maturation. Having late budburst it could be grown safely on low and flat sites which are too frosty for other varieties. It should ripen with only moderate sugar and high acid levels, suitable for making premium brandy. The literature also indicates Trebbiano to be vigorous, fertile, high-yielding and disease-resistant, and as giving reasonably good, fresh, though rather neutral, wines. It could be very useful in the Manjimup-Pemberton area for producing bulk wines to blend with full-bodied, low-acid wines of greater character from warm areas for the cask and flagon trade, or directly for making bulk sparkling wine.

Further areas of potential viticultural soils exist to the north-west of Manjimup in the **Nannup-Carlotta** area, at the extreme southern end of the Darling Scarp. Climatically as well as geographically, conditions here are probably intermediate between those of Manjimup and Margaret River. There are similar gravelly soils, together with reddish loams formed directly from the gneissic and granitic rocks of the scarp. Nannup could undoubtedly produce excellent medium to full-bodied table wines from most grape varieties.

Margaret River

The Margaret River region, strictly defined, takes in the slopes of the low ridge that extends from Cape Naturaliste, in the north, to Cape Leeuwin at Western Australia's extreme south-western tip. The ridge comprises granitic and gneissic rocks over which laterite has formed. The vineyard soils are derived either from laterite or from the underlying country rock at lower levels in the valleys, and are found chiefly in the drainage basins of creeks and small rivers running north-east or north into Geographe Bay, west to the Indian Ocean, and south-east into the lower reaches of the Blackwood River.

Developed commercially for vine growing at the same time as the Lower Great Southern, the Margaret River region provides an interesting comparison. Whereas the Lower Great Southern has average ripening month mean temperatures ranging from about the cooler limit of Bordeaux down to only a little warmer than the Loire, the established **Margaret River** (Table 22) and **Wilyabrup** (Table 30) vineyards, in the central and north-central parts of the region respectively, are appreciably warmer during ripening than the Médoc, and similar to Pomerol (Chapter 16, Table 136) and Saint Emilion (Table 138). Sunshine hours during the growing season and ripening period are marginally more than in all Bordeaux areas and the Lower Great Southern, while relative humidities are intermediate and probably about optimal. Summer and ripening period rainfall totals are very low. Apart from perhaps too little summer rain for solely rain-fed vines on some soils, these are conditions which in Bordeaux would typify a 'great year'. They doubtless help to account for the intense colour (in red wines) and the intensity of flavour and varietal character which are features of Margaret River wines.

In temperature variability, especially of the maximum, the whole of the Margaret River region has clear advantages over the Lower Great Southern. This can be ascribed to its predominantly west coastal aspect and stronger maritime influence. It does, however, carry the disadvantage of probably greater windiness, particularly in spring when the young growth and flowers readily suffer damage (Chater 3). Against that, spring frosts are uncommon. The raw variability indices are only marginally higher than in Bordeaux and most other European viticultural areas.

A shallow downward gradient of temperature and sunshine hours exists from **Yallingup** (Table 32) in the far north of the region to **Karridale** (Table 17) in the far south. Karridale probably differs most from the rest. From personal observation over many years, I suspect that the published maps of sunshine hours do not fully indicate the rather abrupt increase in cloudiness that can be seen southwards from about Witchcliffe, resulting from the summer on-shore south-easterly winds. However these seldom bring rain, and the sunshine hours remain ample for table wines right to the south coast.

Karridale is an interesting potential locality for winegrapes. Moderate areas of suitable soils are present. They include, firstly, the brown loamy sands on the fringe of the karri forest along the lower eastern slopes of the Leeuwin-Naturaliste ridge, close to the old Karridale townsite. These are limited in extent, and would need close attention to vine nutrition, but could be suitable for making delicate table wines. The second group of soils are the lateritic gravelly loamy sands and sandy loams, such as constitute the main soil type for growing winegrapes throughout south-western Australia. They were first typified in this area, as the Forest Grove series of Smith (1948, 1951). The best carried dominant marri in their native state, and the viticulturally more marginal soils a mixture of marri and jarrah (*Eucalyptus marginata*). Here they occur between Kudardup and Alexandra Bridge, mainly in the drainage basins of creeks flowing east into the Blackwood River and south into Hardy Inlet. Some further gravelly soils and quite large areas of deep, well drained, yellow loamy sands (the Calgardup sands of Smith 1948, 1951) occur east of Forest Grove and Witchcliffe, in the drainage basins of McLeod Creek and of the Chapman and Upper Chapman brooks. These also carry vigorous marri. The Calgardup sands are relatively untried at present, but constitute a further potentially major soil resource for viticulture in the Witchcliffe to Karridale area.

The wines of Karridale-Kudardup-Alexandra Bridge, and to a lesser degree of Forest Grove and Witchcliffe, might be expected to be less full-bodied than those from further north, and should be well balanced. Although cooler and less sunny in summer than Margaret River and areas further north, Karridale shares with the rest of the southern west coastal region an almost unique prolongation of mild, sunny weather into late autumn. It might therefore be particularly suited to making sweet table wines from late-picked grapes. The most likely viticultural problem will be that of erratic budburst and consequent rank growth, associated with the area's marked lack of continentality.

Moving north, the estimates show Wilyabrup (Table 30) to be slightly warmer and sunnier than Margaret River (Table 22), but the climates of both are consistent with the robust, full-flavoured, but well-balanced table wines for which Margaret River is best known. Wilyabrup enjoys an additional reputation for the light, fruity port-style wines that can be made in warm seasons.

The viticultural soils of Margaret River and Wilyabrup are predominantly Forest Grove gravelly sandy loams. At lower altitudes in the valleys, and scattered along the western fringe of the area, there occur as well the gritty sandy loams and loams of the Keenan soil series, formed directly from the underlying granite and gneissic rock. These are also suitable for viticulture, but some drain less readily than the gravels; the surfaces can also become powdery and erosion-prone if cultivated dry.

Quite large further areas of potential vineyard soils, so far little exploited, lie just east of Margaret River, in a strip running northwards from Rosa Glen, through Rosa Brook and Mowen to Osmington. Although further inland than the main existing vineyards, and having a little more winter rain, these areas seem unlikely to differ greatly in their growing season and ripening climates. Wine styles should thus be comparable. The principal difference is that more care may be needed in site selection for soil drainage and frost avoidance. This will tend automatically to offset the effects of greater distance inland on temperature variability patterns.

Yallingup (Table 32) represents the most northerly part of the Margaret River region. This is another area with great potential, which is only now beginning to be exploited. The estimated climatic data are for the western, northern and eastern slopes of the Leeuwin-Naturaliste ridge, and show them to be similar to Wilyabrup in most respects, but to have still more equable temperatures and outstanding air drainage. The area from about Abbey Farm Road northwards to Dunsborough has the unique advantage, for Western Australia, of facing north across Geographe Bay. This means that north-easterly and northerly winds must come across the bay, resulting on average in slightly warmer nights, cooler days and higher relative humidities than in the more land-dominated locations; but there are still very generous sunshine hours. Potential vineyard soils are mainly lateritic gravelly sandy loams, with a few outcrops of country rock resulting in gritty loams and transitional soils. There is reasonable scope for surface water storage, including the drainage systems of Gunyulgup Creek, its tributaries, and other small creeks. Like Wilyabrup, Yallingup has already established a reputation for its light, refined fortified wines as well as for mainly full-bodied table wines. Those from Frontignac show exceptional promise.

Further still to the north, along the Naturaliste Peninsula northwards from Dunsborough, small patches of fair soils exist on the slopes facing east across Geographe Bay. Viticultural potential is small because of limited suitable soils, the unlikely availability of land at economic prices, and strong winds. Nevertheless the temperature variability index here should be even lower than at Yallingup, giving outstanding climatic advantages for winegrape quality in any sheltered pockets where vines can thrive.

A notable feature of the Margaret River region as a whole has been the good adaptation of Bordeaux grape varieties. One exception has been Malbec, which suffers badly from excessive vigour and poor fruit set; however that also happens in Bordeaux. The Cabernet Sauvignon wines of Margaret River established a fine reputation from the start, and the vines are reliable yielders. Wines from Merlot show outstanding promise, although some question remains over setting in years with a wet or windy late spring. Wines from the first plantings of Cabernet Franc also show good promise. Sémillon and Sauvignon Blanc have been very successful for producing both dry and sweet white wines of marked character. The Loire variety Chenin Blanc also grows and yields particularly well, perhaps benefiting from its tolerance of strong winds. It makes mildly fruity wines which are useful for blending with those of stronger flavours, to produce early-drinking commercial styles.

Cool-climate varieties of inland origin, such as Riesling, Traminer and Chardonnay, have as a rule been viticulturally less happy. Their problems stem largely from the region's winds and lack of continentality, as discussed in Chapter 3. Nevertheless the excellence of the full-bodied wines from Chardonnay continues amply to justify that variety, despite its wind

susceptibility and somewhat erratic yield; while full-bodied, sweet wines from late-picked Riesling have also been very good. Late pruning helps to overcome the viticultural problems of these varieties, as does supplementary watering to avoid the late summer stress which contributes to their premature budburst (Hamilton 1986; Dr Michael Peterkin, personal communication).

A notable feature of Margaret River is the wide range of grape varieties which produce wines of individuality and power. Proved varieties in addition to those already mentioned include Shiraz, Zinfandel and Verdelho. This seems consistent with the region's favourable relative humidities and generous but equable warmth and sunshine, which in theory should allow the greatest combined expression of yield and fruit quality. It should be equally so whether high yield is due to variety, or to the management of lower-yielding 'noble' varieties for maximum yield.

An extension of the argument is that such regions are the logical places for future production of Australia's 'commercial' and superior bulk wines, as well as of its premium wines. Padthaway, in South Australia (Chapter 10) is an established step in that direction. Margaret River and similar near-coastal areas, including Manjimup-Pemberton, would appear to be particularly suited to any such future developments in Western Australia.

A final comment on Margaret River is that it would be a logical region in which to try the Petit Verdot variety. The Petit Verdot is highly esteemed in Bordeaux, being described as 'a super Cabernet Sauvignon', and 'the best vat of the grand vin ... its finesse and powerful aromatic quality are inimitable' (Dubourdieu 1990). However early budburst and rather late maturity limit it to the warmest sites of Bordeaux, and, at least in modern times, full maturity is reached only in warm years (see discussion of historical climatic changes, Chapter 23). For these reasons the Petit Verdot now forms only a very small part of the Bordeaux red wine blend, although its contribution, when ripe enough, is greatly valued. The climate and gravelly soils of Margaret River to Yallingup should theoretically provide an ideal environment for its perfect ripening, perhaps to give an extra dimension of quality to the region's Bordeaux-style red wines.

South West Coast (Busselton-Harvey)

The brown and reddish alluvial loamy sands of the coastal plain around **Jindong** and **Carbunup** would have climatic conditions similar to those of **Busselton** (Table 10). These should support good vine growth where well enough drained. In addition there are small areas of alluvial deep, red sandy loams at **Marybrook**, just inland from the coast west of Busselton, which look especially interesting for viticulture. They are closely related to the Houghton and Belhus series of alluvial soils, which are the prime viticultural soils of the

Swan Valley (Bettenay, McArthur and Hingston 1960). Limited experience to date has shown that vines on the true sandy loams here attain great vigour, necessitating special attention to trellising. Temperature variability at Marybrook should be a little lower than at Busselton, perhaps approaching that of Yallingup.

Whether the coastal plain west of Busselton would be better included in the South West Coastal region, or alternatively with Margaret River (broadly defined), is a question which only time and experience will show. Soil and topography are more those of the south west coastal plain, as is the climate in many respects; but geographically, it is hard to separate from Margaret River and Yallingup.

Moving north to areas of indisputably west-coastal aspect, significant amounts of alluvial loams exist along the lower **Capel River**. Like the Marybrook sandy loams and better red loamy sands south of Busselton, these carried vigorous stands of marri in their native state. They have already proved to be very suitable for viticulture.

Brownish sands over limestone, characterized by tuart (*Eucalyptus gomphocephala*), form a further distinctive soil environment along the coast, from a little west of Busselton, through Capel and **Bunbury** (Table 9), northwards to and beyond Harvey. These soils have ample good-quality underground water for supplementary watering. Their low natural fertility, and the readiness with which nutrients are lost by leaching, mean that very close attention is needed to vine nutrition, and to building up soil organic matter levels. Given these, however, there are substantial areas of cleared tuart soils in the Bunbury-Harvey area which could be seriously considered for viticultural expansion.

Spring frosts are rare along the south-west coastal strip, and the growing season climate is warm and sunny. Proximity to the ocean and early sea breezes maintain mostly equable temperatures and high afternoon relative humidities. Wines from existing vineyards at Capel and Bunbury show that the warmest parts remain well within the climatic range for high-quality table wines, which are full-flavoured and soft.

The whole of the Busselton-Harvey coastal strip has a climate closely resembling that of Frontignan, on the south coast of France, apart from having more winter rain and a drier summer. This would appear to be an ideal climate for producing aromatic, lightly fortified dessert wines from early-maturing grape varieties such as Frontignac (= Muscat à petits grains, = Muscat of Frontignan, = Rutherglen brown Muscat), Muscadelle, and very probably Verdelho.

Chapman's Hill and adjacent locations on the lower northern slopes of the Whicher Range, south of Busselton, have suitable gravelly soils, temperatures and sunshine hours for full-bodied table wines and light-bodied fortified wines. The area is in the rain shadows of both the Leeuwin-Naturaliste ridge to the west and the Whicher Range to the south, and is therefore warm, dry and sunny like the coastal plain. Temperature variability is probably greater than near the west coast. The particular environment of the slopes and gravelly soils might nevertheless arguably be classified with the northern part of the Leeuwin-Naturaliste ridge, and with Margaret River in its broad sense, rather than with the Busselton-Harvey coastal plain.

Bunbury Hills

The foothills and hills of the Darling Range between Boyanup and Harvey, just east of Bunbury, form a further distinct environment, of which the warmer parts are typified climatically by **Wokalup** (Table 31). Despite being only a short distance inland, this is clearly a hotter environment (at low altitudes) than the coast itself. The hilly topography and proximity to the coast nevertheless still result in early sea breezes, with moderately high afternoon relative humidities and low temperature variability considering the hot, sunny climate. The foothills and immediate scarp region equally benefit from ocean-warmed air descending as part of the land-breeze convection cycle at night (Chapter 4, Figure 7).

The data in Table 31 represent the low foothills and valleys from Harvey southwards to the Ferguson Valley. Good slopes of excellent gravelly loam soils exist from altitudes of about 50 m to 200 m or more. With variation in latitude, altitude and exposure, temperatures could vary by as much as 1.0°C above or below the estimates. Some of the best sites are in the valleys of **Henty Brook**, south-east of Burekup, and of the **Ferguson River** and **Crooked Brook**, south-east of Dardanup. Good surface catchment water is available.

Climatically, the lower and middle altitudes appear perfectly suited for both full-bodied table wines and fortified wines of the port style. Temperatures and sunshine hours through the growing and ripening season closely match those of the lower and central Douro Valley in Portugal: see Pinhão (Chapter 20, Table 165) and Régua (Table 166). Winter-spring rainfall is very reliable, and somewhat higher than in the Douro; as is probably ripening period relative humidity, due to greater proximity to the coast and the regular occurrence of afternoon sea breezes.

In the higher-altitude and southern parts of the area, e.g. the upper Ferguson Valley, temperatures fall into the mainstream for medium to full-bodied table wines while still in the optimum range for aromatic fortified wines from early-maturing grape varieties such as Frontignac, Muscadelle and Verdelho.

The foothills east of Bunbury find their closest climatic analogies within Australia in the Adelaide-Southern Vales and Clare regions of South Australia, and Rutherglen-Milawa in Victoria. However the Bunbury foothills combine, to a degree not found in those environments, substantial freedom from spring frosts, reliable and ample winter-spring rainfall, and a ripening period which is regularly dry and sunny, but with only moderate atmospheric saturation deficits. A recent trial planting by the Western Australian Department of Agriculture has shown the area to produce winegrapes of outstanding quality (Ian Cameron, personal communication 1992). Despeissis (1897) long ago extolled the viticultural virtues of the Bunbury hinterland.

Donnybrook

The farming lands south of **Donnybrook** (Table 13) have large areas of soils and topography very suitable for vine growing. Potential vineyard soils include both gravelly sandy loams, mainly at the higher levels, and loams formed directly from the gneissic country rock in the valleys, with some transitional soils in between. There are ample good surface water catchments. Ripening temperatures and sunshine hours are in the upper range for good quality in table wines, with the warmer parts very suitable for light fortified wines. The climate closely resembles those of Bendigo and Rutherglen-Milawa in Victoria, but with more winter-spring rainfall and less summer rain.

As in northern Victoria, the main problem of Donnybrook lies in its temperature variability and low relative humidities, due to being cut off from maritime influences by the Darling and Whicher ranges. Conditions for classic table wine quality are therefore inferior to those closer to the coast. Against that, vines should grow and yield well wherever frost can be avoided, which automatically selects the best site climates for quality; and the dry atmosphere in summer should at least be conducive to vine health.

The plentiful sunshine and lack of rainfall and humidity during ripening provide conditions especially suitable for making sweet fortified styles which depend on the natural raisining of early-maturing grape varieties, such as Frontignac (on sufficiently frost-free sites) and Muscadelle. This is also an environment where Grenache could be expected to thrive on the richer soils with adequate air drainage, to make good, light red or rosé wines. Cabernet Sauvignon, Cabernet Franc, Shiraz and possibly Malbec may be other well-adapted red varieties.

Kirup-Bridgetown

Potential vinegrowing locations around **Kirup** and **Bridgetown** (Table 8) are mostly at greater altitudes and significantly cooler than those of Donnybrook. Ripening period mean temperatures and sunshine hours are within the optimum range for full-bodied table wines; and ample good gravelly and loamy soils exist, similar to those at Donnybrook and with good potential for surface water catchment. The main problem, even with rigorous site selection, is wide temperature variability and a severe spring frost risk. Relative humidities are low.

Good table wines can undoubtedly be made on sites at Bridgetown that are free enough from frost, with the advantage that the winters are probably cold enough to avoid budburst problems. Like Donnybrook (above), this is an area not to be discounted entirely for commercial viticulture, given careful selection of mesoclimates and of grape varieties best suited to the conditions. Nevertheless the more humid and equable Margaret River, Manjimup-Pemberton and Mount Barker-Albany regions almost certainly remain preferable for fruit quality, and for large-scale planting.

Perth Coast (Mandurah-Yanchep)

The Perth coastal vine region extends along a strip of limestone-based tuart sands within about 10 km of the coast, from Mandurah northwards through **Baldivis** to **Spearwood**, and then just north of Perth from **Wanneroo** to **Yanchep**. It is a northward extension of the coastal Busselton-Bunbury-Harvey viticultural area.

Mandurah (Table 19) should adequately represent the climate from Baldivis south, and **Perth** (Table 25) that from Spearwood to Yanchep. The Fremantle weather recording site is too coastal and exposed to typify any of the potential vinegrowing sites.

The Perth coastal strip has a more favourable climate for viticulture than the Swan Valley, which is immediately inland from it. Temperature variability and monthly highest temperatures in summer are significantly lower, and the average afternoon relative humidity higher, due to an earlier and more regular arrival of the south-westerly summer sea breeze (usually by about noon). This, together with a mostly ready availability of good-quality groundwater, makes it still a reasonably good environment for making table wines, despite high temperature means. The climate suggests that it should be excellent for fortified wines. Infertility of the deep, sandy soils is the main limitation, together with occasional heat waves from which even the coast is not immune.

Experience has shown that this area produces wines appreciably different from those of the Swan Valley. The red wines, in particular, are softer and have more finesse, and show less obviously hot-climate characteristics.

Shiraz has proved to be especially well adapted, producing soft, fruity wines with an often pronounced berry character. Cabernet Sauvignon has also given excellent wines, but its long-term performance is yet to be assessed – as are those of Cabernet Franc and Merlot. The soil is probably too sandy for Grenache, which in other respects might be well adapted for making light red and rosé wines. White wines have not, in general, so far stood out to the same degree as the red.

Swan Valley

The data for **Guildford** (Table 15) show the Swan Valley to be on the climatic borderline between medium-bodied fortified wines on the one hand, and drying grapes, together with 'liqueur'-type fortified wines, on the other. Some of its most famous wines are of the latter type, while the currants of the Swan Valley and nearby Bindoon are reputed to be among the best in the world.

Some good, full-bodied table wines are nevertheless also made. They tend to lack the fresh, fruity acid and specific aromas of cool-climate wines, but if the grapes ripen fully without undue stress they compensate with good body and flavour. In general white wines are the more successful. Verdelho is regularly outstanding, with often very good wines from Riesling, Chenin Blanc and Chardonnay. Red wines show the effects of the heat more readily, and in hot years can take on a cooked, 'porty' character. Cabernet Sauvignon, Merlot and sometimes Shiraz can nevertheless be good, giving wines with more middle-palate fruit and more suitable for early drinking than many of those from cooler climates.

The best Swan Valley table wines are made in cool to average summers, characterized by regular afternoon sea breezes and no heat waves during ripening. The Swan Valley in its tidal reaches extends inland from south-west to north-east, a fact which enhances its natural role as a funnel for the south-west sea breezes. In a cool summer with mainly on-shore winds, ripening conditions in the Swan Valley can in fact be very favourable for full-bodied table wines, with an average mean temperature about 22°C, low variability, and afternoon relative humidities around 50 per cent. The warm nights favour physiological ripening and full flavour.

Some local environments in the Swan Valley are more favoured than others. A strip immediately along the river (which is tidal to Upper Swan) benefits from its permanent proximity to water, and from the deep, well-drained, reddish sandy loams of the Houghton, Belhus and Swan series (Pym 1955) which constitute the terraces immediately above the river. These young alluvial soils are highly suitable for vines, and can sustain good growth and yields with little stress even under solely natural rainfall. The old weather recording station in Stirling Square at Guildford, from which the present climatic estimates for the Swan Valley were derived, was not far from the river and would represent them well. These soils are fringed on the eastern side by the Herne series of grey-brown sands and sandy loams overlying clay, which are regarded as moderately suitable for viticulture.

The flat, less well-drained soils of Pym's Mongin series, lying eastwards again from the Great Northern Highway to the foot of the Darling Range, are much less suitable for viticulture and have progressively gone out of production. The Upper Swan Viticultural Research Station has largely similar topography to these, and is away from the river. The recent (Bureau of Meteorology, Australia, 1975 and 1988) climatic data for Upper Swan show a significantly harsher and hotter climate than at Guildford.

The lower slopes of the Darling Range foothills skirt the Swan Valley's eastern edge, and provide its other major group of sought-after soils for grape growing. The soils are gravelly sands to gravelly sandy loams overlying brown clay, and belong to the Lotons, Range and Oakover series of Pym (1955). Vine growth is only moderate and in good balance, and the grapes are highly reputed for their quality. Experience has shown that the grapes ripen some days earlier than on the valley flats, consistent with a 'thermal zone' effect of the slopes along the scarp (Chapter 4, Figures 4 and 7), together with a night warming influence of the gravelly, well-drained soils. These factors reduce temperature variability to well below that on the adjacent flats, while the reddish brown colour of the gravels should be favourable for light quality in the vine canopies.

Gingin-Bindoon

No published climatic data are available specifically for **Gingin** and **Bindoon**, which are in foothills about 60 km north of the Swan Valley. Greater altitudes (between 100 and 200 m) should offset the distance north in respect to average mean temperature: see Figure 11. The hilly nature of the country probably also reduces temperature variability, at least to an equivalent of the best parts of the Swan Valley. The vineyards at Gingin on the Moonda Brook, at about 100 to 150 m, almost certainly derive special benefit from cool south-westerly afternoon sea breezes funnelling up the valley in summer, together with some warming influence of the reverse air convection at night and in the early morning, as elsewhere on the exposed face of the Darling Scarp.

What cannot be questioned is that the grapes grown on the brown to red loamy and gravelly loam soils at Gingin and Bindoon, with supplementary watering, are of surprisingly high quality for table wines for so hot a climate. As in the Swan Valley, the white varieties are the more successful overall. Red wines mostly do not match those made in the State's cooler southern areas, but early-picked Cabernet Sauvignon is used for dry rosé wines with great success.

Another area which might be mentioned is Dandaragan, to the north of Gingin-Bindoon. No directly relevant temperature figures are available, but given its altitude (mostly about 200–250 m) and undulating topography, it might be guessed to have temperatures similar to those of Gingin, Bindoon and Guildford (Figure 11). The attraction of Dandaragan is its large area of deep, reddish loamy sands, which in the native state carried marri woodland and should support good vine growth. Those along the western edge of the hills would benefit from early sea breezes and night air convection as at Gingin. However rainfall here is less, about 650 mm, and the prospects for satisfactory water supplies less certain.

Perth Hills

Valleys in the hills of the Darling Range close to Perth (see **Kalamunda**, Table 16) have moderate areas of ironstone-gravelly sandy loams and gravelly loams overlying clay, which carried vigorous marri in their native state and are similar to those used for vine-growing elsewhere in Western Australia. There are also some clay loams formed directly from gneissic and doleritic country rock in the valley bottoms. Some of these originally carried wandoo (*Eucalyptus wandoo*), and are perhaps more doubtfully suitable for viticulture.

Several vineyards were established in the Hills area around the turn of the century, but were later gradually abandoned for mainly economic reasons. There have nevertheless been several new plantings in recent years, including one major vineyard on the brow of the scarp at **Chittering**.

With altitudes up to about 300 m, temperatures in the hills are significantly lower than in the Swan Valley, Gingin, Bindoon and probably nearby coastal regions. Sites close to the scarp (represented by Kalamunda) have the most favourable temperature relations, because of their free air drainage and relative warmth at night, together with full exposure to afternoon sea breezes. These conditions extend a short distance up the valleys breaking through the scarp, but tend to be lost as the valleys change direction and become cut off from coastal influences.

I estimate that grapes on the brow of the Darling Scarp should ripen about a week later than on the Swan. Minimum temperatures then fall quite rapidly with greater distance inland, as is reflected in the growing season mean temperatures, adjusted to sea level (Figure 11). Sea-warmed convectional air at night and in the early morning (Chapter 4, Figure 7) does not reach much beyond the scarp and its immediate valleys, whereas afternoon sea breezes do, albeit progressively later with distance inland. The result is a strip of lowest mean temperature not far inland from the scarp. Despeissis (1902) wrote that in the **Mundaring-Chidlows** area grapes ripen two to three weeks later than on the Swan, which suggests effective mean temperatures not much above those estimated here for Collie (Table 11). Eastwards from about Chidlows, minimum as well as maximum temperatures again start to rise. Ripening at **Bakers Hill** (altitude 305 m; Table 7) is nevertheless still estimated to be 10–12 days later than in the Swan Valley, and at temperatures 1.2–2.1°C lower.

Annual rainfall is 900–1200 mm within 20–40 km east of the Darling Scarp, with ample potential for surface storage of good-quality water; but thereafter both rainfall and storage possibilities diminish rapidly inland. Below 750 mm annual rainfall there are increasing problems of salt, in both surface water storage and in the very limited underground supplies. The 750 mm isohyet (a little west of Bakers Hill) marks the approximate inland limit of viticultural suitability at Perth latitudes. By all criteria the better sites are within 20–30 km inland from the scarp, and the best sites along or very close to the scarp itself.

Wine styles are not yet fully defined, but clearly this is a cooler viticultural environment than the adjacent coastal plain and Gingin-Bindoon. Red wines made so far from the newer plantings have had intense colour and flavour. The climatic figures suggest it to be most truly an area for fortified wines, with full-bodied table wines as well in the more coastal parts. The climate closely resembles that of the Douro Valley in Portugal (Chapter 20, Tables 165 and 166), and of Wokalup (Table 29).

Inland areas

Early inland settlers grew grapes for fresh fruit, drying and some winemaking, as part of their locally-sufficient economy. Late in the 19th century Despeissis (1897) reported that there were vineyards producing excellent wine at Toodyay. The estates of F.H. and C.A. Piesse had 45 hectares under vineyards and orchards at Wagin, Arthur River and Katanning, in the Great Southern area; while A. Piesse, a younger brother, was clearing new land to extend his existing 10-hectare vineyard at Katanning. The census for 1909 showed areas (hectares) of bearing winegrapes as follows: Victoria Plains (New Norcia-Bolgart area) 12; Toodyay 111; Avon 24; York 1; Wagin 9; Katanning 21; and Tambellup 2. Older vineyards (though mostly of newer plantings) still exist at Toodyay and Katanning, and did until recently at Bakers Hill. Small new vineyards are in production at Wandering, south of Bakers Hill, and at Kulikup, much further south between Boyup Brook and Kojonup.

Spring frosts are the greatest hazard through most of the inland region. Rainfall and relative humidities are also seriously sub-optimal and deteriorate from west to east, together with the capacity to store good-quality water.

Ripening-period temperatures are highest in the north, around Bakers Hill-Toodyay. Those estimated for **Bakers Hill** (Table 7) probably represent typical vineyard sites in that area, with temperatures perhaps a little higher towards Toodyay and substantially higher with descent into the Avon Valley. The temperatures for York and Northam (not presented), on the floor of the valley, are similar to those for the Swan Valley; but these localities, and to a lesser extent Bakers Hill, differ from the Swan Valley in that the sea breezes in summer arrive late in the afternoon, if at all. In any case, by that time the breezes have become warmer and drier than near the coast.

The data for **Wandering** (Table 29), **Narrogin** (Table 24), **Katanning** (Table 18) and **Kojonup** (Table 19), considered with those for **Collie** (Table 11), suggest that the central region west of the Great Southern railway can be treated more or less as a unit. The frost risk in spring is great. Ripening mean temperatures and sunshine hours are appreciably below those for Bakers Hill-Toodyay.

The main requirement throughout, apart from supplementary water to reduce stresses in summer, is very careful site selection to minimize temperature variability and frost. The best sites, indeed the only sites on which viticulture can in any way be contemplated, will be the slopes or brows of isolated hills having very good air drainage, together with good marri soils. Wandoo (E. wandoo) soils are especially to be avoided because of their mostly limited drainage.

Late budburst and heat tolerance are desirable characteristics of grape varieties for such environments. Adapted varieties might include Cabernet Sauvignon, Carignan, Mataro and Ruby Cabernet, and perhaps Shiraz, Cinsaut and Grenache, among the reds; and Trebbiano, and perhaps Riesling, Muscadelle and Taminga among the whites. With proper selection of site, soil and grape varieties it should be possible to produce fair table wines, and reasonably good, light to medium-bodied fortified wines. Sweet, fortified wines from naturally raisined Muscadelle (and from Frontignac if frosting can be avoided) could be the most successful because of the dry ripening period, as at Rutherglen and Donnybrook. Generally, however, this is not a region to be recommended for commercial-scale viticulture.

Esperance

This isolated south coastal area has climatic characteristics which in several respects are quite favourable for table wines. Ripening average mean temperatures, sunshine hours and average afternoon relative humidities are all optimal for producing medium to full-bodied styles.

Rainfall is sufficient only very near **Esperance** (Table 14). The best area climatically is just east of the town, where the low hills result in most rainfall and provide a range of sites and altitudes with temperatures ranging from a little warmer to a little cooler than those represented in the table. Rainfall distribution is favourable on average, but varies greatly from season to season and within seasons. Temperature variability index and the average ripening month extreme maxima are both rather high.

The Esperance Plain has notably abrupt changes in temperature within a summer day. Normally the morning is warm and dry, with a northerly wind. Then, about midday, the wind changes to southerly, with a sudden and marked drop in temperature and rise in relative humidity (hence the high average 1500 hours relative humidity, column 18). The effect of this on grape and wine quality remains to be seen, but may not be unfavourable because average afternoon and evening conditions fall well within the optimum range for ripening table winegrapes. However failure of the afternoon sea breeze (or wind) can periodically result in serious stresses, and is the main reason for the great variability of the maximum temperature. In these respects Esperance typifies Australian south coastal climates to the ultimate degree. Therefore substantial vintage variation can be expected.

Soil selection is critical at Esperance, because most soils there are either deep, erodible sands, or else flat, shallow and poorly drained. Really suitable soils occur, at most, only in small pockets. The aim would be to combine a gravelly sand to gravelly loam surface, good depth and drainage, but a retentive subsoil, and some slope. Wind breaks would be extremely important, because strong winds can sweep across the plain from several directions.

TABLE 6

ALBANY (Airport), WESTERN AUSTRALIA. 34°57'S, 117°48'E. ALT. 69m. (For Redmond and Lower Kalgan)

	1	2	3	4	5	6	7	8	9	10	11	12	13	14	15	16	17	18
	Temperature °C									Summation of Day Degrees above 10° with 19° Cut-Off			Temperature Variability Index					
PERIOD	Lowest Recorded Min.	No.Mths with <0.3°	No.Mths with <2.5°	Av.Mthly Lowest Min.	Av.Min. Minus Col.4	Daily Mean	Daily Range /2	Av.Mthly Highest Max.	Col.8 Minus Av.Max.	Raw	Adj.1 for Lat. & Daily Temp Rng	Adj.2 for Vine Sites	Raw	Adj. for Vine Sites	Av. Sunshine Hrs	Av. Number Rain Days	Av. Rainfall mm (43YRS)	Av. % R.H. 1500 HRS
OCTOBER				4.2	4.4	13.5	4.9	27.9	9.5	109	109	124	33.5	31.1	205	16	71	66
NOVEMBER				5.8	4.7	15.8	5.3	32.1	11.0	174	169	180	36.9	34.5	225	11	39	63
DECEMBER				7.0	4.9	17.7	5.8	35.1	11.6	239	231	237	39.7	37.3	245	10	30	58
JANUARY				7.8	5.0	19.0	6.2	37.3	12.1	279	270	276	41.9	39.5	232	9	27	54
FEBRUARY				8.3	5.1	19.2	5.8	36.2	11.2	252	252	252	39.5	37.1	212	8	28	56
MARCH				7.4	5.2	18.2	5.6	35.5	11.7	254	253	259	39.3	36.9	205	11	42	59
APRIL				6.3	5.1	16.1	4.7	29.3	8.5	183	190	205	32.4	30.0	180	16	66	63
OCT-APRIL				—	4.9	17.1	5.5	—	10.8	1,490	1,474	1,533	37.6	35.2	1,504	81	303	60
MAY-SEPT				—	—	12.2	—	—	—	—	—	—	—	—	—	—	497	—
DATA FOR MONTH TO MATURITY																		
MATURITY GROUP	EST. MATURITY DATE					*	*	*						*				
1	FEBRUARY 26					19.4	5.2	35.9	11.3					37.3	229	8	30	56
2	MARCH 4					19.3	5.2	35.8	11.4					37.3	226	8	31	56
3	MARCH 9					19.2	5.1	35.7	11.5					37.2	222	9	32	56
4	MARCH 15					19.1	5.1	35.6	11.6					37.1	217	9	34	57
5	MARCH 21					18.9	5.0	35.4	11.7					37.0	212	10	37	58
6	MARCH 28					18.6	5.0	35.2	11.7					36.9	207	10	40	59
7	APRIL 4					18.1	4.9	34.8	11.4					36.5	201	11	45	60
8	APRIL 11					17.6	4.7	34.0	10.8					35.5	196	12	51	61

* Adjusted for vine sites.

TABLE 7

BAKERS HILL (CSIRO), WESTERN AUSTRALIA. 31°45'S, 116°28'E. ALT. 305m.

	1	2	3	4	5	6	7	8	9	10	11	12	13	14	15	16	17	18
	Temperature °C									Summation of Day Degrees above 10° with 19° Cut-Off			Temperature Variability Index					
PERIOD	Lowest Recorded Min. (12YRS)	No.Mths with <0.3° (12YRS)	No.Mths with <2.5° (12YRS)	Av.Mthly Lowest Min. (12YRS)	Av.Min. Minus Col.4	Daily Mean	Daily Range /2	Av.Mthly Highest Max. (12YRS)	Col.8 Minus Av.Max.	Raw	Adj.1 for Lat. & Daily Temp Rng	Adj.2 for Vine Sites	Raw	Adj. for Vine Sites	Av. Sunshine Hrs	Av. Number Rain Days	Av. Rainfall mm	Av. % R.H. 1500 HRS
OCTOBER	0.0	1	5	3.0	5.4	14.7	6.3	29.5	8.5	146	142		39.1		258	11	36	47
NOVEMBER	1.6	0	1	5.0	6.0	18.3	7.3	35.4	9.8	249	227		45.0		306	5	11	36
DECEMBER	4.8	0	0	7.6	5.6	21.3	8.1	38.2	8.8	279	279		46.8		337	3	10	32
JANUARY	7.1	0	0	9.4	5.6	22.9	7.9	39.7	8.9	279	279		46.1		338	2	7	30
FEBRUARY	6.5	0	0	9.4	5.8	22.9	7.7	39.8	9.2	252	252		45.8		285	2	13	30
MARCH	5.9	0	0	8.3	5.7	20.9	6.9	36.5	8.7	279	279		42.0		270	4	19	36
APRIL	4.1	0	0	6.5	5.7	17.7	5.5	32.2	9.0	231	236		36.7		229	8	37	46
OCT-APRIL	—	1	6	—	5.7	19.8	7.1	—	9.0	1,715	1,694		43.1		2,023	35	133	37
MAY-SEPT	—	—	—	—	—	11.8	—	—	—	—	—		—		—	—	485	—
DATA FOR MONTH TO MATURITY																		
MATURITY GROUP	EST. MATURITY DATE																	
1	FEBRUARY 15					23.0	7.8	39.8	9.0				46.0		325	2	10	30
2	FEBRUARY 20					23.0	7.8	39.8	9.0				45.9		320	2	11	30
3	FEBRUARY 25					22.9	7.7	39.8	9.0				45.8		315	2	12	30
4	MARCH 3					22.9	7.7	39.7	9.0				45.6		310	2	13	30
5	MARCH 8					22.7	7.6	39.4	9.0				45.2		304	2	14	31
6	MARCH 14					22.4	7.4	38.8	9.0				44.8		297	3	15	32
7	MARCH 19					22.1	7.2	38.0	9.0				44.0		289	3	16	33
8	MARCH 25					21.6	7.0	37.0	9.0				43.0		280	4	18	35

TABLE 8
BRIDGETOWN, WESTERN AUSTRALIA. 33°57'S, 116°10'E. ALT. 154m.

	1	2	3	4	5	6	7	8	9	10	11	12	13	14	15	16	17	18
	Temperature °C									Summation of Day Degrees above 10° with 19° Cut-Off			Temperature Variability Index					
PERIOD	Lowest Recorded Min. (61YRS)	No.Mths with <0.3° (30YRS)	No.Mths with <2.5° (30YRS)	Av.Mthly Lowest Min. (30YRS)	Av.Min. Minus Col.4	Daily Mean (54YRS)	Daily Range /2 (54YRS)	Av.Mthly Highest Max. (30YRS)	Col.8 Minus Av.Max.	Raw	Adj.1 for Lat. & Daily Temp Rng	Adj.2 for Vine Sites	Raw	Adj. for Vine Sites	Av. Sunshine Hrs	Av. Number Rain Days	Av. Rainfall mm (85YRS)	Av. % R.H. 1500 HRS
OCTOBER	−2.2	16	29	0.4	6.1	13.4	6.9	29.2	8.9	105	98	105	42.6	38.0	205	14	67	53
NOVEMBER	−0.9	5	21	1.7	6.6	16.4	8.1	33.3	8.8	192	163	181	47.8	43.2	240	8	30	44
DECEMBER	0.0	1	7	3.1	6.9	18.8	8.8	36.6	9.0	273	228	247	51.1	46.5	282	5	19	39
JANUARY	0.9	0	0	5.1	6.4	20.6	9.1	38.4	8.7	279	278	279	51.5	46.9	284	3	14	36
FEBRUARY	0.5	0	3	4.3	7.0	20.4	9.1	38.1	8.6	252	248	252	52.0	47.4	245	4	14	36
MARCH	−0.9	3	14	2.4	7.8	18.6	8.4	36.5	9.5	267	236	255	50.9	46.3	230	5	29	42
APRIL	−2.2	9	21	1.3	6.6	15.7	7.8	30.7	7.2	171	154	173	45.0	40.4	190	9	46	51
OCT-APRIL	—	34	95	—	6.8	17.7	8.3	—	8.7	1,539	1,405	1,492	48.7	44.1	1,676	48	219	43
MAY-SEPT	—	—	—	—	—	11.1	—	—	—	—	—	—	—	—	—	—	637	—

DATA FOR MONTH TO MATURITY

MATURITY GROUP	EST. MATURITY DATE		*	*	*						*				
1	FEBRUARY 27		20.5	8.0	37.0	8.6					47.4	266	4	15	36
2	MARCH 4		20.3	7.9	36.8	8.7					47.3	261	4	16	37
3	MARCH 10		20.0	7.8	36.6	8.9					47.2	256	4	18	38
4	MARCH 16		19.7	7.7	36.3	9.1					47.0	250	4	20	39
5	MARCH 22		19.3	7.6	36.0	9.3					46.8	243	5	23	40
6	MARCH 28		18.9	7.4	35.5	9.4					46.4	235	5	27	41
7	APRIL 6		18.1	7.2	34.5	9.1					45.6	226	6	31	43
8	APRIL 16		17.1	7.0	32.8	8.4					44.0	212	7	36	46

* Adjusted for vine sites.

TABLE 9
BUNBURY, WESTERN AUSTRALIA. 33°18'S, 115°38'E. ALT. 5m.

	1	2	3	4	5	6	7	8	9	10	11	12	13	14	15	16	17	18
	Temperature °C									Summation of Day Degrees above 10° with 19° Cut-Off			Temperature Variability Index					
PERIOD	Lowest Recorded Min. (65YRS)	No.Mths with <0.3° (30YRS)	No.Mths with <2.5° (30YRS)	Av.Mthly Lowest Min. (30YRS)	Av.Min. Minus Col.4	Daily Mean (56YRS)	Daily Range /2 (56YRS)	Av.Mthly Highest Max. (30YRS)	Col.8 Minus Av.Max.	Raw	Adj.1 for Lat. & Daily Temp Rng	Adj.2 for Vine Sites	Raw	Adj. for Vine Sites	Av. Sunshine Hrs	Av. Number Rain Days	Av. Rainfall mm (96YRS)	Av. % R.H. 1500 HRS (44YRS)
OCTOBER	0.6	0	2	4.8	5.3	15.0	4.9	26.5	6.6	155	154	149	31.5	32.7	230	11	55	65
NOVEMBER	4.0	0	0	7.0	4.9	17.5	5.6	30.4	7.3	225	218	215	34.6	35.8	275	6	25	60
DECEMBER	3.6	0	0	8.4	5.2	19.6	6.0	33.4	7.8	279	279	279	37.0	38.2	320	4	14	59
JANUARY	5.6	0	0	9.7	5.1	21.2	6.4	34.9	7.3	279	279	279	38.0	39.2	315	3	10	57
FEBRUARY	5.2	0	0	9.9	5.2	21.4	6.3	34.8	7.1	252	252	252	37.5	38.7	260	3	12	56
MARCH	4.1	0	0	8.7	5.3	20.0	6.0	32.9	6.9	279	279	279	36.2	37.4	255	4	23	58
APRIL	2.6	0	0	6.8	5.3	17.7	5.6	28.0	4.7	231	235	232	32.4	33.6	210	8	46	63
OCT-APRIL	—	0	2	—	5.2	18.9	5.8	—	6.8	1,700	1,696	1,685	35.3	36.5	1,865	39	185	60
MAY-SEPT	—	—	—	—	—	13.6	—	—	—	—	—	—	—	—	—	—	696	—

DATA FOR MONTH TO MATURITY

MATURITY GROUP	EST. MATURITY DATE		*	*	*						*				*
1	FEBRUARY 15		21.4	6.6	35.0	7.2					39.0	298	3	11	54
2	FEBRUARY 20		21.4	6.6	34.9	7.2					38.9	291	3	12	54
3	FEBRUARY 26		21.3	6.6	34.7	7.1					38.7	283	3	13	54
4	MARCH 3		21.2	6.6	34.5	7.1					38.5	278	3	14	54
5	MARCH 9		21.1	6.5	34.3	7.0					38.3	274	3	15	55
6	MARCH 14		20.6	6.5	34.0	7.0					38.1	269	3	17	55
7	MARCH 20		20.6	6.4	33.7	6.9					37.8	264	4	19	56
8	MARCH 26		20.3	6.3	33.4	6.9					37.5	259	4	21	56

*Adjusted for vine sites.

TABLE 10 BUSSELTON, WESTERN AUSTRALIA. 33°38′S, 115°22′E. ALT. 3m.

	1	2	3	4	5	6	7	8	9	10	11	12	13	14	15	16	17	18
	Temperature °C									Summation of Day Degrees above 10° with 19° Cut-Off			Temperature Variability Index					
PERIOD	Lowest Recorded Min. (61YRS)	No.Mths with <0.3° (30YRS)	No.Mths with <2.5° (30YRS)	Av.Mthly Lowest Min. (30YRS)	Av.Min. Minus Col.4	Daily Mean (56YRS)	Daily Range /2 (56YRS)	Av.Mthly Highest Max. (30YRS)	Col.8 Minus Av.Max.	Raw	Adj.1 for Lat. & Daily Temp Rng	Adj.2 for Vine Sites	Raw	Adj. for Vine Sites	Av. Sunshine Hrs	Av. Number Rain Days	Av. Rainfall mm (92YRS)	Av. % R.H. 1500 HRS (7YRS)
OCTOBER	0.1	0	3	4.1	4.9	14.5	5.5	26.8	6.8	140	137	131	33.7	34.9	215	13	55	64
NOVEMBER	1.7	0	0	5.7	5.1	17.2	6.4	30.7	7.1	215	209	200	37.8	39.0	260	7	23	55
DECEMBER	4.4	0	0	7.0	5.2	19.3	7.1	34.0	7.6	279	267	257	41.2	42.4	300	4	14	54
JANUARY	4.4	0	0	8.0	5.5	20.9	7.4	35.8	7.5	279	279	279	42.6	43.8	298	3	10	51
FEBRUARY	4.3	0	0	7.9	5.7	20.9	7.3	35.6	7.4	252	252	252	42.3	43.5	248	3	11	50
MARCH	2.6	0	0	6.3	6.2	19.3	6.8	33.3	7.2	279	279	271	40.6	41.8	247	4	23	54
APRIL	0.1	0	1	4.9	5.6	16.7	6.2	28.1	5.2	202	205	199	35.6	36.8	205	9	43	62
OCT-APRIL	—	0	4	—	5.5	18.4	6.7	—	7.0	1,646	1,628	1,589	39.1	40.3	1,773	42	179	56
MAY-SEPT	—	—	—	—	—	12.7	—	—	—	—	—	—	—	—	—	—	659	—
DATA FOR MONTH TO MATURITY																		
MATURITY GROUP	EST. MATURITY DATE					*	*	*						*				
1	FEBRUARY 21					20.8	7.6	35.7	7.4					43.6	278	3	12	50
2	FEBRUARY 26					20.8	7.6	35.7	7.4					43.5	270	3	12	50
3	MARCH 4					20.7	7.5	35.5	7.4					43.3	264	3	13	50
4	MARCH 9					20.6	7.5	35.1	7.3					43.0	260	3	14	51
5	MARCH 15					20.3	7.4	34.7	7.3					42.7	256	3	16	52
6	MARCH 21					19.9	7.3	34.3	7.3					42.4	253	4	18	53
7	MARCH 27					19.5	7.2	33.8	7.2					42.0	249	4	21	54
8	APRIL 2					18.9	7.1	33.3	7.1					41.6	245	4	24	55

* Adjusted for vine sites.

TABLE 11

COLLIE, WESTERN AUSTRALIA. 33°21′S, 116°08′E. ALT. 184m.

	1	2	3	4	5	6	7	8	9	10	11	12	13	14	15	16	17	18
	Temperature °C									Summation of Day Degrees above 10° with 19° Cut-Off			Temperature Variability Index					
PERIOD	Lowest Recorded Min. (61YRS)	No.Mths with <0.3° (30YRS)	No.Mths with <2.5° (30YRS)	Av.Mthly Lowest Min. (30YRS)	Av.Min. Minus Col.4	Daily Mean (56YRS)	Daily Range /2 (56YRS)	Av.Mthly Highest Max. (30YRS)	Col.8 Minus Av.Max.	Raw	Adj.1 for Lat. & Daily Temp Rng	Adj.2 for Vine Sites	Raw	Adj. for Vine Sites	Av. Sunshine Hrs	Av. Number Rain Days	Av. Rainfall mm (73YRS)	Av. % R.H. 1500 HRS
OCTOBER	−0.6	3	24	1.4	5.9	13.9	6.6	29.8	9.3	120	117	125	41.6	38.0	225	14	71	50
NOVEMBER	0.3	0	5	3.7	5.9	17.2	7.6	34.3	9.5	215	193	212	45.8	42.2	270	8	29	42
DECEMBER	1.7	0	0	5.8	5.7	19.8	8.3	37.5	9.4	279	264	279	48.3	44.7	312	5	17	37
JANUARY	3.2	0	0	7.3	5.6	21.6	8.7	39.4	9.1	279	279	279	49.5	45.9	312	4	14	34
FEBRUARY	1.8	0	0	7.3	5.4	21.3	8.6	39.0	9.1	252	252	252	48.9	45.3	273	3	14	35
MARCH	0.2	0	1	4.9	6.4	19.3	8.0	36.8	9.5	279	263	279	47.9	44.3	250	5	26	41
APRIL	−1.3	2	13	2.6	5.8	15.9	7.5	30.6	7.2	177	165	185	43.0	39.4	207	10	50	52
OCT-APRIL	—	5	43	—	5.8	18.4	7.9	—	9.0	1,601	1,533	1,611	46.4	42.8	1,849	49	221	42
MAY-SEPT	—	—	—	—	—	11.0	—	—	—	—	—	—	—	—	—	—	767	—
DATA FOR MONTH TO MATURITY																		
MATURITY GROUP	EST. MATURITY DATE					*	*	*						*				
1	FERUARY 18					21.9	7.7	38.5	9.1					45.5	302	3	14	34
2	FEBRUARY 23					21.7	7.7	38.4	9.1					45.4	299	3	15	35
3	MARCH 1					21.5	7.7	38.2	9.1					45.3	294	3	15	35
4	MARCH 6					21.3	7.6	37.9	9.2					45.2	286	3	16	36
5	MARCH 12					21.0	7.5	37.5	9.3					45.0	279	4	18	37
6	MARCH 17					20.6	7.4	37.1	9.3					44.8	271	4	20	38
7	MARCH 22					20.2	7.3	36.6	9.4					44.6	263	4	22	39
8	MARCH 28					19.8	7.2	36.1	9.5					44.4	255	5	25	40

* Adjusted for vine sites.

TABLE 12

DENMARK (Agricultural Research Station), WESTERN AUSTRALIA. 34°56′S, 117°20′E. ALT. 18m.

	1	2	3	4	5	6	7	8	9	10	11	12	13	14	15	16	17	18
	Temperature °C									Summation of Day Degrees above 10° with 19° Cut-Off			Temperature Variability Index					
PERIOD	Lowest Recorded Min. (23YRS)	No.Mths with <0.3° (23YRS)	No.Mths with <2.5° (23YRS)	Av.Mthly Lowest Min. (23YRS)	Av.Min. Minus Col.4	Daily Mean	Daily Range /2	Av.Mthly Highest Max. (23YRS)	Col.8 Minus Av.Max.	Raw	Adj.1 for Lat. & Daily Temp Rng	Adj.2 for Vine Sites	Raw	Adj. for Vine Sites	Av. Sunshine Hrs	Av. Number Rain Days	Av. Rainfall mm	Av. % R.H. 1500 HRS
OCTOBER	1.2	0	17	2.5	6.1	13.8	5.2	27.9	8.9	118	116	127	35.8	31.8	187	18	88	68
NOVEMBER	−1.1	1	4	3.8	6.4	15.8	5.6	31.6	10.2	174	169	174	39.0	35.0	205	13	50	67
DECEMBER	2.3	0	1	5.3	5.8	17.4	6.3	34.6	10.9	229	222	220	41.9	37.9	230	10	37	65
JANUARY	0.6	0	1	5.8	6.2	18.7	6.7	37.9	12.5	270	258	260	45.5	41.5	222	7	21	65
FEBRUARY	4.0	0	0	5.4	6.8	18.7	6.5	36.6	11.4	244	238	237	44.2	40.2	200	8	27	66
MARCH	1.7	0	2	5.1	6.6	17.9	6.2	34.7	10.6	245	244	242	42.0	38.0	195	11	43	67
APRIL	1.7	0	5	3.9	7.0	16.4	5.5	29.7	7.8	192	195	201	36.8	32.8	166	16	88	68
OCT-APRIL	—	1	30	—	6.4	17.0	6.0	—	10.3	1,472	1,442	1,461	40.7	36.7	1,405	83	354	67
MAY-SEPT	—	—	—	—	—	12.3	—	—	—	—	—	—	—	—	—	—	658	—
	DATA FOR MONTH TO MATURITY																	
MATURITY GROUP	EST. MATURITY DATE					*	*	*						*				
1	MARCH 4					18.6	5.5	35.4	11.3					39.9	213	8	30	66
2	MARCH 10					18.5	5.4	35.2	11.2					39.7	210	8	32	66
3	MARCH 17					18.3	5.4	34.9	11.1					39.3	206	9	35	66
4	MARCH 23					18.1	5.3	34.4	10.9					38.8	201	10	39	67
5	MARCH 30					17.9	5.2	33.7	10.6					38.0	195	11	43	67
6	APRIL 6					17.6	5.1	32.7	10.1					37.0	189	12	49	67
7	APRIL 14					17.2	4.9	31.4	9.4					35.8	182	13	59	67
8	APRIL 22					16.7	4.7	30.1	8.6					34.3	175	15	72	68

* Adjusted for vine sites.

TABLE 13

DONNYBROOK, WESTERN AUSTRALIA. 33°33′S, 115°49′E. ALT. 63m.

	1	2	3	4	5	6	7	8	9	10	11	12	13	14	15	16	17	18
	Temperature °C									Summation of Day Degrees above 10° with 19° Cut-Off			Temperature Variability Index					
PERIOD	Lowest Recorded Min. (60YRS)	No.Mths with <0.3° (30YRS)	No.Mths with <2.5° (30YRS)	Av.Mthly Lowest Min. (30YRS)	Av.Min. Minus Col.4	Daily Mean (55YRS)	Daily Range /2 (55YRS)	Av.Mthly Highest Max. (30YRS)	Col.8 Minus Av.Max.	Raw	Adj.1 for Lat. & Daily Temp Rng	Adj.2 for Vine Sites	Raw	Adj. for Vine Sites	Av. Sunshine Hrs	Av. Number Rain Days	Av. Rainfall mm (72YRS)	Av. % R.H. 1500 HRS
OCTOBER	−0.5	3	25	1.8	6.4	14.4	6.2	29.2	8.6	136	134	134	39.8	36.2	225	13	70	44
NOVEMBER	0.4	0	3	3.9	6.3	17.6	7.4	33.6	8.6	227	207	220	44.5	40.9	265	8	30	37
DECEMBER	1.7	0	2	5.4	6.7	20.0	7.9	37.0	9.1	279	276	279	47.4	43.8	310	5	16	32
JANUARY	3.3	0	0	6.4	6.9	21.7	8.4	38.8	8.7	279	279	279	49.2	45.6	310	3	11	31
FEBRUARY	0.8	0	0	7.0	6.5	21.8	8.3	38.6	8.6	252	252	252	48.2	44.6	260	3	15	31
MARCH	0.6	0	2	4.6	7.5	19.7	7.6	36.8	9.5	279	279	279	47.4	43.8	250	5	28	35
APRIL	−0.2	1	9	3.3	6.5	17.0	7.2	31.1	6.9	211	203	214	42.2	38.6	210	9	51	44
OCT-APRIL	—	4	41	—	6.7	18.9	7.6	—	8.6	1,663	1,630	1,657	45.5	41.9	1,830	46	221	36
MAY-SEPT	—	—	—	—	—	12.1	—	—	—	—	—	—	—	—	—	—	798	—
	DATA FOR MONTH TO MATURITY																	
MATURITY GROUP	EST. MATURITY DATE					*	*	*						*				
1	FEBRUARY 16					22.0	7.5	37.8	8.6					45.0	293	3	14	31
2	FEBRUARY 21					22.0	7.4	37.8	8.6					44.8	288	3	15	31
3	FEBRUARY 27					21.8	7.4	37.7	8.6					44.6	282	3	16	31
4	MARCH 4					21.6	7.3	37.5	8.7					44.4	276	3	18	31
5	MARCH 10					21.3	7.2	37.2	8.8					44.3	271	4	20	32
6	MARCH 16					20.9	7.0	36.9	9.0					44.2	265	4	22	33
7	MARCH 21					20.5	6.8	36.5	9.2					44.1	260	4	24	34
8	MARCH 27					20.0	6.7	36.1	9.4					44.0	254	5	27	35

* Adjusted for vine sites.

TABLE 14

ESPERANCE, WESTERN AUSTRALIA. 33°50'S, 121°55'E. ALT. 4m.

	1	2	3	4	5	6	7	8	9	10	11	12	13	14	15	16	17	18
	Temperature °C									Summation of Day Degrees above 10° with 19° Cut-Off			Temperature Variability Index					
PERIOD	Lowest Recorded Min. (73YRS)	No.Mths with <0.3° (30YRS)	No.Mths with <2.5° (30YRS)	Av.Mthly Lowest Min. (30YRS)	Av.Min. Minus Col.4	Daily Mean (30YRS)	Daily Range /2 (30YRS)	Av.Mthly Highest Max. (30YRS)	Col.8 Minus Av.Max.	Raw	Adj.1 for Lat. & Daily Temp Rng	Adj.2 for Vine Sites	Raw	Adj. for Vine Sites	Av. Sunshine Hrs	Av. Number Rain Days	Av. Rainfall mm	Av. % R.H. 1500 HRS (44YRS)
OCTOBER	1.1	0	3	4.6	5.6	15.2	5.0	32.2	12.0	161	158	137	37.6	40.0	253	12	55	64
NOVEMBER	3.3	0	0	5.9	6.5	17.3	4.9	35.3	13.1	219	213	191	39.2	41.6	278	7	28	64
DECEMBER	4.4	0	0	8.3	5.9	18.9	4.7	37.5	13.9	276	269	243	38.6	41.0	302	6	23	63
JANUARY	4.9	0	0	9.6	5.9	20.1	4.6	39.7	15.0	279	279	279	39.3	41.7	288	5	18	63
FEBRUARY	4.9	0	0	9.2	6.5	20.5	4.8	37.9	12.6	252	252	252	38.3	40.7	240	5	20	64
MARCH	3.9	0	0	7.9	7.0	19.5	4.6	37.3	13.2	279	279	273	38.6	41.0	230	7	28	64
APRIL	3.3	0	0	6.1	6.4	17.4	4.9	32.3	10.0	222	227	205	36.0	38.4	200	10	45	65
OCT-APRIL	—	0	3	—	6.3	18.4	4.8	—	12.8	1,688	1,677	1,580	38.2	40.6	1,791	52	217	64
MAY-SEPT	—	—	—	—	—	13.3	—	—	—	—	—	—	—	—	—	—	458	—

		DATA FOR MONTH TO MATURITY																
MATURITY GROUP	EST. MATURITY DATE					*	*	*						*				*
1	FEBRUARY 23					19.9	5.5	38.4	13.3					41.1	265	5	19	61
2	FEBRUARY 28					19.8	5.5	38.1	13.0					41.0	260	5	20	61
3	MARCH 5					19.7	5.5	37.9	12.9					41.0	255	5	21	61
4	MARCH 11					19.6	5.4	37.7	12.9					41.0	250	5	22	61
5	MARCH 16					19.4	5.4	37.6	12.8					41.0	245	6	23	61
6	MARCH 22					19.2	5.3	37.5	12.8					40.9	239	6	25	61
7	MARCH 28					18.9	5.3	37.3	12.8					40.8	233	7	27	61
8	APRIL 4					18.5	5.3	36.9	12.6					40.5	226	7	31	61

* Adjusted for vine sites.

TABLE 15

GUILDFORD, WESTERN AUSTRALIA. 31°53'S, 116°01'E. ALT. 8m. (For the Swan Valley, Gingin, Bindoon)

	1	2	3	4	5	6	7	8	9	10	11	12	13	14	15	16	17	18
	Temperature °C									Summation of Day Degrees above 10° with 19° Cut-Off			Temperature Variability Index					
PERIOD	Lowest Recorded Min. (70YRS)	No.Mths with <0.3° (26YRS)	No.Mths with <2.5° (26YRS)	Av.Mthly Lowest Min. (26YRS)	Av.Min. Minus Col.4	Daily Mean (46YRS)	Daily Range /2 (46YRS)	Av.Mthly Highest Max. (26YRS)	Col.8 Minus Av.Max.	Raw	Adj.1 for Lat. & Daily Temp Rng	Adj.2 for Vine Sites	Raw	Adj. for Vine Sites	Av. Sunshine Hrs	Av. Number Rain Days	Av. Rainfall mm (78YRS)	Av. % R.H. 1500 HRS
OCTOBER	1.6	0	1	5.0	5.3	16.5	6.2	31.8	9.1	202	197		39.2		248	11	56	52
NOVEMBER	3.6	0	0	6.5	6.3	19.8	7.0	35.5	8.7	270	270		43.0		290	8	20	45
DECEMBER	5.2	0	0	8.6	6.5	22.4	7.3	38.5	8.8	279	279		44.5		328	4	13	41
JANUARY	6.0	0	0	9.9	6.7	24.1	7.5	40.5	8.9	279	279		45.6		325	3	8	40
FEBRUARY	5.6	0	0	10.6	6.1	24.3	7.6	39.9	8.0	252	252		44.5		280	2	10	39
MARCH	5.5	0	0	8.6	6.7	22.5	7.2	38.4	8.7	279	279		44.2		270	4	18	42
APRIL	1.1	0	2	5.7	7.2	19.7	6.8	32.5	6.0	270	270		40.4		221	8	43	48
OCT-APRIL	—	0	3	—	6.4	21.3	7.1	—	8.3	1,831	1,826		43.1		1,962	40	168	44
MAY-SEPT	—	—	—	—	—	14.0	—	—	—	—	—		—		—	—	697	—

		DATA FOR MONTH TO MATURITY																
MATURITY GROUP	EST. MATURITY DATE					*	*	*					*					
1	FEBRUARY 3					24.2	7.5	40.5	8.8				45.5		323	3	8	40
2	FEBRUARY 9					24.4	7.5	40.4	8.7				45.3		320	3	9	40
3	FEBRUARY 14					24.5	7.5	40.3	8.5				45.1		315	2	9	39
4	FEBRUARY 20					24.5	7.6	40.2	8.3				44.9		310	2	10	39
5	FEBRUARY 25					24.4	7.6	40.0	8.1				44.7		304	2	11	39
6	MARCH 3					24.2	7.6	39.8	8.1				44.5		299	2	12	39
7	MARCH 9					24.0	7.5	39.4	8.2				44.4		293	3	13	40
8	MARCH 14					23.7	7.4	39.2	8.4				44.3		287	3	14	40

* Adjusted for vine sites.

TABLE 16

KALAMUNDA, WESTERN AUSTRALIA. 31°59'S, 116°04'E. ALT. 282m. (For Perth Hills)

	1	2	3	4	5	6	7	8	9	10	11	12	13	14	15	16	17	18
	Temperature °C									Summation of Day Degrees above 10° with 19° Cut-Off			Temperature Variability Index					
PERIOD	Lowest Recorded Min. (60YRS)	No.Mths with <0.3° (30YRS)	No.Mths with <2.5° (30YRS)	Av.Mthly Lowest Min. (30YRS)	Av.Min. Minus Col.4	Daily Mean (38YRS)	Daily Range /2 (38YRS)	Av.Mthly Highest Max. (30YRS)	Col.8 Minus Av.Max.	Raw	Adj.1 for Lat. & Daily Temp Rng	Adj.2 for Vine Sites	Raw	Adj. for Vine Sites	Av. Sunshine Hrs	Av. Number Rain Days	Av. Rainfall mm (75YRS)	Av. % R.H. 1500 HRS
OCTOBER	3.3	0	0	5.4	4.7	15.4	5.3	29.9	9.2	167	164		35.1		246	11	70	60
NOVEMBER	3.3	0	0	7.6	5.1	18.8	6.1	34.5	9.6	264	252		39.1		288	6	27	55
DECEMBER	6.1	0	0	9.5	5.3	21.5	6.7	37.5	9.3	279	279		41.4		326	4	20	47
JANUARY	5.7	0	0	10.7	5.3	23.1	7.1	39.2	9.0	279	279		42.7		325	2	10	45
FEBRUARY	7.2	0	0	11.1	5.2	23.2	6.9	39.1	9.0	252	252		41.8		277	2	14	44
MARCH	5.0	0	0	10.2	5.0	21.3	6.1	36.6	9.2	279	279		38.6		270	4	23	47
APRIL	3.9	0	0	7.8	5.7	18.7	5.2	31.1	7.2	261	267		33.7		217	7	56	61
OCT-APRIL	—	0	0	—	5.2	20.3	6.2	—	8.9	1,781	1,772		38.9		1,949	36	220	51
MAY-SEPT	—	—	—	—	—	13.1	—	—	—	—	—		—		—	—	849	—

		DATA FOR MONTH TO MATURITY																
MATURITY GROUP	EST. MATURITY DATE																	
1	FEBRUARY 9					23.4	7.0	39.2	9.0				42.4		318	2	12	45
2	FEBRUARY 14					23.4	7.0	39.2	9.0				42.2		314	2	13	44
3	FEBRUARY 20					23.4	6.9	39.1	9.0				42.0		309	2	14	44
4	FEBRUARY 26					23.3	6.9	39.1	9.0				41.9		303	2	15	44
5	MARCH 3					23.1	6.8	39.0	9.0				41.5		297	2	16	44
6	MARCH 9					22.8	6.7	38.7	9.1				41.0		292	2	17	45
7	MARCH 14					22.5	6.6	38.2	9.1				40.4		287	3	18	45
8	MARCH 20					22.1	6.4	37.5	9.1				39.7		281	3	20	46

TABLE 17

KARRIDALE, WESTERN AUSTRALIA. 34°12'S, 115°03'E. ALT. 48m.

	1	2	3	4	5	6	7	8	9	10	11	12	13	14	15	16	17	18
	Temperature °C									Summation of Day Degrees above 10° with 19° Cut-Off			Temperature Variability Index					
PERIOD	Lowest Recorded Min.	No.Mths with <0.3°	No.Mths with <2.5°	Av.Mthly Lowest Min.	Av.Min. Minus Col.4	Daily Mean (34YRS)	Daily Range /2 (34YRS)	Av.Mthly Highest Max.	Col.8 Minus Av.Max.	Raw	Adj.1 for Lat. & Daily Temp Rng	Adj.2 for Vine Sites	Raw	Adj. for Vine Sites	Av. Sunshine Hrs	Av. Number Rain Days	Av. Rainfall mm (56YRS)	Av. % R.H. 1500 HRS
OCTOBER				4.8	4.6	13.9	4.5	26.5	8.1	121	127		30.7		188		84	67
NOVEMBER				6.2	4.7	16.1	5.2	30.6	9.3	183	178		34.8		218		38	61
DECEMBER				7.5	5.0	17.9	5.4	33.9	10.6	245	236		37.2		248		27	57
JANUARY				8.3	5.3	19.1	5.5	34.8	10.2	279	272		37.5		250		19	55
FEBRUARY				7.8	5.7	19.1	5.6	34.3	9.6	252	249		37.7		225		22	55
MARCH				6.6	5.9	18.1	5.6	33.4	9.7	251	250		38.0		220		36	59
APRIL				5.1	5.7	16.4	5.6	29.5	7.5	192	195		35.6		175		70	66
OCT-APRIL		—	—	—	5.3	17.2	5.3	—	9.3	1,523	1,507		35.9		1,524		296	60
MAY-SEPT		—	—	—	—	12.9	—	—	—	—	—		—		—		904	—

		DATA FOR MONTH TO MATURITY																
MATURITY GROUP	EST. MATURITY DATE																	
1	FEBRUARY 27					19.1	5.6	34.3	9.6				37.7		243		24	55
2	MARCH 5					19.0	5.6	34.2	9.6				37.7		240		25	55
3	MARCH 11					18.8	5.6	34.0	9.6				37.8		236		26	56
4	MARCH 17					18.6	5.6	33.8	9.7				37.9		232		28	57
5	MARCH 23					18.4	5.6	33.6	9.7				37.9		227		31	58
6	MARCH 30					18.1	5.6	33.4	9.7				38.0		221		35	59
7	APRIL 6					17.7	5.6	33.0	9.5				37.7		214		41	60
8	APRIL 13					17.4	5.6	32.0	8.9				37.1		205		50	62

TABLE 18

KATANNING, WESTERN AUSTRALIA. 33°42'S, 117°35'E. ALT. 312m.

	1	2	3	4	5	6	7	8	9	10	11	12	13	14	15	16	17	18
	Temperature °C									Summation of Day Degrees above 10° with 19° Cut-Off			Temperature Variability Index					
PERIOD	Lowest Recorded Min. (74YRS)	No.Mths with <0.3° (30YRS)	No.Mths with <2.5° (30YRS)	Av.Mthly Lowest Min. (30YRS)	Av.Min. Minus Col.4	Daily Mean (44YRS)	Daily Range /2 (44YRS)	Av.Mthly Highest Max. (30YRS)	Col.8 Minus Av.Max.	Raw	Adj.1 for Lat. & Daily Temp Rng	Adj.2 for Vine Sites	Raw	Adj. for Vine Sites	Av. Sunshine Hrs	Av. Number Rain Days	Av. Rainfall mm (77YRS)	Av. % R.H. 1500 HRS (56YRS)
OCTOBER	−0.6	1	22	2.1	5.5	14.0	6.4	30.3	9.9	124	122	125	41.0	39.8	235	10	38	51
NOVEMBER	1.7	0	1	4.3	5.6	17.6	7.7	34.6	9.3	228	203	210	45.7	44.5	280	6	20	40
DECEMBER	3.1	0	0	6.7	5.3	20.3	8.3	37.9	9.6	279	279	279	47.8	46.6	310	4	17	36
JANUARY	5.0	0	0	8.3	5.0	21.7	8.4	39.8	9.7	279	279	279	48.3	47.1	310	3	11	35
FEBRUARY	3.3	0	0	8.4	4.9	21.4	8.1	39.1	9.6	252	252	252	46.9	45.7	260	4	16	37
MARCH	1.7	0	0	6.6	5.6	19.3	7.1	36.5	10.3	279	277	279	44.1	42.9	230	5	25	43
APRIL	0.6	0	5	4.2	5.8	16.4	6.4	30.5	7.6	192	195	199	39.1	37.9	203	7	31	49
OCT-APRIL	—	1	28	—	5.4	18.7	7.5	—	9.4	1,633	1,607	1,623	44.7	43.5	1,828	39	158	42
MAY-SEPT	—	—	—	—	—	11.0	—	—	—	—	—	—	—	—	—	—	332	—

		DATA FOR MONTH TO MATURITY						
MATURITY GROUP	EST. MATURITY DATE	*	*	*			*	
1	FEBRUARY 18	21.8	7.9	39.2	9.6	46.2	293	4	15	36
2	FEBRUARY 23	21.7	7.8	39.0	9.6	46.0	290	4	16	36
3	MARCH 1	21.5	7.7	38.7	9.6	45.7	285	4	17	37
4	MARCH 7	21.2	7.5	38.3	9.7	45.3	269	4	18	38
5	MARCH 12	20.9	7.3	37.8	9.9	44.9	262	4	19	39
6	MARCH 18	20.5	7.1	37.2	10.1	44.4	254	5	20	40
7	MARCH 23	20.1	7.0	36.5	10.2	43.8	244	5	22	41
8	MARCH 29	19.6	6.9	35.8	10.3	43.1	233	5	24	43

* Adjusted for vine sites.

TABLE 19

KOJONUP, WESTERN AUSTRALIA. 33°50'S, 117°09'E. ALT. 305m.

	1	2	3	4	5	6	7	8	9	10	11	12	13	14	15	16	17	18
	Temperature °C									Summation of Day Degrees above 10° with 19° Cut-Off			Temperature Variability Index					
PERIOD	Lowest Recorded Min. (15YRS)	No.Mths with <0.3° (15YRS)	No.Mths with <2.5° (15YRS)	Av.Mthly Lowest Min. (15YRS)	Av.Min. Minus Col.4	Daily Mean (12YRS)	Daily Range /2 (12YRS)	Av.Mthly Highest Max. (15YRS)	Col.8 Minus Av.Max.	Raw	Adj.1 for Lat. & Daily Temp Rng	Adj.2 for Vine Sites	Raw	Adj. for Vine Sites	Av. Sunshine Hrs	Av. Number Rain Days	Av. Rainfall mm (43YRS)	Av. % R.H. 1500 HRS
OCTOBER	0.8	0	11	2.0	5.1	13.3	6.2	28.2	8.7	102	100	105	38.6	36.8	225	11	43	49
NOVEMBER	3.1	0	0	4.4	4.8	16.6	7.3	33.6	9.7	197	179	191	43.8	42.0	263	7	22	40
DECEMBER	3.1	0	0	5.9	5.5	19.2	7.8	37.8	10.8	279	253	265	47.5	45.7	304	4	16	34
JANUARY	5.0	0	0	7.3	6.3	21.7	8.1	39.1	9.3	279	279	279	48.0	46.2	298	3	12	35
FEBRUARY	5.0	0	0	8.2	5.4	21.1	7.4	38.0	9.5	252	252	252	44.6	42.8	254	3	15	35
MARCH	5.8	0	0	7.2	5.4	19.4	6.9	36.3	10.0	279	279	279	42.9	41.1	229	5	24	45
APRIL	2.2	0	0	5.0	5.2	15.7	5.6	29.9	8.6	172	174	179	36.1	34.3	198	8	32	53
OCT-APRIL	—	0	13	—	5.4	18.1	7.0	—	9.5	1,560	1,516	1,550	43.1	41.3	1,771	41	164	42
MAY-SEPT	—	—	—	—	—	10.8	—	—	—	—	—	—	—	—	—	—	383	—

		DATA FOR MONTH TO MATURITY						
MATURITY GROUP	EST. MATURITY DATE	*	*	*			*	
1	FEBRUARY 24	21.4	7.2	37.9	9.5	43.3	279	3	14	35
2	MARCH 1	21.3	7.0	37.7	9.5	42.8	275	3	15	35
3	MARCH 7	21.0	6.9	37.4	9.6	42.4	269	3	16	36
4	MARCH 12	20.7	6.8	37.1	9.6	42.1	261	4	17	37
5	MARCH 18	20.4	6.7	36.8	9.7	41.8	251	4	19	39
6	MARCH 23	20.1	6.6	36.5	9.8	41.5	240	4	21	42
7	MARCH 29	19.7	6.5	36.1	9.9	41.2	232	5	23	44
8	APRIL 3	19.3	6.4	35.6	9.8	39.8	225	5	25	46

* Adjusted for vine sites.

TABLE 20

MANDURAH, WESTERN AUSTRALIA. 32°31'S, 115°03'E. ALT. 48m.

	1	2	3	4	5	6	7	8	9	10	11	12	13	14	15	16	17	18
	Temperature °C									Summation of Day Degrees above 10° with 19° Cut-Off			Temperature Variability Index					
PERIOD	Lowest Recorded Min. (61YRS)	No.Mths with <0.3° (30YRS)	No.Mths with <2.5° (30YRS)	Av.Mthly Lowest Min. (30YRS)	Av.Min. Minus Col.4	Daily Mean (56YRS)	Daily Range /2 (56YRS)	Av.Mthly Highest Max. (30YRS)	Col.8 Minus Av.Max.	Raw	Adj.1 for Lat. & Daily Temp Rng	Adj.2 for Vine Sites	Raw	Adj. for Vine Sites	Av. Sunshine Hrs	Av. Number Rain Days	Av. Rainfall mm (79YRS)	Av. % R.H. 1500 HRS
OCTOBER	2.1	0	0	5.8	4.8	15.8	5.2	29.1	8.1	178	176		33.7		240	11	55	61
NOVEMBER	2.7	0	0	8.1	4.8	18.7	5.8	33.2	8.7	261	251		36.7		285	6	20	55
DECEMBER	3.9	0	0	10.3	4.8	21.0	5.9	36.2	9.3	279	279		37.7		320	3	12	50
JANUARY	5.2	0	0	11.4	5.1	22.8	6.3	38.1	9.0	279	279		39.3		320	2	7	47
FEBRUARY	5.3	0	0	11.7	4.8	22.8	6.3	37.5	8.4	252	252		38.4		270	2	12	47
MARCH	3.8	0	0	10.5	4.8	21.4	6.1	35.7	8.2	279	279		37.4		262	4	20	49
APRIL	2.7	0	0	7.1	5.5	18.5	5.9	30.3	5.9	255	260		35.0		215	8	45	57
OCT-APRIL	—	0	0	—	4.9	20.1	5.9	—	8.2	1,783	1,776		36.9		1,912	36	171	52
MAY-SEPT	—	—	—	—	—	13.7	—	—	—	—	—		—		—	—	726	—
	DATA FOR MONTH TO MATURITY																	
MATURITY GROUP	EST. MATURITY DATE																	
1	FEBRUARY 8					22.9	6.3	37.9	8.8				39.1		312	2	9	47
2	FEBRUARY 13					23.0	6.3	37.9	8.7				39.0		307	2	9	47
3	FEBRUARY 19					23.0	6.3	37.8	8.6				38.8		301	2	10	47
4	FEBRUARY 24					22.9	6.3	37.6	8.5				38.6		295	2	11	47
5	MARCH 2					22.8	6.3	37.4	8.4				38.4		290	2	12	47
6	MARCH 7					22.7	6.3	37.1	8.4				38.2		284	3	13	48
7	MARCH 13					22.5	6.2	36.7	8.3				38.0		279	3	15	48
8	MARCH 19					22.3	6.2	36.3	8.2				37.8		273	3	17	48

TABLE 21

MANJIMUP, WESTERN AUSTRALIA. 34°14'S, 116°09'E. ALT. 279m. (Notes also for Nannup-Carlotta)

	1	2	3	4	5	6	7	8	9	10	11	12	13	14	15	16	17	18
	Temperature °C									Summation of Day Degrees above 10° with 19° Cut-Off			Temperature Variability Index					
PERIOD	Lowest Recorded Min. (30YRS)	No.Mths with <0.3° (30YRS)	No.Mths with <2.5° (30YRS)	Av.Mthly Lowest Min. (30YRS)	Av.Min. Minus Col.4	Daily Mean (19YRS)	Daily Range /2 (19YRS)	Av.Mthly Highest Max. (30YRS)	Col.8 Minus Av.Max.	Raw	Adj.1 for Lat. & Daily Temp Rng	Adj.2 for Vine Sites	Raw	Adj. for Vine Sites	Av. Sunshine Hrs	Av. Number Rain Days	Av. Rainfall mm	Av. % R.H. 1500 HRS #
OCTOBER	0.1	1	12	2.8	5.0	12.9	5.1	27.2	9.2	90	88	108	34.6	32.2	190	14	81	58
NOVEMBER	1.7	0	3	4.7	4.8	15.4	5.9	31.4	10.1	162	157	169	38.5	36.1	225	10	45	49
DECEMBER	4.4	0	0	6.2	4.6	17.3	6.5	34.7	10.9	226	218	230	41.5	39.1	250	7	26	43
JANUARY	5.6	0	0	7.5	4.8	19.2	6.9	36.6	10.5	279	269	279	42.9	40.5	250	5	20	42
FEBRUARY	4.4	0	0	7.3	5.2	19.4	6.9	36.2	9.9	252	251	252	42.7	40.3	230	5	19	44
MARCH	3.3	0	0	6.1	5.8	18.0	6.1	34.6	10.5	248	247	259	40.7	38.3	213	7	33	49
APRIL	1.7	0	6	4.5	5.7	15.5	5.3	28.8	8.0	165	168	185	34.9	32.5	175	11	64	61
OCT-APRIL	—	1	21	—	5.1	16.8	6.1	—	9.9	1,422	1,398	1,482	39.4	37.0	1,533	59	288	49
MAY-SEPT	—	—	—	—	—	11.1	—	—	—	—	—	—	—	—	—	—	767	—
	DATA FOR MONTH TO MATURITY																	
MATURITY GROUP	EST. MATURITY DATE					*	*	*						*				#
1	MARCH 2					19.8	6.2	36.0	10.0					40.2	247	5	22	44
2	MARCH 7					19.7	6.1	35.7	10.1					39.9	241	6	23	45
3	MARCH 13					19.6	6.0	35.5	10.2					39.6	234	6	25	46
4	MARCH 19					19.3	5.9	35.1	10.3					39.2	227	6	27	47
5	MARCH 25					19.0	5.8	34.8	10.4					38.7	220	7	30	48
6	APRIL 1					18.6	5.6	34.3	10.3					38.0	212	7	34	49
7	APRIL 8					18.1	5.4	33.7	10.0					37.0	203	8	40	51
8	APRIL 16					17.3	5.0	32.4	9.4					35.7	193	8	48	54

* Adjusted for vine sites. # Raw data from "Climatic Averages Australia", 1975. All figures period-adjusted by +2%.

TABLE 22

MARGARET RIVER, WESTERN AUSTRALIA. 33°57'S, 115°04'E. ALT. 90m.

	1	2	3	4	5	6	7	8	9	10	11	12	13	14	15	16	17	18
	Temperature °C									Summation of Day Degrees above 10° with 19° Cut-Off			Temperature Variability Index					
PERIOD	Lowest Recorded Min. (6YRS)	No.Mths with <0.3° (6YRS)	No.Mths with <2.5° (6YRS)	Av.Mthly Lowest Min. (6YRS)	Av.Min. Minus Col.4	Daily Mean	Daily Range /2	Av.Mthly Highest Max. (6YRS)	Col.8 Minus Av.Max.	Raw	Adj.1 for Lat. & Daily Temp Rng	Adj.2 for Vine Sites	Raw	Adj. for Vine Sites	Av. Sunshine Hrs	Av. Number Rain Days	Av. Rainfall mm	Av. % R.H. 1500 HRS
OCTOBER	1.7	0	2	3.4	4.9	13.5	5.1	25.0	6.5	108	107	120	31.8	29.4	194	14	76	61
NOVEMBER	4.2	0	0	5.0	5.0	15.8	5.7	29.8	8.4	174	169	175	36.2	33.8	232	9	40	55
DECEMBER	3.5	0	0	6.2	5.2	18.0	6.6	34.2	9.6	248	237	244	41.2	38.8	272	5	22	54
JANUARY	6.6	0	0	7.8	5.5	20.0	6.7	36.1	9.4	279	279	279	41.7	39.3	274	4	14	53
FEBRUARY	6.3	0	0	7.8	5.8	20.3	6.7	36.0	9.0	252	252	252	41.6	39.2	238	4	12	52
MARCH	6.0	0	0	7.0	5.6	18.8	6.2	34.0	9.0	273	271	277	39.4	37.0	226	6	34	56
APRIL	4.0	0	0	5.6	5.6	16.5	5.2	28.9	7.3	195	198	210	33.7	31.3	190	10	76	61
OCT-APRIL	—	0	2	—	5.4	17.6	6.0	—	8.5	1,529	1,513	1,557	37.9	35.5	1,626	52	274	56
MAY-SEPT	—	—	—	—	—	13.0	—	—	—	—	—	—	—	—	—	—	918	—

DATA FOR MONTH TO MATURITY

MATURITY GROUP	EST. MATURITY DATE				6*	7*	8*	9					14*	15	16	17	18
1	FEBRUARY 26				20.5	6.1	35.6	9.0					39.2	258	4	13	52
2	MARCH 4				20.4	6.1	35.4	9.0					39.1	255	4	14	52
3	MARCH 9				20.2	6.0	35.1	9.0					38.8	251	4	16	53
4	MARCH 15				20.0	5.9	34.7	9.0					38.4	245	5	19	53
5	MARCH 21				19.7	5.8	34.3	9.0					37.9	239	5	23	54
6	MARCH 26				19.4	5.7	33.8	9.0					37.4	232	6	28	55
7	APRIL 1				19.0	5.6	33.2	8.9					36.8	224	6	35	56
8	APRIL 8				18.4	5.5	32.4	8.6					36.0	216	7	43	58

* Adjusted for vine sites.

TABLE 23

MOUNT BARKER, WESTERN AUSTRALIA. 34°36'S, 117°38'E. ALT. 253m.

	1	2	3	4	5	6	7	8	9	10	11	12	13	14	15	16	17	18
	Temperature °C									Summation of Day Degrees above 10° with 19° Cut-Off			Temperature Variability Index					
PERIOD	Lowest Recorded Min. (63YRS)	No.Mths with <0.3° (30YRS)	No.Mths with <2.5° (30YRS)	Av.Mthly Lowest Min. (30YRS)	Av.Min. Minus Col.4	Daily Mean (50YRS)	Daily Range /2 (50YRS)	Av.Mthly Highest Max. (30YRS)	Col.8 Minus Av.Max.	Raw	Adj.1 for Lat. & Daily Temp Rng	Adj.2 for Vine Sites	Raw	Adj. for Vine Sites	Av. Sunshine Hrs	Av. Number Rain Days	Av. Rainfall mm (75YRS)	Av. % R.H. 1500 HRS
OCTOBER	0.6	0	8	3.3	4.5	13.1	5.3	29.0	10.5	96	95	107	36.3	35.1	208	16	74	62
NOVEMBER	1.1	0	0	4.8	4.8	15.6	6.0	32.9	11.1	168	163	175	40.1	38.9	225	11	41	57
DECEMBER	1.1	0	0	6.3	4.8	17.6	6.5	36.7	12.6	236	227	239	43.4	42.2	245	10	30	53
JANUARY	1.7	0	0	7.3	5.1	19.0	6.6	38.8	13.2	279	268	279	44.7	43.5	235	8	22	49
FEBRUARY	3.9	0	0	7.3	7.3	19.0	6.7	37.9	12.1	252	243	252	44.0	42.8	215	7	24	48
MARCH	3.6	0	0	6.3	6.3	17.7	5.8	35.8	12.4	239	238	250	41.1	39.9	205	10	37	51
APRIL	2.2	0	1	5.0	5.0	15.7	5.4	29.7	8.8	171	174	186	35.5	34.3	185	13	57	59
OCT-APRIL	—	0	9	—	5.0	16.8	6.0	—	11.5	1,441	1,408	1,488	40.7	39.5	1,518	75	285	54
MAY-SEPT	—	—	—	—	—	11.1	—	—	—	—	—	—	—	—	—	—	471	—

DATA FOR MONTH TO MATURITY

MATURITY GROUP	EST. MATURITY DATE				6*	7*	8*	9					14*	15	16	17	18
1	FEBRUARY 28				19.4	6.4	38.0	12.1					42.8	232	7	26	48
2	MARCH 6				19.3	6.3	37.8	12.1					42.5	230	7	27	48
3	MARCH 12				19.1	6.1	37.5	12.2					42.0	225	8	29	49
4	MARCH 18				18.8	5.9	37.1	12.3					41.4	218	8	31	49
5	MARCH 24				18.5	5.7	36.6	12.4					40.7	211	9	34	50
6	MARCH 31				18.1	5.5	35.9	12.4					39.9	205	10	37	51
7	APRIL 8				17.6	5.4	34.8	11.9					38.8	200	11	41	53
8	APRIL 16				17.0	5.3	33.0	11.0					37.4	195	12	46	55

* Adjusted for vine sites.

TABLE 24

NARROGIN (Agricultural School), WESTERN AUSTRALIA. 32°55'S, 117°09'E. ALT. 340m.

	1	2	3	4	5	6	7	8	9	10	11	12	13	14	15	16	17	18
	Temperature °C									Summation of Day Degrees above 10° with 19° Cut-Off			Temperature Variability Index					
PERIOD	Lowest Recorded Min. (55YRS)	No.Mths with <0.3° (30YRS)	No.Mths with <2.5° (30YRS)	Av.Mthly Lowest Min. (30YRS)	Av.Min. Minus Col.4	Daily Mean (39YRS)	Daily Range /2 (39YRS)	Av.Mthly Highest Max. (30YRS)	Col.8 Minus Av.Max.	Raw	Adj.1 for Lat. & Daily Temp Rng	Adj.2 for Vine Sites	Raw	Adj. for Vine Sites	Av. Sunshine Hrs	Av. Number Rain Days	Av. Rainfall mm (77YRS)	Av. % R.H. 1500 HRS
OCTOBER	−1.7	8	25	1.2	5.8	13.8	6.8	30.1	9.4	119	111	122	42.5	40.1	243	9	34	52
NOVEMBER	0.0	2	10	3.2	6.2	17.5	8.1	35.4	9.5	226	193	208	48.4	46.0	290	4	15	42
DECEMBER	1.8	0	1	5.4	6.3	20.3	8.6	38.2	9.2	279	279	279	50.0	47.6	330	3	13	38
JANUARY	4.3	0	0	7.5	6.1	22.2	8.6	39.4	8.7	279	279	279	49.1	46.7	322	2	9	35
FEBRUARY	3.9	0	0	7.7	5.9	21.9	8.3	39.5	9.2	252	252	252	48.4	46.0	280	3	17	36
MARCH	3.3	0	0	6.3	6.1	19.8	7.4	36.4	9.4	279	279	279	44.9	42.5	245	4	22	45
APRIL	−0.4	2	9	3.4	6.6	16.5	6.5	30.8	7.6	194	199	199	40.4	38.0	215	6	29	54
OCT–APRIL	—	12	45	—	6.1	18.9	7.8	—	9.0	1,628			46.2	43.8	1,925	31	139	43
MAY–SEPT	—	—	—	—	—	11.0	—	—	—	—			—	—	—	—	368	—
	DATA FOR MONTH TO MATURITY																	
MATURITY GROUP	EST. MATURITY DATE			*	*	*								*				
1	FEBRUARY 19					22.4	7.8	39.0	9.1					46.2	311	2	15	36
2	FEBRUARY 24					22.3	7.8	39.1	9.1					46.1	307	3	16	36
3	MARCH 2					22.0	7.7	39.1	9.2					46.0	302	3	17	36
4	MARCH 7					21.7	7.6	38.9	9.2					45.8	294	3	18	37
5	MARCH 13					21.4	7.5	38.5	9.2					45.4	284	3	19	38
6	MARCH 18					21.0	7.3	38.0	9.3					44.8	272	4	20	40
7	MARCH 23					20.6	7.1	37.3	9.3					43.9	260	4	21	42
8	MARCH 29					20.2	6.9	36.4	9.4					42.9	250	4	22	44

* Adjusted for vine sites.

TABLE 25

PEMBERTON, WESTERN AUSTRALIA. 34°27'S, 116°01'E. ALT. 171m.

	1	2	3	4	5	6	7	8	9	10	11	12	13	14	15	16	17	18
	Temperature °C									Summation of Day Degrees above 10° with 19° Cut-Off			Temperature Variability Index					
PERIOD	Lowest Recorded Min. (26YRS)	No.Mths with <0.3° (26YRS)	No.Mths with <2.5° (26YRS)	Av.Mthly Lowest Min. (26YRS)	Av.Min. Minus Col.4	Daily Mean (11YRS)	Daily Range /2 (11YRS)	Av.Mthly Highest Max. (26YRS)	Col.8 Minus Av.Max.	Raw	Adj.1 for Lat. & Daily Temp Rng	Adj.2 for Vine Sites	Raw	Adj. for Vine Sites	Av. Sunshine Hrs	Av. Number Rain Days	Av. Rainfall mm	Av. % R.H. 1500 HRS #
OCTOBER	1.4	0	8	3.3	4.3	12.6	5.0	27.3	9.7	81	79	90	34.0	32.8	185	16	94	65
NOVEMBER	2.1	0	2	5.0	4.3	15.0	5.7	30.9	10.2	150	146	152	37.3	36.1	206	12	56	58
DECEMBER	3.9	0	0	6.3	4.8	16.8	5.7	34.2	11.7	211	203	209	39.3	38.1	228	9	39	52
JANUARY	4.4	0	0	7.4	5.2	19.2	6.6	36.8	11.0	279	276	279	42.6	41.4	238	6	22	48
FEBRUARY	4.4	0	0	7.4	5.6	19.3	6.3	35.7	10.1	252	252	252	40.9	39.7	214	6	17	49
MARCH	3.9	0	0	6.4	6.2	18.3	5.7	34.4	10.4	256	256	262	39.4	38.2	195	8	43	55
APRIL	2.8	0	0	4.7	5.7	15.5	5.1	28.4	7.8	165	167	177	33.9	32.7	157	13	90	63
OCT–APRIL	—	0	10	—	5.2	16.7	5.7	—	10.1	1,394	1,379	1,421	38.2	37.0	1,423	70	361	56
MAY–SEPT	—	—	—	—	—	11.5	—	—	—	—	—	—	—	—	—	—	894	—
	DATA FOR MONTH TO MATURITY																	
MATURITY GROUP	EST. MATURITY DATE			*	*	*								*				#
1	MARCH 8					19.5	5.9	35.4	10.2					39.3	221	6	22	50
2	MARCH 14					19.3	5.8	35.2	10.3					39.0	215	7	25	51
3	MARCH 19					19.1	5.7	35.0	10.3					38.7	209	8	30	52
4	MARCH 25					18.9	5.5	34.8	10.4					38.5	202	8	36	53
5	APRIL 1					18.6	5.4	34.5	10.3					38.1	195	9	43	55
6	APRIL 8					18.1	5.3	33.8	9.9					37.2	187	9	51	56
7	APRIL 15					17.4	5.1	32.3	9.3					36.0	177	10	61	58
8	APRIL 25					16.4	4.9	30.6	8.5					34.2	165	12	76	61

* Adjusted for vine sites. # Raw data from 'Climatic Averages Australia', 1975; all figures period-adjusted by +2%.

TABLE 26

PERTH (Observatory), WESTERN AUSTRALIA. 31°57'S, 115°51'E. ALT. 60m. (For Spearwood, Wanneroo, Yanchep)

	1	2	3	4	5	6	7	8	9	10	11	12	13	14	15	16	17	18
	Temperature °C									Summation of Day Degrees above 10° with 19° Cut-Off			Temperature Variability Index					
PERIOD	Lowest Recorded Min. (25YRS)	No.Mths with <0.3° (25YRS)	No.Mths with <2.5° (25YRS)	Av.Mthly Lowest Min. (25YRS)	Av.Min. Minus Col.4	Daily Mean (68YRS)	Daily Range /2 (68YRS)	Av.Mthly Highest Max. (25YRS)	Col.8 Minus Av.Max.	Raw	Adj.1 for Lat. & Daily Temp Rng	Adj.2 for Vine Sites	Raw	Adj. for Vine Sites	Av. Sunshine Hrs	Av. Number Rain Days	Av. Rainfall mm (88YRS)	Av. % R.H. 1500 HRS (30YRS)
OCTOBER	4.6	0	0	6.7	4.5	16.1	4.9	30.6	9.6	189	186		33.7		246	11	54	54
NOVEMBER	6.6	0	0	8.8	4.9	19.1	5.4	34.8	10.3	270	270		36.8		289	6	21	47
DECEMBER	8.6	0	0	10.8	5.1	21.6	5.7	37.9	10.6	279	279		38.5		325	4	15	46
JANUARY	9.7	0	0	12.3	5.2	23.5	6.0	39.4	9.9	279	279		39.1		324	3	8	43
FEBRUARY	10.6	0	0	12.9	4.8	23.7	6.0	38.8	9.1	252	252		37.9		276	3	11	43
MARCH	8.4	0	0	11.5	4.9	22.1	5.7	37.6	9.8	279	279		37.5		270	4	21	46
APRIL	4.6	0	0	8.6	5.4	19.3	5.3	32.1	7.5	270	270		34.1		219	8	46	48
OCT-APRIL	—	0	0	—	5.0	20.8	5.6	—	9.5	1,818	1,815		36.8		1,949	39	176	47
MAY-SEPT	—	—	—	—	—	14.2	—	—	—	—	—		—		—	—	707	—

MATURITY GROUP	DATA FOR MONTH TO MATURITY																	
	EST. MATURITY DATE																	
1	FEBRUARY 4					23.6	6.0	39.3	9.8					39.0	320	3	8	43
2	FEBRUARY 10					23.7	6.0	39.2	9.6					38.8	314	3	9	43
3	FEBRUARY 16					23.8	6.0	39.1	9.4					38.6	309	3	10	43
4	FEBRUARY 21					23.8	6.0	39.0	9.2					38.3	304	3	11	43
5	FEBRUARY 27					23.7	6.0	38.8	9.1					38.0	299	3	12	43
6	MARCH 4					23.6	6.0	38.6	9.2					37.9	295	3	13	43
7	MARCH 10					23.4	5.9	38.4	9.3					37.8	291	3	14	44
8	MARCH 15					23.2	5.9	38.2	9.4					37.7	286	3	16	44

TABLE 27

PORONGURUP, WESTERN AUSTRALIA. 34°40'S, 117°54'E. ALT. approx. 260m. (Estimated from Mount Barker)

	1	2	3	4	5	6	7	8	9	10	11	12	13	14	15	16	17	18
	Temperature °C									Summation of Day Degrees above 10° with 19° Cut-Off			Temperature Variability Index					
PERIOD	Lowest Recorded Min.	No.Mths with <0.3°	No.Mths with <2.5°	Av.Mthly Lowest Min.	Av.Min. Minus Col.4	Daily Mean	Daily Range /2	Av.Mthly Highest Max.	Col.8 Minus Av.Max.	Raw	Adj.1 for Lat. & Daily Temp Rng	Adj.2 for Vine Sites	Raw	Adj. for Vine Sites	Av. Sunshine Hrs	Av. Number Rain Days	Av. Rainfall mm (40YRS)	Av. % R.H. 1500 HRS
OCTOBER				3.3	4.5	13.1	5.3	29.0	10.5	96	95	105	36.3	33.1	208	16	70	62
NOVEMBER				4.8	4.8	15.6	6.0	32.9	11.1	168	163	166	40.1	36.9	225	11	43	57
DECEMBER				6.3	4.8	17.6	6.5	36.7	12.6	236	227	230	43.4	40.2	245	10	31	53
JANUARY				7.3	5.1	19.0	6.6	38.8	13.2	279	268	272	44.7	41.5	235	8	28	49
FEBRUARY				7.3	5.0	19.0	6.7	37.9	12.1	252	243	249	44.0	40.8	215	7	27	48
MARCH				6.3	5.6	17.7	5.8	35.8	12.4	239	238	241	44.1	37.9	205	10	46	51
APRIL				5.0	5.3	15.7	5.4	29.7	8.8	171	174	183	35.5	32.3	185	13	65	59
OCT-APRIL				—	5.0	16.8	6.0	—	11.5	1,441	1,408	1,446	40.7	37.5	1,518	75	310	54
MAY-SEPT				—	—	11.1	—	—	—	—	—	—	—	—	—	—	438	—

MATURITY GROUP	DATA FOR MONTH TO MATURITY																	
	EST. MATURITY DATE			*	*	*								*			#	*
1	MARCH 4					19.0	5.8	36.9	12.1					40.4	228	7	30	50
2	MARCH 10					18.8	5.6	36.6	12.2					40.0	224	8	33	50
3	MARCH 16					18.6	5.4	36.2	12.3					39.5	218	8	37	51
4	MARCH 23					18.3	5.2	35.7	12.4					38.8	212	9	41	52
5	MARCH 30					17.9	5.0	35.1	12.4					38.0	206	10	45	53
6	APRIL 6					17.4	4.9	34.2	12.0					37.0	200	10	49	54
7	APRIL 14					16.9	4.8	33.0	11.0					35.6	195	11	54	55
8	APRIL 23					16.3	4.7	30.7	9.9					33.8	190	12	59	57

* Adjusted for vine sites. # Average and ripening month rainfall from direct recording for Porongurup.

TABLE 28

ROCKY GULLY/FRANKLAND, WESTERN AUSTRALIA. Approx. 34°22'S, 117°00'E. ALT. approx. 200m. (All data estimated)

PERIOD	Lowest Recorded Min.	No.Mths with <0.3°	No.Mths with <2.5°	Av.Mthly Lowest Min.	Av.Min. Minus Col.4	Daily Mean	Daily Range /2	Av.Mthly Highest Max.	Col.8 Minus Av.Max.	Raw	Adj.1 for Lat. & Daily Temp Rng	Adj.2 for Vine Sites	Raw	Adj. for Vine Sites	Av. Sunshine Hrs	Av. Number Rain Days	Av. Rainfall mm	Av. % R.H. 1500 HRS
OCTOBER				3.1	4.8	13.3	5.4	28.3	9.6	102	101	104	36.0	34.8	210	14	56	57
NOVEMBER				4.9	4.8	15.9	6.2	32.5	10.4	177	172	175	40.0	38.8	232	10	29	50
DECEMBER				6.2	5.0	18.0	6.8	36.4	11.6	248	234	242	43.8	42.6	265	8	22	44
JANUARY				7.5	5.3	19.8	7.0	38.1	11.3	279	279	279	44.6	43.4	260	6	17	43
FEBRUARY				7.6	5.3	19.8	6.9	37.4	10.7	252	252	252	43.6	42.4	238	6	16	43
MARCH				6.7	5.6	18.4	6.1	35.7	11.2	260	259	262	41.2	40.0	218	8	30	48
APRIL				5.0	5.4	15.8	5.4	29.6	8.4	174	177	180	35.4	34.2	188	11	45	58
OCT-APRIL			—	5.2	17.3	6.3	—		10.5	1,492	1,474	1,494	40.7	39.5	1,611	63	215	49
MAY-SEPT			—	—	11.1	—	—	—		—	—	—	—	—	—	—	472	—

DATA FOR MONTH TO MATURITY

MATURITY GROUP	EST. MATURITY DATE		*	*	*						*				
1	FEBRUARY 28		19.9	6.6	37.2	10.7					42.4	256	6	16	43
2	MARCH 6		19.8	6.5	37.0	10.8					42.1	250	6	18	44
3	MARCH 11		19.6	6.3	36.7	10.9					41.6	243	7	21	45
4	MARCH 17		19.4	6.2	36.4	11.1					41.1	236	7	24	46
5	MARCH 23		19.1	6.0	36.0	11.2					40.6	228	8	27	47
6	MARCH 30		18.6	5.8	35.5	11.2					40.0	220	8	30	48
7	APRIL 6		17.9	5.7	34.8	10.9					39.1	213	9	33	50
8	APRIL 14		17.3	5.5	33.0	10.1					37.8	205	9	37	52

* Adjusted for vine sites.

TABLE 29

WANDERING, WESTERN AUSTRALIA. 32°40'S, 116°41'E. ALT. 335m.

PERIOD	Lowest Recorded Min. (68YRS)	No.Mths with <0.3° (30YRS)	No.Mths with <2.5° (30YRS)	Av.Mthly Lowest Min. (30YRS)	Av.Min. Minus Col.4	Daily Mean (38YRS)	Daily Range /2 (38YRS)	Av.Mthly Highest Max. (30YRS)	Col.8 Minus Av.Max.	Raw	Adj.1 for Lat. & Daily Temp Rng	Adj.2 for Vine Sites	Raw	Adj. for Vine Sites	Av. Sunshine Hrs	Av. Number Rain Days	Av. Rainfall mm (80YRS)	Av. % R.H. 1500 HRS
OCTOBER	−2.2	18	29	0.1	6.2	13.5	7.2	30.6	9.7	109	95	116	44.9	39.7	245	11	44	47
NOVEMBER	−1.7	2	17	2.4	6.6	17.4	8.4	35.6	9.6	223	185	214	50.0	44.8	295	6	17	36
DECEMBER	1.7	0	4	5.0	6.7	20.6	8.9	38.3	8.9	279	279	279	51.1	45.9	330	4	14	32
JANUARY	3.3	0	0	6.8	6.6	22.3	8.9	40.2	9.0	279	279	279	51.2	46.0	325	3	9	30
FEBRUARY	2.8	0	0	6.9	6.3	22.0	8.8	39.9	9.0	252	252	252	50.6	45.4	284	3	14	30
MARCH	−0.6	0	2	4.9	6.8	19.8	8.1	37.2	9.4	279	277	279	48.5	43.3	248	5	22	36
APRIL	−2.2	8	19	1.8	6.7	16.2	7.7	31.3	7.5	185	172	199	44.9	39.7	221	7	33	46
OCT-APRIL	—	28	71	—	6.6	18.8	8.3	—	9.0	1,606	1,539	1,618	48.7	43.5	1,948	39	153	37
MAY-SEPT	—	—	—	—	—	10.7	—	—	—	—	—	—	—	—	—	—	480	—

DATA FOR MONTH TO MATURITY

MATURITY GROUP	EST. MATURITY DATE		*	*	*						*				
1	FEBRUARY 19		22.6	7.6	39.2	9.1					45.6	315	3	12	30
2	FEBRUARY 24		22.5	7.6	39.1	9.1					45.5	312	3	13	30
3	MARCH 2		22.3	7.5	39.0	9.2					45.3	308	3	14	30
4	MARCH 7		22.0	7.4	38.8	9.2					45.1	302	3	15	31
5	MARCH 13		21.7	7.3	38.5	9.2					44.8	294	4	16	32
6	MARCH 18		21.3	7.2	38.0	9.2					44.4	284	4	17	33
7	MARCH 23		20.8	7.1	37.3	9.2					44.0	272	4	19	34
8	MARCH 29		20.3	6.9	36.5	9.2					43.5	258	5	21	36

* Adjusted for vine sites.

TABLE 30

WILLYABRUP, WESTERN AUSTRALIA. 33°48′S, 115°00′E. ALT. approx. 90m. (All data estimated)

	1	2	3	4	5	6	7	8	9	10	11	12	13	14	15	16	17	18
	Temperature °C									Summation of Day Degrees above 10° with 19° Cut-Off			Temperature Variability Index					
PERIOD	Lowest Recorded Min.	No.Mths with <0.3°	No.Mths with <2.5°	Av.Mthly Lowest Min.	Av.Min. Minus Col.4	Daily Mean	Daily Range /2	Av.Mthly Highest Max.	Col.8 Minus Av.Max.	Raw	Adj.1 for Lat. & Daily Temp Rng	Adj.2 for Vine Sites	Raw	Adj. for Vine Sites	Av. Sunshine Hrs	Av. Number Rain Days	Av. Rainfall mm	Av. %R.H. 1500 HRS
OCTOBER				4.4	4.8	13.9	4.7	25.0	6.4	121	123	135	30.0	28.2	196	14	73	62
NOVEMBER				6.0	4.9	16.2	5.3	29.4	7.9	186	180	187	34.0	32.2	238	8	36	56
DECEMBER				7.3	5.0	18.3	6.0	33.3	9.0	257	247	251	38.0	36.2	280	5	20	54
JANUARY				8.6	5.3	20.2	6.3	35.4	8.9	279	279	279	39.4	37.6	284	4	14	54
FEBRUARY				8.6	5.5	20.4	6.3	35.2	8.5	252	252	252	39.2	37.4	239	4	11	53
MARCH				7.5	5.6	18.9	5.8	33.1	8.4	276	274	279	37.2	35.4	229	6	31	56
APRIL				6.3	5.5	16.7	4.9	28.0	6.4	201	206	218	31.5	29.7	195	10	68	62
OCT-APRIL			—	5.2	17.8	5.6	—	7.9		1,572	1,561	1,601	35.6	33.8	1,661	51	253	57
MAY-SEPT			—	—	13.1	—	—	—		—	—	—	—	—	—	—	885	—

					DATA FOR MONTH TO MATURITY													
MATURITY GROUP	EST. MATURITY DATE				*	*	*							*				
1	FEBRUARY 23					20.7	5.9	35.0	8.6					37.5	265	4	12	53
2	FEBRUARY 28					20.6	5.8	34.9	8.5					37.4	261	4	12	53
3	MARCH 6					20.4	5.8	34.7	8.5					37.2	256	4	13	54
4	MARCH 11					20.2	5.7	34.4	8.4					36.9	250	5	15	54
5	MARCH 17					20.0	5.6	34.0	8.4					36.5	244	5	18	55
6	MARCH 22					19.7	5.5	33.6	8.4					36.1	238	5	22	55
7	MARCH 27					19.3	5.4	33.1	8.4					35.6	232	6	28	56
8	APRIL 3					18.8	5.3	32.4	8.3					35.0	226	6	34	57

* Adjusted for vine sites.

TABLE 31

WOKALUP, WESTERN AUSTRALIA. 33°08′S, 115°53′E. ALT. 116m.

	1	2	3	4	5	6	7	8	9	10	11	12	13	14	15	16	17	18
	Temperature °C									Summation of Day Degrees above 10° with 19° Cut-Off			Temperature Variability Index					
PERIOD	Lowest Recorded Min. (23YRS)	No.Mths with <0.3° (23YRS)	No.Mths with <2.5° (23YRS)	Av.Mthly Lowest Min. (23YRS)	Av.Min. Minus Col.4	Daily Mean	Daily Range /2	Av.Mthly Highest Max. (23YRS)	Col.8 Minus Av.Max.	Raw	Adj.1 for Lat. & Daily Temp Rng	Adj.2 for Vine Sites	Raw	Adj. for Vine Sites	Av. Sunshine Hrs	Av. Number Rain Days	Av. Rainfall mm	Av. %R.H. 1500 HRS
OCTOBER	1.9	0	6	4.0	5.3	15.1	5.8	28.3	7.4	158	155	152	35.9	33.5	230	11	67	53
NOVEMBER	2.2	0	1	5.5	5.9	17.8	6.4	33.0	8.8	234	226	223	40.3	37.9	278	8	27	46
DECEMBER	3.8	0	0	7.2	6.3	20.8	7.3	36.9	8.8	279	279	279	44.3	41.9	325	4	17	43
JANUARY	5.5	0	0	9.0	6.0	22.6	7.6	39.1	8.9	279	279	279	45.3	42.9	322	3	10	41
FEBRUARY	6.1	0	0	9.3	6.3	22.9	7.3	38.9	8.7	252	252	252	44.2	41.8	272	3	19	41
MARCH	5.0	0	0	7.9	6.1	20.9	6.9	36.8	9.0	279	279	279	42.7	40.3	258	4	28	45
APRIL	2.9	0	0	6.2	6.3	18.0	5.5	31.8	8.3	240	245	243	36.6	34.2	210	8	56	53
OCT-APRIL	—	0	7	—	6.0	19.7	6.7	—	8.6	1,721	1,715	1,707	41.3	38.9	1,895	41	224	46
MAY-SEPT	—	—	—	—	—	13.1	—	—	—	—	—	—	—	—	—	—	786	—

					DATA FOR MONTH TO MATURITY													
MATURITY GROUP	EST. MATURITY DATE				*	*	*							*				
1	FEBRUARY 14					23.0	6.8	38.3	8.8					42.3	310	3	16	41
2	FEBRUARY 19					23.0	6.8	38.3	8.9					42.2	304	3	17	41
3	FEBRUARY 25					22.9	6.8	38.2	9.0					42.0	298	3	19	41
4	MARCH 2					22.7	6.7	38.1	9.0					41.8	292	3	21	41
5	MARCH 8					22.4	6.7	37.9	8.9					41.6	286	3	22	42
6	MARCH 13					22.1	6.6	37.7	8.8					41.3	280	3	23	42
7	MARCH 19					21.7	6.5	37.4	8.6					41.0	273	4	25	43
8	MARCH 24					21.3	6.4	36.9	8.4					40.7	266	4	27	44

* Adjusted for vine sites.

TABLE 32

YALLINGUP, WESTERN AUSTRALIA. Approx. 33°42′S, 115°05′E. ALT. approx 80m. (All data estimated)

	1	2	3	4	5	6	7	8	9	10	11	12	13	14	15	16	17	18
	Temperature °C									Summation of Day Degrees above 10° with 19° Cut-Off			Temperature Variability Index					
PERIOD	Lowest Recorded Min.	No.Mths with <0.3°	No.Mths with <2.5°	Av.Mthly Lowest Min.	Av.Min. Minus Col.4	Daily Mean	Daily Range /2	Av.Mthly Highest Max.	Col.8 Minus Av.Max.	Raw	Adj.1 for Lat. & Daily Temp Rng	Adj.2 for Vine Sites	Raw	Adj. for Vine Sites	Av. Sunshine Hrs	Av. Number Rain Days	Av. Rainfall mm	Av. % R.H. 1500 HRS
OCTOBER				5.0	4.7	14.2	4.5	25.0	6.3	130	135	147	29.0	27.2	198	14	70	65
NOVEMBER				6.5	4.8	16.5	5.2	28.7	7.0	195	188	196	32.6	30.8	242	8	32	59
DECEMBER				8.0	4.8	18.5	5.7	32.0	7.8	264	252	257	35.4	33.6	286	5	17	58
JANUARY				9.0	5.0	20.0	6.0	33.8	7.8	279	279	279	36.8	35.0	290	4	13	55
FEBRUARY				9.0	5.2	20.2	6.0	33.7	7.5	252	252	252	36.7	34.9	240	4	11	55
MARCH				7.8	5.5	18.9	5.6	31.7	7.2	276	274	279	35.1	33.3	231	6	28	57
APRIL				6.5	5.4	16.8	4.9	26.3	4.6	204	209	219	29.6	27.8	198	11	61	63
OCT-APRIL				—	5.1	17.9	5.4	—	6.9	1,600	1,589	1,629	33.6	31.8	1,685	52	232	59
MAY-SEPT				—	—	13.2	—	—	—	—	—	—	—	—	—	—	851	—
DATA FOR MONTH TO MATURITY																		
MATURITY GROUP	EST. MATURITY DATE					*	*	*						*				
1	FEBRUARY 19					20.4	5.5	33.4	7.6					34.9	270	4	12	55
2	FEBRUARY 25					20.4	5.5	33.4	7.6					34.9	263	4	12	55
3	MARCH 3					20.3	5.5	33.2	7.5					34.8	256	4	13	55
4	MARCH 8					20.2	5.4	32.9	7.5					34.7	251	5	14	55
5	MARCH 14					20.0	5.4	32.6	7.4					34.5	246	5	16	56
6	MARCH 19					19.8	5.3	32.3	7.4					34.2	241	5	19	56
7	MARCH 24					19.5	5.2	31.9	7.3					33.8	236	6	23	57
8	MARCH 30					19.1	5.1	31.4	7.2					33.3	231	6	28	57

* Adjusted for vine sites.

Commentaries and vineyard site adjustments

Albany (Airport) (Table 6)

Vineyard site adjustments, °C

Min.	Max.	
–	–	Moderate slopes, vs flats, but further inland
+0.8	–0.4	Gravelly soils
–	–	Altitudes ~ as at airport
+0.8	–0.4	Mean +0.2°, variability index –2.4

Notes

The figures are calculated for the valley of the lower Kalgan and its tributaries, from Kalgan up to about Takenup; also for the scattered sites with suitable topography and soils westward from the airport to Blue Gum Creek and the lower Hay River valley. Throughout an area where most soils are extremely poor, better gravelly soils occur on scattered rises and on some slopes along the creeks and rivers. These should be conducive to good vine balance and grape quality, if not necessarily to high yields.

Bakers Hill (Table 7)

Vineyard site adjustments: None

Notes

The data, from the CSIRO 'Yalanbee' research station, are from a limited period but probably represent the viticultural environments of Bakers Hill-Toodyay well enough. This being among the warmest of the Western Australian inland viticultural climates, selection of sites for frost freedom is less critical than further south, so that relatively slight slopes are usable. Careful site selection for air drainage would still be significant for quality, however.

The data for relative humidity are estimates, based on those of comparable stations further south.

Bridgetown (Table 8)

Vineyard site adjustments, °C

Min.	Max.	
+1.2	−0.5	Moderate slopes, vs frosty valley recording site
+0.4	−0.2	Some soils stony or gravelly
−0.4	−0.4	Alt. 200–240 m, vs Bridgetown 154 m
+1.2	−1.1	Mean +0.05°, variability index −4.6

Notes

Bridgetown is in a frosty valley, so a substantial adjustment is needed for the slopes where vines are most likely to be grown. The potential vine sites are here assumed to be on a level with or higher than existing apple orchards. The lower sites have loamy soils, some stony, derived directly from the mainly gneissic country rock that is exposed in the Blackwood Valley. Outside and above the valley proper the loams merge into the gravelly lateritic soils of the old plateau. In terms of effective temperature and its variability, the gravelly nature of these possibly offsets their greater altitude and flatter topography.

Spring frosts remain a serious hazard in the Blackwood Valley despite the most careful site selection. This reflects the valley's topographic isolation from both west and south coasts. The climatic contrast is especially strong with Manjimup, only 35 km to the south, where the coastal influence is clearly felt.

Bunbury (Table 9)

Vineyard site adjustments, °C

Min.	Max.	
−0.8	+0.4	Flat topography, further inland
+0.4	−0.2	Sandy soils over limestone
−0.4	+0.2	Mean −0.1°, variability index +1.2 Relative humidity reduced 2 per cent.

Notes

The Bunbury weather recording site is close to the coast and not fully representative. Present and potential vineyard sites on the tuart sands are mostly a little further inland, and fairly flat.

The Bunbury coastal strip as a whole is nevertheless notable for its temperature equability, with little variation of the maximum (column 9) and quite moderate summer extremes (column 8) for such a warm climate. This and the high average 1500 hours relative humidity are caused by the very regular early appearance of the south-west sea breeze in summer. That the coastal influence diminishes rapidly inland is shown by the much higher summer extremes and lower afternoon relative humidities at Wokalup (Table 31). The reason is that only a little way inland, the south-west sea breezes here merge into semi-land winds, similar to those in the afternoon at Collie (Table 11) and Donnybrook (Table 13), and to a smaller extent Busselton (Table 10).

Busselton (Table 10)

Vineyard site adjustments, °C

Min.	Max.	
−0.4	+0.2	Likely vineyard sites further inland
−0.1	−0.1	Vineyard altitudes ~ 20 m altitude, vs Busselton 3 m
−	−	Flat terrain, soils loamy sands or sandy loams
−0.5	+0.1	Mean −0.2°, variability index +1.2

Notes

The figures are estimates for the flat hinterland of Busselton, extending in an arc from Marybrook and Carbanup in the west through Jindong and Yoongarillup to Wonnerup and Ruabon in the east.

The rather high figure for temperature variability at Busselton is due solely to a wide diurnal temperature range (column 7), resulting from flat topography and the fact that the afternoon south-westerly winds are partly cut off by the Leeuwin-Naturaliste ridge. However in the absence of spring frosts or marked summer temperature extremes (column 8), and with summer nights still fairly warm, the variability index is probably not excessive.

The physical factor most likely to limit vine growing at Busselton is soil type; and in particular, drainage. The most extensive soil group of the coastal plain is that described by Bettenay, McArthur and Hingston (1960) as the Abba association of meadow podzolics, characterized by poor drainage and seasonal swamps. Interspersed are sandy rises of low fertility. Additionally, an arc of alluvial brown to reddish loamy sands and sandy loams extends through Jindong and Carbanup to Marybrook. The best of these are the recent deposits of deep, well-drained reddish sandy loams at Marybrook, as described in the main text. The reddish loamy sands further inland, for instance around Jindong, have their drainage to varying degrees restricted by a subsoil ironstone layer; nevertheless good stands of marri suggest that vines might also succeed with proper attention to deep drainage. Finally, the strip of tuart yellow sands over limestone which parallels the coast eastwards from Marybrook, through Wonnerup to Ludlow and beyond, is also largely suitable for viticulture. However much of the best of this is forest reserve.

Collie (Table 11)

Vineyard site adjustments, °C

Min.	Max.	
+0.6	−0.3	Moderate slopes, vs Collie approaching flat
+0.6	−0.3	Moderately gravelly soils
−0.1	−0.1	Vineyard altitudes ~ 200 m, vs Collie 184 m
+1.1	−0.7	Mean +0.2°, variability index −3.6

Notes

The figures represent a strip north and south of Collie, at similar distances inland from the Darling Scarp and at similar altitudes. The sea-level isotherms here more or less parallel the coast (Figure 11).

The ripening temperatures and generous sunshine hours indicate full-bodied table wines and lighter styles of fortified wine. Temperature variability and extreme maxima are higher than desirable, however, and relative humidities quite low, showing occlusion from coastal influences behind the Darling Scarp. Collie thus has an essentially inland climate, apart from an ample winter and spring rainfall. Deep, retentive soils are therefore especially needed in this environment, or else supplementary watering during ripening, for which there is good scope from surface water storage. Soil types are varied, with limited areas of marri soils suitable for viticulture. Careful site selection is essential to avoid spring frosts, with the best sites closest to the Darling Scarp and having free air drainage to the coastal plain.

Denmark (Agricultural Research Station) (Table 12)

Vineyard site adjustments, °C

Min.	Max.	
+1.2	−0.5	Moderate slopes of isolated hill range
+0.45	+0.15	Mainly north and north-east facing slopes
–	–	Loamy karri soils
−0.7	−0.7	Altitude ~ 130 m, vs Denmark 18 m
+0.95	−1.05	Mean −0.05°, variability index −4.0

Notes

The inland, north-facing slopes of the Bennett Range, immediately west of Denmark, are sheltered from ocean winds and have moderately fertile karri loam soils. They constitute an interesting environment for viticulture, with unique features for Western Australia because of their close proximity to the south coast. The assumed altitude of about 130 m represents some of the intermediate and higher slopes suitable for viticulture. Lower slopes towards Scotsdale Brook probably have similar biologically effective temperatures, losing through their site-climate characteristics approximately what they gain in lower altitude.

Temperature variability is greater than might be expected so close to the sea, because of the marked variability of the maximum (column 8). This is typical of the south coast. Average afternoon relative humidities are nevertheless high and sunshine hours fairly low, because of the normally regular summer afternoon on-shore winds and frequent associated cloudiness.

Donnybrook (Table 13)

Vineyard site adjustments, °C

Min.	Max.	
+0.8	−0.4	Moderate slopes, vs Donnybrook in valley
−0.3	−0.3	Vineyard altitudes ~ 100–120 m, vs Donnybrook 63 m
+0.4	−0.2	Some soils gravelly
+0.9	−0.9	Mean unchanged, variability index −3.6

Notes

The adjusted data are for the lower valley-side slopes of the Preston and upper Capel (north branch) rivers, rather than for the valley floors. The latter would have about the same mean temperatures, but greater temperature variability and more frost (as shown by the raw data for Donnybrook).

Ripening conditions are similar to those in the Collie area, the more southerly location of Donnybrook being offset by its lower altitude. The low 1500 hours relative humidity at Donnybrook is notable. The reason is that any afternoon winds from the south are here still essentially land winds, while the lie of the valleys means that sea breezes from the west are cut off by intervening hills.

Esperance (Table 14)

Vineyard site adjustments, °C

Min.	Max.	
−0.8	+0.4	Inland, undulating, vs coastal, flat
−0.5	−0.5	Altitude 60–120 m, vs sea level
−1.3	−0.1	Mean −0.7°, variability index + 2.4

Notes

I have assumed that the sites most likely to be chosen for viticulture are inland but not greatly distant from the coast, and in moderately undulating country (few real hills are present). No adjustment is made for soil type, because most are sandy-surfaced; but some reduction in the temperature variability index might be achieved by selection of gravelly-surfaced soils.

Guildford (for Swan Valley, Gingin and Bindoon) (Table 15)

Vineyard site adjustments: None

Notes

The main data for Guildford, from Climatic Averages Australia (1956), are from the old recording site in Stirling Square. Column 18 is from the new Perth Airport at South Guildford, period-adjusted.

Other aspects of the Swan Valley, and of Gingin, Bindoon and Dandaragan, are covered in the main text.

Kalamunda (Table 16)

Vineyard site adjustments, °C

Min.	Max.	
−0.8	+0.4	Moderate slopes but further from scarp than Kalamunda
+0.8	−0.4	Gravelly soils
0.0	0.0	Mean and variability index unchanged

Notes

Kalamunda is on the brow of the Darling Scarp, and lies clearly in its thermal zone. At 282 m altitude its average mean October-April temperature is only 1.0°C lower than at Guildford, with lower maximum temperatures but similar minima. Thus the climate is distinctly more favourable for vines and winegrape quality than on the adjacent flats of the coastal plain.

Potential hills vine sites are likely in the main to be further inland than Kalamunda. Some climatic deterioration relative to Kalamunda is allowed for in the site adjustments, but the adjustment for gravelly soils offsets this.

Karridale (Table 17)

Vineyard site adjustments: None

Notes

The old Karridale weather station, from which the data are taken, was on an east-facing slight to moderate slope, at an altitude representative of potential vine sites on the karri loamy sands there. The estimates would also apply broadly to the Alexandra Bridge and Kudardup areas.

Katanning (Table 18)

Vineyard site adjustments, °C

Min.	Max.	
+0.4	−0.2	Slight to moderate slopes (Katanning flat)
+0.4	−0.2	Mean +0.1°, variability index −1.2

Notes

Although Katanning is at the cooler and wetter south-western edge of the Western Australian wheatbelt, in the absence of readily available water for supplementary watering the conditions for vines are at best marginal. The area has few redeeming soil or topographic features.

Kojonup (Table 19)

Vineyard site adjustments, °C

Min.	Max.	
+0.4	−0.2	Selected slopes (Kojonup moderately sloping)
+0.2	−0.1	Some moderately gravelly soils
+0.6	−0.3	Mean +0.15°, variability index −1.8

Notes

Kojonup is not much nearer the coast than Katanning, and the recorded afternoon relative humidities are similarly low. However the country to the south and south-west of Kojonup, where vine growing is most likely to be contemplated, is more hilly and offers better opportunity for site selection and water conservation. It is nevertheless still marginal for viticulture.

Most soils are loamy, but some moderately gravelly soils occur in combination with suitable topography. Soil depth and drainage are crucial in this country, with soils characterized by wandoo (*Eucalyptus wandoo*), flooded gum (*E. rudis*) and probably jam (*Acacia acuminata*) unsuitable. As elsewhere, the indicator for good vineyard soils is vigorous dominant marri (*E. calophylla*).

Mandurah (Table 20)

Vineyard site adjustments, °C

Min.	Max.	
−0.4	+0.2	Vines further inland; similar altitude and terrain
+0.4	−0.2	Sandy soils over limestone
0.0	0.0	Mean and variability index unchanged

Notes

Compared with Bunbury, Mandurah has higher temperatures throughout growth and ripening, and quite substantially higher extreme maxima (column 8), associated with greater variability of the maximum (column 9). The latter is in turn probably associated with a later average onset of the afternoon sea breeze, together with the fact that land winds here, and increasingly to the north, are more likely to come direct from the continental interior than from the Great Australian Bight.

One special environment deserves mention. The narrow strip of land between Peel Inlet and the sea, and to a lesser extent between the Harvey Estuary and the sea, would have slightly lower maxima and a reduced temperature variability index compared with Mandurah because land winds would be appreciably moderated by passage across estuarine waters. Potential viticultural sites here are rare, but there are a few possibly suitable tuart-marri sandy soils.

Manjimup (Table 21)

Vineyard site adjustments, °C

Min.	Max.	
+0.4	−0.2	Slight to moderate slopes (Manjimup ~ flat)
+0.4	−0.2	Soils variably gravelly
+0.2	+0.2	Vineyard altitude ~ 240 m (Manjimup 279 m)
+1.0	−0.2	Mean +0.4°, variability index −2.4 Relative humidity period-adjusted up by 2 per cent

Notes

The full-season average mean temperature and sunlight hours at Manjimup are almost identical with those of Mount Barker. However the seasonal pattern differs subtly, Manjimup being cooler than Mount Barker up to the end of December, but a little warmer and sunnier through January, February and March. Estimated maturities at Manjimup are a day or two later than in the main Mount Barker and Rocky Gully-Frankland vineyards, but maturation still takes place under slightly greater warmth and sunshine hours.

A reservation concerning Manjimup is that the temperature records are fairly short-term. Some period adjustment was necessary, of which the accuracy is open to question. The conclusion that ripening conditions will be intermediate between those of Mount Barker and Margaret River (but closer to Mount Barker) is probably justified, but the precise placing remains uncertain.

Margaret River (Table 22)

Vineyard site adjustments, °C

Min.	Max.	
–	–	Vineyards and recording station on moderate slopes
+0.4	−0.2	Vineyards closer to the coast
+0.4	−0.2	Some soils gravelly
+0.8	−0.4	Mean +0.2°, variability index −2.4

Notes

The published temperature data for Margaret River in Climatic Averages Australia (1975) are from a very short period, at a time when temperatures – especially in summer – were well above the 1911–1940 average in south-western Australia. A downward period adjustment averaging 0.4°C over the growing season, together with smoothing of the individual months, was applied to bring the estimate into line with the longer-term records for Busselton and other south-western centres.

The estimate is for vineyards close to and just south of Margaret River, as far as Witchcliffe. See Wilyabrup (Table 30) for those nearest to the north.

Mount Barker (Table 23)

Vineyard site adjustments, °C

Min.	Max.	
–	–	Vineyards and Mount Barker both on moderate slopes
+0.4	−0.2	Some soils gravelly
+0.3	+0.3	Vineyard altitudes ~ 200 m, vs Mt Barker 253 m
+0.7	+0.1	Mean +0.4°, variability index −1.2

Notes

The main established vineyard sites are in valleys at lower altitudes than Mount Barker town. (The foothills of the Porongurups are treated separately: see Table 26.) In the Narrikup-Denbarker areas, present and potential vineyard sites are at 150 m or less. Probably these are sufficiently far south of Mount Barker to offset the altitude difference when estimating temperature means. Conversely, vine sites north of Mount Barker tend to be at greater altitudes than the town, with a similar result. The Mount Barker area (excluding the Porongurups) can therefore reasonably be considered as thermally uniform, apart from a progressively widening diurnal temperature range and increasing spring frost risk from south to north. There is also a fairly marked trend from south to north towards greater sunshine hours but decreasing rainfall and relative humidity.

Narrogin (Table 24)

Vineyard site adjustments, °C

Min.	Max.	
+0.4	−0.2	Selected moderate slopes, vs undulating
+0.4	−0.2	A few gravelly soils
+0.8	−0.4	Mean +0.2°, variability index −2.4

Notes

This represents the extreme eastern limit of potential viticulture, with rainfall well below optimum and limited potential for surface storage of good-quality water. With much of the country carrying wandoo (*Eucalyptus wandoo*) or jam (*Acacia acuminata*) in the native state, most soils are shallow and/or indifferently drained; few are suitable for viticulture.

Pemberton (Table 25)

Vineyard site adjustments, °C

Min.	Max.	
+0.4	−0.2	Moderate to steep, vs moderate slopes
+0.2	+0.2	Vineyard altitudes ~ 140 m (recording station 171 m)
+0.6	0.0	Mean +0.3°, variability index −1.2 Relative humidity period-adjusted up by 2 per cent

Notes

The climatic figures used for Pemberton were from an 11 year period only, probably between the early 1950s and early 1960s. Those published in Climatic Averages Australia (1975) covered a period starting in 1957 and showed a growing season mean temperature fully 0.5°C higher. Comparison of the 1975 versus long-term records for other southern stations suggests that the earlier 11 year period was more truly representative of Pemberton's long-term climate. The estimate based on it is nevertheless very tentative. For what it is worth, the estimate suggests that grapes will ripen under temperatures very similar to those at Mount Barker, but with less sunshine hours. Wine styles should therefore be perceptibly lighter than at Manjimup, and perhaps a little lighter than at Mount Barker.

Topography in the Pemberton area is favourable for viticulture, with many good exposures on moderate slopes. Typical soils are brown loams, formed directly from the gneissic country rock and carrying karri (Eucalyptus diversicolor) in the native state. There is excellent scope for surface water storage.

Perth (Table 26)

Vineyard site adjustments, °C

Min.	Max.	
−0.4	+0.2	Slightly closer to sea but further from river
+0.4	−0.2	Sandy soils over limestone
0.0	0.0	Mean and variability index unchanged

Notes

The raw data are from the old weather station at the observatory, near King's Park. This has a favourable site climate for viticulture, being on a hill close to the Swan Estuary, which acts as a funnel for the summer afternoon sea breezes. Present and potential vineyard sites on tuart (*Eucalyptus gomphocephala*) soils to the south in the Fremantle-Spearwood area, and to the north from Wanneroo to Yanchep, are mostly at similar altitudes in mildly hilly country. They lack the river influence, but are closer to the sea. The old Perth climatic records probably represent them well.

Porongurup (estimate based on Mount Barker) (Table 27)

Vineyard site adjustments, °C

Min.	Max.	
+0.7	−0.1	Slopes of isolated range, vs normal moderate slopes
+0.3	+0.1	Predominantly northerly aspects
+0.4	−0.2	Soils variably stony or gravelly
−0.5	−0.5	Vineyard altitude ~ 330 m, vs Mt Barker 253 m
+0.9	−0.7	Mean +0.1°, variability index −3.2 Relative humidity adjusted up by 2 per cent for greater altitude

Notes

The Porongurups are a small, isolated, intrusive granite range, offering site climate advantages on their lower slopes to a degree that is perhaps unique in Western Australia. The calculated site data are for 330 m altitude on the northern slope, which is about the altitude of existing plantings there.

Similar figures would apply to land not facing north, but at about 250 to 300 m altitude. Some climatically comparable sites are probably also available at these altitudes on the slopes of the smaller hills Mount Barrow and Mount Barker, close to the town of Mount Barker.

A slightly higher rainfall throughout summer compared with Mount Barker, as indicated by the long-term records for Woodburn, on the north-east Porongurups foothills, is not in itself disadvantageous; but being often associated with thunderstorms, it may entail some added hail risk.

Rocky Gully-Frankland (estimated) (Table 28)

Vineyard site adjustments, °C

Min.	Max.	
+0.4	−0.2	Moderate slopes (sources average slight slopes)
+0.4	−0.2	Mean +0.1°, variability index −1.2

Notes

The figures for Rocky Gully-Frankland are very tentative, because of the amount of estimation involved. They show ripening a day or two earlier than at Mount Barker, at about 0.5°C higher average mean temperature. Ripening sunshine hours are greater, and relative humidities lower. Rainfall is less throughout the growing season. The temperature variability index is estimated to be similar to that at Mount Barker, the greater distance inland of the Rocky Gully-Frankland area being offset by its more westerly location.

Wandering (Table 29)

Vineyard site adjustments, °C

Min.	Max.	
+1.2	−0.5	Moderate slopes, vs frosty hollow
+0.6	−0.3	Gravelly soils
−0.2	−0.2	Vineyards 30 m higher
+1.6	−1.0	Mean +0.3°, variability index −5.2

Notes

The weather-recording station at Wandering is in a notoriously frosty basin, so substantial adjustments are needed for potential vineyard sites. These will still have a very significant risk of heavy frost after budburst, but probably little more than at comparable sites such as Narrogin. Rainfall and relative humidities are very marginal for rain-fed viticulture, but rainfall is sufficient to allow some surface water storage.

Small-scale viticulture is possible in this area with very careful site selection. Commercial-scale development cannot be recommended.

Wilyabrup (estimated) (Table 30)

Vineyard site adjustments, °C

Min.	Max.	
−	−	Estimated directly for vine site topography
+0.6	−0.3	Many soils gravelly
+0.6	−0.3	Mean +0.15°, variability index −1.8

Notes

Wilyabrup has the oldest of the existing vineyards in the Margaret River region sensu lato, with plantings dating from 1967. The wine styles have broadly confirmed the initial predictions based on climate. Both red and white table wines have been predominantly full-bodied. Late picking for full-bodied sweet white wines of the Sauternes and auslese Riesling styles has been very successful. In warm seasons there have also been some attractive light-bodied port-style wines. Grape varietal characteristics are expressed to a marked degree.

Wokalup (Table 31)

Vineyard site adjustments, °C

Min.	Max.	
−	−	Recording site at typical altitude and topography
−0.3	−0.3	Latitude adjustment: main potential areas south of Wokalup
+0.8	−0.4	Well drained gravelly soils
+0.5	−0.7	Mean −0.1°, variability index −2.4

Yallingup (estimated) (Table 32)

Vineyard site adjustments, °C

Min.	Max.	
−	−	Data estimated for moderate slopes and altitude 80 m
+0.6	−0.3	Many soils gravelly
+0.6	−0.3	Mean +0.15°, variability index −1.8

Notes

The Yallingup figures are estimated for the slopes of the Leeuwin-Naturaliste ridge, northwards from the Abbeys Farm road to about Dunsborough.

10

Viticultural Environments of South Australia

South Australia's first vines were planted in 1837, the year after the State's inauguration. The suitability of the Adelaide area for vines quickly became apparent, followed soon after by the Barossa Valley just to the north.

Intensive viticultural development of the Barossa Valley began in 1842, with the first settlement there of German farmers from Silesia, sponsored by George Fife Angas. The new settlers played a major part in the expansion of the industry that followed over the next quarter of a century, and their descendants are today still a dominant force in the South Australian wine industry. Good accounts of various parts of the South Australian wine industry include those of Rankine (1971), Burden (1976) and Halliday (1981, 1983, 1985a, 1985b, 1991a).

South Australia remained free from the Phylloxera plague that crippled the European wine industry, and destroyed that of Victoria, in the 1870s and 1880s. This, combined with a shift in consumer tastes towards sweet fortified wines – for which the Adelaide and Barossa climates were well suited – resulted in a renewed growth and rapidly growing exports from the mid 1880s onwards. The robust dry red wines of South Australia also came into strong demand at this time, for blending with, and strengthening, the weak vintages of Europe's newly reconstructed vineyards. The fact that the world was then in a cool climatic phase (see Chapter 23) would certainly have added: both to the need for strengthening the European wines, and to the quality of the table wines from South Australia's warm areas. The emergence of Clare as an important wine-producing area also dates from this period.

The removal of internal tariffs following federation of the Australian States in 1901 further consolidated South Australia's viticultural dominance. Its firmly established industry was able to continue expanding at the expense of the smaller and higher-cost industries of New South Wales and Western Australia, and of recovery from Phylloxera in Victoria.

It is only now, after nearly a century, that the viticultural dominance of South Australia is again starting to be challenged. The change in tastes back to table wines, and in part to the styles of table wine made in cooler areas, has been a major contributing factor. As a response to this challenge, South Australia's wine industry is itself now evolving into cooler production areas, including some where vines were planted in the early years, but later abandoned.

The coolest viticultural region of South Australia is the Mount Lofty Range, immediately east of Adelaide and of the Barossa Valley. Coonawarra, in the far south-east of the State, is a little warmer. These produce medium-bodied table wines of relatively 'cool climate' character. Plantings in them have increased dramatically in recent years, as have those at Padthaway (Keppoch Valley), which is north of Coonawarra and a little warmer again.

The other premium-quality vineyards of South Australia are in mainly warm regions that have traditionally produced both full-bodied table wines and fortified wines. They comprise the coastal plain and foothills of the Mount Lofty Range, south as far as McLaren Vale; the Barossa Valley, just north of Adelaide; and Clare, on the western edge of the ranges further north again. There is also a small, long-established industry at Langhorne Creek, over the hills to the south-east of Adelaide near Strathalbyn. In the early years all these areas, with the exception of Langhorne Creek, practised almost exclusively dryland viticulture. Now the increasing trend is to give supplementary watering in summer wherever possible. This, combined with other developments in the vineyard and in winemaking technology, has resulted in dramatically improved in table wine quality.

Finally there are the extensive hot-climate, fully irrigated Riverland plantings on the Murray River, from Renmark down to Waikerie and beyond. Mostly developed since the beginning of the present century, these plantings originally produced mainly drying grapes. Currant and Muscat Gordo Blanco were the principal varieties at first, with Sultana becoming more popular in later years. Around World War I the region diversified into distillation wines, and between the wars into inexpensive fortified wines. It is now Australia's biggest producer of bulk table wines for the cask and flagon trade, largely from the 'dual purpose' varieties Muscat Gordo Blanco and Sultana. More recent plantings have included superior wine varieties which, with appropriate vineyard management, can

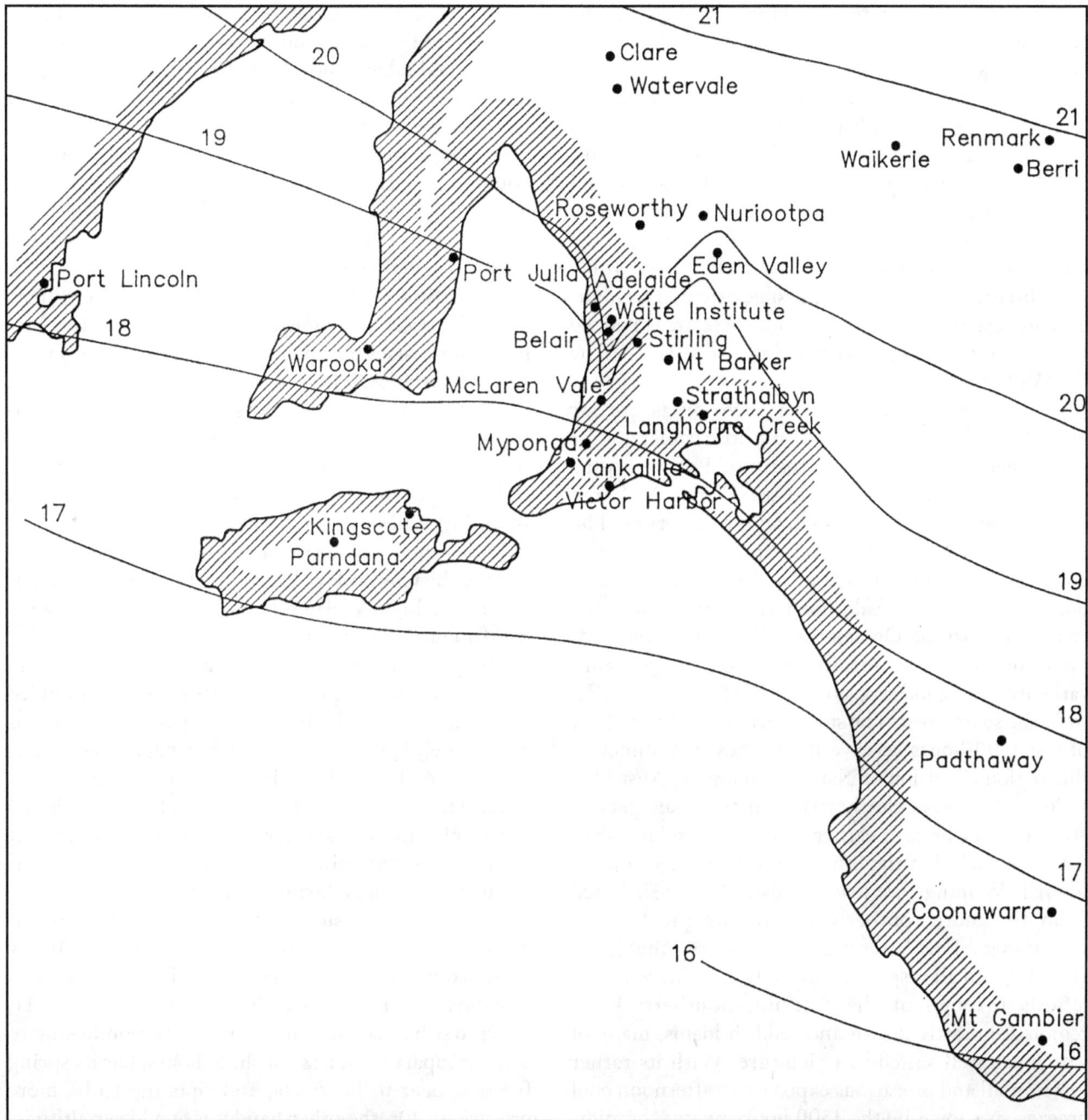

Figure 12. South Australian vinegrowing regions, with isotherms of October-April average mean temperatures, reduced to sea level.

achieve reasonable and sometimes good quality in the lower and middle-priced bottled wine market.

Figure 12 shows the South Australian locations mentioned in the text, together with the growing season (October-April) isotherms, reduced to sea level. The altitude-adjusted temperatures fall with distance eastwards from the coast and from the Mount Lofty Range scarp. They do so for the same reason as inland from the west coast of Western Australia. Minimum temperatures fall quickly inland from the scarp, as the night and early morning cycle of air drainage to the coast and overhead return of sea-warmed air diminishes; but maximum temperatures do not rise as fast inland because the summer afternoon sea breezes and

the normal winter and spring on-shore winds penetrate deeper. As in Western Australia, this means that the viticulturally most favourable temperature (and relative humidity) regimes are those closest to the coast, and along the slopes, crest and proximal valleys of the scarp.

Adelaide-Southern Vales

The climate of the lower foothills is typified by **Adelaide, Waite Institute** (Table 35). **Adelaide City** (Table 34) is on the flat, central part of the coastal plain. Its recorded temperatures are almost certainly higher than those of the main vineyards, owing to the combination of lower altitude and, perhaps especially, urban warming (Chapter 23). However they may be reasonably indicative of the Angle Vale area to the north.

Whether the Waite Institute, at 123 m, adequately represents the main Adelaide foothills and the Southern Vales vineyards is debatable. The Magill vineyards, just to the north, are (or were) at 150 to 250 m, and mostly more sloping. Similarly those at Tea Tree Gully are at 140 to over 200 m, and in a valley opening to the south-west to south which should pick up cool afternoon breezes to the greatest possible extent. The highest and best-exposed parts of both these areas may be better represented by **Belair** (308 m, Table 37) than by the Waite Institute.

The Reynella and Southern Vales areas are far enough south to be probably also a little cooler and/or more equable than the Waite Institute. Reynella is at a similar altitude but closer to the sea. Within McLaren Vale the situation differs according to location. The northern parts around Blewett Springs reach nearly 200 m altitude. Being on a ridge with good air drainage, both south into the vale itself and north down the steep slopes to the Onkaparinga River, the vineyards undoubtedly enjoy locally minimal temperature variability and good exposure to cool afternoon breezes from the south-west. The steep north to south gradient of the 1500 hours relative humidities in summer in this region (Division of National Mapping, Australia, 1986, p. 35) shows clearly that McLaren Vale gets its afternoon sea breezes earlier and more regularly than Adelaide itself. On the other hand the southern parts towards Willunga are low and flat, and are sheltered from the more southerly breezes by the Sellicks Hill Range. This latter area appears less suitable for viticulture in all respects, especially for table wines.

Soils throughout the Adelaide-Southern Vales region are mostly brown and reddish loams, many of which are well suited to viticulture. With its rather low rainfall and precarious exposure to afternoon cool breezes, as shown by the 1500 hours relative humidities, this is an area which can be expected to respond to site selection and above all to supplementary watering in summer; and for table winemaking, to modern oenological methods. Its vintage port styles remain famous, in line with theoretical expectations based on climatic similarities to the Douro Valley of Portugal (compare Tables 165 and 166, Chapter 20); but as well it is now known equally for the increasingly reliable high quality of its full-bodied table wines, most notably those from Shiraz, Chardonnay and Sauvignon Blanc (Halliday 1991a).

Adelaide Hills

A transect from Adelaide across the Mount Lofty Range to Mount Barker illustrates well the role of altitude in South Australia. The Waite Institute, enjoying the benefits of foothills topography, is climatically fairly close to optimal for fortified wines. By Belair, at 308 m (Table 37), temperature has fallen and relative humidity risen to approach optimal conditions for full-bodied table wines, apart from a still undesirably marked variability of the maximum temperature: a factor that is more or less universal in South Australia. The much higher 1500 hours relative humidities compared with the adjacent coastal plain suggest better exposure to the south-westerly and southerly sea breezes. Rainfall is nevertheless still below optimum for unirrigated vines.

By **Stirling**, at altitude 496 m (Table 47), the climate is much cooler and moister, and approaches the cool limit for satisfactory ripening of Maturity Group 3 and 4 grape varieties. These do not ripen until mid to late April. Because temperature variability falls, and relative humidities rise, rapidly through late March and April, early-maturing grapes ripen on average under close to optimum conditions for light table wines and sparkling wines.

The estimates based on Stirling are for the Petaluma and nearby vineyards at Piccadilly, as being representative of the high Adelaide Hills. Their altitudes range from about 450 to 550 m.

The planting of Pinot Noir and Chardonnay here for champagne-style wines appears to be well-founded on climatic grounds. Estimated ripening conditions quite closely approach those in Champagne (see Reims, Chapter 16, Table 137). The main differences are that relative humidity at Stirling is less and sunshine hours fairly substantially greater. Stirling also has greater temperature variability, with higher extreme maximum temperatures during ripening.

The differences suggest that it may not be possible to get quite the same combination of delicacy with strength of character in the base wines as in Champagne. Drumborg, in Victoria (Chapter 11, Table 63) has not dissimilar ripening conditions to Stirling, apart from less sunshine hours; but its spring frosts appear to be worse, and ripening to be more precarious. On the other hand it is at a lower altitude, and would therefore have an appreciably greater atmospheric carbon dioxide concentration (see Chapter 23). In terms of potential quality, if not of safety, the two areas are probably comparable for making champagne-style wines.

In as cool an area as Stirling, ripening is very sensitive to altitude, aspect and varietal maturity grading. This is because all months of the growing season are below 19°C average mean temperature, so the vines remain phenologically responsive to temperature differences through the whole season (Chapter 3). Theoretically an altitude difference of 50 m, resulting in a temperature difference of 0.3°C, should in this environment change ripening time by the equivalent of one vine maturity group, or 10–12 days for a given variety. Season to season differences in maturity date will therefore be magnified compared with warmer areas. Given that rainfall increases very rapidly in late autumn here, occasional wet vintages can be expected even with early-maturing grape varieties.

Mount Barker (Table 42) is well inland from the Mount Lofty Range's western scarp, and has a growing season average mean temperature 1.7°C less than at Belair despite its very similar altitude. As in comparable situations in Western Australia (Chapter 9), this is due mainly to differences in the minimum temperatures; which combined with more variability of the minimum (Column 5), results in a much greater incidence of frosts at Mount Barker. It shows clearly the deterioration in viticultural climate with distance east from the scarp, and from the marine influence of the Gulf of Saint Vincent.

Barossa Valley

I have made two estimates for the Barossa Valley, one based on medium-term **Nuriootpa** data (Table 43) and the other on temperature records from the vine trials of Boehm (1970) (Table 36). Long-term climatic data are lacking, but the two estimates show reasonable agreement.

The figures show the Barossa Valley to be cooler than is generally believed. Its *biologically effective* temperature summation and ripening mean temperatures are almost identical with those of Bordeaux and Margaret River. Raw temperatures are a little higher, however: the lower biological effectiveness of the Barossa's temperatures being due to its wide diurnal temperature range, typical of valley-floor location. There are big differences in other respects, too. The Barossa has more sunshine, lower relative humidities and rainfall, and much greater variability of the maximum temperature and higher average extreme maxima than either Bordeaux or (to a smaller extent) Margaret River. Nevertheless the Barossa is still close enough to the coast to get afternoon sea breezes which, although too late to influence greatly the 1500 hours relative humidity figures, are probably still enough to ameliorate the viticultural climate quite substantially. As a result, the Barossa remains capable of making good to very good table wines, typically of full-bodied character. The best mesoclimates are those of the foothills fringing the valley.

Low rainfall and stresses during ripening have been the principal limitations of the Barossa, necessitating severe pruning and yield reduction to achieve good grape quality. Supplementary watering by drip irrigation has now largely changed this, and thereby ensured the Valley's continued role as a leading producer of some of Australia's best wines in traditional robust style. The area nevertheless remains vulnerable to its fairly extreme temperature variability. Being towards the inland margin of regular sea breezes, the latter quite often fail. Hot northerly winds, without respite, can occasionally still do considerable damage to the vines and grapes regardless of irrigation.

Barossa Hills

As in the latitude of Adelaide, a transect through the Barossa Valley and adjoining hills of the Mount Lofty Range reveals major viticultural differences. To the west **Roseworthy**, on the plain near Gawler, is hot and arid (Table 46). The Barossa Valley is cooler by an amount slightly greater than would be expected from the normal lapse rate with altitude, but (at least at Nuriootpa) has a similar, fairly wide, temperature variability pattern reflecting its valley-floor situation.

The **Eden Valley-Springton** area (Table 41), averaging about 450 m in the hills east of the valley, is cooler again by about the amount expected from its altitude. Relative to the valley it has a slightly reduced diurnal temperature range, and probably less overall variability of the temperatures.

Lower seasonal temperatures in the hills mean that ripening (in late March to early May) is later and under markedly cooler and more equable conditions than in the Barossa proper. The main white varieties (maturity groups 2–5) ripen under optimal average mean temperatures for light to medium-bodied table wines, with not greatly excessive highest maxima during ripening. The main midseason red varieties, on the other hand, need the warmest situations to ripen reliably. Only at Keyneton do their wines regularly achieve full Bordeaux-style, medium body. This is far enough north to be appreciably warmer than most of the hills at given altitudes, while the best-known Keyneton vineyards (Hill of Grace, Mount Edelstone) are at only 380–400 m altitude. Their temperatures are probably intermediate between those indicated for the Barossa Valley (Tables 36, 43) and for average Barossa Hills locations (Table 41), so the proven performance of Shiraz (Maturity Group 5) in them conforms well to expectations.

Recently developed major vineyards further south in the Barossa Hills, such as at Pewsey Vale, Heggies and High Eden, average nearly 500 m in altitude, and in the cases of the latter two reach as high as 530 m. They would be appreciably cooler than the average shown for Eden Valley-Springton (Table 41), and temperatures in the coolest parts probably approximate to those for Stirling (Table 47) in the Adelaide Hills. This is true white wine country, with ripening mean temperatures for grape varieties such as Chardonnay (Maturity Group 3) and Riesling (Group 4) very close to those experienced in Burgundy (Table 130, Chapter 16) and the Rhine Valley (Tables 143 and 145, Chapter 17) respectively.

In other respects, however, these Australian locations differ from their European counterparts in important ways. They have greater temperature variability, more sunshine hours, less summer rain and lower relative humidities. They are also at greater altitude. These differences perhaps help to explain the firmer, more 'steely' character of classic South Australian wines, particularly those from the Riesling grape, as compared with those from the much 'softer' European ripening climates. The tougher growing season conditions also help to explain why the Riesling vine naturally achieves a reasonable balance between growth and fruiting here, without necessarily having the vertical training that is practised in northern Europe. Pinot Noir (Maturity Group 3) and other early red varieties can ripen well, but are at risk of fruit damage from stresses and temperature extremes. Cabernet Sauvignon (Group 6) is also grown, but its wines often show signs of grape immaturity.

Clare-Watervale

The published climatic data for **Clare** (Table 39) pose a problem in matching wine style against climate. Some have suggested that the 'official' temperature figures are atypical, or at least that they do not represent many of the vineyards (e.g. Halliday 1985a, 1985b). Certainly it is true that despite having moderately high recorded temperatures, which point more to fortified wines, Clare remains essentially a table wine area. It enjoys a good reputation for its firm, full-bodied, deeply coloured red wines, and in particular for its dry rieslings. For these reasons we need to look more closely than usual at the topography of the area, and possible mesoclimatic differences within it.

From a recent (1991) inspection I can confirm that the Clare weather recording station is not only at the bottom of the valley and in the heart of the town, but also close to roadways and buildings. A valley-bottom situation could be expected to maximize the recorded diurnal temperature range; but at the same time, for its altitude, it would probably reduce the mean temperature slightly compared with average undulating sites (Chapter 4). This may be more than offset by urban warming (Chapter 23), since Clare was already a well-established town by the 1911–40 period when the figures used here were recorded. Warming could well have been especially marked in Clare's case because of the thermometer's unusual closeness to buildings and roads.

To distort significantly the perceived relationship to local viticulture, the effect would have to be greater than that embodied in the average of the climatic records used to establish the broad relationship between climate and vine phenology (Chapter 6). Here we are on more difficult grounds. To the extent that this does happen, the net result would be to raise artificially the effective temperature estimates, both for the valley floor, and for the more sloping sites after adjusting for average vineyard topography, as I have done in Table 39.

But equally or perhaps more importantly, there exist quite large differences among the vineyards in their altitudes, latitudes and individual topographies, as Halliday (1985b) discusses in some detail.[1] My estimate

in Table 39 is for altitudes and latitudes similar to those of the town. While that is true for most of Clare's old-established vineyards, there are also many, particularly the newer ones, at higher altitudes or some way south of the town; or both. For instance Petaluma's Hanlin Hill vineyard, just east of Clare, is at 470 to 500 m, compared with 398 m for the town weather station. The vineyards further south at Sevenhill and Penwortham also reach 500 m or more. Those further south still, around Watervale and Leasingham, average only about 400 m; but the general trends of temperature with latitude (Figure 12) suggest that this area should be appreciably cooler than Clare at comparable altitudes.

A likely contributing factor is the incidence of afternoon cool breezes from the south. While those coming early enough in the afternoon would affect the recorded maximum temperatures, and therefore the calculated means and temperature summations, those coming about after 1500 hours would not. Nor would they affect the recorded afternoon relative humidities. Yet they could still have profound importance for viticulture (Chapter 3).

Halliday (1985b, p. 179) quotes Tim Knappstein as saying that sea breezes penetrate to Clare sufficiently to cool the summer evenings. Reactions to my own limited enquiries have been that such breezes are at least not a marked or regular feature of the Clare summer afternoons. A specific description, from the township, was that the predominant summer winds are from the north, through much of the day. About 1600 hours, however, they usually slacken or cease, with a marked drop in temperature which tends to be accompanied by indefinite breezes from the south-west or south.

Both descriptions are consistent with the recorded 1500 hours relative humidity averages for Clare (Table 39), which are typical of the dry Australian inland. But some sea breezes evidently do penetrate to there later in the afternoon; and this may well be enough to benefit the vines substantially and enhance their end-of-day photosynthesis and ripening conditions, as compared with more truly interior regions. Most importantly, locally differing exposure to the breezes might contribute critically to differences in grape quality.

Let us therefore now examine how different parts of the Clare region are exposed to winds and breezes from the various quarters. The Clare Valley itself falls in a shallow gradient to the north. It and the town are fully exposed to the hot northerly winds, and somewhat sheltered from the cooler southerly breezes.

---

1 Halliday is mistaken in one point, saying that the heat summation method takes only day temperatures into

account. The figures used in the published Australian debate, as in the USA but unlike those often used in Europe, have to the best of my knowledge all been means. That is, they are the mid-point averages of the daily maximum and minimum temperatures. The means almost certainly predict vine phenology, and probably adaptation, more accurately than the day temperatures or maxima: see Chapter 6.

Sevenhill and Penwortham, on the watershed at the top of the valley, would have similar exposure in both directions. Watervale, Leasingham and Auburn are in the opposing valley grading to the south. They would be the most favoured to receive cool and humid breezes funnelling up the valley from the south, and would certainly tend to receive them earlier than Clare.

The high country immediately east of Clare deserves attention in a different way, because topographically it constitutes a truly isolated ridge, with free air drainage away in all directions. As documented in Chapter 4, such isolated hills and ridges have uniquely favourable mesoclimates, and provide the sites of many of the world's greatest vineyards. For inland Australian locations, they give the best available means to offset broader regional climatic shortcomings for viticulture.

The Clare ridge extends from some kilometres northeast of the town, southwards nearly to Auburn. As the highest point of the North Mount Lofty Range's western arm, it probably forms a dividing line between air convections influenced by Spencer and Saint Vincent gulfs on the west side, and those of purely interior air on the east side. The western face of the ridge would be directly exposed to any sea breezes from the southwest that reach that far, whereas its inland-facing slope would tend to remain in the opposing convection as illustrated in Figure 7, Chapter 4. To the extent that this occurs, it gives the western slopes a clear advantage for viticulture.

It is probably fair to say that most of the table wines for which Clare now enjoys its highest reputation come from grapes grown either in the higher central or southern parts of the region, or if from elsewhere, from higher west-facing slopes with good exposure to cool breezes.

Originally, vines at Clare were planted mostly in the valley for the good reason that the deep, water-retentive soils there were excellent for fully dry-land viticulture. In the absence of supplementary watering, and with marginal and mainly winter-spring rainfall, they were in fact essential. As in the Barossa Valley and Hills, the climate-imposed limitation on growth (even, later, with some supplementary watering) was enough to bring vines into natural fruiting balance and reasonable canopy light relations with simple or no trellising.

Varieties to respond best appear to have been those which in 'softer' climates (at least at low latitudes) tend to become too vegetative, unless accorded advanced trellising and canopy management. Riesling is a typical example. It is also notable that Malbec, a notoriously rampant and difficult variety in coastal climates, is reckoned to be well adapted at Clare (Halliday 1985b). This situation parallels that, not only of the Barossa Valley and Hills, but also of the dry inland fringe of the Lower Great Southern viticultural region in Western Australia (Chapter 9). It exemplifies the compensatory interactions of climate and vine management in determining vine balance, and therefore to some degree

wine quality, as discussed in Chapter 5.

Of particular importance to Clare is the fact that surface water conservation and supplementary watering now make possible the use of shallower soils on the middle and higher slopes, with better air drainage and temperature relations than before. By careful control over water supply, simple trellising can continue to be appropriate. Alternatively, nutrition and water can be increased to give higher yields with more complex trellising. Either way, there are clear quality advantages over the old valley floor sites, both in greater control over vegetative growth and in the lesser temperature variability that these sites enjoy.

The question must nevertheless still be asked, whether such plantings in semi-inland areas of Australia can be fully competitive with those in theoretically more favoured coastal regions, once the latter have come fully to terms *with the most appropriate methods of trellising and canopy management*.

I suggest that the environmental equation will ultimately favour the coastal regions for most table wines, as will the economic equation – provided that suitable coastal and near-coastal land remains available for viticulture as against alternative uses. That in itself could become a pressing problem in the longer term, as indeed it has already in some regions.

The viticultural prospects for 'semi inland' localities with still fairly moderate climates, such as Clare, are thus by no means entirely gloomy. Cheaper land availability, combined with optimal site selection and management practices, will probably enable them to continue making very good wines competitively, in traditional Australian styles for which there are established markets. Such areas and wine styles are unlikely ever to be displaced completely.

Langhorne Creek

The climate of Langhorne Creek is estimated from **Strathalbyn** (Table 48). Ripening average mean temperatures are in the upper optimum range for medium to full-bodied table wines, with higher relative humidities than in the Barossa, Clare, and even Adelaide/Southern Vales. This is because of the predominantly on-shore, southerly winds in summer. Sunshine hours are less, though still ample, resulting in a ratio of growing season sunshine hours to effective temperature summation of only a little over one (see Chapter 3).

Because of these factors, together, perhaps, with the deep alluvial soil and irrigation from the Bremer River, the wines at their best are reputedly softer and fruitier than those from further inland. Langhorne Creek produces some of Australia's most highly regarded dry red wines. However it still suffers occasionally from very hot, dry northerly winds which can dominate over the southerlies for prolonged periods, and which are reflected in the variability of the maximum temperatures (column 9).

South-East

Both **Coonawarra** (Table 40) and **Padthaway** (Table 44) share Strathalbyn's variability of the maximum temperature, a result of their 'south coast' location, together with a lack of backing hills to act as barriers against hot winds from the north. To this must be added a wide diurnal temperature range and marked variability of the minimum, due to their flat terrain. Frosts are a serious hazard in spring, while in summer the very hot, dry, northerly winds can occasionally stress the vines regardless of irrigation.

Against this, the two areas have average ripening mean temperatures that are optimal for light to medium, and medium to full-bodied table wines respectively; appropriate sunshine hours; and reasonably regular afternoon sea breezes throughout summer which provide optimum temperature and relative humidity conditions once they have arrived, and temper the sporadic extremes. Stresses in mid-summer, provided that the vines do not suffer unduly, can perhaps on balance be beneficial in this cool climate, by helping to stop unwanted vegetative growth. Both areas have well-drained, red, limestone-based soils with good water for supplementary watering. With sprinkler irrigation the same water can be used, when necessary, for frost control in spring.

Coonawarra is known best for its often outstanding medium-bodied red wines: particularly from Shiraz, and more recently from Cabernet Sauvignon. Halliday (1983) has described the area and its vintages to that date. White wines, despite some very good ones, are overall less highly reputed. The difference may be related to the respective conditions under which the grapes ripen. Shiraz and Cabernet do not normally ripen until April, by which time conditions are mild and equable. In good seasons the temperatures are very close to optimal for quality. The warmth and red colour of the better soils may also be a factor in allowing full ripening and flavour development to continue under relatively cool autumn conditions. Full ripening can be precarious, however, in cool seasons with early autumn rains. Bunch rot is also a problem in some seasons. But in the mostly warm and dry autumns of recent decades, both varieties have fairly regularly given wines of excellent balance and character.

The earlier-ripening white varieties, and early red varieties such as Pinot Noir, ripen at Coonawarra when temperatures are on average still too variable for fully reliable quality. Day temperatures are still above the ideal, but night temperatures are already too low for optimum physiological ripening.

The Keppoch Valley at Padthaway, with irrigation, has become established predominantly as a volume producer of good-average quality table wines. Average mean ripening temperatures are optimal for medium to full-bodied wines, but afternoon relative humidities are often too low, and the variability of the temperatures during ripening of all maturity types too great,

to bring out the best grape qualities reliably.

Interestingly, the best Keppoch wines have usually been white, or else sometimes red wines made from Pinot Noir, for which the grapes ripen early and under fairly warm conditions. Perhaps midseason red grapes fall into the same climatic hole as early white grapes at Coonawarra (above): that is, day temperatures during ripening are still high and variable, but night temperatures have meantime fallen too low for optimum physiological ripening. A remedy, as at Clare, might theoretically be to find hillside sites with better air drainage and temperature relations; but unfortunately the Keppoch Valley lacks hills, and the best soils for viticulture are confined to the lowest flats.

The Riverland

The climate of the Murray River irrigation area is fairly uniform, and is typified by Berri (Table 38).

Of the fully irrigated viticultural areas of Australia, this is the coolest during ripening. Ripening average mean temperatures are nearly 2°C lower than at **Griffith** (Chapter 12, Table 91) and **Leeton** (Table 95) in the Murrumbidgee irrigation area of New South Wales. On the other hand the average extreme maxima during ripening at Berri are about 1.5°C higher. The greater variability of the maximum at Berri reflects its marginally south coast/dry mediterranean climatic type, whereas Leeton and Griffith have a slight sub-tropical influence which moderates the summer extremes. Consistent with their sub-tropical influence, Leeton and Griffith are more subject to early spring frosts.

On balance no clear distinction can be drawn between the two areas. Both are hot and arid, though less so than the inland valleys of California (see Fresno, Chapter 22, Table 174). Ripening is within the optimum mean temperature range for fortified wines, and not inconsistent with good full-bodied table wines, given the right grape varieties and vineyard management. However the combination of rather high means with wide temperature variation, and very low relative humidities, results in a total ripening environment that in most seasons is clearly adverse to table wine quality.

Soils in the Riverland range from sands to well-drained red sandy loam over limestone, and are generally favourable for viticulture.

Potential new viticultural areas

Several areas remain in South Australia which, climatically, appear at least as good as those already planted. The most interesting is the lower Fleurieu Peninsula, for which I have attempted climatic estimates based on Victor Harbor (Table 49).

The Fleurieu Peninsula is a south-westerly extension of the Mount Lofty Range. Altitudes are less than in the range near Adelaide, but for average temperatures the distance south offsets this. The great climatic advantage of the area lies in its strong maritime influences,

Table 33. Supplementary ripening-month estimates for the lower Fleurieu Peninsula, South Australia, based on Victor Harbor

Grape maturity group	Sea level				300 metres			
	Maturity date	Mean temp. °C	Highest max. °C	Sunshine hrs	Maturity date	Mean temp. °C	Highest max. °C	Sunshine hrs
1	Feb. 17	19.6	38.0	255	March 21	17.4	34.0	232
2	Feb. 22	19.5	37.5	250	March 29	17.1	33.5	227
3	Feb. 28	19.4	37.1	245	April 7	16.3	32.5	218
4	March 5	19.3	36.7	241	April 17	15.5	31.0	203
5	March 11	19.2	36.4	238	April 29	14.5	28.7	180
6	March 16	19.1	36.1	235	May 15	13.1	25.6	160
7	March 22	19.0	35.8	232	–	–	-	-
8	March 29	18.8	35.4	228	–	–	-	-

resulting in greater humidity and a moderation of the temperature extremes experienced further east and inland. Average 1500 hours relatively humidity in January reaches 60 per cent in the south (Division of National Mapping, Australia, 1986), which is optimal for table wines and greatly exceeds that of Adelaide and areas further north. Sunshine hours remain fully adequate for table wines.

The estimates based on Victor Harbor in Table 49 are for potential vineyard sites a little way inland, at about 100–150 m altitude. Table 33 gives supplementary estimates for sites close to sea level, using the raw Victor Harbor data, and at 300 m altitude. Values for other altitudes can be interpolated as desired. The tables assume a normal temperature lapse rate of 0.6°C per 100 m altitude, and that average temperature variability and sunshine hours are uniform across the region.

Short-term figures for Myponga, towards the west coast of the peninsula (Climatic Averages Australia, 1975) show 1323 raw day degrees at 226 m altitude, which is 25 day degrees below the regression with altitude based on Victor Harbor records. Myponga is in a valley, which would depress mean temperature because of cold nights. The correspondence suggests that the Victor Harbor-based estimates are probably representative enough of the Fleurieu Peninsula as a whole.

The Myponga figures for 1500 hours relative humidity (mean of 48 per cent for the whole growing period, with a minimum of 40 per cent in December rising to 55 per cent in April) nevertheless indicate that sheltered pockets in valleys towards the west coast can be distinctly less humid in summer than the south coast, as would be expected from the predominantly south to south-westerly afternoon winds in summer. The peninsula's west coast (actually facing north-west, into the Gulf of Saint Vincent) could also be expected to have more summer sunshine, for the same reason.

I believe this to be a region with exciting potential for viticulture. The temperature and sunshine data point to a wide range of table wine styles, from full-bodied Bordeaux and Rhône styles at low altitudes to classical cool-climate styles from about 200 m up.

The fully exposed south coast is probably too windy, apart from a few very sheltered valley locations. The slopes descending north-westwards from the main peninsula ridge, on the other hand, should be reasonably sheltered from the full force of the south-west and southerly winds, while still benefiting from their equable temperatures and humidity. Suitable, moderately hilly, topography exists through much of the Hindmarsh, Inman and Yankalilla valleys, and in the basins of smaller streams running into the Gulf of Saint Vincent between Yankalilla and Cape Jervis.

The western part of the peninsula has the further advantage that hot winds from the north and north-west are moderated by their passage over the Gulf of Saint Vincent. The highest maximum temperatures experienced in summer and during ripening might therefore be appreciably lower here, for given altitudes and grape maturity groups, than estimated in Tables 33 and 49 from the Victor Harbor data.

Annual rainfall across the region is mostly between 500 and 600 mm. This would necessitate surface water conservation and supplementary drip irrigation more or less universally. The soils are podsolic and generally less deep or fertile than in neighbouring established viticultural regions, such as the Southern Vales and Langhorne Creek. Nevertheless many do appear suitable for viticulture. Combined with supplementary drip irrigation and the region's dry summer, they should allow good control over the growth and fruiting balance of most grape varieties, as discussed above for Clare.

At least in climatic terms, the lower Fleurieu Peninsula has arguably the best conditions of all in mainland South Australia for table wine production.

Another region which appears *climatically* promising is **Kangaroo Island**. Unfortunately the only long-term data are from Kingscote, which is on a small peninsula and atypically maritime. Climatic Averages Australia (1975) has more recent data for both Kingscote and Parndana East, the latter being at 155 m altitude in the centre of the island. Raw temperature summations (with 19°C cut-off) are 1571 and 1362 respectively for Kingscote and Parndana East, compared with 1483 for Kingscote from Climatic Averages Australia (1956).

Because all northerly winds must pass over the ocean, variability of the maximum temperature should be restricted compared with the mainland, especially along the north coast. Temperature variability indices for the growing season and ripening month can be estimated roughly to be in the low 30s, which is highly favourable for winegrape quality. Growing season relative humidities are also well within the optimum range.

Soils and/or drainage are admittedly unsuitable for viticulture over much of Kangaroo Island. With its low relief, winds would also be a problem necessitating wind breaks and very careful canopy management in most areas. Lack of continentality, combined with the temperature summations, suggest that any plantings should be confined initially to Bordeaux grape varieties. But within those constraints, the indications for quality are excellent if sheltered pockets of well-drained gravelly or calcareous soils can be found.

A further area of interest is the southern tip of Eyre Peninsula, represented by Port Lincoln (Table 45). Patches of terra rossa soil over limestone, resembling that found at Coonawarra and in the Keppoch Valley, occur between Port Lincoln and Dutton Bay, which is on the southern west coast of the peninsula. As at Coonawarra and Keppoch, these soils overlie readily available water of good quality.

The tip of Eyre Peninsula appears climatically to be well suited to producing full-bodied table wines, given some supplementary watering and shelter from southerly and westerly winds. Temperature variability and overall conditions deteriorate rapidly to the north and inland. As on Kangaroo Island, Bordeaux grape varieties are indicated.

A final region of South Australia which could warrant some consideration for viticulture is the south-east of **Yorke Peninsula**, on the terra rossa sandy loams that extend from about Port Julia in the north to Warooka

in the west (Hubble 1970). No long-term climatic data are available for this area, but those from Climatic Averages Australia (1975) for Warooka show a growing season raw temperature summation (with 19°C cut-off) of 1691; diurnal temperature range/2, 5.7; and 1500 hours relative humidity, 47 per cent. The figures point to the cooler southern part of Yorke Peninsula being in the upper optimum temperature range for table wines. This is a flat area, but reasonably moderated by maritime influences. Given suitable soils, supplementary water and adequate wind breaks, it might have potential for full-bodied table wines.

Addendum (*added in original second printing*)

One area overlooked in the first printing is the coastal strip west and south of Coonawarra, in the immediate hinterlands of Beachport and Robe and extending south to the Victorian border. Hubble (1970) indicates shallow red sandy soils over limestone, with some sandy yellow solodic soils in the north, both merging into black clay soils or bleached sands at the inland margin.

Date for Robe in *Climatic Averages Australia* (1956, 1975, 1988) indicate a growing season raw temperature summation almost identical with that of Coonawarra, but with a warmer spring, cooler summer and warmer autumn – as might be expected from its coastal location. However, a narrower diurnal temperature range makes the summation for Robe more biologically effective. Taking these factors into account, together with the fact that any vineyards would be inland from Robe (and therefore, climatically, intermediate between it and Coonawarra), it can be predicted that early-maturing grape varieties will ripen at about the same time and under the same mean temperatures as at Coonawarra. Later-maturing red varieties, such as Shiraz and Cabernet Sauvignon, should ripen up to a few days earlier: a beneficial difference. Ripening potential would diminish towards the southern end of the strip, with areas near Mt Gambier probably confined to early grape varieties (up to Maturity Group 4).

A smaller diurnal temperature range and higher afternoon relative humidities are clear quality advantages for all varieties. However, the vines would need good natural or artificial protection from wind, together with careful selection of the best soils. Given these, the viticultural potential appears very favourable, with the best spots possibly in Category 1 as defined in Chapter 24.

TABLE 34

ADELAIDE (city), SOUTH AUSTRALIA. 34°56'S, 138°35'E. ALT. 43m.

	Temperature °C								Summation of Day Degrees above 10° with 19° Cut-Off			Temperature Variability Index					
	1	2	3	4	5	6	7	8	9	10	11	12	13	14	15	16	17	18
PERIOD	Lowest Recorded Min. (113YRS)	No.Mths with <0.3° (30YRS)	No.Mths with <2.5° (30YRS)	Av.Mthly Lowest Min. (30YRS)	Av.Min. Minus Col.4	Daily Mean (30YRS)	Daily Range /2 (30YRS)	Av.Mthly Highest Max. (30YRS)	Col.8 Minus Av.Max.	Raw	Adj.1 for Lat. & Daily Temp Rng	Adj.2 for Vine Sites	Raw	Adj. for Vine Sites	Av. Sunshine Hrs	Av. Number Rain Days	Av. Rainfall mm	Av. % R.H. 1500 HRS (30YRS)
OCTOBER	2.2	0	0	6.0	4.9	16.7	5.8	31.5	9.0	208	205		37.1		225	11	44	41
NOVEMBER	4.9	0	0	7.8	5.2	19.3	6.3	36.0	10.4	270	270		40.8		260	8	31	35
DECEMBER	6.1	0	0	9.6	5.3	21.5	6.6	38.6	10.5	279	279		42.2		300	6	27	32
JANUARY	7.3	0	0	11.1	5.0	22.7	6.6	40.3	11.0	279	279		42.4		300	4	20	31
FEBRUARY	7.5	0	0	11.3	5.3	23.2	6.6	38.2	8.4	252	252		40.1		250	4	21	32
MARCH	6.6	0	0	10.2	4.8	21.2	6.2	35.9	8.5	279	279		38.1		245	5	24	34
APRIL	4.2	0	0	8.2	4.2	17.6	5.2	30.6	7.8	228	231		32.8		185	9	45	44
OCT-APRIL	—	0	0	—	5.0	20.3	6.2	—	9.4	1,795	1,795		39.1		1,765	47	212	36
MAY-SEPT	—	—	—	—	—	13.0	—	—	—	—	—		—		—	—	321	—

DATA FOR MONTH TO MATURITY

MATURITY GROUP	EST. MATURITY DATE																	
1	FEBRUARY 2					22.9	6.6	40.2	10.8				42.3		298	4	20	31
2	FEBRUARY 8					23.2	6.6	40.0	10.2				41.8		292	4	20	31
3	FEBRUARY 13					23.4	6.6	39.6	9.7				41.3		286	4	20	31
4	FEBRUARY 19					23.4	6.6	39.2	9.2				40.8		280	4	21	32
5	FEBRUARY 25					23.4	6.6	38.7	8.7				40.4		275	4	21	32
6	MARCH 2					23.1	6.6	38.2	8.5				40.0		270	4	21	32
7	MARCH 8					22.8	6.5	37.7	8.5				39.7		265	4	21	32
8	MARCH 13					22.5	6.4	37.2	8.5				39.3		260	4	22	33

TABLE 35

ADELAIDE (Waite Institute), SOUTH AUSTRALIA. 34°58'S, 138°38'E. ALT. 123m. (Adelaide metropolitan, Southern Vales)

	Temperature °C								Summation of Day Degrees above 10° with 19° Cut-Off			Temperature Variability Index					
	1	2	3	4	5	6	7	8	9	10	11	12	13	14	15	16	17	18
PERIOD	Lowest Recorded Min. (45YRS)	No.Mths with <0.3° (30YRS)	No.Mths with <2.5° (30YRS)	Av.Mthly Lowest Min. (30YRS)	Av.Min. Minus Col.4	Daily Mean (26YRS)	Daily Range /2 (26YRS)	Av.Mthly Highest Max. (30YRS)	Col.8 Minus Av.Max.	Raw	Adj.1 for Lat. & Daily Temp Rng	Adj.2 for Vine Sites	Raw	Adj. for Vine Sites	Av. Sunshine Hrs	Av. Number Rain Days	Av. Rainfall mm	Av. % R.H. 1500 HRS
OCTOBER	3.4	0	0	5.6	5.0	15.5	4.9	30.2	9.8	171	170		34.4		225	13	52	47
NOVEMBER	4.4	0	0	6.9	5.8	18.1	5.4	34.2	10.7	243	237		38.1		260	9	38	41
DECEMBER	5.4	0	0	8.5	6.0	20.3	5.8	37.3	11.2	279	279		40.4		300	8	31	37
JANUARY	8.3	0	0	10.0	5.6	21.5	5.9	38.8	11.4	279	279		40.6		300	5	24	35
FEBRUARY	6.7	0	0	10.3	5.6	21.5	5.6	37.2	10.1	252	252		38.1		250	5	28	36
MARCH	6.0	0	0	9.6	5.5	20.4	5.3	34.8	9.1	279	279		35.8		245	6	21	39
APRIL	5.3	0	0	7.5	4.9	16.8	4.4	30.5	9.3	204	216		31.8		185	11	57	50
OCT-APRIL	—	0	0	—	5.5	19.2	5.3	—	10.2	1,707	1,712		37.0		1,765	57	251	41
MAY-SEPT	—	—	—	—	—	12.2	—	—	—	—	—		—		—	—	376	—

DATA FOR MONTH TO MATURITY

MATURITY GROUP	EST. MATURITY DATE																	
1	FEBRUARY 9					21.6	5.8	38.3	11.0				39.9		290	5	26	35
2	FEBRUARY 15					21.6	5.8	38.0	10.8				39.4		285	5	27	35
3	FEBRUARY 21					21.6	5.7	37.7	10.5				38.8		279	5	28	36
4	FEBRUARY 26					21.6	5.7	37.4	10.2				38.3		273	5	29	36
5	MARCH 4					21.5	5.6	37.0	10.0				37.9		268	5	29	36
6	MARCH 10					21.4	5.6	36.5	9.8				37.5		264	5	28	37
7	MARCH 15					21.2	5.5	36.1	9.6				37.1		260	5	26	37
8	MARCH 21					21.0	5.4	35.7	9.4				36.7		255	6	24	38

TABLE 36

BAROSSA VALLEY, SOUTH AUSTRALIA (estimate). Approx. 34°35′S, 139°00′E. ALT. approx. 250m. (See also Table 43.)

	1	2	3	4	5	6	7	8	9	10	11	12	13	14	15	16	17	18
	Temperature °C									Summation of Day Degrees above 10° with 19° Cut-Off			Temperature Variability Index					
PERIOD	Lowest Recorded Min.	No.Mths with <0.3°	No.Mths with <2.5°	Av.Mthly Lowest Min.	Av.Min. Minus Col.4	Daily Mean	Daily Range /2	Av.Mthly Highest Max.	Col.8 Minus Av.Max.	Raw	Adj.1 for Lat. & Daily Temp Rng	Adj.2 for Vine Sites	Raw	Adj. for Vine Sites	Av. Sunshine Hrs	Av. Number Rain Days	Av. Rainfall mm	Av. % R.H. 1500 HRS
OCTOBER				1.7	6.1	13.6	5.8	29.5	10.1	112	110		39.4		240	11	45	46
NOVEMBER				2.8	7.2	16.5	6.5	34.5	11.5	195	190		44.7		272	8	28	39
DECEMBER				4.8	7.0	19.0	7.2	37.0	10.8	279	258		46.6		300	6	26	36
JANUARY				6.6	6.6	21.0	7.8	39.6	10.8	279	279		48.6		300	5	19	36
FEBRUARY				7.2	5.9	20.1	7.0	37.0	9.9	252	252		43.8		250	3	25	37
MARCH				5.5	6.0	18.7	7.2	35.3	9.3	270	258		44.1		245	5	19	41
APRIL				2.9	5.8	14.6	5.9	30.6	10.1	138	139		39.5		195	9	42	51
OCT-APRIL				—	6.4	17.6	6.8	—	10.4	1,525	1,486		43.8		1,802	47	204	41
MAY-SEPT				—	—	10.0	—	—	—	—	—		—		—	—	302	—

DATA FOR MONTH TO MATURITY

MATURITY GROUP	EST. MATURITY DATE																	
1	FEBRUARY 24					20.6	7.1	37.3	10.1				44.3		275	3	25	37
2	MARCH 2					20.5	7.0	36.9	9.9				43.8		269	3	25	37
3	MARCH 7					20.3	7.0	36.5	9.7				43.8		264	3	24	38
4	MARCH 13					20.0	7.1	36.1	9.6				43.9		259	4	23	38
5	MARCH 19					19.6	7.1	35.8	9.5				44.0		255	4	22	39
6	MARCH 25					19.2	7.2	35.4	9.4				44.1		250	5	21	40
7	APRIL 1					18.6	7.2	35.0	9.3				44.0		243	5	20	41
8	APRIL 9					17.6	7.0	34.0	9.5				43.5		231	3	25	43

TABLE 37

BELAIR, SOUTH AUSTRALIA. 35°00′S, 138°38′E. ALT. 308m.

	1	2	3	4	5	6	7	8	9	10	11	12	13	14	15	16	17	18
	Temperature °C									Summation of Day Degrees above 10° with 19° Cut-Off			Temperature Variability Index					
PERIOD	Lowest Recorded Min. (43YRS)	No.Mths with <0.3° (25YRS)	No.Mths with <2.5° (25YRS)	Av.Mthly Lowest Min. (30YRS)	Av.Min. Minus Col.4	Daily Mean (30YRS)	Daily Range /2 (30YRS)	Av.Mthly Highest Max. (30YRS)	Col.8 Minus Av.Max.	Raw	Adj.1 for Lat. & Daily Temp Rng	Adj.2 for Vine Sites	Raw	Adj. for Vine Sites	Av. Sunshine Hrs	Av. Number Rain Days	Av. Rainfall mm	Av. % R.H. 1500 HRS (26YRS)
OCTOBER	1.7	0	0	5.0	5.0	14.6	4.6	30.4	11.2	143	147		34.6		220	10	60	59
NOVEMBER	4.4	0	0	6.3	5.8	17.4	5.3	35.0	12.3	222	216		39.3		256	7	42	52
DECEMBER	5.7	0	0	7.8	5.7	19.1	5.6	36.4	11.7	279	279		39.8		297	5	35	47
JANUARY	7.7	0	0	9.4	5.1	20.4	5.9	38.6	12.3	279	279		41.0		298	4	25	44
FEBRUARY	4.3	0	0	9.5	5.7	21.0	5.8	37.8	11.0	252	252		39.9		248	4	28	43
MARCH	4.7	0	0	8.7	5.4	19.0	4.9	35.4	11.5	279	279		36.5		242	4	30	50
APRIL	4.2	0	0	6.9	4.9	15.7	3.9	29.4	9.8	171	190		30.3		181	9	62	62
OCT-APRIL	—	0	0	—	5.4	18.2	5.1	—	11.4	1,625	1,642		37.3		1,742	43	282	51
MAY-SEPT	—	—	—	—	—	11.2	—	—	—	—	—		—		—	—	473	—

DATA FOR MONTH TO MATURITY

MATURITY GROUP	EST. MATURITY DATE																	
1	FEBRUARY 15					21.2	5.8	38.2	11.6				40.4		283	4	27	43
2	FEBRUARY 20					21.2	5.8	38.1	11.3				40.2		278	4	27	43
3	FEBRUARY 26					21.1	5.8	37.9	11.1				40.0		273	4	28	43
4	MARCH 3					20.9	5.7	37.6	11.1				39.8		268	4	28	44
5	MARCH 9					20.6	5.6	37.3	11.1				39.5		263	4	28	45
6	MARCH 15					20.3	5.5	36.9	11.2				39.0		258	4	29	46
7	MARCH 20					19.9	5.3	36.5	11.3				38.3		253	4	29	47
8	MARCH 26					19.4	5.1	36.0	11.4				37.3		248	4	30	49

TABLE 38
BERRI, SOUTH AUSTRALIA. 34°17'S, 140°38'E. ALT. 31m.

	1	2	3	4	5	6	7	8	9	10	11	12	13	14	15	16	17	18
	Temperature °C									Summation of Day Degrees above 10° with 19° Cut-Off			Temperature Variability Index					
PERIOD	Lowest Recorded Min. (39YRS)	No.Mths with <0.3° (30YRS)	No.Mths with <2.5° (30YRS)	Av.Mthly Lowest Min. (30YRS)	Av.Min. Minus Col.4	Daily Mean (25YRS)	Daily Range /2 (25YRS)	Av.Mthly Highest Max. (30YRS)	Col.8 Minus Av.Max.	Raw	Adj.1 for Lat. & Daily Temp Rng	Adj.2 for Vine Sites	Raw	Adj. for Vine Sites	Av. Sunshine Hrs	Av. Number Rain Days	Av. Rainfall mm	Av. % R.H. 1500 HRS (22YRS)
OCTOBER	1.9	0	1	4.8	5.0	16.7	6.9	34.2	10.6	208	198		43.2		265	5	22	39
NOVEMBER	4.3	0	0	6.3	5.9	19.6	7.4	37.9	10.9	270	266		46.4		292	5	24	27
DECEMBER	5.4	0	0	8.2	6.1	22.1	7.8	39.9	10.0	279	279		47.3		328	5	20	25
JANUARY	5.3	0	0	9.7	5.4	23.0	7.9	41.8	10.9	279	279		47.9		325	4	20	24
FEBRUARY	6.7	0	0	9.3	5.7	22.6	7.6	39.4	9.2	252	252		45.3		282	3	22	28
MARCH	3.9	0	0	7.5	5.8	20.6	7.3	37.1	9.2	279	279		44.2		275	3	11	30
APRIL	−1.1	1	2	4.4	5.6	16.3	6.3	31.6	9.0	189	192		39.8		235	6	20	40
OCT-APRIL	—	1	3	—	5.6	20.1	7.3	—	10.0	1,756	1,745		44.9		2,002	31	139	30
MAY-SEPT	—	—	—	—	—	12.1	—	—	—	—	—		—		—	—	130	—
DATA FOR MONTH TO MATURITY																		
MATURITY GROUP	EST. MATURITY DATE																	
1	FEBRUARY 3					23.0	7.9	41.8	10.8				47.6		323	4	20	24
2	FEBRUARY 9					23.0	7.8	41.3	10.4				47.1		319	4	21	25
3	FEBRUARY 15					22.9	7.8	40.8	9.8				46.6		315	3	22	26
4	FEBRUARY 20					22.8	7.7	40.2	9.5				46.0		311	3	23	27
5	FEBRUARY 26					22.7	7.6	39.6	9.3				45.5		307	3	24	28
6	MARCH 3					22.5	7.6	39.2	9.2				45.2		302	3	22	28
7	MARCH 9					22.2	7.5	38.8	9.2				45.0		296	3	20	29
8	MARCH 14					21.9	7.5	38.4	9.2				44.8		291	3	18	29

TABLE 39
CLARE, SOUTH AUSTRALIA. 33°50'S, 138°37'E. ALT. 398m.

	1	2	3	4	5	6	7	8	9	10	11	12	13	14	15	16	17	18
	Temperature °C									Summation of Day Degrees above 10° with 19° Cut-Off			Temperature Variability Index					
PERIOD	Lowest Recorded Min. (92YRS)	No.Mths with <0.3° (30YRS)	No.Mths with <2.5° (30YRS)	Av.Mthly Lowest Min. (30YRS)	Av.Min. Minus Col.4	Daily Mean (30YRS)	Daily Range /2 (30YRS)	Av.Mthly Highest Max. (30YRS)	Col.8 Minus Av.Max.	Raw	Adj.1 for Lat. & Daily Temp Rng	Adj.2 for Vine Sites	Raw	Adj. for Vine Sites	Av. Sunshine Hrs	Av. Number Rain Days	Av. Rainfall mm	Av. % R.H. 1500 HRS (67YRS)
OCTOBER	−1.3	5	25	1.8	5.5	14.4	7.1	30.5	9.0	136	125	143	42.9	39.3	250	10	55	44
NOVEMBER	0.0	1	11	3.6	6.4	17.5	7.5	34.6	9.6	225	203	225	46.0	42.4	280	7	36	37
DECEMBER	2.6	0	0	4.9	7.4	20.1	7.8	37.6	9.7	279	279	279	48.3	44.7	310	5	30	33
JANUARY	4.4	0	0	6.7	6.7	21.3	7.9	39.1	9.9	279	279	279	48.2	44.6	310	4	26	32
FEBRUARY	3.3	0	0	6.9	6.9	21.6	7.8	38.0	8.6	252	252	252	46.7	43.1	255	4	26	33
MARCH	2.7	0	0	4.7	6.9	19.2	7.6	35.1	8.3	279	267	279	45.6	42.0	255	5	24	37
APRIL	0.4	0	16	2.3	5.9	14.8	6.6	30.3	8.9	144	145	156	41.2	37.6	210	8	48	46
OCT-APRIL	—	6	52	—	6.5	18.4	7.5	—	9.1	1,594	1,550	1,613	45.6	42.0	1,870	43	245	37
MAY-SEPT	—	—	—	—	—	9.9	—	—	—	—	—	—	—	—	—	—	387	—
DATA FOR MONTH TO MATURITY																		
MATURITY GROUP	EST. MATURITY DATE					*	*	*					*					
1	FEBRUARY 14					22.1	6.9	38.0	9.2				43.7		296	5	27	33
2	FEBRUARY 20					22.1	6.9	37.8	9.0				43.5		290	5	27	33
3	FEBRUARY 25					22.0	6.9	37.5	8.8				43.3		282	5	28	33
4	MARCH 3					21.8	6.9	37.2	8.6				43.1		274	5	28	33
5	MARCH 8					21.5	6.9	36.8	8.5				42.9		270	5	27	34
6	MARCH 14					21.1	6.8	36.4	8.5				42.7		267	5	26	34
7	MARCH 19					20.7	6.8	35.9	8.4				42.5		264	5	25	35
8	MARCH 24					20.2	6.7	35.3	8.3				42.2		260	5	24	36

* Adjusted for vine sites.

115

TABLE 40
COONAWARRA, SOUTH AUSTRALIA. 37°18'S, 140°50'E. ALT. 59m.

	1	2	3	4	5	6	7	8	9	10	11	12	13	14	15	16	17	18
	Temperature °C									Summation of Day Degrees above 10' with 19° Cut-Off			Temperature Variability Index					
PERIOD	Lowest Recorded Min.	No.Mths with <0.3°	No.Mths with <2.5°	Av.Mthly Lowest Min.	Av.Min. Minus Col.4	Daily Mean	Daily Range /2	Av.Mthly Highest Max.	Col.8 Minus Av.Max.	Raw	Adj.1 for Lat. & Daily Temp Rng	Adj.2 for Vine Sites	Raw	Adj. for Vine Sites	Av. Sunshine Hrs	Av. Number Rain Days	Av. Rainfall mm	Av. % R.H. 1500 HRS
OCTOBER				0.2	5.9	12.6	6.5	28.7	9.6	81	80	86	41.5	39.1	213	14	55	58
NOVEMBER				1.1	6.7	14.8	7.0	33.7	11.9	144	135	148	46.6	44.2	220	12	39	53
DECEMBER				2.6	6.9	17.0	7.5	37.3	12.8	217	198	213	49.7	47.3	255	10	33	46
JANUARY				4.0	6.8	19.3	8.5	40.6	12.8	279	253	268	53.6	51.2	275	7	24	40
FEBRUARY				4.4	7.0	19.6	8.2	39.0	11.2	252	242	252	51.0	48.6	235	5	26	40
MARCH				2.9	6.7	17.4	7.8	35.8	10.6	229	209	224	48.5	46.1	220	8	29	45
APRIL				1.2	6.4	14.5	6.9	30.2	8.8	135	130	142	42.8	40.4	175	12	51	54
OCT-APRIL				—	6.6	16.5	7.5	—	11.1	1,337	1,247	1,333	47.7	45.3	1,593	68	257	48
MAY-SEPT				—	—	10.2	—	—	—	—	—	—	—	—	—	—	371	—
	DATA FOR MONTH TO MATURITY																	
MATURITY GROUP	EST. MATURITY DATE					*	*	*						*				
1	MARCH 11					19.1	7.4	37.4	11.0					47.8	245	6	28	42
2	MARCH 17					18.7	7.4	36.8	10.8					47.3	237	6	28	43
3	MARCH 24					18.1	7.3	36.0	10.6					46.7	227	7	29	44
4	APRIL 2					17.4	7.2	35.0	10.4					45.9	216	8	31	46
5	APRIL 10					16.8	6.9	33.6	10.1					44.5	203	9	36	48
6	APRIL 22					15.4	6.5	31.2	9.5					42.3	188	11	45	51
7	MAY 7					13.9	6.1	28.5	8.5					39.3	162	13	56	56
8	—					—	—	—	—					—	—	—	—	—

* Adjusted for vine sites.

TABLE 41
EDEN VALLEY-SPRINGTON, SOUTH AUSTRALIA. Approx. 34°35'S, 139°02'E. ALT. approx. 450m.

	1	2	3	4	5	6	7	8	9	10	11	12	13	14	15	16	17	18
	Temperature °C									Summation of Day Degrees above 10' with 19° Cut-Off			Temperature Variability Index					
PERIOD	Lowest Recorded Min.	No.Mths with <0.3°	No.Mths with <2.5°	Av.Mthly Lowest Min.	Av.Min. Minus Col.4	Daily Mean	Daily Range /2	Av.Mthly Highest Max.	Col.8 Minus Av.Max.	Raw	Adj.1 for Lat. & Daily Temp Rng	Adj.2 for Vine Sites	Raw	Adj. for Vine Sites	Av. Sunshine Hrs	Av. Number Rain Days	Av. Rainfall mm	Av. % R.H. 1500 HRS
OCTOBER				1.4	6.0	12.3	4.9	27.0	9.8	71	72		35.4		235			50
NOVEMBER				2.0	6.7	14.8	6.1	32.4	11.5	144	140		42.6		270			43
DECEMBER				3.7	6.6	17.3	7.0	35.4	11.1	226	211		45.7		295			40
JANUARY				4.9	6.5	19.3	7.9	37.2	10.0	279	258		48.1		295			39
FEBRUARY				6.1	5.7	18.9	7.1	36.0	10.0	249	235		44.1		245			40
MARCH				4.6	6.0	17.5	6.9	33.9	9.5	232	225		43.1		238			44
APRIL				2.9	5.6	13.6	5.1	28.6	9.9	108	110		35.9		186			55
OCT-APRIL				—	6.2	16.2	6.4	—	10.3	1,309	1,251		42.1		1,764			44
MAY-SEPT				—	—	9.0	—	—	—	—	—		—		—			—
	DATA FOR MONTH TO MATURITY																	
MATURITY GROUP	EST. MATURITY DATE																	
1	MARCH 20					18.2	7.0	34.9	10.0				43.4		248			42
2	MARCH 26					17.9	6.9	34.5	10.0				43.2		242			43
3	APRIL 2					17.2	6.8	34.1	9.9				42.9		236			44
4	APRIL 13					15.8	6.0	32.0	9.9				40.8		218			47
5	APRIL 30					13.6	5.1	28.5	9.8				35.9		186			55
6	—					—	—	—	—				—		—			—
7	—					—	—	—	—				—		—			—
8	—					—	—	—	—				—		—			—

TABLE 42

MOUNT BARKER, SOUTH AUSTRALIA. 35°04'S, 138°52'E. ALT. 330m.

PERIOD	1 Lowest Recorded Min. (108YRS)	2 No.Mths with <0.3° (30YRS)	3 No.Mths with <2.5° (30YRS)	4 Av.Mthly Lowest Min. (30YRS)	5 Av.Min. Minus Col.4	6 Daily Mean (30YRS)	7 Daily Range /2 (30YRS)	8 Av.Mthly Highest Max. (30YRS)	9 Col.8 Minus Av.Max.	10 Raw	11 Adj.1 for Lat. & Daily Temp Rng	12 Adj.2 for Vine Sites	13 Raw	14 Adj. for Vine Sites	15 Av. Sunshine Hrs	16 Av. Number Rain Days	17 Av. Rainfall mm (30YRS)	18 Av. % R.H. 1500 HRS (30YRS)
OCTOBER	-1.2	15	27	0.5	6.6	13.1	6.0	28.6	9.5	96	95	95	40.1	37.7	230	13	69	51
NOVEMBER	-1.1	4	20	1.8	6.9	15.5	6.8	33.3	11.0	165	156	161	45.1	42.7	245	9	41	45
DECEMBER	0.4	0	8	3.5	7.0	17.7	7.2	36.2	11.3	239	220	229	47.1	44.7	292	7	34	41
JANUARY	1.7	0	1	5.6	5.8	18.8	7.4	38.3	12.1	273	251	260	47.5	45.1	293	6	27	39
FEBRUARY	1.7	0	0	5.5	6.4	19.3	7.4	36.7	10.3	252	243	251	46.0	43.6	245	5	28	40
MARCH	0.6	0	5	3.9	6.3	17.3	7.1	34.0	9.6	226	216	225	44.3	41.9	240	7	31	42
APRIL	-1.2	7	21	1.6	6.5	13.8	5.7	28.6	9.1	114	116	116	38.4	36.0	185	11	62	54
OCT-APRIL	—	26	81	—	6.5	16.5	6.8	—	10.4	1,365	1,297	1,337	44.1	41.7	1,730	58	292	45
MAY-SEPT	—	—	—	—	—	9.8	—	—	—	—	—	—	—	—	—	—	489	—

DATA FOR MONTH TO MATURITY

MATURITY GROUP	EST. MATURITY DATE			6	*7	*8						*14	15	16	17	18
1	MARCH 7			19.0	6.7	35.5	10.1					43.3	260	5	30	40
2	MARCH 13			18.7	6.7	35.1	9.9					43.0	254	6	30	41
3	MARCH 20			18.3	6.6	34.5	9.8					42.6	248	6	31	41
4	MARCH 28			17.7	6.5	33.7	9.7					42.1	242	7	31	42
5	APRIL 6			16.6	6.3	32.7	9.5					41.5	235	8	35	44
6	APRIL 18			15.2	5.8	31.0	9.3					39.0	208	9	45	47
7	MAY 13			14.0	4.8	27.0	8.8					34.0	170	13	71	57
8	—			—	—	—	—					—	—	—	—	—

* Adjusted for vine sites.

TABLE 43

NURIOOTPA (BAROSSA VALLEY), SOUTH AUSTRALIA. 34°29'S, 139°01'E. ALT. 274m. (See also Table 36.)

PERIOD	1 Lowest Recorded Min. (16YRS)	2 No.Mths with <0.3° (16YRS)	3 No.Mths with <2.5° (16YRS)	4 Av.Mthly Lowest Min. (16YRS)	5 Av.Min. Minus Col.4	6 Daily Mean (24YRS)	7 Daily Range /2 (24YRS)	8 Av.Mthly Highest Max. (16YRS)	9 Col.8 Minus Av.Max.	10 Raw	11 Adj.1 for Lat. & Daily Temp Rng	12 Adj.2 for Vine Sites	13 Raw	14 Adj. for Vine Sites	15 Av. Sunshine Hrs	16 Av. Number Rain Days	17 Av. Rainfall mm	18 Av. % R.H. 1500 HRS
OCTOBER	-0.3	2	14	1.2	6.1	13.4	6.1	29.6	10.1	105	104		40.6		245	11	45	46
NOVEMBER	-1.0	1	9	2.0	7.5	16.3	6.8	35.0	11.9	189	179		46.6		277	8	28	39
DECEMBER	0.8	0	3	4.0	7.4	18.8	7.4	36.7	10.5	273	250		47.5		302	6	26	36
JANUARY	2.2	0	1	5.6	7.3	20.8	7.9	38.8	10.1	279	279		49.0		300	5	19	36
FEBRUARY	4.5	0	0	7.4	5.6	20.2	7.2	37.7	10.3	252	252		44.7		250	3	25	37
MARCH	0.8	0	2	4.8	6.3	18.3	7.2	34.4	8.9	257	245		44.0		247	5	19	41
APRIL	0.2	1	8	2.5	5.9	14.4	6.0	30.9	10.5	132	134		40.4		196	9	42	51
OCT-APRIL	—	4	37	—	6.6	17.5	6.9	—	10.3	1,487	1,443		44.7		1,817	47	204	41
MAY-SEPT	—	—	—	—	—	9.9	—	—	—	—	—		—		—	—	302	—

DATA FOR MONTH TO MATURITY

MATURITY GROUP	EST. MATURITY DATE			6	7	8	9					14	15	16	17	18
1	FEBRUARY 27			20.5	7.2	37.8	10.3					44.9	272	3	27	37
2	MARCH 4			20.3	7.2	37.5	10.1					44.6	268	3	25	37
3	MARCH 10			20.0	7.2	37.0	9.8					44.4	263	4	23	38
4	MARCH 16			19.6	7.2	36.3	9.5					44.2	258	4	21	39
5	MARCH 23			19.0	7.2	35.5	9.2					44.1	253	5	20	40
6	MARCH 30			18.4	7.2	34.5	8.9					44.0	248	5	19	41
7	APRIL 7			17.4	7.0	33.5	9.2					43.5	240	6	23	43
8	APRIL 17			16.1	6.6	32.4	9.7					42.2	220	7	30	46

TABLE 44

PADTHAWAY (KEPPOCH VALLEY), SOUTH AUSTRALIA. 36°37'S, 140°28'E. ALT. approx. 50m. (All data estimated.)

	1	2	3	4	5	6	7	8	9	10	11	12	13	14	15	16	17	18
	Temperature °C									Summation of Day Degrees above 10° with 19° Cut-Off			Temperature Variability Index					
PERIOD	Lowest Recorded Min.	No.Mths with <0.3°	No.Mths with <2.5°	Av.Mthly Lowest Min.	Av.Min. Minus Col.4	Daily Mean	Daily Range /2	Av.Mthly Highest Max.	Col.8 Minus Av.Max.	Raw	Adj.1 for Lat. & Daily Temp Rng	Adj.2 for Vine Sites	Raw	Adj. for Vine Sites	Av. Sunshine Hrs	Av. Number Rain Days	Av. Rainfall mm	Av. % R.H. 1500 HRS
OCTOBER				0.8	5.9	13.5	6.8	29.9	9.6	109	103	114	42.7	40.3	230	12	46	52
NOVEMBER				1.9	6.7	15.9	7.3	35.1	11.9	177	162	177	47.8	45.4	240	10	32	47
DECEMBER				3.6	6.9	18.1	7.6	38.5	12.8	251	229	244	50.1	47.7	279	8	27	41
JANUARY				4.9	6.8	20.2	8.5	41.5	12.8	279	279	279	53.6	51.2	290	5	20	35
FEBRUARY				5.4	7.0	20.6	8.2	40.0	11.2	252	252	252	51.0	48.6	247	5	21	36
MARCH				3.9	6.7	18.3	7.7	36.6	10.6	257	238	253	48.1	45.7	250	6	23	41
APRIL				1.8	6.4	15.1	6.9	30.8	8.8	153	148	160	42.8	40.4	194	10	39	50
OCT-APRIL				—	6.6	17.4	7.6	—	11.1	1,478	1,411	1,479	48.0	45.6	1,720	56	208	43
MAY-SEPT				—	—	10.8	—	—	—	—	—	—	—	—	—	—	301	—
DATA FOR MONTH TO MATURITY																		
MATURITY GROUP	EST. MATURITY DATE					*	*	*						*				
1	FEBRUARY 27					20.7	7.6	39.7	11.2					48.7	268	5	23	36
2	MARCH 4					20.5	7.5	39.3	11.1					48.2	266	5	23	36
3	MARCH 10					20.2	7.4	38.7	11.0					47.7	264	5	23	37
4	MARCH 16					19.8	7.3	37.9	10.9					47.2	260	6	23	38
5	MARCH 22					19.3	7.2	37.1	10.8					46.6	256	6	23	39
6	MARCH 29					18.7	7.1	36.3	10.7					45.9	251	6	23	40
7	APRIL 4					18.0	7.1	35.5	10.4					45.0	244	6	25	42
8	APRIL 12					17.2	7.0	34.3	10.0					43.8	235	7	28	44

* Adjusted for vine sites.

TABLE 45

PORT LINCOLN, SOUTH AUSTRALIA. 34°44'S, 135°51'E. ALT. 4m.

	1	2	3	4	5	6	7	8	9	10	11	12	13	14	15	16	17	18
	Temperature °C									Summation of Day Degrees above 10° with 19° Cut-Off			Temperature Variability Index					
PERIOD	Lowest Recorded Min. (45YRS)	No.Mths with <0.3° (30YRS)	No.Mths with <2.5° (30YRS)	Av.Mthly Lowest Min.	Av.Min. Minus Col.4	Daily Mean (59YRS)	Daily Range /2 (59YRS)	Av.Mthly Highest Max. (30YRS)	Col.8 Minus Av.Max.	Raw	Adj.1 for Lat. & Daily Temp Rng	Adj.2 for Vine Sites	Raw	Adj. for Vine Sites	Av. Sunshine Hrs	Av. Number Rain Days	Av. Rainfall mm	Av. % R.H. 1500 HRS (30YRS)
OCTOBER	3.9	0	0	6.4	3.7	15.1	5.0	30.9	10.8	158	156	144	34.5	35.7	220	11	35	61
NOVEMBER	5.8	0	0	7.8	4.1	17.0	5.1	33.6	11.5	210	204	193	36.0	37.2	232	7	23	57
DECEMBER	5.6	0	0	9.7	3.9	18.8	5.2	35.5	11.5	273	264	252	36.2	37.4	270	5	18	55
JANUARY	8.9	0	0	11.1	3.6	20.0	5.3	38.3	13.0	279	279	279	37.8	39.0	265	4	14	53
FEBRUARY	8.3	0	0	11.7	3.5	20.3	5.1	36.2	10.8	252	252	252	34.7	35.9	225	4	15	55
MARCH	8.4	0	0	10.6	3.6	19.0	4.8	34.6	10.8	279	279	265	33.6	34.8	220	5	19	58
APRIL	6.1	0	0	8.9	3.4	16.8	4.5	30.2	8.9	204	215	198	30.3	31.5	170	10	37	63
OCT-APRIL	—	0	0	—	3.7	18.1	5.0	—	11.0	1,655	1,649	1,583	34.7	35.9	1,602	46	161	57
MAY-SEPT	—	—	—	—	—	12.9	—	—	—	—	—	—	—	—	—	—	325	—
DATA FOR MONTH TO MATURITY																		
MATURITY GROUP	EST. MATURITY DATE					*												*
1	FEBRUARY 21					20.1	5.2	36.3	11.2					35.3	248	4	15	52
2	FEBRUARY 26					20.0	5.1	36.1	11.0					34.8	246	4	15	53
3	MARCH 4					19.8	5.1	35.8	10.8					34.5	242	4	15	53
4	MARCH 9					19.6	5.0	35.5	10.8					34.2	237	4	16	54
5	MARCH 15					19.4	5.0	35.2	10.8					34.0	232	4	17	54
6	MARCH 21					19.2	4.9	34.8	10.8					33.8	227	4	18	55
7	MARCH 27					18.9	4.9	34.4	10.8					33.6	222	5	19	55
8	APRIL 2					18.5	4.8	34.0	10.6					33.3	216	5	20	56

* Adjusted for vine sites.

TABLE 46

ROSEWORTHY (Agricultural College), SOUTH AUSTRALIA. 34°31'S, 138°41'E. ALT. 114m.

	1	2	3	4	5	6	7	8	9	10	11	12	13	14	15	16	17	18
	Temperature °C									Summation of Day Degrees above 10° with 19° Cut-Off			Temperature Variability Index					
PERIOD	Lowest Recorded Min. (62YRS)	No.Mths with <0.3° (30YRS)	No.Mths with <2.5° (30YRS)	Av.Mthly Lowest Min. (30YRS)	Av.Min. Minus Col.4	Daily Mean (41YRS)	Daily Range /2 (41YRS)	Av.Mthly Highest Max. (30YRS)	Col.8 Minus Av.Max.	Raw	Adj.1 for Lat. & Daily Temp Rng	Adj.2 for Vine Sites	Raw	Adj. for Vine Sites	Av. Sunshine Hrs	Av. Number Rain Days	Av. Rainfall mm	Av. % R.H. 1500 HRS (31YRS)
OCTOBER	−0.9	1	15	2.5	6.4	15.6	6.7	32.4	10.1	174	168		43.3		240	10	41	46
NOVEMBER	0.0	1	1	4.5	7.0	18.6	7.1	36.5	10.8	258	242		46.2		275	6	27	40
DECEMBER	2.9	0	0	6.7	7.0	21.1	7.4	39.2	10.7	279	279		47.3		302	5	24	38
JANUARY	5.2	0	0	8.4	6.1	22.1	7.6	41.4	11.7	279	279		48.2		300	4	22	36
FEBRUARY	5.6	0	0	8.5	6.4	22.3	7.4	39.3	9.6	252	252		45.6		253	3	20	36
MARCH	4.2	0	0	7.3	6.0	20.4	7.1	36.7	9.2	279	279		43.6		250	4	19	39
APRIL	0.6	0	1	5.0	5.6	16.7	6.1	31.7	8.9	201	204		38.9		200	8	38	48
OCT-APRIL	—	2	17	—	6.4	19.5	7.1	—	10.1	1,722	1,703		44.7		1,820	40	191	40
MAY-SEPT	—	—	—	—	—	11.9	—	—	—	—	—		—		—	—	248	—
DATA FOR MONTH TO MATURITY																		
MATURITY GROUP	EST. MATURITY DATE																	
1	FEBRUARY 10					22.5	7.5	41.0	11.0				47.1		292	3	21	36
2	FEBRUARY 15					22.5	7.5	40.5	10.5				46.7		288	3	21	36
3	FEBRUARY 21					22.5	7.4	40.0	10.1				46.3		284	3	20	36
4	FEBRUARY 26					22.4	7.4	39.5	9.8				45.9		279	3	20	36
5	MARCH 4					22.2	7.3	39.0	9.6				45.5		274	3	20	36
6	MARCH 9					21.9	7.3	38.5	9.5				45.1		269	3	20	37
7	MARCH 15					21.6	7.2	38.0	9.4				44.7		264	4	20	37
8	MARCH 20					21.3	7.2	37.5	9.3				44.3		259	4	20	38

TABLE 47

STIRLING, SOUTH AUSTRALIA. 35°00'S, 138°42'E. ALT. 496m.

	1	2	3	4	5	6	7	8	9	10	11	12	13	14	15	16	17	18
	Temperature °C									Summation of Day Degrees above 10° with 19° Cut-Off			Temperature Variability Index					
PERIOD	Lowest Recorded Min. (19YRS)	No.Mths with <0.3° (19YRS)	No.Mths with <2.5° (19YRS)	Av.Mthly Lowest Min. (19YRS)	Av.Min. Minus Col.4	Daily Mean (30YRS)	Daily Range /2 (30YRS)	Av.Mthly Highest Max. (19YRS)	Col.8 Minus Av.Max.	Raw	Adj.1 for Lat. & Daily Temp Rng	Adj.2 for Vine Sites	Raw	Adj. for Vine Sites	Av. Sunshine Hrs	Av. Number Rain Days	Av. Rainfall mm	Av. % R.H. 1500 HRS (30YRS)
OCTOBER	0.3	0	17	1.7	5.6	12.3	5.0	26.6	9.3	71	70	81	34.9	33.7	230	11	79	59
NOVEMBER	1.4	0	3	3.3	5.4	14.4	5.7	31.3	11.2	132	129	134	39.4	38.2	260	9	64	53
DECEMBER	2.7	0	0	5.2	5.1	16.5	6.2	34.5	11.8	202	195	201	41.7	40.5	300	9	49	50
JANUARY	3.5	0	0	5.6	5.5	17.7	6.6	34.2	9.9	239	230	237	41.8	40.6	298	7	35	46
FEBRUARY	5.4	0	0	6.5	5.3	18.4	6.6	35.0	10.0	235	229	236	41.7	40.5	255	5	45	45
MARCH	3.1	0	0	5.0	5.5	16.7	6.2	32.7	9.8	208	207	213	40.1	38.9	238	8	41	49
APRIL	1.0	0	1	3.5	4.9	13.2	4.8	27.1	9.1	96	100	111	33.2	32.0	190	11	99	62
OCT-APRIL	—	0	21	—	5.4	15.6	5.9	—	10.2	1,183	1,160	1,213	39.0	37.8	1,771	60	412	52
MAY-SEPT	—	—	—	—	—	9.2	—	—	—	—	—	—	—	—	—	—	709	—
DATA FOR MONTH TO MATURITY																		
MATURITY GROUP	EST. MATURITY DATE					*	*	*					*					
1	MARCH 22					17.6	6.0	33.5	9.9				39.4		250	7	49	48
2	MARCH 31					16.9	5.9	32.6	9.8				38.9		238	8	49	49
3	APRIL 11					15.6	5.5	31.1	9.6				37.0		219	9	58	52
4	APRIL 25					14.6	4.8	28.0	9.2				33.0		198	11	88	60
5	—					—	—	—	—				—		—	—	—	—
6	—					—	—	—	—				—		—	—	—	—
7	—					—	—	—	—				—		—	—	—	—
8	—					—	—	—	—				—		—	—	—	—

* Adjusted for vine sites.

TABLE 48

STRATHALBYN, SOUTH AUSTRALIA. 35°16'S, 138°55'E. ALT. 71m. (For Langhorne Creek)

	1	2	3	4	5	6	7	8	9	10	11	12	13	14	15	16	17	18
	Temperature °C									Summation of Day Degrees above 10° with 19° Cut-Off			Temperature Variability Index					
PERIOD	Lowest Recorded Min. (92YRS)	No.Mths with <0.3° (30YRS)	No.Mths with <2.5° (30YRS)	Av.Mthly Lowest Min. (30YRS)	Av.Min. Minus Col.4	Daily Mean (30YRS)	Daily Range /2 (30YRS)	Av.Mthly Highest Max. (30YRS)	Col.8 Minus Av.Max.	Raw	Adj.1 for Lat. & Daily Temp Rng	Adj.2 for Vine Sites	Raw	Adj. for Vine Sites	Av. Sunshine Hrs	Av. Number Rain Days	Av. Rainfall mm	Av. % R.H. 1500 HRS (30YRS)
OCTOBER	−0.1	1	15	2.3	6.1	14.7	6.3	31.1	9.8	146	144	153	41.4		225	11	44	52
NOVEMBER	1.7	0	0	4.1	6.1	17.0	6.8	35.7	11.9	210	200	209	45.2		250	8	29	47
DECEMBER	1.8	0	0	5.7	6.3	18.9	6.9	38.2	12.4	276	261	270	46.3		285	6	25	45
JANUARY	4.6	0	0	7.2	5.6	19.9	7.1	40.6	13.6	279	279	279	47.6		290	4	21	43
FEBRUARY	4.4	0	0	7.1	6.2	20.3	7.0	38.9	11.9	252	252	252	45.8		250	5	22	44
MARCH	2.2	0	0	5.3	6.5	18.6	6.8	36.4	11.0	267	261	270	44.7		240	6	24	46
APRIL	−1.1	0	11	3.3	6.2	15.4	5.9	31.4	10.1	162	164	173	39.9		190	10	39	54
OCT-APRIL	—	1	26	—	6.1	17.8	6.7	—	11.5	1,592	1,561	1,606	44.4		1,730	50	204	47
MAY-SEPT	—	—	—	—	—	11.6	—	—	—	—	—	—	—		—	—	291	—
	DATA FOR MONTH TO MATURITY																	
MATURITY GROUP	EST. MATURITY DATE					*		*										
1	FEBRUARY 15					20.7	7.0	40.1	12.7				46.7		280	4	22	43
2	FEBRUARY 21					20.7	7.0	39.9	12.3				46.3		276	5	23	44
3	FEBRUARY 27					20.6	7.0	39.6	12.0				45.9		271	5	24	44
4	MARCH 5					20.5	7.0	39.3	11.8				45.7		266	5	24	44
5	MARCH 10					20.3	6.9	38.9	11.7				45.5		261	5	24	45
6	MARCH 16					20.0	6.9	38.4	11.5				45.3		255	5	24	45
7	MARCH 22					19.6	6.9	37.7	11.3				45.1		249	6	24	45
8	MARCH 28					19.2	6.8	36.9	11.1				44.8		243	6	24	46

* Adjusted for vine sites.

TABLE 49

VICTOR HARBOR, SOUTH AUSTRALIA. 35°33'S, 138°37'E. ALT. 6m. (For Lower Fleurieu Peninsula)

	1	2	3	4	5	6	7	8	9	10	11	12	13	14	15	16	17	18
	Temperature °C									Summation of Day Degrees above 10° with 19° Cut-Off			Temperature Variability Index					
PERIOD	Lowest Recorded Min. (37YRS)	No.Mths with <0.3° (30YRS)	No.Mths with <2.5° (30YRS)	Av.Mthly Lowest Min. (30YRS)	Av.Min. Minus Col.4	Daily Mean (18YRS)	Daily Range /2 (18YRS)	Av.Mthly Highest Max. (30YRS)	Col.8 Minus Av.Max.	Raw	Adj.1 for Lat. & Daily Temp Rng	Adj.2 for Vine Sites	Raw	Adj. for Vine Sites	Av. Sunshine Hrs	Av. Number Rain Days	Av. Rainfall mm	Av. % R.H. 1500 HRS
OCTOBER	2.8	0	0	4.8	5.2	15.2	5.2	30.7	10.3	161	159	137	36.3		230	11	44	56
NOVEMBER	4.4	0	0	6.6	5.0	16.9	5.0	35.1	13.2	207	202	181	38.5		248	7	29	58
DECEMBER	5.6	0	0	8.2	5.3	18.8	5.3	37.8	13.7	273	265	243	40.2		277	6	24	55
JANUARY	6.1	0	0	9.5	5.3	19.6	5.3	39.6	14.7	279	279	268	40.7		273	4	22	60
FEBRUARY	6.1	0	0	10.0	4.9	19.4	4.9	37.1	12.8	252	252	239	36.9		225	4	22	61
MARCH	5.3	0	0	8.4	5.1	18.8	5.1	35.2	11.3	273	273	250	37.0		225	6	22	56
APRIL	1.7	0	0	5.9	4.5	16.1	4.5	30.3	9.7	183	193	165	33.4		177	10	44	54
OCT-APRIL	—	0	0	—	5.2	17.8	5.0	—	12.2	1,628	1,623	1,483	37.6		1,655	48	207	57
MAY-SEPT	—	—	—	—	—	12.6	—	—	—	—	—	—	—		—	—	331	—
	DATA FOR MONTH TO MATURITY																	
MATURITY GROUP	EST. MATURITY DATE					*		*										
1	FEBRUARY 26					18.7	4.9	36.6	12.9				37.0		247	4	22	61
2	MARCH 4					18.6	4.9	36.2	12.6				36.9		243	4	22	61
3	MARCH 10					18.5	5.0	35.8	12.3				36.9		239	4	22	60
4	MARCH 16					18.4	5.0	35.4	12.0				36.9		235	5	22	59
5	MARCH 23					18.3	5.1	35.0	11.7				36.9		231	5	22	58
6	MARCH 29					18.1	5.1	34.6	11.4				36.9		227	6	22	57
7	APRIL 5					17.9	5.0	34.0	11.1				36.6		220	6	24	56
8	APRIL 12					17.7	4.9	33.2	10.7				36.0		210	7	28	55

* Adjusted for vine sites.

Commentaries and vineyard site adjustments

Adelaide (city) (Table 34)

Vineyard site adjustments: None

Notes

Adelaide city is on the flat coastal plain. The data show it to be quite appreciably hotter throughout the growing season than the Waite Institute (Table 35) in the lower foothills. A heating effect of the city itself probably contributes, suggesting at least that the figures represent the hottest parts of the coastal plain. On the other hand Adelaide city also shows a substantially lower average 1500 hours relative humidity compared with the Waite Institute, and even more so compared with Belair (Table 37), which is on the scarp at a greater altitude again. The coastal plain thus appears largely to miss the early benefit of afternoon cool breezes enjoyed by the foothills and scarp.

Adelaide (Waite Institute) (Table 35)

Vineyard site adjustments: None

Barossa Valley (estimate) (Table 36)

Vineyard site adjustments: None

Notes

Columns 6 and 7 are averages for a four-year experimental period (1959–63) from three actual vineyard sites (Boehm 1970), with smoothing and period adjustment based on Nuriootpa temperatures over the same period vs Nuriootpa 1942–65 (Table 43). The remaining columns are estimated using data from Nuriootpa and Kapunda.

This is an alternative estimate for the Barossa Valley to that provided directly by Nuriootpa, which is on the valley floor and perhaps climatically unfavourable. No adjustment is made for soil types, which vary from sands to clay loams.

The foothill slopes fringing the valley should have a superior mesoclimate, with less temperature variability and (at the lower altitudes) higher biologically effective temperature summations than on the valley floor. The foothills have probably the best climates in the Barossa region for full-bodied red wines from mid-season grape varieties.

Belair (Table 37)

Vineyard site adjustments: None

Notes

At 308 m altitude, Belair enjoys the good mesoclimatic characteristics of the Mount Lofty Range's western scarp. The diurnal temperature range and temperature variability index are similar to those for the Waite Institute on the lower slope, but the average mean growing season temperature is 1.0°C lower, consistent

with Belair's 185 m greater altitude. Relative humidity at 1500 hours averages a full 10 per cent higher than at the Waite Institute, indicating better exposure to afternoon sea breezes. Rainfall is also substantially greater. This represents probably the best climate for full-bodied table wines in the immediate Adelaide area.

Berri (Table 38)

Vineyard site adjustments: None

Clare (Table 39)

Vineyard site adjustments, °C

Min.	Max.	
+0.8	−0.4	Moderate slopes, vs Clare in valley
+0.4	−0.2	Limestone subsoils fairly common
+1.2	−0.6	Mean temperature +0.3°, variability index −3.6

Notes

The estimate assumes altitudes and latitudes similar to those of Clare township. Questions of their relevance to the region's viticulture are canvassed fully in the main text.

Coonawarra (Table 40)

Vineyard site adjustments, °C

Min.	Max.	
–	–	Flat terrain
+0.8	−0.4	Limestone-based soil
+0.8	−0.4	Mean temperature +0.2°, variability index −2.4

Notes

Only short-term published data are available for Coonawarra, and are from a period when temperatures appear to have been higher than the long-term average. I have used the figures from Climatic Averages Australia (1975), with period adjustments based on nearby stations (particularly Mount Gambier and Kybybolite) which have long-term as well as shorter-term records. The resulting 0.2°C reduction compared with the 1975 figures for the growing season is possibly conservative. Several of the columns in the table required smoothing. The estimates for Coonawarra are therefore only tentative, but they do agree reasonably with its observed vine phenology.

Eden Valley-Springton (estimate) (Table 41)

Vineyard site adjustments: None

Notes

The estimates for columns 6 and 7 are directly from experimental data for four vineyard sites, averaging about 450 m altitude (Boehm 1970), with period adjustment parallel to that for the Barossa Valley (above) versus Nuriootpa. Columns 5 and 9 are estimated from Nuriootpa and Stirling. Again, therefore, these estimates can at best only be tentative.

Mount Barker (South Australia) (Table 42)

Vineyard site adjustments, °C

Min.	Max.	
+0.8	−0.4	Selected moderate slopes vs Mount Barker semi valley
−0.2	−0.2	Altitude 30 m higher than Mount Barker
+0.6	−0.6	Mean temperature unchanged, variability index −2.4

Notes

The data for Mount Barker show clearly the climatic deterioration inland from the western scarp of the Mount Lofty Range, particularly in greater variability of the minimum temperature and therefore greater frost risk. Sufficiently frost-free sites could produce reasonable table wines, but much better conditions exist in the hills closer to Adelaide and along the lower Fleurieu Peninsula.

Nuriootpa (Barossa Valley) (Table 43)

Vineyard site adjustments: None

Notes

The data in columns 6 and 7 are from Boehm (1970), and those in column 18 from Climatic Averages Australia (1975), period-adjusted by comparison with Kapunda and slightly smoothed. The figures probably represent the less favourable, valley-floor sites of the Barossa.

Padthaway (estimate) (Table 44)

Vineyard site adjustments, °C

Min.	Max.	
–	–	Flat terrain
+0.8	−0.4	Calcareous soil
+0.8	−0.4	Mean temperature +0.2°, variability index −2.4

Notes

The estimates for Padthaway are based on data from Keith, Lucindale, Naracoorte, Kybybolite and Coonawarra, smoothed and period-adjusted.

Padthaway has many of the climatic and soil characteristics of Coonawarra, but with moderately higher temperatures, more sunshine hours, less rainfall and lower relative humidities. Ripening of midseason to late varieties is assured, in contrast to Coonawarra where it can be incomplete in cool, wet autumns. Temperature variability and low relative humidities during ripening appear to be the main environmental factors limiting potential for grape and wine quality.

Port Lincoln (Table 45)

Vineyard site adjustments, °C

Min.	Max.	
–	–	Predominantly flat terrain
+0.8	−0.4	Limestone-based soils
−0.8	+0.4	Further inland, less insular than Port Lincoln
−0.4	−0.4	Altitude of 50–75 m assumed
−0.4	−0.4	Mean temperature −0.4°, variability index unchanged

Notes

The estimates apply directly to the limestone country west and north-west of Port Lincoln. The low, non-calcareous hills that lie parallel to the east coast just north of Port Lincoln are a further possible area for vines, with a probably similar climate. Surface catchment water would be needed in the latter area, because ground water is unlikely to be available and the natural rainfall is hardly sufficient.

Roseworthy (Agricultural College) (Table 46)

Vineyard site adjustments: None

Notes

Roseworthy is on flat terrain west of the Barossa Valley. It is included for comparison, being a leading centre for viticultural and oenological research and teaching. Clearly it is very marginal for viticulture.

Stirling (Table 47)

Vineyard site adjustments, °C

Min.	Max.	
+0.4	−0.2	Selected moderate slopes
+0.15	+0.05	East to north-east aspects
+0.55	−0.15	Mean temperature +0.2°, variability index −1.2

Notes

Columns 5 and 9 are based on raw data for monthly extremes for 1957–76, versus average minima and maxima from 1957 to 1974 from Climatic Averages Australia (1975). Columns 4 and 10 were calculated from these, and applied to the temperature means from Climatic Averages Australia (1956). Columns 1–3 were period-adjusted in parallel.

Strathalbyn (for Langhorne Creek) (Table 48)

Vineyard site adjustments, °C

Min.	Max.	
–	–	Vineyards closer to Lake Alexandrina but very flat
+0.3	+0.3	Vineyards at altitude 20 m or less
+0.3	+0.3	Mean temperature +0.3°, variability index unchanged

Victor Harbor (for lower Fleurieu Peninsula) (Table 49)

Vineyard site adjustments, °C

Min.	Max.	
−1.0	+0.5	More inland locations
+1.0	−0.5	Slopes, possibly stony soils
−0.7	−0.7	Altitude 100–150 m (but see Notes and main text)
−0.7	−0.7	Mean temperature −0.7°, variability index unchanged

Notes

Victor Harbor is the only place on the lower Fleurieu Peninsula with published long-term climatic records, but its coastal location is not representative of most potential vineyard sites. The detailed estimates in Table 49 are for altitudes of 100–150 m, with the assumption that slopes and/or stony soils offset greater distance inland in determining temperature variability patterns. Table 33, in the main text, shows supplementary estimates for sites at sea level and 300 m.

11

Viticultural Environments of Victoria

Grape vines follow history, in Victoria as elsewhere. Early plantings around Melbourne soon showed the suitability of the Port Phillip Bay region for making table wines.

Swiss settlers from Neuchatel planted small vineyards at Geelong from 1842 onwards. They were followed a little later by more wealthy compatriots such as the brothers Paul and Hubert de Castella, and Guillaume, Baron de Pury and his brother Samuel, who established the famous Yarra Valley vineyards at Lilydale in the 1850s and 1860s. Experienced vine dressers were brought out from Switzerland to work the Yarra vineyards. Not surprisingly, Chasselas and Pinot Noir vines were prominent among the early plantings. These varieties proved to be well suited to the region, and produced good table wines. The early viticulture of Geelong and Lilydale has been described by de Castella (1886), and Peel (1965).

Vines followed the gold rushes of the 1850s to Avoca, Bendigo, Great Western, Ballarat and Rutherglen. Many of the gold-bearing soils proved to be ideally suited to viticulture, and some of the diggers, and those who followed them, stayed to plant vines.

Phylloxera appeared at Geelong in 1875, resulting in all vines there being uprooted in a vain attempt to stop its spread. By then, in any case, most of the Geelong vineyards were already decrepit, following the death, departure or loss of interest of the original settlers. Elsewhere similar factors were at work, together with public indifference to the light wines that the cooler areas produced. In the Yarra Valley an evident increase in the already serious problem of frost, thought to be caused by extensive clearing of the natural vegetation, was a further factor; while Peel (1972) has written of the problems of drought and wet vintages at Sunbury, just north of Melbourne, in the 1880s. In fact it is not unlikely that a world-wide cooling of the climate about that time was a contributing factor: see Chapter 23. In any case the combined result was a progressive abandonment of cool-climate viticulture in Victoria. By the mid–20th century the State's only remaining substantial plantings were in the north-east around Rutherglen, and in the Murray River irrigation areas of the north-west.

The Royal Commission on Vegetable Products, in its *Handbook of Viticulture* (1891), defined cool, intermediate and hot viticultural areas of Victoria, as I have described in Chapter 2. The present study confirms early reports that the main limitation in the cool Region 1 was, and is, frost. The flat topography of much of southern Victoria offers little opportunity for favourable site selection. Moreover, lack of continentality in the southern areas means that frost and other climatic inclemencies can recur well into late spring. The south coast suffers as well from a fairly marked variability of the maximum temperatures in summer, as elsewhere in south coastal Australia.

Local factors modify and improve conditions for viticulture around Port Phillip Bay. Gippsland also has some reasonably favourable sites. The Great Dividing Range offers further environments for cool-climate viticulture, with careful site selection; but distance inland produces increasing atmospheric saturation deficits and moisture stresses.

The viticultural regions of Victoria will be discussed in roughly clockwise order, starting with Port Phillip Bay. Figure 13 shows the main locations mentioned, and isotherms of the average mean temperatures for the October-April growing season, reduced to sea level. For general accounts of Victorian viticulture and wines, see Benwell (1976) and Halliday (1982a, 1985a, 1991a).

Port Phillip Bay region

Probably more than any other part of mainland Australia, this region resembles climatically the classic vineyards of western France. It differs mainly in having more variable maximum temperatures, less continentality, and lower rainfall and relative humidities around the northern and western shores of the bay.

The climatic figures for **Melbourne** (Table 66) probably over-estimate temperatures because of urban warming. **Werribee** (Table 76) gives an alternative, lower estimate for temperature; and because it is very flat, a 'worst possible' estimate for frost risk in areas close to the bay.

The original Melbourne vineyards would have resembled Bordeaux quite closely in their growing season and ripening period average mean temperatures, rainfall and sunshine hours. Average afternoon

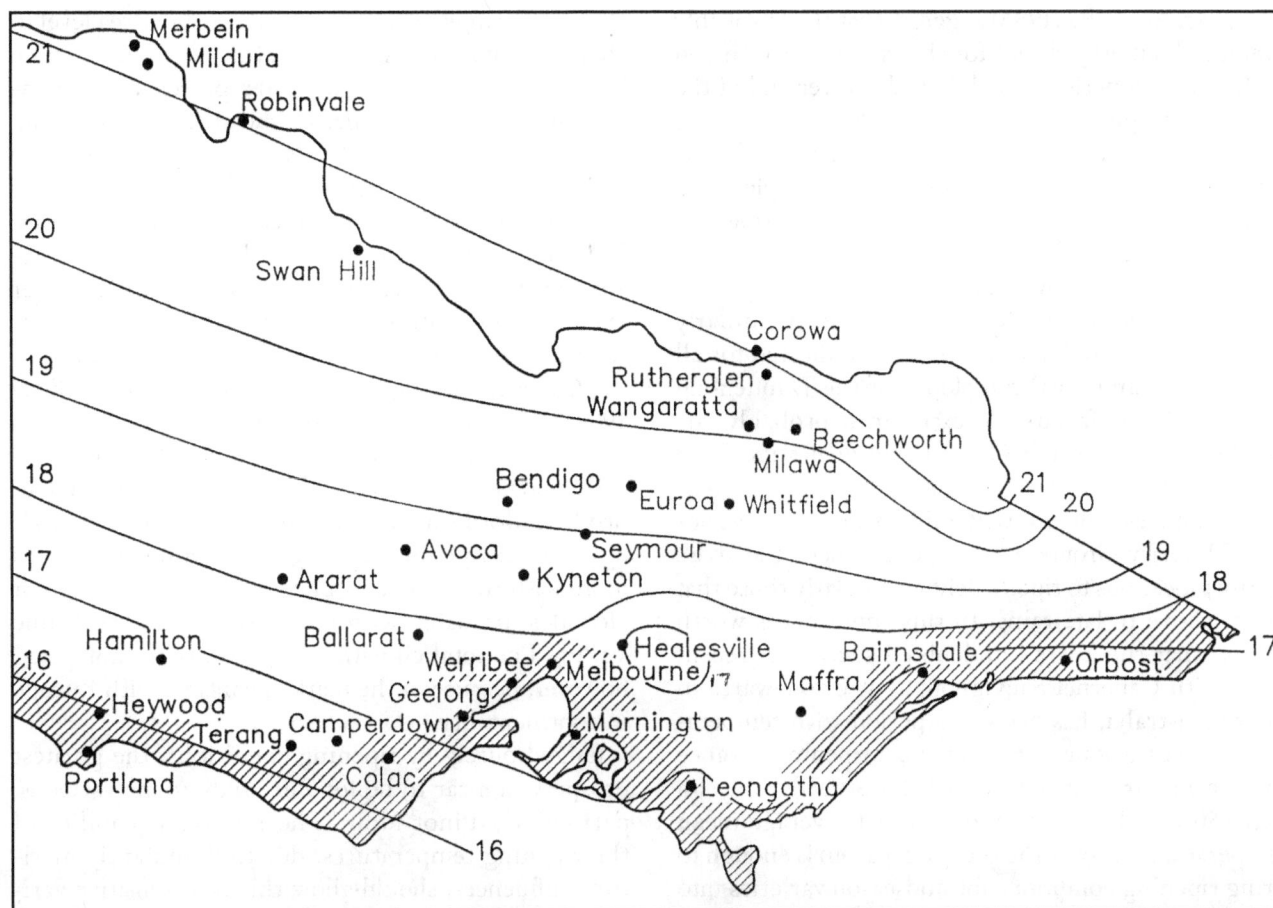

Figure 13. Victoria, with isotherms of October-April average mean temperatures, reduced to sea level.

relative humidities would have been a little lower than in Bordeaux, but were still probably optimal for vine health and wine quality. Melbourne suffers in the comparison mainly through its liability to occasional extremes of high temperature and low relative humidity in summer; and in some years, through insufficient total rainfall for solely rain-fed vines.

The present vineyards of **Geelong** (Table 60) are mostly west of that city, and are to varying degrees more cut off from maritime influences. This and their greater altitude make them appreciably cooler than Geelong, and more liable to frost (depending on slope etc.).

In its ripening temperatures and sunshine hours, Geelong is about half way between Bordeaux and Burgundy. Midseason Bordeaux varieties ripen fairly safely because of the prolonged autumn, usually without too much rain; but they do so on average at lower mean temperatures and sunshine hours than in Bordeaux. On the other hand total growing season sunshine hours equal those of Bordeaux. The result is light to medium, fresh Bordeaux-style red wines of good finesse and quality. The Burgundy varieties (Maturity Group 3) ripen at temperatures and sunshine hours slightly above those in Burgundy, which in itself is not unfavourable; but there is a much greater risk, for susceptible varieties such as Pinot Noir and Gamay, of damaging high temperatures during ripening.

Two other viticultural problems are predictable at Geelong. With only moderate total rainfall, restricted growth and yields can be expected without supplementary watering. This probably assists with vine fruiting balance and canopy light relations in the absence of special trellising, but supplementary water to reduce stress during ripening could assist quality greatly in some seasons. Secondly, lack of continentality, and often the persistence of rain until Christmas, mean that flowering and setting are vulnerable to bad weather.

The **Yarra Valley**, east of Melbourne, is represented by **Healesville** (Table 62). The vineyards are mostly west and downstream from Healesville, but are on average at comparable altitudes, on selected slopes to avoid frost. There could be very significant temperature differences above and below those estimated in the table, depending on vineyard altitude, slope and exposure. Importantly, the Yarra Valley has a number of small, isolated or projecting hills which provide specially favourable mesoclimates on their slopes, including Yeringberg and the Warramate Hills.

The estimated ripening temperatures and sunshine hours in the Yarra Valley are marginally lower than at Geelong, but sunshine hours in spring and early summer, and total sunshine for the growing season, are a little greater. Rainfall is substantially greater in all seasons. I have no figures for 1500 hours relative humidities, but the maps of the Division of National

Mapping, Australia (1986) suggests that they probably average about 50 per cent for the growing season and early ripening period, which is at the lower end of the optimum range.

The Yarra Valley's climate shows that its reputation for Burgundy grape varieties and wine styles is no accident. High temperature extremes during ripening are still a hazard, but they are less so than at Geelong and especially Coonawarra, which in average ripening mean temperatures is a close equivalent. A similarly favourable climate for wine quality is indicated for all the western and northern slopes of the Dandenong Hills, and, for favourable exposures, probably the southern slopes of the Great Dividing Range from Healesville west to about Sunbury.

The figures suggest that midseason grape varieties (Maturity Groups 5–6) in general need the area's warmer locations to ripen safely, particularly those that are sensitive to late rains. In this context it is worth noting that the Yarra's undoubted success in recent years with Cabernet Sauvignon, as at Coonawarra in South Australia, has been in a period with temperatures above the long-term average. A ripening advance by one maturity group (which here needs only an extra 50 raw day degrees, or a 0.24°C average mean temperature rise over the growing season) is enough to bring ripening conditions for midseason varieties into the optimum range for medium-bodied wines. Were temperatures to continue to rise, Cabernet Sauvignon and other Bordeaux varieties would be established as widely and fully adapted in the Yarra Valley, while Pinot Noir may need to retreat further into the hills for best results. However, as I argue in Chapter 23, this seems unlikely to occur at least until several decades into the 21st century. The more immediate prospect is for a return to long-term average temperatures. In that case the Yarra Valley will remain as an area of special adaptation (by Australian standards) for Pinot Noir, and will continue so for some decades to come.

The **Mornington Peninsula**, as estimated from the data for **Mornington** (Table 68), has arguably the best ripening climate in Victoria for light to medium-bodied table wines. Its peninsular geography reduces temperature variability, including that of the maximum (column 9) – a unique feature for south coastal Victoria. Hot, dry winds from the north and north-west must here traverse the full width of Port Phillip Bay. The peninsula is also largely frost-free and has optimal average relative humidities, together with suitable totals and seasonal distributions of both rainfall and sunshine hours. Two potential drawbacks are lack of continentality, resulting in budburst and setting problems in susceptible varieties, and strong winds, requiring careful attention to vineyard aspect, canopy management and windbreaks. For more detailed accounts of the Mornington Peninsula and its viticulture, see Anon. (1987) and Halliday (1991a).

Actual and potential vineyard sites on the Mornington Peninsula range in altitude from just above sea level to about 200 m in the South Red Hill area. The estimate based on Mornington (Table 68) represents the warmest part of the peninsula, at low altitude and facing inland towards Port Phillip Bay. Such sites appear very well suited, climatically, for producing medium-bodied red Bordeaux styles from midseason grape varieties. In all respects the ripening month compares favourably with that of the Médoc, with marginally lower ripening temperatures being offset by a longer potential ripening period. The reported great refinement of the Cabernet Sauvignon wines (Anon. 1987; Halliday 1991a) conforms with climatic expectation.

Locations much above 100 m, unless having atypically warm site climates, are probably too cool to produce Bordeaux styles regularly. They find analogy in their ripening capacity and ripening conditions more with the Loire and Burgundy; or, at the highest altitudes, probably with Champagne and the Rhine Valley. The total climatic comparisons are not exact, however, because of the marked contrast with Europe in continentality.

Judged purely by ripening conditions, the greatest prospects appear to be for Burgundy grape varieties, particularly Pinot Noir. The relative equability of the ripening temperatures, due to dominantly maritime influences, should allow this heat-sensitive variety to ripen without too great a risk of scorching. Potential Pinot styles should theoretically range from full-bodied on the low-altitude sites to light red or sparkling champagne styles at the highest altitudes. Provided that viticultural problems associated with lack of continentality can be overcome, this (together with selected sites in the Yarra Valley) would appear to be one of the few regions of mainland Australia where the precise characteristics of the great Burgundy wines (both red and white) might reasonably be aspired to.

Western Districts

The nearer Western districts are represented by **Camperdown** (Table 57), **Colac** (Table 58) and **Terang** (Table 74).

Colac and Terang have the flat, or nearly flat, frosty environments that typify much of Victoria's Region 1. The fact that Colac town is close to Lake Colac does not save it from an extreme frost risk in spring. Terang is a little cooler in average recorded temperatures, but less frosty, and therefore perhaps approaches more the kind of site on which vine growing might be attempted.

Camperdown's much more favourable figures reflect its unique topographical location: on the lower northern slope of the isolated Mount Leura. It is also close to lakes, though these are probably not large enough, or close enough, to influence its climate greatly. This must be as favourable a climate as is available in the region, but it remains cooler and frostier than Geelong, and has less capacity for prolonged autumn ripening if needed.

One environment worth closer examination is the lower, north-facing slopes of the Otway Range to the south-east of Colac. If frost-free enough sites can be found, they might produce true cool-climate wine styles from early-maturing grape varieties (up to Maturity Group 4).

The far-western region of Victoria is represented by **Hamilton** (Table 61), **Heywood** (Table 63) and **Portland** (Table 70).

The figures for Heywood have been adjusted to represent **Drumborg**, which has existing vines on sloping gravelly, red-brown volcanic soil. The resulting estimates must be treated with caution, because period as well as site adjustments were needed (see commentary on Heywood); but they agree with practical experience as reported by Heinze (1977). Heinze says that the latest-maturing variety normally to ripen is Riesling. Traminer, Sylvaner and Pinot Noir achieve full ripeness more reliably, which Table 63 shows they still do under conditions of quite low temperature and sunshine hours; but this follows a growing season with very ample sunshine hours for so cool a climate. The growing and ripening conditions are thus appropriate to German-style and sparkling wines (Table 2, Chapter 3). Comparisons with the Adelaide Hills for production of champagne-type sparkling wines have already been noted under Stirling, South Australia (Chapter 10). Occasional seasons successfully ripen Cabernet Sauvignon, to produce light red wines. According to Heinze, fruit set is a problem because of low temperatures at flowering, and bunch rot occurs readily towards the end of ripening. The latter is predictable from the rapid rise in rainfall and relative humidity in late autumn (Table 61).

Hamilton, further inland, has a ripening capacity similar to that of Drumborg and Terang. Compared with Drumborg, however, there is a worse frost problem and little hope of finding superior sites, because of the flatness of the country and its mostly clay soils.

One area with some climatic potential is just inland from the south coastal town of **Portland** (Table 70). The data for Portland are adjusted for the sheltered, north and north-east facing slopes of small hills close to the town. Suitability would depend on soil types, and on using grape varieties and management practices adapted to a highly maritime temperature regime.

Western Great Dividing Range

This region takes in the cool, high-altitude viticultural areas of the main divide from Great Western eastwards to the Goulburn Valley. It includes **Ararat** (Table 51), **Ballarat** (Table 54) and **Kyneton**, for which I have not attempted a detailed estimate for reasons noted below.

Temperatures along the range are moderate to low, depending on altitude and latitude. Other climatic characteristics in common are generous sunshine hours, resulting in very high ratios of growing-season sunshine hours to raw and effective temperature

summations; intermediate temperature variability indices and relative humidities; and in general barely sufficient rainfall for rain-fed viticulture. Exposed situations experience damagingly strong winds (Hamilton 1988). The result, historically, has been very good, clean, well balanced table wines of brilliant colour and great aging capacity, but (especially in the absence of supplementary watering) low yields and doubtful economics of production.

The old-established vineyards of Great Western, a little to the north of Ararat, are at 240 to 320 m altitude. The higher vineyards here are on mostly north-east facing slopes of the Ararat Hills. Slope, aspect and soil factors probably cancel the altitude differences out to a large extent in determining effective temperatures, which one might guess to be warmer than calculated for Ararat by the equivalent of one to two maturity groups.

Newer vineyards just to the south-west and south of Ararat, and to the east beyond Mount Langi Ghiran, average about 350 m in altitude. The figures for Ararat should represent these well. The presence of prominent isolated hills such as Mount Langi Ghiran and Mount Ararat suggests also that considerable further potential exists for selecting viticulturally favoured mesoclimates in this area, with similar ripening times and mean temperatures to those calculated for Ararat. With greater exposure to maritime influences than on the inland slopes at Great Western, and somewhat (but not excessively) later ripening, average climatic conditions on suitably sheltered slopes appear to be very suitable for producing medium-bodied wines from early to midseason grape varieties. Optimum maturity groups are probably 3–4 for white wines, and 4–5 for red wines.

Ballarat, to the south-east at 437 m altitude, is much cooler. Even given optimal air drainage and a predominantly northern aspect, the figures show that only very early-maturing varieties can be expected to ripen, and then at low temperatures. The temperature variability is greater than desirable for flavour development. There is a strong risk of killing frosts. This is clearly a marginal environment at best.

The published climatic data for Kyneton (altitude 511 m) present problems of interpretation. Climatic Averages Australia (1956) give a growing season average mean temperature of 15.5°C, and a raw temperature summation of 1171 day degrees. However the average diurnal temperature range/2 for the growing season is 7.7, which is extremely wide for a mountain location and suggests an anomalous recording site in a frosty hollow. Site adjustments based on Kyneton for potential vineyards must therefore be very large and speculative. My guess is that sites with a low enough diurnal temperature range and spring frost risk to allow viticulture would have 1250–1300 effective day degrees at this latitude and altitude; which agrees with the practical record of midseason grape varieties such as Shiraz ripening fully, but late and rather precariously (Anon. 1990). The crucial factor is site selection.

North-central Victoria

The 'north-central' region takes in several vine-growing areas which have much in common in their wines. Most characteristic are their deeply coloured and robust red wines from Cabernet Sauvignon and Shiraz. These areas correspond to the warm, intermediate Region 2 of the Royal Commission on Vegetable Products (1891), and to the warmer parts of Region II together with Region III of Amerine and Winkler (1944). Representative viticultural climates are estimated from the published data for **Avoca** (Table 52) and **Euroa** (Table 59), for the northern slopes of the Pyrenees and Strathbogie ranges respectively; from **Bendigo** (Table 56), for that region; and from **Seymour** (Table 72), for Tabilk in the central Goulburn Valley.

The **Pyrenees** estimate is adjusted to 350–400 m, which is the altitude of the Taltarni vineyards. These have excellent site characteristics, being on the slopes of a projecting ridge within an isolated hill range, and having mainly easterly aspects and gravelly soils. The raw temperature variability index for Avoca itself indicates a semi-valley situation, so that large site adjustments are needed for the vineyards.

Average ripening mean temperatures on the higher Pyrenees slopes now used for viticulture are optimal for full-bodied table wines; and although the area has some fairly stressful temperatures and low relative humidities in mid summer, both temperature and relative humidity are surprisingly moderate during actual ripening. This is a characteristic of the whole north-central region of Victoria, and is associated with a marked drop in the variability of maximum temperatures from January to February. It results from a late southward extension of the north-east monsoonal influence, which, with distance north, comes increasingly to dominate the summer climates of New South Wales and Queensland (see Chapters 12 and 13). Mid and late afternoon relative humidities are also tempered here by some penetration of afternoon southerly winds across the low western arm of the Great Dividing Range, in contrast to north-eastern Victoria, which is effectively sheltered from the south by the high Australian Alps. Sunshine hours are ample, however, verging into the fortified wine range. Rainfall is fairly evenly spread but limited, so supplementary watering is almost universally desirable.

Avoca has substantial plantings of Trebbiano and Doradillo for brandy/sparkling wine production. The ripening conditions for these late-maturing varieties are very suitable for light to medium bulk table wines; but at least by Cognac and Champagne standards, they appear to be warmer than optimum for brandy and sparkling wines.

The **Strathbogie Range** estimate, based on Euroa, is for the Mount Helen vineyard at 420–460 m. Because this necessitated a large adjustment from Euroa, both for altitude and, to a smaller extent, distance south from the town, the estimate is more than usually tentative. It suggests that growing and ripening conditions resemble those of the Pyrenees vineyards.

Other (some possibly more) favourable mesoclimates are available in the region, most notably on the slopes of isolated or projecting outliers of the Strathbogie Range. Temperatures would depend on altitude, but the means for sites on the lower slopes (say, up to 300 m in the south or 400 m in the north) should not greatly differ from what I have calculated for Seymour (below). Their temperature variabilities, however, should be much less.

Bendigo is the most northerly part of the north-central viticultural region, and fairly low with a flat to undulating topography. These features explain its higher temperature summation and ripening mean temperatures compared with the ranges. Relative humidities are lower, and sunshine hours greater. These characteristics place it climatically a little beyond the optimum for table wines, and more into the range for light fortified wines. Nevertheless temperature variability remains fairly moderate by Australian inland standards. Bendigo, like other parts of north-central Victoria, benefits from the drop in variability of the maximum temperature between January and February, so that the highest maxima during ripening are not greatly excessive for midseason and later grapes. It is perhaps for that reason, together with the occurrence of quartz gravel subsoils in which water is available to deep-penetrating roots right through the summer, that Bendigo's reputation rests primarily on its robust, deeply coloured and fully flavoured red table wines.

Seymour, in the Goulburn Valley, has the typical wide diurnal range and pronounced temperature variability of a flat valley floor: perhaps mitigated to some extent close to the river and to Lake Nagambie. Despite raw temperature summations similar to those of Bendigo, vine development is probably retarded by the wide diurnal temperature range, and I estimate that maturation takes place later and at lower average temperatures than at Bendigo. In terms of extreme maxima during ripening this benefits late-maturing grape varieties, but as at Bendigo does little to help those that ripen early. Midseason to late varieties are therefore the best adapted and ripen within the optimum range of mean temperatures for full-bodied table wines, with not greatly excessive extremes during ripening. But in other respects this is an often stressful climate, inferior for table wines to the nearby hills and foothills as discussed above for the Strathbogie Range.

North-west irrigation areas

Merbein (Table 67, also representing **Mildura**) and **Swan Hill** (Table 73) typify the hot, fully irrigated viticultural areas of north-western Victoria. The 'dual purpose' grape varieties Muscat Gordo Blanco and

Sultana supply much of the production. Originally planted for drying, these varieties are now much used for bulk wine production according to the relative strengths of the alternative markets. However some true winegrapes are now also grown.

Both places have ripening mean temperatures in the middle to upper optimum range for fortified wines; marked temperature variability; long sunshine hours; and low relative humidities. Spring is warm enough for frosts to be a minimal problem, despite the temperature variability.

This is a region of bulk fortified and table wine production, with some medium-grade bottled wines from the best grapes. Flor sherry styles from Mildura have a deserved market popularity.

North-east Victoria

Rutherglen (Table 71) and **Wangaratta** (Table 75), together with **Corowa**, just over the border in New South Wales (Chapter 12, Table 86), represent the low-altitude vineyards of north-eastern Victoria. The Rutherglen figures appear to come from a frosty recording site which is almost certainly unrepresentative of the vineyards, so fairly large site adjustments are needed. Wangaratta is probably more indicative, with a moderate diurnal temperature range and variability for the region, although the reasons for this are obscure because it is flat and lies in a valley. Urban night-warming is one possible cause. Most of the vineyards have the advantage of being on at least slight slopes or rises, or close to the Murray River.

Some of the older Rutherglen plantings extended up to the gravelly hill tops; but with dependence on natural rainfall, and low yields, these have now mostly been abandoned (Halliday 1985a). Such sites would have enjoyed significantly more favourable site climates than indicated by the tables, with ripening times and mean temperatures much as in the tables, but higher night temperatures and reduced variability indices (probably about 40 or less).

A locality with remaining hill-slope vineyards is that along the lower western slopes of the isolated Warby Range at Glenrowan, just west of Wangaratta. It is uniquely favoured, for inland Australia, in also abutting Lake Mokoan, which covers a substantial area immediately to the west. The vines are on deep, granite-derived friable red soils which form a narrow strip between the hills and the lake. With a temperature variability index perhaps significantly below 40, the strip is renowned for its red wines, mainly from Shiraz, which have enormous body, depth of colour and flavour. It also produces some of the best of the district's (and the world's) fortified, sweet white wines from Frontignac and Muscadelle: very sweet, with intense aroma and flavour.

Average mean ripening temperatures for early-maturing grapes in the Rutherglen-Wangaratta region, at between 22 and 23°C, are more or less identical with those of Frontignan and other localities in southern France which produce sweet, lightly fortified muscats from the Frontingnac variety (see Montpellier, Chapter 16, Table 132); also with those of Setubal, in Portugal (see Lisbon, Chapter 20, Table 164). The corresponding figure for Constantia in South Africa, the source of the sweet Constantia muscats which enjoyed great fame in the 19th century, is about 21°C. Grapes for the sweet but unfortified muscats (spumante style) of Asti, in northern Italy, ripen at between 22 and 23°C (Turin, Chapter 19, Table 159). As already noted in Chapter 3, this quite remarkable correspondence of ripening temperatures across the world points to around 22°C as an optimum ripening mean temperature for sweet, non-acid wines from Frontignac and probably other early-maturing aromatic grape varieties. The difference between Rutherglen and the other regions mentioned lies in Rutherglen's low relative humidities through and after ripening. This leads to natural partial raisining in the absence of noble rot infection, and an intense, very sweet style.

The outstanding climatic feature of north-eastern Victoria as a wine-producing region is its more or less complete lack of afternoon sea breezes or other maritime influences, being cut off from both the east and south coasts by the Australian Alps. This helps to explain why it is essentially an area for sweet, fortified wines, and experiences difficulty in producing consistently good table wines despite average temperature which are not greatly excessive. Individual vineyards producing the best (full-bodied) table wines generally have special advantages of isolated hill slopes (Glenrowan) and/or very close proximity to substantial water bodies such as Lake Mokoan or the Murray River.

The vineyards of Milawa, a little south of Wangaratta, are generally thought to be cooler than Wangaratta and Rutherglen, and have deep, red alluvial loam soils over river gravel which give the vines naturally favourable moisture relations right through summer (Halliday 1985a). This area is now best known for its table wines.

In recent years the Milawa-based plantings have extended to increasingly greater altitudes, in search of 'cool climate' characteristics in table wines. Table 55, based on **Beechworth**, gives estimates for possible vineyard sites in that area at 400 m. The outstanding feature of Beechworth's climate is the great variability of its minimum temperatures (column 5), resulting in a very serious frost risk. This appears to be a common feature of the west-facing slopes of the Great Dividing Range in far north-eastern Victoria and in New South Wales (see Chapter 12). The most likely cause is periodic tongues of chilled night air descending from the alpine snowfields. These might be localized, or follow distinct paths. Careful selection of sites with by-passing air-drainage channels is therefore essential in this region, especially if viticulture is to extend to any great altitude.

A good example of this being apparently achieved is the very recent plantings at Whitlands near Whitfield, south of Milawa, at heights reaching nearly 800 m. The average altitude of the vines is approximately 770 m, on a north-projecting ridge between the King River on the east side, and its tributary Boggy Creek on the west. Because the King River Valley curves away to the south-west at the southern end of the ridge, it might be expected to intercept and divert dense chilled air descending from the snowfields to the south and south-east.

Table 50 gives Wangaratta-derived estimates for Whitfield, at 100 m intervals from altitude 200 m (valley floor) to 800 m. Topography is assumed to change progressively from flat, at 200 m, to steep slopes (by vineyard standards) at 500 m and above; and from all aspects up to 400 m, to increasingly northern aspects between 500 and 800 m. Soils are assumed to be loamy (therefore no temperature adjustment for soil type) throughout. A 0.6°C reduction in average mean temperature is allowed for the more southerly latitude of Whitfield as compared with Wangaratta (cf. Figure 13), in addition to the standard 0.6°C reduction per 100 m increase in altitude. In line with an evident general lapse rate of diurnal temperature range with altitude in north-east Victoria (e.g. Beechworth vs Wangaratta, Rutherglen and Corowa) I have also made a further altitude adjustment of 0.2°C reduction in the diurnal range/2 per 100 m altitude increase, additional to the normal topographic adjustments for slope and aspect per se (Chapter 4). (With small altitude differences this would be normally small enough to be neglected, as I have done elsewhere; but with large differences in altitude, any such trend needs to be taken into account if present.) The table then follows through the standard calculation of biologically effective temperature summation at each altitude, and thence estimates the average dates of maturity and ripening month average mean temperatures for each maturity group.

The calculations suggest about 800 m as the upper limit for reliable ripening of the earliest high-quality grape varieties (Maturity Groups 2/3). Between 700 and 800 m should be optimal for true cool-climate white wines (German and Champagne styles); about 700 m for Burgundy-style red wines; 300 to 700 m for full to medium-bodied white wines generally, from grape varieties of Maturity Groups 3–5; and 300–600 m for full to medium-bodied red wines, of Bordeaux or Rhône styles, from Maturity Groups 4–6. Provided that full ripening takes place before excessive autumn rains, quality should improve with altitude because of rising relative humidities and falling temperature variability indices, which I estimate for the middle and upper slopes to be between 33 and 36 for the growing season, and slightly lower during ripening. Greater altitudes and lower average temperatures will not necessarily increase the danger of killing frosts after budburst, because of the reduced temperature variability combined with later budburst.

The regional data nevertheless confirm that nowhere will relative humidity reach as high as that normally experienced in coastal regions; nor will afternoon sea breezes bring relief from stress, as in coastal regions. Additionally, the highest locations are at altitudes which call into question the direct effects of reduced carbon dioxide partial pressure on photosynthesis and water use efficiency, and hence on must and wine pH (Chapters 3, 23).

Whilst it is too early yet for an accurate practical check of these estimates, some published comments are now available on the initial vintages from Whitlands (Clarke 1991). He refers to Cabernet Franc and Merlot (Maturity Group 5) as making excellent wines 'when they ripen'; and to Cabernet Sauvignon (Maturity Group 6) as sometimes not being picked until June. Wines from Pinot Noir (Maturity Group 3) are reported to be deep-coloured and rich.

Such behaviour seems best consistent with my predictions for about 650–700 m: that is, some 100 m lower than the actual vineyard height. It needs to be remembered, however, that the prediction was based on early to mid 20th century temperatures, whereas the last decade is generally recognized as having been unusually warm (see Chapter 23). My estimate of the temperature difference between the two periods (Figure 17) is about 0.3°C, or half of the equivalent of a 100 m altitude difference. The suggested, very tentative, conclusion is therefore that the method used to project from Wangaratta (150 m) to Whitlands (770 m) was conservative in its estimate of ripening capacity at the higher altitude by the equivalent of 0.3°C, or 50 m in altitude, or one grape maturity group. Alternatively the topographical adjustments may have been accurate, but the allowance for distance south from Wangaratta too great by 0.3°C.

Either way, the exercise seems broadly to confirm the methods developed in Chapters 4, 6, 7 and 8, and that they can be safely used to make conservative adjustments for smaller differences of topography and altitude.

In summary, the crude calculations support the case for high-altitude viticulture in central and north-east Victoria, but with reservations compared with more coastal regions having similar temperatures. The best sites in the region are along the north-facing slopes of the Great Dividing Range, wherever there are isolated or projecting hills enjoying superior local air drainage and protection from the incursion of cold air masses from snowfields on nearby mountains. Individual site selection is crucial, if unacceptable frost risks are to be avoided and ripening period temperature variability reduced to a minimum. With a range of altitudes it should be possible to produce table wines ranging from full-bodied at intermediate altitudes to light, true cool-climate styles towards the upper limit. Comparable site

Table 50. Estimates (°C) for different altitudes in the King River Valley, Victoria. Based on Wangaratta, adjusted for latitude (–0.6°) and increasing altitude (–0.6° per 100 m):

a. for raw monthly average mean temperatures;
b. for average maturity dates, and
c. for average mean temperatures of the respective final 30 days of ripening, for the eight grape maturity groups; with
d. assumed topographies and corresponding adjustments at the different altitudes

Month	Wangaratta, 150 m		King River Valley altitudes						
	Mean temp.	Daily range/2	200 m	300 m	400 m	500 m	600 m	700 m	800 m
			a. Monthly average mean temperature						
Oct.	14.7	6.5	13.8	13.2	12.6	12.0	11.4	10.8	10.2
Nov.	18.2	7.4	17.3	16.7	16.1	15.5	14.9	14.3	13.7
Dec.	21.2	7.7	20.3	19.7	19.1	18.5	17.9	17.3	16.7
Jan.	22.6	7.8	21.7	21.1	20.5	19.9	19.3	18.7	18.1
Feb.	23.0	7.8	22.1	21.5	20.9	20.3	19.7	19.1	18.5
March	19.7	7.5	18.8	18.2	17.6	17.0	16.4	15.8	15.2
April	15.1	6.8	14.2	13.6	13.0	12.4	11.8	11.2	10.6
Average	19.2	7.4	18.3	17.7	17.1	16.5	15.9	15.3	14.7
Maturity group			b. Average date of maturity						
1			20/2	22/2	24/2	27/2	3/3	8/3	15/3
2			25/2	27/2	1/3	4/3	8/3	15/3	25/3
3			3/3	5/3	7/3	10/3	15/3	23/3	7/4
4			8/3	10/3	12/3	16/3	23/3	3/4	-
5			14/3	16/3	18/3	23/3	3/4	22/4	-
6			20/3	22/3	25/3	31/3	16/4	-	-
7			27/3	29/3	2/4	11/4	–	–	–
8			3/4	5/4	12/4	4/5	–	–	–
Maturity group			c. Average mean temperature of ripening month						
1			22.3	21.9	21.5	21.0	20.3	19.3	18.2
2			22.2	21.7	21.2	20.6	19.8	18.7	16.9
3			22.0	21.5	20.8	20.1	19.2	17.5	15.1
4			21.6	21.0	20.3	19.4	18.3	16.1	-
5			21.1	20.4	19.6	18.6	16.8	13.3	-
6			20.4	19.7	18.7	17.6	14.7	-	-
7			19.5	18.8	17.6	16.1	–	–	–
8			18.4	17.6	16.2	12.4	–	–	–
			d. Assumed topographies and topographic adjustments						
Alt. adj. to daily range/2			-	- 0.2	- 0.4	- 0.6	- 0.8	- 1.0	- 1.2
Slope			Flat	Mod.	Mod.+	Steep	Steep	Steep	Steep
Slope adj. to mean temp.			-	+ 0.2	+ 0.4	+ 0.5	+ 0.5	+ 0.5	+ 0.5
Slope adj. to daily range/2			-	- 0.6	- 0.8	- 1.0	- 1.0	- 1.0	- 1.0
Aspect			All	All	All	SE-N-SW	E-N-W	NE-N-NW	N
Aspect adj. to mean temp.			–	–	–	+ 0.1	+ 0.2	+ 0.3	+ 0.4
Aspect adj. to daily range/2			–	–	–	–	- 0.1	- 0.1	- 0.2
Total adj. to mean temp.			0	+ 0.2	+ 0.4	+ 0.6	+ 0.8	+ 1.0	+ 1.2
Total adj. to daily range/2			0	- 0.8	- 1.2	- 1.6	- 1.9	- 2.1	- 2.4

selection for cool ripening becomes less feasible north-wards from north-east Victoria, because of greater spring temperature variability and frost risk.

East Gippsland

The data for **Bairnsdale** (Table 53) and **Orbost** (Table 69) show an environment suitable *on average* for making medium-bodied table wines. The climate compares closely with that of the upper Yarra Valley, except that East Gippsland has less rainfall throughout the year, and more sunshine at all times except mid-summer. East Gippsland also has much less spring frost risk, due to a smaller variability of the minimum temperature.

In many respects the climate of East Gippsland resembles that of north-east Tasmania (see Launceston, Chapter 14, Table 114). The frost risk is less than at Launceston, but may not differ from that closer to the Tasmanian coast. However the Victorian coast, because of its southern aspect, has much greater variability of the maximum temperature, and greater summer extremes. To the east in Victoria there is also increasing variability of the annual rainfall, although nowhere does the variability equal that experienced in coastal southern New South Wales (Chapter 12).

Irregularity of rainfall within seasons is another feature of the East Gippsland climate, with a strong tendency for it to come in large amounts interspersed by long dry spells. Water conservation is therefore desirable, and there is a risk of unseasonal heavy rain during ripening.

In summary, East Gippsland has climatic conditions which on average should yield good table wines of medium body; but much vintage variation can be expected, because of the variability of rainfall and of maximum temperatures. A further probable limiting factor is scarcity of topographically favourable sites, owing to the flatness of the coastal plain and the predominantly southern aspect of the hill slopes. The estimates for both Bairnsdale and Orbost are for the low foothills, although slight to moderate slopes on the plain, closer to the ocean, are probably not much different.

Where suitable sites can be found, East Gippsland shows promise for small-scale viticulture. Large-scale investment should perhaps await further experience.

La Trobe Valley

The La Trobe Valley, as represented by **Maffra** (Table 65) is a little warmer and drier than East Gippsland, with a wider diurnal temperature range but less variability of the maximum. The foothills along the northern edge of the valley, like those of East Gippsland, have predominantly southern aspects. Favourable east to north exposures do nevertheless exist along the rivers Avon, Macallister and Thomson, and on the upper reaches of the La Trobe River and its northern tributaries. East-facing foothills to the west of Maffra offer further possibilities.

The rather low rainfall of the valley itself indicates a need for supplementary watering. Some of the foothills presumably have more rain, but its likely variability suggests that a capacity for supplementary watering is still desirable. In other respects the foothills represented by Table 65 are climatically not dissimilar to the Yarra Valley. They could have nearly as good potential for table wine production.

South Gippsland

Coastal southern Gippsland, as exampled by **Leongatha** (Table 64), is mild and reasonably free from frosts; but it is cooler and less sunny than East Gippsland, with more rainfall in all seasons. Estimated ripening mean temperatures, sunshine hours, and probably relative humidities are comparable with those of Burgundy and the central Loire; total seasonal sunshine is similar to that of the central Loire, but a little less than in Burgundy; and total seasonal and ripening period rainfall are more than in either. On the other hand ripening period sunshine hours are fully comparable with those of the Loire and Burgundy. Early-maturing grape varieties, up to Maturity Group 4, should ripen fully, to make mainly light wines. However as elsewhere near the south coast, the variability of the maximum temperature is very marked. Ripening period extremes could be excessive in some seasons for delicate styles of wine.

This is not an environment which could on present information be recommended for large-scale commercial viticulture. Nevertheless selected sites might be very satisfactory for small-scale development. Varieties such as Chasselas, Müller-Thurgau, Traminer, Sauvignon Blanc, Chardonnay, Meunier and the Pinots are logical choices in the first instance. Halliday (1991a) has reported highly promising results with Pinot Noir and Chardonnay grown south of Leongatha. The main viticultural need is to select warm, sheltered aspects, and to train the vines for good canopy light relations and disease control.

The combined figures for Maffra and Leongatha suggest interesting prospects also on the inland-facing slopes of the South Gippsland hills just north of Leongatha, running down to the La Trobe Valley. The foothills of the Morwell River Valley look especially interesting. With less rainfall and more sunshine than at Leongatha due to rain shadow, and shelter from coastal winds and the good air drainage of a relatively isolated hill range, it seems likely that excellent climatic conditions exist for winegrapes to make light to medium-bodied table wines. Given suitable soils, the only apparently adverse factor is a still rather marked variability of the maximum temperature. Early-midseason varieties, such as Cabernet Franc and Merlot, should ripen well on the lowest and warmest slopes; while the main white varieties, together with early-maturing red varieties such as Pinot Noir and Gamay, should ripen up to about 200 m.

Western Port

The Western Port area of west Gippsland is another which, on climatic grounds, appears well suited to viticulture. Typical of the environments warranting consideration would be the low hills adjoining the Bass and Lang Lang rivers.

Detailed climatic figures are lacking, but from the general trends one may surmise that conditions are intermediate between those of the Mornington Peninsula and perhaps the cooler parts of the Yarra Valley. Major soil groups (Hubble 1970) also resemble those of the Mornington Peninsula and Yarra Valley.

These characteristics point to Western Port as an area for Burgundy and Loire grape varieties and wine styles, especially Chardonnay, the Pinots, and (on soils not too conducive to vine vigour) perhaps Sauvignon Blanc.

TABLE 51

ARARAT, VICTORIA. 37°17'S, 142°57'E. ALT. 332m. (For Great Western)

	1	2	3	4	5	6	7	8	9	10	11	12	13	14	15	16	17	18
	Temperature °C									Summation of Day Degrees above 10° with 19° Cut-Off			Temperature Variability Index					
PERIOD	Lowest Recorded Min. (46YRS)	No.Mths with <0.3° (30YRS)	No.Mths with <2.5° (30YRS)	Av.Mthly Lowest Min. (30YRS)	Av.Min. Minus Col.4	Daily Mean (29YRS)	Daily Range /2 (29YRS)	Av.Mthly Highest Max. (30YRS)	Col.8 Minus Av.Max.	Raw	Adj.1 for Lat. & Daily Temp Rng	Adj.2 for Vine Sites	Raw	Adj. for Vine Sites	Av. Sunshine Hrs	Av. Number Rain Days	Av. Rainfall mm	Av. % R.H. 1500 HRS (20YRS)
OCTOBER	−1.7	4	19	1.2	6.3	12.9	5.4	27.2	8.9	90	89	92	36.8	35.6	240	12	61	53
NOVEMBER	1.1	0	6	3.5	5.6	15.3	6.2	32.2	10.7	159	157	160	41.1	39.9	265	9	48	49
DECEMBER	3.3	0	0	5.0	5.7	17.5	6.8	35.6	11.3	233	224	232	44.2	43.0	290	7	38	43
JANUARY	3.9	0	0	6.3	5.1	18.7	7.3	38.6	12.6	270	253	261	46.9	45.7	305	5	31	39
FEBRUARY	3.9	0	0	6.3	6.2	19.4	6.9	36.3	10.0	252	252	252	43.8	42.6	255	6	34	42
MARCH	2.2	0	1	5.3	5.9	17.3	6.1	33.5	10.1	226	226	229	40.4	39.2	235	6	39	46
APRIL	−1.1	2	15	2.7	6.0	13.5	4.8	27.9	9.6	105	109	116	34.8	33.6	190	9	44	59
OCT-APRIL	—	6	41	—	5.8	16.4	6.2	—	10.5	1,335	1,310	1,342	41.1	39.9	1,780	54	295	47
MAY-SEPT	—	—	—	—	—	9.3	—	—	—	—	—	—	—	—	—	—	321	—
DATA FOR MONTH TO MATURITY																		
MATURITY GROUP	EST. MATURITY DATE					*	*	*						*				
1	MARCH 7					19.3	6.4	35.7	10.0					41.8	270	6	34	42
2	MARCH 13					19.1	6.3	35.4	10.0					41.3	260	6	35	43
3	MARCH 20					18.6	6.1	34.8	10.0					40.5	250	6	36	44
4	MARCH 28					17.8	5.9	34.0	10.0					39.5	240	6	38	45
5	APRIL 4					16.9	5.6	33.0	10.0					38.2	230	6	40	48
6	APRIL 16					15.5	5.3	30.8	9.8					36.0	212	8	42	53
7	MAY 14					11.8	4.6	26.3	9.4					32.3	178	11	48	61
8	—					—	—	—	—					—	—	—	—	—

* Adjusted for vine sites.

TABLE 52
AVOCA, VICTORIA. 37°05′S, 143°29′E. ALT. 242m. (For Pyrenees Range)

	1	2	3	4	5	6	7	8	9	10	11	12	13	14	15	16	17	18
						Temperature °C				Summation of Day Degrees above 10° with 19° Cut-Off			Temperature Variability Index					
PERIOD	Lowest Recorded Min. (70YRS)	No.Mths with <0.3° (30YRS)	No.Mths with <2.5° (30YRS)	Av.Mthly Lowest Min. (30YRS)	Av.Min. Minus Col.4	Daily Mean (30YRS)	Daily Range /2 (30YRS)	Av.Mthly Highest Max. (30YRS)	Col.8 Minus Av.Max.	Raw	Adj.1 for Lat. & Daily Temp Rng	Adj.2 for Vine Sites	Raw	Adj. for Vine Sites	Av. Sunshine Hrs	Av. Number Rain Days	Av. Rainfall mm	Av. % R.H. 1500 HRS *
OCTOBER	−2.8	9	26	1.1	5.8	13.4	6.5	28.7	8.8	105	105	105	40.6	34.2	250	10	49	51
NOVEMBER	0.0	3	10	3.1	5.9	16.2	7.2	33.3	9.9	186	173	183	44.6	38.2	275	7	38	47
DECEMBER	1.9	0	1	5.1	6.2	18.8	7.5	37.0	10.7	273	252	268	46.9	40.5	310	6	36	41
JANUARY	4.4	0	0	6.9	5.3	20.0	7.8	39.3	11.5	279	279	279	48.0	41.6	310	4	27	37
FEBRUARY	1.7	0	1	6.5	6.4	20.6	7.7	37.1	8.8	252	252	252	46.0	39.6	260	4	38	40
MARCH	1.7	0	1	4.8	6.0	17.9	7.1	34.2	9.2	245	235	244	43.6	37.2	250	5	30	44
APRIL	−2.2	7	20	1.6	6.4	13.9	5.9	28.7	8.9	117	118	118	38.9	32.5	200	7	40	57
OCT-APRIL	—	19	59	—	6.0	17.3	7.1	—	9.7	1,457	1,414	1,449	44.1	37.7	1,855	43	258	45
MAY-SEPT	—	—	—	—	—	9.2	—	—	—	—	—	—	—	—	—	—	284	—
						DATA FOR MONTH TO MATURITY												
MATURITY GROUP	EST. MATURITY DATE					*	*	*						*				*
1	FEBRUARY 24					20.7	6.1	35.8	9.1					39.9	284	4	40	39
2	MARCH 2					20.6	6.1	35.5	8.8					39.5	279	4	40	40
3	MARCH 7					20.4	6.0	35.1	8.9					39.1	274	4	38	41
4	MARCH 13					20.0	5.9	34.5	9.0					38.6	268	4	36	42
5	MARCH 20					19.3	5.8	33.8	9.1					38.1	262	5	33	43
6	MARCH 27					18.5	5.6	33.0	9.2					37.5	255	5	31	44
7	APRIL 4					17.3	5.3	32.2	9.1					36.5	245	5	31	46
8	APRIL 14					16.1	4.9	31.1	9.1					35.2	232	6	34	49

* Adjusted for vine sites. Raw relative humidities estimated as Ararat minus 2%.

TABLE 53
BAIRNSDALE, VICTORIA. 37°49′S, 147°37′E. ALT. 14m. (East Gippsland)

	1	2	3	4	5	6	7	8	9	10	11	12	13	14	15	16	17	18
						Temperature °C				Summation of Day Degrees above 10° with 19° Cut-Off			Temperature Variability Index					
PERIOD	Lowest Recorded Min. (66YRS)	No.Mths with <0.3° (30YRS)	No.Mths with <2.5° (30YRS)	Av.Mthly Lowest Min. (30YRS)	Av.Min. Minus Col.4	Daily Mean (30YRS)	Daily Range /2 (30YRS)	Av.Mthly Highest Max. (30YRS)	Col.8 Minus Av.Max.	Raw	Adj.1 for Lat. & Daily Temp Rng	Adj.2 for Vine Sites	Raw	Adj. for Vine Sites	Av. Sunshine Hrs	Av. Number Rain Days	Av. Rainfall mm	Av. % R.H. 1500 HRS
OCTOBER	−2.8	0	8	2.9	5.0	13.8	5.9	29.6	9.9	118	117	112	38.5	36.7	230	13	70	56
NOVEMBER	−1.1	0	0	4.4	5.0	15.4	6.0	33.5	12.1	162	160	155	41.1	39.3	253	11	64	54
DECEMBER	0.0	0	0	6.0	5.3	17.3	6.0	35.6	12.3	226	222	217	41.6	39.8	280	10	68	52
JANUARY	1.7	0	0	7.0	4.9	18.0	6.1	37.7	13.6	248	245	240	42.9	41.1	268	8	60	50
FEBRUARY	3.9	0	0	7.0	5.5	18.5	6.0	35.7	11.2	238	236	231	40.7	38.9	230	7	50	52
MARCH	0.0	0	0	5.8	5.2	16.9	5.9	34.0	11.2	214	213	208	40.0	38.2	231	9	66	53
APRIL	−1.7	4	11	2.7	5.6	14.0	5.7	29.2	9.5	120	121	116	37.9	36.1	205	9	50	54
OCT-APRIL	—	4	19	—	5.2	16.3	5.9	—	11.4	1,326	1,314	1,279	40.4	38.6	1,697	67	428	53
MAY-SEPT	—	—	—	—	—	10.1	—	—	—	—	—	—	—	—	—	—	268	—
						DATA FOR MONTH TO MATURITY												
MATURITY GROUP	EST. MATURITY DATE					*	*	*						*				
1	MARCH 14					17.9	5.5	34.2	11.2					38.6	239	8	60	52
2	MARCH 21					17.5	5.5	33.9	11.2					38.4	236	8	63	53
3	MARCH 29					16.9	5.4	33.5	11.2					38.2	232	9	66	53
4	APRIL 8					16.0	5.3	32.8	10.8					37.5	226	9	60	53
5	APRIL 21					14.8	5.2	30.6	10.1					36.8	212	9	54	54
6	MAY 10					13.1	5.1	27.2	9.0					35.6	195	10	50	55
7	—					—	—	—	—					—	—	—	—	—
8	—					—	—	—	—					—	—	—	—	—

* Adjusted for vine sites.

TABLE 54

BALLARAT, VICTORIA. 37°35'S, 143°50'E. ALT. 437m.

	1	2	3	4	5	6	7	8	9	10	11	12	13	14	15	16	17	18
				Temperature °C						Summation of Day Degrees above 10° with 19° Cut-Off			Temperature Variability Index					
PERIOD	Lowest Recorded Min. (66YRS)	No.Mths with <0.3° (30YRS)	No.Mths with <2.5° (30YRS)	Av.Mthly Lowest Min. (30YRS)	Av.Min. Minus Col.4	Daily Mean (30YRS)	Daily Range /2 (30YRS)	Av.Mthly Highest Max. (30YRS)	Col.8 Minus Av.Max.	Raw	Adj.1 for Lat. & Daily Temp Rng	Adj.2 for Vine Sites	Raw	Adj. for Vine Sites	Av. Sunshine Hrs	Av. Number Rain Days	Av. Rainfall mm	Av. % R.H. 1500 HRS (30YRS)
OCTOBER	−1.7	15	29	0.2	6.3	11.7	5.2	25.3	8.4	53	52	69	35.5	33.3	214	16	68	56
NOVEMBER	−0.3	3	17	2.0	5.8	13.7	5.9	29.8	10.2	111	110	120	39.6	37.4	232	13	56	53
DECEMBER	1.1	0	10	3.2	6.5	16.1	6.4	34.0	11.5	189	186	197	43.6	41.4	264	11	52	47
JANUARY	2.2	0	2	4.1	6.2	17.3	7.0	36.8	12.5	226	215	234	46.7	44.5	279	8	38	41
FEBRUARY	2.2	0	3	4.4	7.2	18.3	6.7	35.0	10.0	232	227	240	44.0	41.8	242	8	50	41
MARCH	−0.6	1	10	3.3	6.8	16.1	6.0	31.9	9.9	189	189	200	40.6	38.4	224	10	47	47
APRIL	−1.7	6	22	1.4	6.2	12.4	4.8	26.4	9.2	72	76	94	34.6	32.4	174	13	56	59
OCT-APRIL	—	25	93	—	6.4	15.1	6.0	—	10.2	1,072	1,055	1,154	40.7	38.5	1,629	79	367	49
MAY-SEPT	—	—	—	—	—	8.2	—	—	—	—	—	—	—	—	—	—	352	—
				DATA FOR MONTH TO MATURITY														
MATURITY GROUP	EST. MATURITY DATE					*	*	*						*				
1	MARCH 29					16.7	5.5	32.0	9.9					38.6	227	10	47	47
2	APRIL 9					15.5	5.2	30.4	9.7					36.9	212	11	49	50
3	APRIL 26					13.3	4.5	27.2	9.3					33.2	181	13	54	57
4	—					—	—	—	—					—	—	—	—	—
5	—					—	—	—	—					—	—	—	—	—
6	—					—	—	—	—					—	—	—	—	—
7	—					—	—	—	—					—	—	—	—	—
8	—					—	—	—	—					—	—	—	—	—

* Adjusted for vine sites.

TABLE 55

BEECHWORTH, VICTORIA. 36°21'S, 146°41'E. ALT. 552m.

	1	2	3	4	5	6	7	8	9	10	11	12	13	14	15	16	17	18
				Temperature °C						Summation of Day Degrees above 10° with 19° Cut-Off			Temperature Variability Index					
PERIOD	Lowest Recorded Min. (96YRS)	No.Mths with <0.3° (30YRS)	No.Mths with <2.5° (30YRS)	Av.Mthly Lowest Min. (30YRS)	Av.Min. Minus Col.4	Daily Mean (29YRS)	Daily Range /2 (29YRS)	Av.Mthly Highest Max. (30YRS)	Col.8 Minus Av.Max.	Raw	Adj.1 for Lat. & Daily Temp Rng	Adj.2 for Vine Sites	Raw	Adj. for Vine Sites	Av. Sunshine Hrs	Av. Number Rain Days	Av. Rainfall mm	Av. % R.H. 1500 HRS (30YRS)
OCTOBER	−1.7	17	28	0.4	7.2	12.8	5.2	25.6	7.6	87	86	130	35.6	32.6	255	11	89	49
NOVEMBER	−0.6	7	21	1.9	7.9	15.8	6.0	30.4	8.6	174	171	204	40.5	37.5	300	9	64	39
DECEMBER	−0.3	1	6	3.4	8.9	18.7	6.4	33.8	8.7	270	263	279	43.2	40.2	325	8	57	37
JANUARY	−0.6	0	3	5.4	8.1	20.1	6.6	35.5	8.8	279	279	279	43.3	40.3	325	6	46	33
FEBRUARY	2.2	0	0	5.5	8.5	20.6	6.6	34.2	7.0	252	252	252	41.9	38.9	275	6	51	36
MARCH	0.0	0	7	4.0	7.7	17.8	6.1	31.1	7.2	242	241	276	39.3	36.3	274	6	62	39
APRIL	−1.1	8	23	1.7	6.5	13.4	5.2	25.4	6.8	102	103	146	34.1	31.1	255	8	68	48
OCT-APRIL	—	33	88	—	7.8	17.0	6.0	—	7.8	1,406	1,395	1,566	39.7	36.7	1,979	54	437	40
MAY-SEPT	—	—	—	—	—	8.3	—	—	—	—	—	—	—	—	—	—	492	—
				DATA FOR MONTH TO MATURITY														
MATURITY GROUP	EST. MATURITY DATE					*	*	*						*				
1	FEBRUARY 18					21.8	5.8	35.1	7.7					39.4	306	6	52	35
2	FEBRUARY 24					21.8	5.7	34.9	7.4					39.2	300	6	54	35
3	MARCH 1					21.7	5.7	34.6	7.0					38.9	294	6	55	36
4	MARCH 6					21.5	5.7	34.2	7.0					38.5	290	6	57	36
5	MARCH 12					21.2	5.6	33.7	7.1					38.1	286	6	58	37
6	MARCH 19					20.8	5.5	32.9	7.1					37.6	282	6	59	38
7	MARCH 26					20.0	5.4	32.0	7.2					36.9	278	6	61	39
8	APRIL 3					18.9	5.3	31.0	7.2					36.0	273	6	62	40

* Adjusted for vine sites.

TABLE 56
BENDIGO, VICTORIA. 36°46′S, 144°17′E. ALT. 223m.

PERIOD	1 Lowest Recorded Min. (110YRS)	2 No.Mths with <0.3° (30YRS)	3 No.Mths with <2.5° (30YRS)	4 Av.Mthly Lowest Min. (30YRS)	5 Av.Min. Minus Col.4	6 Daily Mean (30YRS)	7 Daily Range /2 (30YRS)	8 Av.Mthly Highest Max. (30YRS)	9 Col.8 Minus Av.Max.	10 Raw	11 Adj.1 for Lat. & Daily Temp Rng	12 Adj.2 for Vine Sites	13 Raw	14 Adj. for Vine Sites	15 Av. Sunshine Hrs	16 Av. Number Rain Days	17 Av. Rainfall mm	18 Av. % R.H. 1500 HRS (30YRS)
OCTOBER	−1.4	1	17	2.3	5.8	14.3	6.2	28.6	8.1	133	132		38.7		250	12	52	40
NOVEMBER	0.8	0	3	4.0	6.5	17.3	6.8	33.7	9.6	219	215		43.3		282	9	37	35
DECEMBER	2.8	0	0	6.2	6.5	19.8	7.1	36.7	9.8	279	279		44.7		310	7	33	32
JANUARY	2.8	0	0	7.6	6.0	21.0	7.4	38.7	10.3	279	279		45.9		317	6	33	30
FEBRUARY	2.2	0	0	7.6	7.0	21.7	7.1	36.6	7.8	252	252		43.2		265	5	35	32
MARCH	3.3	0	0	5.9	5.9	18.9	6.7	33.4	7.8	276	272		40.5		258	6	36	37
APRIL	0.8	0	8	5.8	5.8	14.7	5.7	27.9	7.5	141	142		36.1		212	8	41	45
OCT-APRIL	—	1	28	—	6.2	18.2	6.7	—	8.7	1,579	1,571		41.8		1,894	53	267	36
MAY-SEPT	—	—	—	—	—	9.8	—	—	—	—	—		—		—	—	279	—

DATA FOR MONTH TO MATURITY

MATURITY GROUP	EST. MATURITY DATE												13		15	16	17	18
1	FEBRUARY 17					21.8	7.2	37.0	8.3				44.2		298	6	36	31
2	FEBRUARY 22					21.8	7.2	36.8	8.0				43.7		292	5	37	32
3	FEBRUARY 28					21.7	7.1	36.6	7.8				43.2		286	5	38	32
4	MARCH 5					21.4	7.1	36.3	7.8				42.7		281	5	38	33
5	MARCH 11					21.0	7.0	35.8	7.8				42.2		276	6	37	34
6	MARCH 16					20.6	6.9	35.2	7.8				41.7		271	6	37	35
7	MARCH 22					20.0	6.8	34.5	7.8				41.2		266	6	36	36
8	MARCH 28					19.3	6.7	33.8	7.8				40.7		261	6	36	37

TABLE 57
CAMPERDOWN, VICTORIA. 38°13′S, 143°09′E. ALT. 165m.

PERIOD	1 Lowest Recorded Min. (70YRS)	2 No.Mths with <0.3° (30YRS)	3 No.Mths with <2.5° (30YRS)	4 Av.Mthly Lowest Min. (30YRS)	5 Av.Min. Minus Col.4	6 Daily Mean (30YRS)	7 Daily Range /2 (30YRS)	8 Av.Mthly Highest Max. (30YRS)	9 Col.8 Minus Av.Max.	10 Raw	11 Adj.1 for Lat. & Daily Temp Rng	12 Adj.2 for Vine Sites	13 Raw	14 Adj. for Vine Sites	15 Av. Sunshine Hrs	16 Av. Number Rain Days	17 Av. Rainfall mm	18 Av. % R.H. 1500 HRS
OCTOBER	−1.1	1	15	2.6	4.8	13.3	5.9	26.7	7.5	102	102		35.9		210	15	73	
NOVEMBER	−0.6	0	4	4.1	4.7	15.1	6.3	30.6	9.2	153	152		39.1		215	13	60	
DECEMBER	0.6	0	0	5.3	5.3	17.4	6.8	35.3	11.1	229	222		43.6		240	10	48	
JANUARY	1.1	0	0	6.5	4.8	18.7	7.4	38.4	12.3	270	253		46.7		275	8	37	
FEBRUARY	0.0	1	1	6.7	5.6	19.5	7.2	35.7	9.0	252	252		43.4		223	8	42	
MARCH	1.7	0	0	6.3	4.5	17.5	6.7	34.0	9.8	233	229		.41.1		210	9	49	
APRIL	−1.7	0	4	4.0	4.5	14.0	5.5	28.2	8.7	120	121		35.2		160	13	62	
OCT-APRIL	—	2	24	—	4.9	16.5	6.5	—	9.7	1,359	1,331		40.7		1,533	76	371	
MAY-SEPT	—	—	—	—	—	9.9	—	—	—	—	—		—		—	—	406	

DATA FOR MONTH TO MATURITY

MATURITY GROUP	EST. MATURITY DATE												13		15	16	17	18
1	MARCH 9					19.1	7.1	35.5	9.3				42.8		232	8	46	
2	MARCH 15					18.8	7.0	35.2	9.4				42.3		224	8	47	
3	MARCH 21					18.1	6.9	34.6	9.6				41.7		217	9	48	
4	MARCH 30					17.5	6.7	33.8	9.8				41.1		210	9	49	
5	APRIL 8					16.5	6.3	32.4	9.6				40.0		200	10	51	
6	APRIL 21					15.0	5.8	30.0	9.0				37.0		180	12	56	
7	MAY 9					13.0	5.3	26.8	8.5				34.0		148	14	64	
8	—					—	—	—	—				—		—	—	—	

TABLE 58

COLAC, VICTORIA. 38°20'S, 143°42'E. ALT. 134m.

	1	2	3	4	5	6	7	8	9	10	11	12	13	14	15	16	17	18
				Temperature °C						Summation of Day Degrees above 10° with 19° Cut-Off			Temperature Variability Index					
PERIOD	Lowest Recorded Min. (70YRS)	No.Mths with <0.3° (30YRS)	No.Mths with <2.5° (30YRS)	Av.Mthly Lowest Min. (30YRS)	Av.Min. Minus Col.4	Daily Mean (36YRS)	Daily Range /2 (36YRS)	Av.Mthly Highest Max. (30YRS)	Col.8 Minus Av.Max.	Raw	Adj.1 for Lat. & Daily Temp Rng	Adj.2 for Vine Sites	Raw	Adj. for Vine Sites	Av. Sunshine Hrs	Av. Number Rain Days	Av. Rainfall mm	Av. % R.H. 1500 HRS
OCTOBER	−3.3	13	27	0.6	6.4	12.8	5.8	26.3	7.7	87	86	94	37.3	33.9	202	16	68	61
NOVEMBER	−2.8	3	20	2.0	6.4	14.8	6.4	30.2	9.0	144	143	150	41.0	37.6	200	13	57	56
DECEMBER	−0.6	3	14	2.5	7.0	16.6	7.1	35.6	11.9	205	193	210	47.3	43.9	228	10	45	50
JANUARY	0.6	0	7	3.6	7.3	18.2	7.3	37.7	12.2	254	239	259	48.7	45.3	265	7	32	41
FEBRUARY	0.6	0	7	3.9	7.1	18.6	7.6	36.1	9.9	241	224	243	47.4	44.0	214	8	37	43
MARCH	0.0	2	7	3.4	6.6	16.8	6.8	33.9	10.3	211	206	218	44.1	40.7	200	9	44	47
APRIL	−3.9	6	22	1.4	6.6	13.7	5.7	27.5	8.1	111	112	119	37.5	34.1	155	13	56	55
OCT-APRIL	—	27	104	—	6.8	15.9	6.7	—	9.9	1,253	1,203	1,293	43.3	39.9	1,464	76	339	50
MAY-SEPT	—	—	—	—	—	9.6	—	—	—	—	—	—	—	—	—	—	382	—
				DATA FOR MONTH TO MATURITY														
MATURITY GROUP	EST. MATURITY DATE			*	*	*								*				
1	MARCH 13					18.3	6.3	34.6	10.1					42.8	217	8	42	45
2	MARCH 20					17.9	6.2	34.2	10.1					42.2	211	9	43	46
3	MARCH 27					17.4	6.0	33.5	10.1					41.2	204	9	44	47
4	APRIL 5					16.6	5.8	32.2	9.9					40.2	194	9	46	49
5	APRIL 16					15.4	5.4	30.4	9.3					37.7	180	11	49	51
6	MAY 4					13.5	4.8	26.5	8.0					33.8	150	13	57	56
7	—					—	—	—	—					—	—	—	—	—
8	—					—	—	—	—					—	—	—	—	—

* Adjusted for vine sites.

TABLE 59

EUROA, VICTORIA. 36°46'S, 145°33'E. ALT. 175m. (For Strathbogie Range)

	1	2	3	4	5	6	7	8	9	10	11	12	13	14	15	16	17	18
				Temperature °C						Summation of Day Degrees above 10° with 19° Cut-Off			Temperature Variability Index					
PERIOD	Lowest Recorded Min. (58YRS)	No.Mths with <0.3° (30YRS)	No.Mths with <2.5° (30YRS)	Av.Mthly Lowest Min. (30YRS)	Av.Min. Minus Col.4	Daily Mean (26YRS)	Daily Range /2 (26YRS)	Av.Mthly Highest Max. (30YRS)	Col.8 Minus Av.Max.	Raw	Adj.1 for Lat. & Daily Temp Rng	Adj.2 for Vine Sites	Raw	Adj. for Vine Sites	Av. Sunshine Hrs	Av. Number Rain Days	Av. Rainfall mm	Av. % R.H. 1500 HRS *
OCTOBER	−0.6	4	15	2.3	6.2	14.6	6.1	28.7	8.0	143	141	98	38.6	36.2	250	9	61	55
NOVEMBER	0.0	1	3	4.3	6.3	17.6	7.0	34.0	9.4	228	217	174	43.7	41.3	282	6	44	47
DECEMBER	3.3	0	0	6.2	7.1	20.4	7.1	37.5	10.0	279	279	273	45.5	43.1	300	5	40	40
JANUARY	0.6	0	0	8.7	5.6	21.8	7.5	39.5	10.2	279	279	279	45.8	43.4	317	4	36	39
FEBRUARY	2.6	0	0	8.1	7.0	22.5	7.4	37.9	8.0	252	252	252	44.6	42.2	265	4	33	41
MARCH	1.7	0	1	6.6	6.8	20.0	6.6	34.2	7.6	279	279	266	40.8	38.4	260	5	46	46
APRIL	−0.6	2	11	3.3	6.4	15.3	5.6	28.5	7.6	159	161	118	36.4	34.0	208	6	48	55
OCT-APRIL	—	7	30	—	6.5	18.9	6.8	—	8.7	1,619	1,608	1,460	42.2	39.8	1,882	39	308	46
MAY-SEPT	—	—	—	—	—	10.1	—	—	—	—	—	—	—	—	—	—	341	—
				DATA FOR MONTH TO MATURITY														
MATURITY GROUP	EST. MATURITY DATE			*	*	*								*				*
1	FEBRUARY 26					21.1	6.8	36.1	8.2					42.3	290			41
2	MARCH 3					21.0	6.7	35.7	8.0					41.6	286			41
3	MARCH 9					20.7	6.5	35.1	7.9					40.9	282			42
4	MARCH 14					20.4	6.4	34.4	7.8					40.3	278			43
5	MARCH 19					19.9	6.3	33.7	7.7					39.7	273			44
6	MARCH 25					19.4	6.1	32.9	7.6					39.1	266			45
7	APRIL 2					18.2	6.0	31.9	7.6					38.2	257			46
8	APRIL 12					16.7	5.8	30.6	7.6					36.9	244			49

* Adjusted for vine sites. All relative humidities adjusted in parallel with Beechworth vs. Wangaratta.

TABLE 60
GEELONG, VICTORIA. 38°09'S, 144°22'E. ALT. 27m.

	1	2	3	4	5	6	7	8	9	10	11	12	13	14	15	16	17	18
				Temperature °C						Summation of Day Degrees above 10° with 19° Cut-Off			Temperature Variability Index					
PERIOD	Lowest Recorded Min. (61YRS)	No.Mths with <0.3° (30YRS)	No.Mths with <2.5° (30YRS)	Av.Mthly Lowest Min. (30YRS)	Av.Min. Minus Col.4	Daily Mean (30YRS)	Daily Range /2 (30YRS)	Av.Mthly Highest Max. (30YRS)	Col.8 Minus Av.Max.	Raw	Adj.1 for Lat. & Daily Temp Rng	Adj.2 for Vine Sites	Raw	Adj. for Vine Sites	Av. Sunshine Hrs	Av. Number Rain Days	Av. Rainfall mm (30YRS)	Av. % R.H. 1500 HRS (30YRS)
OCTOBER	−0.3	1	8	3.4	5.2	14.1	5.5	29.0	9.4	127	126	110	36.6	37.8	190	12	50	58
NOVEMBER	1.7	0	1	5.2	5.0	15.7	5.5	33.3	12.1	171	169	154	39.1	40.3	195	10	51	57
DECEMBER	3.2	0	0	6.5	5.5	17.6	5.6	36.4	13.2	236	233	218	41.1	42.3	220	8	47	56
JANUARY	3.9	0	0	7.9	5.1	18.8	5.8	39.4	14.8	273	270	254	43.1	44.3	260	6	26	52
FEBRUARY	5.0	0	0	8.7	5.1	19.5	5.7	37.2	12.0	252	252	250	39.9	41.1	220	6	40	55
MARCH	3.1	0	0	7.4	5.2	17.9	5.3	34.8	11.6	245	244	229	38.0	39.2	210	7	45	57
APRIL	1.7	0	2	4.9	5.4	15.1	4.8	29.3	9.4	153	154	139	34.0	35.2	160	10	42	62
OCT-APRIL	—	1	11	—	5.2	17.0	5.5	—	11.8	1,457	1,448	1,354	38.8	40.0	1,455	59	301	57
MAY-SEPT	—	—	—	—	—	11.0	—	—	—	—	—	—	—	—	—	—	240	—

DATA FOR MONTH TO MATURITY

MATURITY GROUP	EST. MATURITY DATE	Daily Mean *	Daily Range /2 *	Av.Mthly Highest Max. *	Col.8 Minus Av.Max.	Adj. for Vine Sites *	Av. Sunshine Hrs	Av. Number Rain Days	Av. Rainfall mm	Av. % R.H. *
1	MARCH 8	18.8	5.9	36.4	11.9	40.6	232	6	43	50
2	MARCH 15	18.6	5.8	36.0	11.8	40.1	226	7	44	51
3	MARCH 22	18.2	5.7	35.5	11.7	39.7	219	7	44	52
4	MARCH 29	17.5	5.6	34.8	11.6	39.4	212	7	45	53
5	APRIL 7	16.7	5.5	33.7	11.2	38.5	203	7	44	54
6	APRIL 17	15.8	5.3	32.0	10.5	37.2	185	8	43	56
7	APRIL 30	14.5	5.1	29.1	9.4	35.2	160	10	42	59
8	MAY 21	12.6	4.8	26.0	8.0	32.0	125	12	44	63

* Adjusted for vine sites. Adjusted relative humidities from means of Geelong and Colac.

TABLE 61
HAMILTON, VICTORIA. 37°45'S, 142°01'E. ALT. 187m.

	1	2	3	4	5	6	7	8	9	10	11	12	13	14	15	16	17	18
				Temperature °C						Summation of Day Degrees above 10° with 19° Cut-Off			Temperature Variability Index					
PERIOD	Lowest Recorded Min. (84YRS)	No.Mths with <0.3° (30YRS)	No.Mths with <2.5° (30YRS)	Av.Mthly Lowest Min. (30YRS)	Av.Min. Minus Col.4	Daily Mean (30YRS)	Daily Range /2 (30YRS)	Av.Mthly Highest Max. (30YRS)	Col.8 Minus Av.Max.	Raw	Adj.1 for Lat. & Daily Temp Rng	Adj.2 for Vine Sites	Raw	Adj. for Vine Sites	Av. Sunshine Hrs	Av. Number Rain Days	Av. Rainfall mm	Av. % R.H. 1500 HRS (28YRS)
OCTOBER	−3.4	6	24	1.4	5.2	12.4	5.8	27.7	9.5	74	74	77	37.9	36.7	218	15	66	53
NOVEMBER	−1.7	0	6	3.3	4.7	14.3	6.3	30.0	9.4	129	127	130	39.3	38.1	233	12	51	50
DECEMBER	−0.6	0	1	4.5	5.0	16.4	6.9	35.6	12.3	198	189	197	44.9	43.7	262	10	46	45
JANUARY	0.3	0	0	5.6	4.8	17.8	7.4	38.5	13.3	242	225	232	47.7	46.5	282	8	33	39
FEBRUARY	1.3	0	0	6.0	5.3	18.6	7.3	36.2	10.3	241	227	234	44.8	43.6	235	7	33	38
MARCH	−1.6	0	1	4.9	5.0	16.7	6.8	34.0	10.5	208	203	210	42.7	41.5	225	9	43	43
APRIL	−2.1	1	11	3.0	4.9	13.5	5.6	28.2	9.1	105	106	109	36.4	35.2	177	13	55	53
OCT-APRIL	—	7	43	—	5.0	15.7	6.6	—	10.6	1,197	1,151	1,189	42.0	40.8	1,632	74	327	46
MAY-SEPT	—	—	—	—	—	9.5	—	—	—	—	—	—	—	—	—	—	365	—

DATA FOR MONTH TO MATURITY

MATURITY GROUP	EST. MATURITY DATE	Daily Mean *	Daily Range /2 *	Av.Mthly Highest Max. *	Col.8 Minus Av.Max.	Adj. for Vine Sites *	Av. Sunshine Hrs	Av. Number Rain Days	Av. Rainfall mm	Av. % R.H. *
1	MARCH 26	17.3	6.6	34.2	10.5	41.8	231	9	42	42
2	APRIL 5	16.3	6.3	33.1	10.3	40.9	219	10	44	44
3	APRIL 16	15.0	6.0	31.4	9.9	38.6	202	11	48	48
4	MAY 11	12.5	5.0	26.0	8.6	33.0	160	15	58	56
5	—	—	—	—	—	—	—	—	—	—
6	—	—	—	—	—	—	—	—	—	—
7	—	—	—	—	—	—	—	—	—	—
8	—	—	—	—	—	—	—	—	—	—

* Adjusted for vine sites.

TABLE 62

HEALESVILLE, VICTORIA. 37°42'S, 145°30'E. ALT. approx. 130m. (Yarra Valley)

PERIOD	1 Lowest Recorded Min. (11YRS)	2 No.Mths with <0.3° (11YRS)	3 No.Mths with <2.5° (11YRS)	4 Av.Mthly Lowest Min. (11YRS)	5 Av.Min. Minus Col.4	6 Daily Mean (10YRS)	7 Daily Range /2 (10YRS)	8 Av.Mthly Highest Max. (11YRS)	9 Col.8 Minus Av.Max.	10 Raw	11 Adj.1 for Lat. & Daily Temp Rng	12 Adj.2 for Vine Sites	13 Raw	14 Adj. for Vine Sites	15 Av. Sunshine Hrs	16 Av. Number Rain Days	17 Av. Rainfall mm (20YRS)	18 Av. % R.H. 1500 HRS
OCTOBER	0.0	4	10	1.0	6.7	13.0	5.3	27.6	9.3	93	92	115	37.2	33.8	200		105	
NOVEMBER	2.2	0	3	3.2	5.9	15.2	6.1	32.5	11.2	156	154	167	41.5	38.1	205		87	
DECEMBER	−1.1	2	2	3.7	6.9	17.0	6.4	33.9	10.5	217	214	227	43.0	39.6	235		82	
JANUARY	−1.1	1	1	4.0	7.0	17.9	6.9	36.6	11.8	245	235	255	46.4	43.0	275		66	
FEBRUARY	2.8	0	0	4.7	6.6	18.2	6.9	35.2	10.1	230	222	240	44.3	40.9	220		77	
MARCH	0.6	0	2	3.5	6.8	16.8	6.5	33.8	10.5	211	210	224	43.3	39.9	205		75	
APRIL	−1.1	2	7	1.3	6.9	13.3	5.1	28.1	9.7	99	100	124	37.0	33.6	150		105	
OCT-APRIL	—	9	25	—	6.7	15.9	6.2	—	10.4	1,251	1,227	1,352	41.8	38.4	1,490		597	
MAY-SEPT	—	—	—	—	—	9.4	—	—	—	—	—	—	—	—	—	—	411	

DATA FOR MONTH TO MATURITY

MATURITY GROUP	EST. MATURITY DATE				6 *	7 *	8 *	9					14 *	15	16	17	18
1	MARCH 6				18.5	6.0	34.5	10.2					40.7	230		79	
2	MARCH 13				18.3	5.9	34.2	10.3					40.5	222		78	
3	MARCH 20				17.9	5.8	33.9	10.4					40.3	215		77	
4	MARCH 27				17.5	5.7	33.5	10.5					40.0	210		75	
5	APRIL 4				16.8	5.5	32.8	10.3					39.5	200		82	
6	APRIL 15				15.5	4.9	31.4	9.9					36.8	180		90	
7	APRIL 30				13.8	4.4	27.7	9.7					33.6	150		105	
8	—				—	—	—	—					—	—		—	

* Adjusted for vine sites.

TABLE 63

HEYWOOD, VICTORIA. 38°08'S, 141°38'E. ALT. 27m. (For Drumborg)

PERIOD	1 Lowest Recorded Min. (27YRS)	2 No.Mths with <0.3° (27YRS)	3 No.Mths with <2.5° (27YRS)	4 Av.Mthly Lowest Min.	5 Av.Min. Minus Col.4	6 Daily Mean	7 Daily Range /2	8 Av.Mthly Highest Max.	9 Col.8 Minus Av.Max.	10 Raw	11 Adj.1 for Lat. & Daily Temp Rng	12 Adj.2 for Vine Sites	13 Raw	14 Adj. for Vine Sites	15 Av. Sunshine Hrs	16 Av. Number Rain Days	17 Av. Rainfall mm	18 Av. % R.H. 1500 HRS
OCTOBER	−2.8	7	23	1.1	6.0	12.7	5.6	27.6	9.3	84	83	92	37.7	32.9	200	16	70	68
NOVEMBER	0.6	0	17	2.8	5.9	14.4	5.7	30.9	10.8	132	131	138	39.5	34.7	210	14	62	60
DECEMBER	0.6	0	9	4.0	6.3	16.2	5.9	35.7	13.6	192	190	195	43.5	38.7	240	13	55	54
JANUARY	1.1	0	5	4.4	6.6	17.4	6.4	36.5	12.7	229	227	227	44.9	40.1	245	10	37	48
FEBRUARY	2.3	0	5	4.4	7.2	18.0	6.4	37.1	12.7	224	222	222	45.5	40.7	225	9	34	49
MARCH	−1.1	5	12	3.3	7.0	16.5	6.2	34.7	12.0	202	201	201	43.8	39.0	210	11	39	54
APRIL	−1.7	9	21	1.1	7.0	13.9	5.8	29.5	9.8	117	118	124	40.0	35.2	160	12	76	57
OCT-APRIL	—	21	92	—	6.6	15.6	6.0	—	11.6	1,180	1,172	1,199	42.1	37.3	1,490	85	373	56
MAY-SEPT	—	—	—	—	—	9.8	—	—	—	—	—	—	—	—	—	—	488	—

DATA FOR MONTH TO MATURITY

MATURITY GROUP	EST. MATURITY DATE				6 *	7 *	8 *	9					14 *	15	16	17	18
1	MARCH 27				16.8	5.0	33.8	12.1					39.2	214	11	38	53
2	APRIL 5				16.0	4.9	32.7	11.7					38.5	204	11	44	54
3	APRIL 17				15.0	4.8	30.9	10.9					37.1	185	11	58	55
4	MAY 1				13.8	4.6	28.2	9.8					35.0	158	12	77	57
5	—				—	—	—	—					—	—	—	—	—
6	—				—	—	—	—					—	—	—	—	—
7	—				—	—	—	—					—	—	—	—	—
8	—				—	—	—	—					—	—	—	—	—

* Adjusted for vine sites.

TABLE 64

LEONGATHA, VICTORIA. 38°38′S, 145°58′E. ALT. 83m. (South Gippsland)

	1	2	3	4	5	6	7	8	9	10	11	12	13	14	15	16	17	18
	Temperature °C									Summation of Day Degrees above 10° with 19° Cut-Off			Temperature Variability Index					
PERIOD	Lowest Recorded Min. (40YRS)	No.Mths with <0.3° (30YRS)	No.Mths with <2.5° (30YRS)	Av.Mthly Lowest Min. (30YRS)	Av.Min Minus Col.4	Daily Mean (28YRS)	Daily Range /2 (28YRS)	Av.Mthly Highest Max. (30YRS)	Col.8 Minus Av.Max.	Raw	Adj.1 for Lat. & Daily Temp Rng	Adj.2 for Vine Sites	Raw	Adj. for Vine Sites	Av. Sunshine Hrs	Av. Number Rain Days	Av. Rainfall mm (30YRS)	Av. % R.H. 1500 HRS
OCTOBER	−0.6	1	13	2.6	5.2	13.1	5.3	27.6	9.2	96	96		35.6		186		99	
NOVEMBER	1.1	0	3	4.0	5.1	14.9	5.8	31.4	10.7	147	146		39.0		195		84	
DECEMBER	1.1	0	1	4.7	5.9	16.9	6.3	35.1	11.9	214	212		43.0		220		78	
JANUARY	2.2	0	1	5.4	6.2	18.1	6.5	37.3	12.7	251	249		44.9		225		55	
FEBRUARY	2.5	0	0	6.0	6.3	18.7	6.4	35.7	10.6	244	242		42.5		205		58	
MARCH	0.6	0	0	4.9	6.4	17.3	6.0	34.3	11.0	226	226		41.4		205		73	
APRIL	0.0	1	12	3.1	6.0	14.1	5.0	27.8	8.7	123	123		34.7		148		83	
OCT-APRIL	—	2	30	—	5.9	16.2	5.9	—	10.7	1,301	1,294		40.2		1,384		530	
MAY-SEPT	—	—	—	—	—	10.3	—	—	—						—		468	
DATA FOR MONTH TO MATURITY																		
MATURITY GROUP	EST. MATURITY DATE																	
1	MARCH 14					18.2	6.2	35.0	10.8				42.0		214		68	
2	MARCH 20					17.9	6.1	34.8	10.9				41.8		211		70	
3	MARCH 27					17.5	6.0	34.5	11.0				41.5		208		72	
4	APRIL 4					17.0	5.9	33.9	10.8				41.0		196		74	
5	MAY 1					14.0	5.0	27.7	8.7				34.6		146		83	
6	—					—	—	—	—				—		—		—	
7	—					—	—	—	—				—		—		—	
8	—					—	—	—	—				—		—		—	

TABLE 65

MAFFRA, VICTORIA. 37°58′S, 146°58′E. ALT. 26m. (Latrobe Valley)

	1	2	3	4	5	6	7	8	9	10	11	12	13	14	15	16	17	18
	Temperature °C									Summation of Day Degrees above 10° with 19° Cut-Off			Temperature Variability Index					
PERIOD	Lowest Recorded Min. (65YRS)	No.Mths with <0.3° (30YRS)	No.Mths with <2.5° (30YRS)	Av.Mthly Lowest Min. (30YRS)	Av.Min Minus Col.4	Daily Mean (30YRS)	Daily Range /2 (30YRS)	Av.Mthly Highest Max. (30YRS)	Col.8 Minus Av.Max.	Raw	Adj.1 for Lat. & Daily Temp Rng	Adj.2 for Vine Sites	Raw	Adj. for Vine Sites	Av. Sunshine Hrs	Av. Number Rain Days	Av. Rainfall mm	Av. % R.H. 1500 HRS
OCTOBER	−1.1	3	20	2.0	6.0	14.1	6.1	29.8	9.6	127	126	112	40.0	37.0	225	12	61	55
NOVEMBER	−0.6	2	6	3.7	6.0	15.9	6.2	33.9	11.8	177	175	162	42.6	39.6	248	10	56	53
DECEMBER	1.4	0	3	5.3	6.2	18.1	6.6	36.4	11.7	251	246	236	44.3	41.3	265	8	59	51
JANUARY	1.9	0	1	6.5	5.8	18.9	6.6	38.7	13.2	276	271	260	45.4	42.4	278	7	52	49
FEBRUARY	1.2	0	1	6.7	6.4	19.6	6.5	36.4	10.3	252	252	252	42.7	39.7	222	6	43	50
MARCH	0.0	0	2	5.1	6.6	17.8	6.1	34.3	10.4	242	241	227	41.4	38.4	205	8	54	52
APRIL	−1.7	5	12	2.4	6.5	14.6	5.7	29.3	9.0	138	139	126	38.3	35.3	200	8	43	55
OCT-APRIL	—	10	45	—	6.2	17.0	6.3	—	10.9	1,463	1,450	1,375	42.1	39.1	1,643	59	368	52
MAY-SEPT	—	—	—	—	—	10.3	—	—	—	—	—	—	—	—	—	—	216	—
DATA FOR MONTH TO MATURITY																		
MATURITY GROUP	EST. MATURITY DATE					*	*	*						*				
1	MARCH 7					18.9	5.7	34.8	10.3					39.4	230	7	47	50
2	MARCH 13					18.7	5.6	34.4	10.3					39.1	224	7	49	51
3	MARCH 19					18.4	5.5	34.0	10.4					38.9	218	8	51	51
4	MARCH 26					17.9	5.4	33.5	10.4					38.6	212	8	53	52
5	APRIL 4					17.1	5.3	32.7	10.2					38.1	207	8	53	52
6	APRIL 13					16.0	5.2	31.5	9.9					37.2	203	8	50	53
7	MAY 13					13.0	5.0	28.0	9.0					34.0	190	9	45	56
8	—					—	—	—	—					—	—	—	—	—

* Adjusted for vine sites.

TABLE 66
MELBOURNE, VICTORIA. 37°49'S, 144°58'E. ALT. 35m.

	1	2	3	4	5	6	7	8	9	10	11	12	13	14	15	16	17	18
	Temperature °C									Summation of Day Degrees above 10° with 19° Cut-Off			Temperature Variability Index					
PERIOD	Lowest Recorded Min. (113YRS)	No.Mths with <0.3° (30YRS)	No.Mths with <2.5° (30YRS)	Av.Mthly Lowest Min. (30YRS)	Av.Min. Minus Col.4	Daily Mean (30YRS)	Daily Range /2 (30YRS)	Av.Mthly Highest Max. (30YRS)	Col.8 Minus Av.Max.	Raw	Adj.1 for Lat. & Daily Temp Rng	Adj.2 for Vine Sites	Raw	Adj. for Vine Sites	Av. Sunshine Hrs	Av. Number Rain Days	Av. Rainfall mm	Av. % R.H. 1500 HRS (30YRS)
OCTOBER	0.0	0	2	4.4	4.9	14.6	5.3	29.0	9.1	143	142	134	35.2	36.2	190	14	67	52
NOVEMBER	2.5	0	0	6.1	4.9	16.4	5.4	33.4	11.6	192	190	182	38.1	39.1	200	12	59	52
DECEMBER	4.4	0	0	8.2	4.7	18.5	5.6	37.2	13.1	264	259	251	40.2	41.2	225	11	58	51
JANUARY	5.6	0	0	9.4	4.4	19.6	5.8	39.5	14.1	279	279	279	41.7	42.7	235	8	48	48
FEBRUARY	4.8	0	0	9.5	5.0	20.2	5.7	37.5	11.6	252	252	252	39.4	40.4	212	8	50	51
MARCH	2.8	0	0	7.5	5.4	18.4	5.5	34.6	10.7	260	260	252	38.1	39.1	206	9	53	51
APRIL	1.1	0	0	4.9	5.5	15.2	4.8	29.3	9.3	156	160	151	34.0	35.0	150	12	59	56
OCT-APRIL	—	0	2	—	5.0	17.6	5.4	—	11.4	1,546	1,542	1,501	38.1	39.1	1,418	74	394	52
MAY-SEPT	—	—	—	—	—	11.1	—	—	—	—	—	—	—	—	—	—	264	—

DATA FOR MONTH TO MATURITY																		
MATURITY GROUP	EST. MATURITY DATE					*	*	*						*				
1	FEBRUARY 23					20.0	6.0	37.9	12.0					40.8	231	8	53	50
2	MARCH 1					19.9	6.0	37.5	11.6					40.4	228	8	54	51
3	MARCH 6					19.8	5.9	37.0	11.4					40.1	224	8	54	51
4	MARCH 12					19.6	5.9	36.5	11.2					39.9	220	8	54	51
5	MARCH 18					19.3	5.8	35.9	11.0					39.7	216	9	53	51
6	MARCH 25					18.8	5.8	35.3	10.9					39.5	211	9	53	51
7	MARCH 31					18.1	5.7	34.6	10.7					39.1	206	9	53	51
8	APRIL 10					16.8	5.5	33.6	10.3					38.2	195	10	54	52

* Adjusted for vine sites.

TABLE 67
MERBEIN, VICTORIA. 34°10'S, 142°04'E. ALT. 56m. (Also for Mildura)

	1	2	3	4	5	6	7	8	9	10	11	12	13	14	15	16	17	18
	Temperature °C									Summation of Day Degrees above 10° with 19° Cut-Off			Temperature Variability Index					
PERIOD	Lowest Recorded Min. (50YRS)	No.Mths with <0.3° (30YRS)	No.Mths with <2.5° (30YRS)	Av.Mthly Lowest Min. (30YRS)	Av.Min. Minus Col.4	Daily Mean (27YRS)	Daily Range /2 (27YRS)	Av.Mthly Highest Max. (30YRS)	Col.8 Minus Av.Max.	Raw	Adj.1 for Lat. & Daily Temp Rng	Adj.2 for Vine Sites	Raw	Adj. for Vine Sites	Av. Sunshine Hrs	Av. Number Rain Days	Av. Rainfall mm	Av. % R.H. 1500 HRS
OCTOBER	0.2	2	8	3.0	6.2	16.7	7.5	33.7	9.5	208	189		45.7		276	7	27	39
NOVEMBER	1.7	0	0	5.0	6.7	19.8	8.1	37.6	9.7	270	261		48.8		300	5	23	27
DECEMBER	4.5	0	0	7.3	6.8	22.3	8.2	40.0	9.5	279	279		49.1		335	4	18	25
JANUARY	5.2	0	0	9.3	5.7	23.3	8.3	41.5	9.9	279	279		48.8		347	3	20	24
FEBRUARY	5.0	0	0	8.8	6.2	23.2	8.2	39.8	8.4	252	252		47.4		298	3	22	28
MARCH	2.5	0	0	6.2	6.8	20.9	7.9	36.5	7.7	279	279		46.1		285	4	21	30
APRIL	1.3	0	10	3.2	5.9	16.2	7.1	31.6	8.3	186	180		42.6		248	5	16	40
OCT-APRIL	—	2	18	—	6.3	20.3	7.9	—	9.0	1,753	1,719		46.9		2,089	31	147	30
MAY-SEPT	—	—	—	—	—	11.4	—	—	—	—	—		—		—	—	128	—

DATA FOR MONTH TO MATURITY																		
MATURITY GROUP	EST. MATURITY DATE																	
1	FEBRUARY 5					23.3	8.3	41.5	9.7					48.8	323	3	20	24
2	FEBRUARY 11					23.4	8.3	41.2	9.3					48.6	315	3	20	24
3	FEBRUARY 15					23.4	8.2	40.8	9.0					48.3	307	3	21	25
4	FEBRUARY 22					23.3	8.2	40.4	8.7					48.0	299	3	22	26
5	FEBRUARY 27					23.2	8.2	40.0	8.5					47.7	291	3	23	27
6	MARCH 5					23.1	8.2	39.6	8.3					47.4	283	3	24	28
7	MARCH 10					22.9	8.1	39.1	8.1					47.1	276	3	23	28
8	MARCH 16					22.7	8.1	38.4	8.0					46.7	269	4	22	29

TABLE 68

MORNINGTON, VICTORIA. 38°13'S, 145°02'E. ALT. 46m. (Mornington Peninsula)

	1	2	3	4	5	6	7	8	9	10	11	12	13	14	15	16	17	18
	Temperature °C									Summation of Day Degrees above 10° with 19° Cut-Off			Temperature Variability Index					
PERIOD	Lowest Recorded Min. (42YRS)	No.Mths with <0.3° (30YRS)	No.Mths with <2.5° (30YRS)	Av.Mthly Lowest Min. (30YRS)	Av.Min. Minus Col.4	Daily Mean (20YRS)	Daily Range /2 (20YRS)	Av.Mthly Highest Max. (30YRS)	Col.8 Minus Av.Max.	Raw	Adj.1 for Lat. & Daily Temp Rng	Adj.2 for Vine Sites	Raw	Adj. for Vine Sites	Av. Sunshine Hrs	Av. Number Rain Days	Av. Rainfall mm	Av. % R.H. 1500 HRS
OCTOBER	1.9	0	1	4.9	4.3	13.6	4.4	24.9	6.9	112	120		28.8		185	13	69	60
NOVEMBER	4.2	0	0	6.0	4.6	15.6	5.0	29.2	8.6	168	166		33.2		190	10	58	56
DECEMBER	4.7	0	0	7.4	4.5	17.5	5.6	33.7	10.6	233	230		37.5		227	8	53	52
JANUARY	4.4	0	0	8.2	4.7	18.8	5.9	36.0	11.3	273	270		39.6		250	7	45	48
FEBRUARY	2.8	0	0	8.4	4.9	19.2	5.9	34.1	9.0	252	252		37.5		210	6	45	53
MARCH	2.5	0	0	7.6	4.9	17.9	5.4	31.3	8.0	245	244		34.5		205	8	52	55
APRIL	2.8	0	0	5.6	4.7	14.8	4.5	26.5	7.2	144	153		29.9		153	11	64	60
OCT-APRIL	—	0	1	—	4.7	16.8	5.2	—	8.8	1,427	1,435		34.4		1,420	63	386	55
MAY-SEPT	—	—	—	—	—	10.8	—	—	—	—	—		—		—	—	350	—

| DATA FOR MONTH TO MATURITY | | | | | | | | | | | | | | | | | | |

MATURITY GROUP	EST. MATURITY DATE																	
1	MARCH 2					19.2	5.9	34.0	9.0					37.3	226	6	49	53
2	MARCH 8					19.0	5.8	33.5	8.8					36.9	221	7	50	54
3	MARCH 14					18.8	5.7	32.9	8.5					36.4	217	7	50	54
4	MARCH 20					18.6	5.6	32.3	8.3					35.7	213	8	51	55
5	MARCH 27					18.3	5.5	31.7	8.1					35.1	209	8	52	55
6	APRIL 3					17.8	5.3	30.8	7.9					34.2	201	8	53	55
7	APRIL 12					17.0	5.0	29.5	7.7					32.8	188	9	56	56
8	APRIL 22					15.7	4.7	27.9	7.4					31.1	170	10	61	58

TABLE 69

ORBOST, VICTORIA. 37°42'S, 148°28'E. ALT. 45m. (East Gippsland)

	1	2	3	4	5	6	7	8	9	10	11	12	13	14	15	16	17	18
	Temperature °C									Summation of Day Degrees above 10° with 19° Cut-Off			Temperature Variability Index					
PERIOD	Lowest Recorded Min. (31YRS)	No.Mths with <0.3° (30YRS)	No.Mths with <2.5° (30YRS)	Av.Mthly Lowest Min. (30YRS)	Av.Min. Minus Col.4	Daily Mean (11YRS)	Daily Range /2 (11YRS)	Av.Mthly Highest Max. (30YRS)	Col.8 Minus Av.Max.	Raw	Adj.1 for Lat. & Daily Temp Rng	Adj.2 for Vine Sites	Raw	Adj. for Vine Sites	Av. Sunshine Hrs	Av. Number Rain Days	Av. Rainfall mm	Av. % R.H. 1500 HRS (11YRS)
OCTOBER	0.6	0	16	2.7	5.0	13.4	5.7	30.1	11.0	105	105	100	38.8	36.6	225	11	78	58
NOVEMBER	1.7	0	1	4.7	5.1	15.5	5.7	33.6	12.4	165	163	158	40.3	38.1	230	9	69	58
DECEMBER	3.3	0	0	5.9	5.2	17.3	6.2	36.6	13.1	226	223	218	43.1	40.9	270	8	76	57
JANUARY	4.4	0	0	7.1	5.3	18.6	6.2	38.3	13.5	267	263	258	43.6	41.4	253	7	70	55
FEBRUARY	3.9	0	0	7.2	5.3	18.4	5.9	37.1	12.8	235	233	229	41.7	39.5	220	7	61	57
MARCH	2.8	0	0	5.7	5.7	17.1	5.7	34.1	11.3	220	220	215	39.8	37.6	220	7	68	57
APRIL	0.0	2	12	2.8	6.2	14.4	5.4	29.2	9.4	132	133	131	37.2	35.0	177	8	71	63
OCT-APRIL	—	2	29	—	5.4	16.4	5.8	—	11.9	1,350	1,340	1,309	40.6	38.4	1,595	57	493	58
MAY-SEPT	—	—	—	—	—	10.4	—	—	—	—	—	—	—	—	—	—	346	—

| DATA FOR MONTH TO MATURITY | | | | | | | | | | | | | | | | | | |

MATURITY GROUP	EST. MATURITY DATE					*	*	*						*				
1	MARCH 12					17.9	5.3	35.2	12.2					38.7	230	7	66	57
2	MARCH 19					17.6	5.2	34.6	11.9					38.3	226	7	67	57
3	MARCH 27					17.1	5.2	33.7	11.5					37.9	222	7	68	57
4	APRIL 5					16.5	5.1	32.5	11.0					37.2	214	7	69	58
5	APRIL 16					15.5	5.0	31.0	10.3					36.3	200	8	70	60
6	APRIL 28					14.3	4.9	28.9	9.5					35.1	178	8	71	63
7	MAY 17					12.7	4.8	26.0	9.0					33.0	152	9	72	64
8	—					—	—	—	—					—	—	—	—	—

* Adjusted for vine sites.

TABLE 70

PORTLAND, VICTORIA. 38°21'S, 141°39'E. ALT. 16m.

	1	2	3	4	5	6	7	8	9	10	11	12	13	14	15	16	17	18
	Temperature °C									Summation of Day Degrees above 10° with 19° Cut-Off			Temperature Variability Index					
PERIOD	Lowest Recorded Min. (77YRS)	No.Mths with <0.3° (30YRS)	No.Mths with <2.5° (30YRS)	Av.Mthly Lowest Min. (30YRS)	Av.Min. Minus Col.4	Daily Mean (58YRS)	Daily Range /2 (58YRS)	Av.Mthly Highest Max. (30YRS)	Col.8 Minus Av.Max.	Raw	Adj.1 for Lat. & Daily Temp Rng	Adj.2 for Vine Sites	Raw	Adj. for Vine Sites	Av. Sunshine Hrs	Av. Number Rain Days	Av. Rainfall mm (30YRS)	Av. % R.H. 1500 HRS (17YRS)
OCTOBER	0.6	0	12	2.8	6.1	13.2	4.3	27.2	9.7	99	110	105	33.0	34.2	210	16	69	66
NOVEMBER	2.2	0	0	5.3	5.1	14.8	4.4	30.7	11.5	144	152	147	34.2	35.4	227	14	56	64
DECEMBER	2.8	0	0	5.9	5.7	16.1	4.5	33.7	13.1	189	195	190	36.8	38.0	240	10	46	63
JANUARY	4.4	0	0	7.1	5.7	17.3	4.5	35.4	13.6	226	232	227	37.3	38.5	240	9	30	62
FEBRUARY	3.3	0	0	7.3	5.8	17.6	4.5	33.9	11.8	213	218	214	35.6	36.8	220	9	45	63
MARCH	2.0	0	0	6.2	5.8	16.4	4.4	32.4	11.6	198	207	203	35.0	36.2	207	11	38	63
APRIL	1.1	0	4	4.3	6.1	14.5	4.1	27.4	8.8	135	149	145	31.3	32.5	158	12	67	66
OCT-APRIL	—	0	16	—	5.8	15.7	4.4	—	11.4	1,204	1,263	1,231	34.7	35.9	1,502	81	351	64
MAY-SEPT	—	—	—	—	—	11.1	—	—	—	—	—	—	—	—	—	—	500	—

DATA FOR MONTH TO MATURITY

MATURITY GROUP	EST. MATURITY DATE					*	*	*						*				*
1	MARCH 25					16.7	4.7	33.0	11.6					36.3	209	11	40	58
2	APRIL 3					16.2	4.7	32.3	11.3					36.0	198	11	40	59
3	APRIL 12					15.6	4.6	30.8	10.5					35.0	183	11	45	60
4	APRIL 24					14.8	4.4	28.8	9.4					33.2	168	12	58	62
5	MAY 6					14.0	4.2	26.4	8.4					32.0	150	12	70	64
6	—					—	—	—	—					—	—	—	—	—
7	—					—	—	—	—					—	—	—	—	—
8	—					—	—	—	—					—	—	—	—	—

* Adjusted for vine sites. Adjusted relative humidities from means of Portland and Mount Gambier.

TABLE 71

RUTHERGLEN, VICTORIA. 36°03'S, 146°28'E. ALT. 169m.

	1	2	3	4	5	6	7	8	9	10	11	12	13	14	15	16	17	18
	Temperature °C									Summation of Day Degrees above 10° with 19° Cut-Off			Temperature Variability Index					
PERIOD	Lowest Recorded Min. (57YRS)	No.Mths with <0.3° (30YRS)	No.Mths with <2.5° (30YRS)	Av.Mthly Lowest Min. (30YRS)	Av.Min. Minus Col.4	Daily Mean (28YRS)	Daily Range /2 (28YRS)	Av.Mthly Highest Max. (30YRS)	Col.8 Minus Av.Max.	Raw	Adj.1 for Lat. & Daily Temp Rng	Adj.2 for Vine Sites	Raw	Adj. for Vine Sites	Av. Sunshine Hrs	Av. Number Rain Days	Av. Rainfall mm (30YRS)	Av. % R.H. 1500 HRS
OCTOBER	−3.4	17	29	0.1	6.6	14.1	7.4	28.6	7.1	127	112	136	43.3	39.1	267	10	55	44
NOVEMBER	−2.8	5	17	1.9	6.9	17.3	8.5	34.6	8.8	219	185	211	49.7	45.5	302	7	39	33
DECEMBER	0.6	0	3	4.1	7.7	20.7	8.9	37.9	8.3	279	279	279	51.6	47.4	335	6	50	27
JANUARY	2.2	0	1	6.9	6.4	22.1	8.8	39.8	8.9	279	279	279	50.5	46.3	340	5	38	26
FEBRUARY	1.2	0	1	6.3	7.4	22.4	8.7	38.4	7.3	252	252	252	49.5	45.3	280	5	44	28
MARCH	−0.9	1	6	4.6	6.3	19.2	8.3	34.1	6.6	279	256	279	46.1	41.9	280	5	39	35
APRIL	−2.9	13	25	1.0	6.2	14.4	7.2	28.8	7.2	132	123	144	42.2	38.0	238	7	46	45
OCT-APRIL	—	36	82	—	6.8	18.6	8.3	—	7.7	1,567	1,486	1,580	47.6	43.4	2,042	45	311	34
MAY-SEPT	—	—	—	—	—	9.1	—	—	—	—	—	—	—	—	—	—	297	—

DATA FOR MONTH TO MATURITY

MATURITY GROUP	EST. MATURITY DATE					*	*	*						*				
1	FEBRUARY 17					22.8	7.7	39.0	7.9					45.7	320	5	43	27
2	FEBRUARY 22					22.8	7.7	38.4	7.6					45.5	312	5	45	27
3	FEBRUARY 28					22.7	7.7	37.7	7.3					45.3	305	5	47	28
4	MARCH 5					22.5	7.6	37.0	7.1					44.9	299	5	46	29
5	MARCH 11					22.2	7.6	36.3	7.0					44.4	295	5	44	30
6	MARCH 16					21.7	7.5	35.6	6.9					43.6	291	5	42	31
7	MARCH 22					20.9	7.4	34.8	6.8					43.0	287	5	41	32
8	MARCH 27					20.2	7.3	34.0	6.7					42.4	283	5	40	34

* Adjusted for vine sites.

TABLE 72

SEYMOUR, VICTORIA. 37°02′S, 145°08′E. ALT. 142m. (Central Goulburn Valley)

	1	2	3	4	5	6	7	8	9	10	11	12	13	14	15	16	17	18
				Temperature °C						Summation of Day Degrees above 10° with 19° Cut-Off			Temperature Variability Index					
PERIOD	Lowest Recorded Min. (71YRS)	No.Mths with <0.3° (30YRS)	No.Mths with <2.5° (30YRS)	Av.Mthly Lowest Min. (30YRS)	Av.Min Minus Col.4	Daily Mean (28YRS)	Daily Range /2 (28YRS)	Av.Mthly Highest Max. (30YRS)	Col.8 Minus Av.Max.	Raw	Adj.1 for Lat. & Daily Temp Rng	Adj.2 for Vine Sites	Raw	Adj. for Vine Sites	Av. Sunshine Hrs	Av. Number Rain Days	Av. Rainfall mm	Av. % R.H. 1500 HRS
OCTOBER	−2.5	8	25	1.0	6.1	14.1	7.0	29.0	7.9	127	118		42.0		245	9	54	50
NOVEMBER	0.0	0	10	3.2	6.0	17.0	7.8	34.1	9.3	210	188		46.5		275	6	45	40
DECEMBER	1.1	0	1	5.1	6.5	19.8	8.2	37.7	9.7	279	272		49.0		290	5	37	37
JANUARY	3.3	0	0	6.4	6.1	20.9	8.4	40.0	10.7	279	279		50.4		310	4	33	35
FEBRUARY	1.7	0	1	6.8	6.0	21.3	8.5	37.9	8.1	252	252		48.1		260	4	37	34
MARCH	1.1	0	3	5.2	5.5	18.6	7.9	34.3	7.8	267	244		44.9		255	5	42	41
APRIL	−1.1	6	17	1.9	5.8	14.5	6.8	28.6	7.3	135	132		40.3		205	7	45	51
OCT-APRIL	—	14	57	—	6.0	18.0	7.8	—	8.7	1,549	1,485		45.9		1,840	40	293	41
MAY-SEPT	—	—	—	—	—	9.6	—	—	—	—	—		—		—	—	303	—

DATA FOR MONTH TO MATURITY

MATURITY GROUP	EST. MATURITY DATE																	
1	FEBRUARY 22					21.4	8.5	38.4	8.6				48.6		287	4	38	34
2	FEBRUARY 27					21.3	8.5	38.0	8.2				48.2		282	4	40	34
3	MARCH 5					21.1	8.4	37.4	8.1				47.7		277	4	40	35
4	MARCH 11					20.8	8.3	36.8	8.0				47.1		272	5	41	36
5	MARCH 16					20.3	8.2	36.1	7.9				46.4		267	5	41	37
6	MARCH 23					19.6	8.1	35.3	7.9				45.7		262	5	42	39
7	MARCH 31					18.6	7.9	34.3	7.8				44.9		255	5	42	41
8	APRIL 8					17.5	7.6	33.1	7.7				43.9		246	5	43	43

TABLE 73

SWAN HILL, VICTORIA. 35°22′S, 143°35′E. ALT. 70m.

	1	2	3	4	5	6	7	8	9	10	11	12	13	14	15	16	17	18
				Temperature °C						Summation of Day Degrees above 10° with 19° Cut-Off			Temperature Variability Index					
PERIOD	Lowest Recorded Min. (75YRS)	No.Mths with <0.3° (30YRS)	No.Mths with <2.5° (30YRS)	Av.Mthly Lowest Min. (30YRS)	Av.Min Minus Col.4	Daily Mean (35YRS)	Daily Range /2 (35YRS)	Av.Mthly Highest Max. (30YRS)	Col.8 Minus Av.Max.	Raw	Adj.1 for Lat. & Daily Temp Rng	Adj.2 for Vine Sites	Raw	Adj. for Vine Sites	Av. Sunshine Hrs	Av. Number Rain Days	Av. Rainfall mm	Av. % R.H. 1500 HRS (13YRS)
OCTOBER	−1.1	1	4	3.7	5.2	16.2	7.3	31.8	8.3	192	177		42.7		268	7	34	37
NOVEMBER	1.7	0	0	5.4	6.3	19.8	8.1	36.7	8.8	270	264		47.5		296	5	26	33
DECEMBER	4.4	0	0	7.6	6.2	22.2	8.4	39.2	8.6	279	279		48.4		341	4	24	29
JANUARY	4.6	0	0	9.3	5.8	23.4	8.3	40.8	9.1	279	279		48.1		340	3	21	29
FEBRUARY	5.6	0	0	9.4	6.0	23.6	8.2	38.7	6.9	252	252		45.7		290	3	24	34
MARCH	3.9	0	0	7.3	5.4	20.6	7.9	35.6	7.1	279	279		44.1		280	3	24	36
APRIL	−1.1	0	0	4.5	5.0	16.4	6.9	30.5	7.2	192	188		39.8		240	4	25	46
OCT-APRIL	—	1	4	—	5.7	20.3	7.9	—	8.0	1,743	1,718		45.2		2,055	29	178	35
MAY-SEPT	—	—	—	—	—	11.0	—	—	—	—	—	—	—		—	—	167	—

DATA FOR MONTH TO MATURITY

MATURITY GROUP	EST. MATURITY DATE																	
1	FEBRUARY 6					23.7	8.3	40.4	8.6				47.5		335	3	26	30
2	FEBRUARY 11					23.8	8.3	40.1	8.1				46.9		330	3	25	31
3	FEBRUARY 17					23.8	8.2	39.7	7.7				46.3		324	3	25	32
4	FEBRUARY 23					23.7	8.2	39.2	7.3				45.7		319	3	24	33
5	FEBRUARY 28					23.6	8.2	38.7	7.0				45.3		313	3	24	34
6	MARCH 6					23.2	8.1	38.1	7.0				45.1		308	3	24	34
7	MARCH 11					22.8	8.1	37.5	7.0				44.9		303	3	24	35
8	MARCH 17					22.4	8.0	36.9	7.1				44.7		298	3	25	35

TABLE 74
TERANG, VICTORIA. 38°14'S, 142°55'E. ALT. 132m.

	1	2	3	4	5	6	7	8	9	10	11	12	13	14	15	16	17	18
	Temperature °C									Summation of Day Degrees above 10° with 19° Cut-Off			Temperature Variability Index					
PERIOD	Lowest Recorded Min. (72YRS)	No.Mths with <0.3° (30YRS)	No.Mths with <2.5° (30YRS)	Av.Mthly Lowest Min. (30YRS)	Av.Min. Minus Col.4	Daily Mean (29YRS)	Daily Range /2 (29YRS)	Av.Mthly Highest Max. (30YRS)	Col.8 Minus Av.Max.	Raw	Adj.1 for Lat. & Daily Temp Rng	Adj.2 for Vine Sites	Raw	Adj. for Vine Sites	Av. Sunshine Hrs	Av. Number Rain Days	Av. Rainfall mm	Av. % R.H. 1500 HRS
OCTOBER	−1.1	7	22	1.7	5.1	12.4	5.6	27.2	9.2	74	74	77	36.7	35.5	218	15	70	
NOVEMBER	0.6	0	9	3.2	5.1	14.1	5.8	31.0	11.1	123	122	125	39.4	38.2	223	13	63	
DECEMBER	0.6	0	1	4.2	5.4	16.0	6.4	35.9	13.5	186	184	187	44.5	43.3	248	10	48	
JANUARY	0.0	1	1	5.4	5.2	17.5	6.9	39.2	14.8	233	224	233	47.6	46.4	274	8	38	
FEBRUARY	−1.1	0	0	5.7	5.9	18.4	6.8	36.6	11.4	235	229	236	44.5	43.3	225	8	42	
MARCH	1.1	0	2	5.0	5.3	16.6	6.3	34.4	11.5	205	204	207	42.0	40.8	223	9	51	
APRIL	−0.6	4	18	2.5	5.8	13.5	5.2	28.1	9.4	105	106	109	36.0	34.8	158	13	66	
OCT-APRIL	—	12	53	—	5.4	15.5	6.1	—	11.6	1,161	1,143	1,174	41.5	40.3	1,569	76	378	
MAY-SEPT	—	—	—	—	—	9.6	—	—	—	—	—	—	—	—	—	—	408	

DATA FOR MONTH TO MATURITY																		
MATURITY GROUP	EST. MATURITY DATE					*	*	*						*				
1	MARCH 29					16.8	6.0	34.4	11.5					40.9	225	9	51	
2	APRIL 8					15.8	5.7	33.2	11.0					39.4	210	10	54	
3	APRIL 22					14.4	5.2	29.8	10.0					36.6	176	12	62	
4	MAY 15					12.3	4.6	25.0	8.8					32.0	140	15	69	
5	—					—	—	—	—					—	—	—	—	
6	—					—	—	—	—					—	—	—	—	
7	—					—	—	—	—					—	—	—	—	
8	—					—	—	—	—					—	—	—	—	

* Adjusted for vine sites.

TABLE 75
WANGARATTA, VICTORIA. 36°21'S, 146°19'E. ALT. 150m.

	1	2	3	4	5	6	7	8	9	10	11	12	13	14	15	16	17	18
	Temperature °C									Summation of Day Degrees above 10° with 19° Cut-Off			Temperature Variability Index					
PERIOD	Lowest Recorded Min. (89YRS)	No.Mths with <0.3° (30YRS)	No.Mths with <2.5° (30YRS)	Av.Mthly Lowest Min. (30YRS)	Av.Min. Minus Col.4	Daily Mean (29YRS)	Daily Range /2 (29YRS)	Av.Mthly Highest Max. (30YRS)	Col.8 Minus Av.Max.	Raw	Adj.1 for Lat. & Daily Temp Rng	Adj.2 for Vine Sites	Raw	Adj. for Vine Sites	Av. Sunshine Hrs	Av. Number Rain Days	Av. Rainfall mm	Av. % R.H. 1500 HRS (27YRS)
OCTOBER	−0.4	0	15	2.8	5.4	14.7	6.5	28.9	7.7	146	144	135	39.1		266	9	63	44
NOVEMBER	1.7	0	1	5.3	5.5	18.2	7.4	34.4	8.8	246	228	219	43.9		300	7	46	33
DECEMBER	3.8	0	0	7.1	6.4	21.2	7.7	37.7	8.8	279	279	279	46.0		335	6	42	27
JANUARY	5.0	0	0	9.5	5.3	22.6	7.8	39.9	9.5	279	279	279	46.0		335	4	38	26
FEBRUARY	5.0	0	0	8.8	6.4	23.0	7.8	38.3	7.5	252	252	252	45.1		278	4	40	28
MARCH	1.6	0	0	6.7	5.5	19.7	7.5	34.5	7.3	279	279	275	42.8		276	5	48	35
APRIL	−0.3	0	11	3.2	5.1	15.1	6.8	29.1	7.2	153	150	141	39.5		233	7	48	45
OCT-APRIL	—	0	27	—	5.7	19.2	7.4	—	8.1	1,634	1,601	1,580	43.2		2,023	42	325	34
MAY-SEPT	—	—	—	—	—	9.9	—	—	—	—	—	—	—		—	—	315	—

DATA FOR MONTH TO MATURITY																		
MATURITY GROUP	EST. MATURITY DATE					*	*	*										
1	FEBRUARY 16					22.8	7.8	38.7	8.4					45.5	316	4	41	27
2	FEBRUARY 21					22.8	7.8	38.4	8.3					45.3	311	4	42	27
3	FEBRUARY 27					22.7	7.8	38.0	7.6					45.1	306	4	43	28
4	MARCH 4					22.6	7.8	37.5	7.5					44.8	301	4	44	28
5	MARCH 10					22.2	7.7	36.9	7.4					44.4	296	5	45	29
6	MARCH 15					21.7	7.6	36.3	7.4					44.0	291	5	46	30
7	MARCH 21					20.9	7.6	35.6	7.4					43.6	286	5	47	32
8	MARCH 26					20.1	7.5	34.8	7.3					43.2	281	5	47	34

* Adjusted for vine sites.

TABLE 76
WERRIBEE (Research Farm), VICTORIA. 37°54′S, 144°38′E. ALT. 23m.

	1	2	3	4	5	6	7	8	9	10	11	12	13	14	15	16	17	18
	Temperature °C									Summation of Day Degrees above 10° with 19° Cut-Off			Temperature Variability Index					
PERIOD	Lowest Recorded Min. (57YRS)	No.Mths with <0.3° (30YRS)	No.Mths with <2.5° (30YRS)	Av.Mthly Lowest Min. (30YRS)	Av.Min. Minus Col.4	Daily Mean (28YRS)	Daily Range /2 (28YRS)	Av.Mthly Highest Max. (30YRS)	Col.8 Minus Av.Max.	Raw	Adj.1 for Lat. & Daily Temp Rng	Adj.2 for Vine Sites	Raw	Adj. for Vine Sites	Av. Sunshine Hrs	Av. Number Rain Days	Av. Rainfall mm (30YRS)	Av. % R.H. 1500 HRS
OCTOBER	−1.4	6	22	1.7	6.3	13.9	5.9	28.1	8.3	121	120		38.1		190	15	49	
NOVEMBER	0.1	1	5	3.3	6.4	15.7	6.0	33.0	11.3	171	169		41.7		196	12	45	
DECEMBER	2.3	0	1	5.3	6.6	18.0	6.1	36.8	12.7	248	245		43.7		222	10	47	
JANUARY	5.0	0	0	6.9	5.9	19.1	6.3	39.0	13.6	279	279		44.7		263	7	30	
FEBRUARY	3.3	0	0	7.4	5.9	19.4	6.1	36.9	11.4	252	252		41.7		218	7	37	
MARCH	1.1	0	1	5.8	6.0	17.6	5.8	33.8	10.4	236	235		39.6		215	9	43	
APRIL	0.0	0	11	3.0	6.5	14.6	5.1	28.8	9.1	138	139		36.0		153	12	40	
OCT-APRIL	—	7	40	—	6.2	16.9	5.9	—	11.0	1,445	1,439		40.8		1,457	72	291	
MAY-SEPT	—	—	—	—	—	10.4	—	—	—	—			—		—	—	203	

	DATA FOR MONTH TO MATURITY																	
MATURITY GROUP	EST. MATURITY DATE																	
1	FEBRUARY 27					19.4	6.1	37.0	11.5				41.9		240	7	37	
2	MARCH 4					19.3	6.1	36.5	11.2				41.4		236	7	39	
3	MARCH 10					19.1	6.0	35.9	11.0				41.0		232	8	40	
4	MARCH 17					18.7	6.0	35.3	10.8				40.6		227	8	41	
5	MARCH 24					18.2	5.9	34.6	10.6				40.1		222	9	42	
6	MARCH 31					17.6	5.8	33.8	10.4				39.6		215	9	43	
7	APRIL 9					16.7	5.6	32.5	10.1				38.8		206	10	42	
8	APRIL 20					15.6	5.4	30.8	9.6				37.6		180	11	41	

Commentaries and vineyard site adjustments

Ararat (Table 51)

Vineyard site adjustments, °C

Min.	Max.	
+0.4	−0.2	Adjusted from slight slope to moderate slope
+0.4	−0.2	Mean temperature +0.1°, variability index −1.2

Avoca (for Pyrenees foothills) (Table 52)

Vineyard site adjustments, °C

Min.	Max.	
+1.3	−0.5	Moderate slopes of isolated hill range
+0.3	+0.1	Slopes facing north-east and north
+0.8	−0.4	Gravelly soils
−0.8	−0.8	Altitude 350–400 m (Avoca 242 m)
+1.6	−1.6	Mean temperature unchanged, variability index −6.4

All relative humidities estimated as Ararat minus 2 per cent.

Bairnsdale East Gippsland (Table 53)

Vineyard site adjustments, °C

Min.	Max.	
+0.6	−0.3	Moderate slopes vs flat, but further from coast and lakes
−0.3	−0.3	Altitude 50–100 m, vs Bairnsdale 14 m
+0.3	−0.6	Mean temperature −0.15°, variability index −1.8

Notes

The estimate is for foothill slopes, without particular aspect. Low altitude sites close to Bairnsdale might be marginally warmer, but otherwise similar.

Ballarat (Table 54)

Vineyard site adjustments, °C

Min.	Max.	
+0.6	−0.3	Moderate slopes
+0.3	+0.1	Predominantly northern aspects
+0.9	−0.2	Mean temperature +0.35°, variability index −2.2

Notes

Comparison of Ballarat with Ballarat Mount Pleasant (Climatic Averages Australia 1956) shows that site climates exist which are superior to that of Ballarat town. The latter still has a fairly narrow diurnal temperature range, however, and comparison with other sites suggests that only a partial topographical adjustment is warranted for potential vineyard sites.

No allowance is made here for soil type. Rocky soils with superior thermal characteristics, if available, could improve temperature relations significantly in this very marginal climate.

Beechworth (Table 55)

Vineyard site adjustments, °C

Min.	Max.	
+0.8	−0.4	Selected moderate slopes (Beechworth ? slight slope)
+0.2	−0.1	Slightly stony soils
+0.9	+0.9	Adjusted to 400 m, vs Beechworth 552 m
+1.9	+0.4	Mean temperature +1.15°, variability index −3.0

Notes

Beechworth is on a small plateau, with the potential vineyard climate estimated for surrounding slopes at 400 m.

The most notable aspect of Beechworth's climate is the variability of the minimum temperature (column 5), probably related to periodic incursions of chilled air from the snowfields to the east. That and the only moderate variability of the maximum (column 9) match the climate more to the western slopes in southern New South Wales than to the northern slopes of the Great Dividing Range in Victoria.

Bendigo (Table 56)

Vineyard site adjustments: None

Notes

The raw data show a quite moderate temperature variability index considering the distance inland, and probably represent vineyard sites adequately for both topography and altitude.

Camperdown (Table 57)

Vineyard site adjustments: None

Colac (Table 58)

Vineyard site adjustments, °C

Min.	Max.	
+1.0	−0.5	Moderate slopes, vs Colac in a frosty site
+0.3	+0.1	Predominantly northern aspects
−0.2	−0.2	Vineyards at 150–200 m (Colac 134 m)
+1.1	−0.6	Mean temperature +0.25°, variability index −3.4

Notes

The adjustments are about as great as could reasonably be made. Better site climates seem unlikely to be available over usable areas except perhaps on the lower northern slopes of the Otway Range.

Euroa (for Strathbogie Range) (Table 59)

Vineyard site adjustments, °C

Min.	Max.	
+0.6	−0.2	Moderate slopes among semi-isolated hills
+0.6	+0.2	Northerly aspect
−0.5	−0.5	Distance south from Euroa
−1.5	−1.5	Altitude 450 m, vs Euroa 175 m
−0.8	−2.0	Mean temperature −1.4°, variability index −2.4

All relative humidities adjusted for altitude in parallel with Beechworth vs Wangaratta.

Notes

Euroa, at the foot of the Strathbogie Range, appears already to enjoy some of the latter's site-climate advantages. Therefore only a part-adjustment is made for slope. The estimate is for the Mount Helen Vineyard, but the Strathbogie Range offers a fairly wide range of potential viticultural environments, perhaps especially on lower and warmer slopes than characterized here.

Geelong (Table 60)

Vineyard site adjustments, °C

Min.	Max.	
−0.4	+0.2	Moderate slopes but inland from Geelong
−0.4	−0.4	Vineyard altitudes ~ 100 m (Geelong 27 m)
−0.8	−0.2	Mean temperature −0.5°, variability index +1.2

Notes

The ripening period relative humidities for vineyard sites are estimated as the average of Geelong and Colac. Adjustments for slope and distance inland take into account the data for Werribee as well.

Hamilton (Table 61)

Vineyard site adjustments, °C

Min.	Max.	
+0.4	−0.2	Slight-moderate slopes, vs Hamilton flat
+0.4	−0.2	Mean temperature +0.1°, variability index −1.2

Healesville (for Yarra Valley) (Table 62)

Vineyard site adjustments, °C

Min.	Max.	
+1.0	−0.5	Selected moderate slopes (Healesville in valley)
+0.3	+0.1	Predominantly northern exposures
+1.3	−0.4	Mean temperature +0.45°, variability index −3.4

Notes

The published climatic data for Healesville are fairly short-term. Although agreeing well enough with observed vine behaviour, they and the resulting vineyard estimates remain open to revision.

Heywood (for Drumborg) (Table 63)

Vineyard site adjustments, °C

Min.	Max.	
+1.0	−0.5	Moderate slopes, vs Heywood flat
+0.6	−0.3	Moderately gravelly soil
−0.4	−0.4	Altitude ~ 100 m, vs Heywood 27 m
+1.2	−1.2	Mean temperature unchanged, variability index −4.8

Notes

The Heywood climatic figures came from Climatic Averages Australia (1975), period-adjusted back to 1911–40 by reference to parallel differences at Hamilton, Portland and Camperdown. Columns 5 and 9 were from raw data for 1965–76, related to means and daily ranges from Climatic Averages Australia (1975). Columns 4 and 8 were then estimated.

Leongatha (South Gippsland) (Table 64)

Vineyard site adjustments: None

Notes

The temperature data for Leongatha should also apply roughly to the lower foothills immediately to the north, where the effects of slope on diurnal temperature range will tend to offset the more inland location. The north-facing inland slopes of the South Gippsland hills are not represented by the table. See main text.

Maffra (La Trobe Valley) (Table 65)

Vineyard site adjustments, °C

Min.	Max.	
+1.0	−0.5	Moderate slopes, vs Maffra flat
−0.7	−0.7	Vine altitude ~ 150 m, vs Maffra 26 m
+0.3	−1.2	Mean temperature −0.45°, variability index −3.0

Notes

The estimate based on Maffra is for foothills along the northern edge of the La Trobe Valley.

Melbourne (Table 66)

Vineyard site adjustments, °C

Min.	Max.	
−0.5	–	To offset urban warming
−0.5	–	Mean temperature −0.25°, variability index +1.0

Notes

The adjustment to offset urban warming may be conservative. That used brings the minima into line with the figures for Kew, which in 1911–40 would have been only moderately urbanized.

I have made no allowance for slope selection in the immediate Melbourne area, assuming that where slopes are available, their site-climate advantages will be offset by greater distance from the Port Phillip Bay. Premium stony soils are unlikely to be available.

Merbein (also for Mildura) (Table 67)

Vineyard site adjustments: None

Mornington (for Mornington Peninsula) (Table 68)

Vineyard site adjustments: None

Notes

The town of Mornington is on Port Phillip Bay, and undoubtedly enjoys the modifying benefit of close proximity to water. Nevertheless, no part of the peninsula is far from water. I have assumed that wider diurnal temperature ranges in the more inland parts can be offset by slope selection, so that the raw variability data for Mornington still apply. Site variation will then depend on altitude (affecting mainly the mean temperature) and aspect, together with soil type. Viticultural differences due to altitude are discussed in the main text, with the Table 68 representing the warm lowlands near Mornington.

Orbost (East Gippsland) (Table 69)

Vineyard site adjustments, °C

Min.	Max.	
+0.6	−0.3	Moderate to steep slopes, but further inland
−0.5	−0.5	Vines at 100–150 m, vs Orbost 45 m
+0.3	+0.1	East to north aspects
+0.4	−0.7	Mean temperature −0.15°, variability index −2.2

Notes

The Snowy River, just inland from Orbost, flows to the south-east, and slopes on the right bank which are sheltered from the south-west by hills should afford some good exposures. Undulating sites around Orbost itself are probably not greatly different.

Orbost has stronger south coastal influences than Bairnsdale, as evidenced by its higher rainfall and relative humidities, and greater variability of the maximum temperature.

Portland (Table 70)

Vineyard site adjustments, °C

Min.	Max.	
−1.5	+1.0	Inland from Portland
+1.2	−0.5	Moderate slopes of small hills
+0.3	+0.1	North-east aspect
−0.3	−0.3	Altitude 50–100 m, vs Portland 16 m
−0.3	+0.3	Mean temperature unchanged, variability index +1.2

Ripening month relative humidity is estimated as the average of Portland and Mount Gambier.

Rutherglen (Table 71)

Vineyard site adjustments, °C

Min.	Max.	
+1.0	−0.5	Slight slopes, vs atypically frosty recording site
+0.4	−0.2	Some soils stony
+1.4	−0.7	Mean temperature +0.35°, variability index −4.2

Notes

Figures (not presented) for the nearby Rutherglen Agricultural Research Institute, at the same altitude as Rutherglen town, show inconsistency between the 1956 and 1975 editions of Climatic Averages Australia, suggesting a change of recording site. The 1956 figures give a growing season mean a full 1.0° warmer than for Rutherglen town, but at the same time, inexplicably, a wider diurnal temperature range. Because of these inconsistencies, together with what appears to be an atypically frosty recording site in Rutherglen itself, I regard the figures from both Rutherglen town and the Research Institute as suspect, and those from Wangaratta and Corowa as being probably more indicative of vineyard climates in the area.

Seymour (for central Goulburn Valley) (Table 72)

Vineyard site adjustments: None

Notes

The main vineyards are on alluvial flats, and therefore attract no topographical or soil-type adjustment. Those very close to the river, and especially to Nagambie Lake, may derive some climatic benefit.

Swan Hill (Table 73)

Vineyard site adjustments: None

Notes

The relative humidities recorded at Swan Hill exceed those from comparable nearby locations, suggesting that they may be influenced by immediate proximity to the river or by some other abnormal factor.

Terang (Table 74)

Vineyard site adjustments, °C

Min.	Max.	
+0.4	−0.2	Slight slopes
+0.4	−0.2	Mean temperature +0.1°, variability index −1.2

Wangaratta (Table 75)

Vineyard site adjustments, °C

Min.	Max.	
−0.3	−0.3	Altitude 200 m, vs Wangaratta 150 m
−0.3	−0.3	Mean temperature −0.3°, variability index unchanged

Notes

I take Wangaratta as directly representative of vineyard site climates on the flatter country of north-east Victoria, with adjustment only for their mainly greater altitude, and as appropriate for distance north and south. Conditions along the foothill strip between the Warby Range and Lake Mokoan would be appreciably more favourable (see text).

Werribee (Table 76)

Vineyard site adjustments: None

Notes

The Werribee area itself is flat, and not a likely prospect for viticulture. However sites with better air drainage exist among the foothills to the north and west, where site advantages probably offset, or more than offset, greater distance from Port Phillip Bay. Werribee provides the basis for an alternative estimate for the nearby Anakie area, to complement that derived from Geelong.

12

Viticultural Environments of New South Wales

Australian viticulture began in New South Wales. Vine cuttings from Rio de Janeiro and the Cape of Good Hope were among the plants brought by the First Fleet in 1788; and it was in Sydney's immediate hinterland that pioneers such as Gregory Blaxland and the Macarthur family made Australia's first wine.

Vine growing soon crossed the Blue Mountains to the Bathurst area, which by the early 1800s had as many vines as Sydney. But the Bathurst vineyards have long gone, as have most of those around Sydney: victims of frost in the first case, and of disease in the second, because of Sydney's warm, humid climate.

When the Hunter Valley was opened up for settlement in the 1820s, an area became available for the first time which was well suited to vines. It was sunnier than Sydney, though still more or less frost-free; there was enough seasonal rainfall for vines to grow well, but not too well; and humidity, on average, was not excessive. Throughout much of the 19th century the valley remained free of today's serious diseases, including downy mildew, powdery mildew and Phylloxera. Moreover, the Hunter had an outstanding pioneer in James Busby, whose writings I have cited elsewhere. Busby's story, with that of the Hunter and of New South Wales' viticulture generally, has been told by Lake (1964, 1970), Cox (1967), Halliday and Jarratt (1979) and Halliday (1980, 1985a, 1991a).

Today's viticulture in New South Wales is nevertheless limited, and widely scattered. Apart from the Hunter Valley, the only established major wine areas at present are on the Murrumbidgee at Griffith, Yenda and Leeton, producing mainly bulk wine with full irrigation, and at Mudgee. New South Wales' share of total Australian wine production remains small, despite the fact that Sydney is the country's biggest market. Why is this so? The answer undoubtedly lies in the climate, which has made it hard to compete economically with the southern States.

As the following analysis shows, many environments in New South Wales provide potentially good ripening conditions for winegrape quality. But mostly this is combined with a grave risk from killing frosts, due to excessive temperature variability in the spring.

Many areas also intermittently have very wet and humid ripening conditions, which can make disease control and harvest difficult or impossible. Some areas have both problems. On top of this, the variability of both seasonal and total annual rainfall is greater than along the continent's southern coast, making drought an extra hazard for those vineyards which depend on natural rainfall or local water catchments.

A further factor has operated in recent years. Because frost risk has largely precluded viticulture in the parts of New South Wales which might, otherwise, have produced true cool-climate styles of wine, the State has been to some degree by-passed by the market fashion of the 1970s and 1980s for cool-climate wines.

But is this justified? It is arguable that the period saw an over-reaction in North America and Australia (following traditional European tastes) against warm areas for wine production. It was based at least in part on factors which are no longer relevant, if they ever were. Wine-making technology can now obviate many of the problems of warm areas. Certainly the common assumption that the best wines are from the coldest areas that will ripen grapes is wrong. In earlier chapters I showed that intermediate ripening temperatures are in general desirable for winemaking, depending to some degree on grape variety and wine style; and that narrow temperature variation about an optimum mean is at least equally important. Restricted temperature variability during ripening is definitely a positive factor in much of New South Wales, particularly in the north. The region will also gain from rising atmospheric carbon dioxide concentrations, as I argue in Chapter 23.

The success of table wines from hot viticultural areas such as the Hunter Valley, both within Australia and on export markets, shows clearly that their styles are now as valid as those from cool areas. In this scenario I can see a number of New South Wales environments which deserve consideration for future expansion, at least from the viewpoint of wine quality. But whether, in the light of the climatic hazards to viticulture, they can be fully cost-competitive with the best southern and western environments is another question.

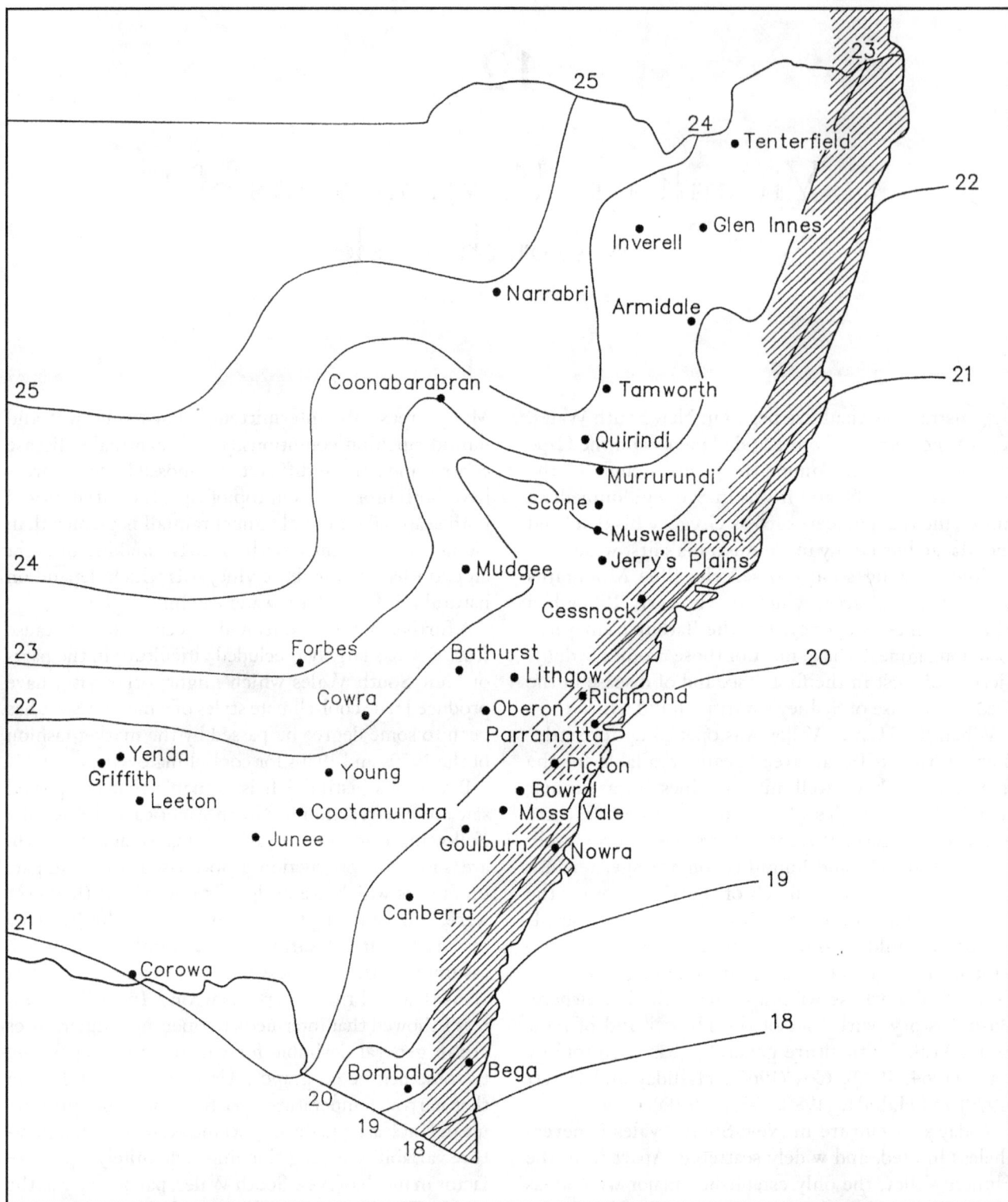

Figure 14. New South Wales vinegrowing regions, with isotherms of October-April average mean temperatures, reduced to sea level.

The Hunter Valley

The climate of **Cessnock** (Table 83) represents the lower Hunter Valley; **Jerry's Plains** (Table 93), the middle Hunter Valley as far upstream as Denman and Sandy Hollow; and **Scone** (Table 104), the extreme upper Hunter Valley beyond Muswellbrook.

The Cessnock data are adjusted to 200 m, an average altitude for the long-established hillside vineyards there. (The uppermost vines on Mount Pleasant vineyard are at about 400 m.) However, because the combined effects of slope and aspect partly offset those of altitude, and because in any case mean temperatures are consistently close to, or above, the 'saturation' level of 19°C for rate of vine phenological development, ripening times in the Hunter Valley are much the same across the full range of altitudes and environments. This is recognized in Table 93, which estimates the same ripening dates for both the valley floor and the hillsides. But the average and extreme temperatures during ripening, and the diurnal temperature range throughout the season, will be greater on the valley floor than shown in the table, to the extent of the net adjustments listed in the commentary.

Figure 14 shows the effects of the Hunter-Goulburn Valley's unique geography on temperatures (adjusted to sea level) inland from its lower reaches. The complete break in the Great Dividing Range allows marine influences to extend much further inland than would otherwise be the case, by letting through afternoon sea breezes which reach the southern slopes of the Liverpool Range and apparently influence temperatures as far inland as Coonabarabran. The converse can be seen in the relative heat (for given altitudes) of the Quirindi-Tamworth area, which is cut off from the ocean by the Liverpool Range to the south and the Hastings Range to the east, but is fully exposed to hot winds from the west and north. The same is apparent in the upper Murray Valley in southern New South Wales, extending into north-east Victoria (Figure 13), where marine influences are entirely cut off by the Australian Alps.

Within the Hunter-Goulburn Valley, the difference with distance inland between Cessnock and Jerry's Plains is in fact very small, whether in average mean temperature, diurnal temperature range or mid-summer afternoon relative humidities, showing that the summer sea breezes penetrate well up the valley before mid afternoon. Rainfall is somewhat less throughout the year at Jerry's Plains, as well as relative humidity in spring and autumn. Generally, however, the figures reveal a similar viticultural climate along the valley floor for some distance inland, probably to Sandy Hollow and beyond.

The northern arm of the Hunter, through Muswellbrook to Scone, shows the effect of being cut off from marine influences by the hills and mountains to the east, resulting in a more inland type of climate. The ripening average mean temperatures, average maxima and average monthly highest maxima here closely approach those of the Murrumbidgee and Murray irrigation areas. Nevertheless in other respects, apart from a greater risk of heavy rain at ripening and vintage, this area has clear advantages over the southern irrigation areas for quality in table wines. In particular, relative humidities are substantially higher, resulting in much less moisture stress on the vines. The ratio of seasonal sunshine hours to effective day degrees is not greatly above that in the lower Hunter, and still lies within the hypothesized range for medium-bodied, fairly soft table wines (Chapter 3).

Many writers have described the haze or partial cloud cover that commonly accompany the summer afternoon sea breezes in the lower and middle Hunter Valley. This and the resulting low ratio of growing season sunshine hours to effective temperature summation (Figure 2, Chapter 3), explain why the

Hunter is purely a table wine area despite its high ripening period mean temperatures. At the same time the warm nights encourage flavour ripening, as opposed to mere sugar accumulation. The combined factors suffice to account for the very good table wine quality of which the Valley is capable, despite its apparently above-optimum mean temperatures. The main problems which can detract from both quality and yield are viticultural, and are related principally to the erratic rainfall. Drought is not uncommon, while the heaviest seasonal rains often coincide with late ripening and harvest in January-February.

It is worth noting that even when conventionally calculated as (maximum plus minimum) /2, the ripening month average mean temperatures in the Hunter Valley are only two to three degrees more than in the warmer classic European table wine areas such as Bordeaux, the southern Rhône, and northern Italy. The true daily average temperatures in mid-summer would be closer still, because of the Hunter's relatively shorter days and longer nights at that time of year. Likewise, the much greater seasonal temperature summation of the Hunter is due mainly to its warm spring and autumn, and is therefore misleading if taken as a direct indication of ripening conditions and potential wine styles. In both these and other respects there are in fact much closer similarities to these European environments than the conventionally-expressed data appear to indicate. The success in the Hunter of Sémillon and Cabernet Sauvignon grapes (from Bordeaux), and of Shiraz (southern Rhône), is thus not entirely surprising. Certain central and northern Italian grape varieties, such as Sangiovese (from Chianti: see Siena, Chapter 19, Table 158) might also be expected to do well.

The Brokenback Range, in the lower Hunter, is an isolated ridge which enjoys outstanding air drainage. Its eastern to northern slopes combine the advantages of soil warmth due to aspect (even in a hot climate: see Chapter 3) with good red soils and full exposure to early sea breezes. In theory as well as in proven practice, this is clearly the Hunter Valley's choicest viticultural environment.

Projection from the Jerry's Plains data suggests that excellent grape quality for table wines could likewise be obtained at greater altitudes (say 300–600 m or more) on the valley slopes adjacent to that area, were viticulturally suitable sites and soils to exist there. Advantages from the inland funnelling of sea breezes through the Hunter/Goulburn river gap extend at least to the exposed north-east-facing slopes towards Mudgee, as well as to the southern slopes of the Liverpool Range along its western arm towards Coolah.

Sydney area
Parramatta (Table 101) is about 25 km from the coast, and represents the near-coastal part of the Sydney

plain. It was the site of the first successful Australian vineyard, planted under Governor Phillip's direction following the substantial failure of the initial planting at Farm Cove (now the Sydney Botanic Garden). Parramatta was also the site of Gregory Blaxland's Brush Farm vineyard, from which samples of wine shipped to England in 1822 and 1827 were reportedly awarded medals by the Royal Society of Arts.

Richmond (Table 103) typifies the more inland parts of the Sydney coastal plain. It is marginally warmer than Parramatta, and drier both in rainfall and in afternoon relative humidities. The diurnal temperature range and temperature variability indices are slightly greater. This would be similar to the environment of John Macarthur's pioneering vineyard at Camden, and also to that of the Minchinbury plantings at Rooty Hill, which survived in commerce until very recently. Some newer plantings exist near Richmond.

The Sydney coastal plain's ratio of growing season sunshine hours to effective temperature summation is still lower than in the lower Hunter Valley, indicating that the wines will be soft, and probably seldom more than medium-bodied.

The area verges on the warm, humid limit for *vinifera* grapes. In that respect it is notable that the vineyards in the Hastings Valley (just inland from Port Macquarie, to the north of Newcastle) have historically depended heavily on direct-producing species hybrid varieties. The new and relatively good-quality hybrid variety Chambourcin is prominent in the Hastings Valley plantings that have been re-established within the last decade (Halliday 1990). This is an area with more humid, and even wetter, summers than the Sydney coastal plain. But to the extent that vine diseases and bunch rot can be avoided, the theoretical potential for making soft, medium-bodied table wines is good in average and drier seasons.

The figures for **Picton** (Table 102) are adjusted to an altitude of 300–350 m (vs Picton 171 m), to estimate climatic conditions on the higher foothills just south of Sydney. There is a risk of spring frosts, but with careful site selection the worst should be avoidable. Ripening period average, average maximum and average highest maximum temperatures are appreciably lower than on the coastal plain or in the Hunter Valley. Assuming that the relative humidities are not much greater than at Picton, and that sunshine hours are much the same, ripening conditions appear quite favourable for table wines. That would continue to as high an altitude as frosts allow. The main hazard, as elsewhere in the Sydney region, is excess rainfall and humidity during ripening. With a likely ratio of growing season sunshine hours to effective temperature summation of around 1.0, the wines should still be of no more than medium body.

Southern New South Wales coast

Bega (Table 79) is some 15 km inland, but still on the coastal lowlands. The vineyard estimate is for moderate slopes above the town at about 100 m altitude, and shows quite good average climate conditions for making medium-bodied table wines. Average and extreme maximum temperatures are very moderate, yet average sunshine hours are fairly generous. The coast here is substantially sunnier in summer than further north. Average relative humidities are excellent for quality; but combined with the heavy summer rainfall, they indicate a still substantial disease risk. The main other potential problem here stems from the very marked variability of the annual and seasonal rainfall: see Division of National Mapping, Australia (1986), pp. 14 and 52.

North-facing slopes south of Candelo, near Bega, have what appears to be the best topography in the area for viticulture, with temperatures suitable for table wines from the lowest slopes to 500 m or more.

Comparable conditions may be available in several other localities along the coast, from Nowra southwards. The soils near the coast between Nowra and Bega are largely red and brown earths, many of which should be suited to viticulture. Sites most to be favoured would be hills or ridges projecting towards the coast (for instance directly inland from Batemans Bay, and from Moruya) which cold inland air can by-pass on its way down the valleys. The principal aim throughout would be to select dry, sunny, elevated sites with good air drainage, and to adopt canopy management methods to offset the summer rain and humidity.

Western plains

The western plains of New South Wales include, from south to north, the viticultural area of **Corowa** (Table 86), just over the Murray River from Rutherglen in Victoria; the Murrumbidgee irrigation areas of **Leeton** (Table 95) and **Griffith** (Table 91); **Forbes** (Table 88), on the Lachlan in the central west; and **Narrabri** (Table 100), to typify potential irrigated viticulture in the north-west.

Corowa is climatically similar to Rutherglen and Wangaratta in Victoria (Chapter 11, tables 71 and 75). The landscape on the new South Wales side of the border is, however, generally flat, and offers little opportunity for superior site selection other then through immediate proximity to the river.

Leeton and Griffith differ little from each other, the Leeton figures being slightly more favourable in diurnal range and relative humidity. Both are appreciably hotter than the South Australian irrigation areas, as typified by Berri (Chapter 10, Table 38), though cooler than the San Joaquin Valley of California: see Fresno (Chapter 22, Table 174). Compared with Berri the extreme maximum temperatures during ripening are a little lower, reflecting the slight subtropical climatic influence that reaches the Riverina in summer, as opposed to Berri's mainly south coastal influences. But in broad viticultural terms the two regions are very similar. Both are hot, flat irrigation areas, with the normal viticultural and oenological problems of such environments.

Forbes (Table 88) on the Lachlan River, has a small area of vines. It is appreciably hotter throughout the season than Leeton and Griffith, but the greater variability of the minimum temperatures in spring results in at least an equal risk of frost after budburst. Conversely, a greater sub-tropical influence in summer results in less variability of both maximum and minimum temperatures at that time of year. The summer monsoonal influence reaches its peak during February, resulting in greater relative humidities but no greater extreme maxima during ripening than at Leeton and Griffith. On balance there seems to be little to choose, climatically, between Forbes and the Murrumbidgee and Murray irrigation areas for viticulture.

Narrabri (Table 100) would reasonably represent such irrigation areas as Wee Waa, on the Namoi, and Moree, on the Gwydir, should they be considered for growing winegrapes. At first sight the risk from spring frosts appears less than further south, but in fact the October-November temperature figures have little relevance because budburst is earlier than in the south. The minimum temperature in spring is very variable, so there is still a fairly substantial frost risk to the young growth.

Average seasonal and ripening period temperatures at Narrabri equal those at Fresno, California, but ripening period extreme maxima remain no higher than at Griffith and Forbes. This is because of the strong summer monsoon influence that reaches Narrabri. The monsoon also brings more summer rain than further south, although sunshine hours are not much reduced. This is perhaps an area for early tablegrapes, but its sub-tropical heat rules it out for satisfactory winegrapes.

South-western slopes
Junee (Table 94), **Cootamundra** (Table 85) and **Young** (Table 107) have climates typical of the south-west slopes. Junee is a previous wine-producing area, and Young has recent plantings.

Growing season and ripening period mean temperatures, and sunshine hours, are virtually identical with those of Rutherglen, Wangaratta and Corowa. Progressively greater altitudes with distance north more or less exactly offset the change in sea-level temperature with latitude (Figure 14). However Cootamundra and Young show increasingly northern climatic influences in their greater summer rainfall and relative humidities, and more moderate highest maxima. The relative humidities nevertheless remain well below optimum for table wine quality, and these areas are too far inland and too much cut off from the coast by the Great Dividing Range to benefit from late afternoon sea breezes.

The greatest problem throughout is frosts, due to the combination in spring of a very wide diurnal temperature range and marked variability of the minimum. The frosts are more severe than in otherwise equivalent environments in Victoria. Careful selection of sites on the slopes of isolated or projecting hills will help, but still cannot achieve optimum temperature variability regimes. Localities with best temperatures for viticulture would include the ridge of hills running from Gundagai to a little west of Cootamundra, together with the slopes of isolated hills east of Young. These include the Douglas Range, Wambanumba Ridge, and Dananbilla Ridge. But water conservation for the summer, and especially for the ripening period, remains essential throughout. The highest altitudes, good air drainage and water conservation appear to be the keys in this region for combining safe yields with best grape quality for table wines.

Central-western slopes
Cowra (Table 87) is at a much lower altitude than Young, and this, together with its latitude, results in a hot and still-dry climate. The altitude of the recording site is listed as 298 m, and existing vine plantings extend up to about 350 m. I have estimated for 350–400 m, but the best climates for table wines would be still further up the slopes. Sugarloaf Mountain (773 m), an isolated volcanic hill west of Cowra, appears especially worth investigating. Its superior air drainage and resulting reduced temperature variability, combined with lower temperatures and higher relative humidities on the upper slopes, should provide climates much better suited for table wines than Cowra itself. The combination of volcanic soils and sheltered eastern slopes could have particular advantages. Water conservation is nevertheless still needed.

By the latitude of **Mudgee** (Table 98), rainfall has become slightly summer-dominant: though still with very ample sunshine hours through the growing season. Relative humidities remain below optimum for table wine quality, but are sufficiently greater than in the south to reduce stresses on the vines and allow balanced vine development under natural rainfall in most seasons. The sub-tropical pattern of consistently moderate maximum temperatures is well established by mid summer, giving good average conditions for ripening.

Mudgee itself is in a valley and has severe spring frosts. The better existing vineyards, by and large, are on slopes at variably greater altitudes. I have estimated for 550–600 m, which represents the higher and cooler vineyards. The figures show ripening temperatures about optimum for fortified wines, and a little above optimum for table wines in the absence of moderating sea breezes.

It is of interest to compare Mudgee with the Hunter Valley. According to my estimates, maturity – of a given grape variety – should be about two weeks later at Mudgee. General report puts the difference in vintage dates at three to four weeks. That may in part reflect traditional differences in maturity stage at picking. Equally, I may have made too little allowance for the much lower night temperatures in spring at Mudgee, which are almost certainly the reason for the later budburst and slower early vine development there.

Ripening period mean and extreme temperatures at Mudgee are appreciably less than in the Hunter, although hardly by enough to justify its occasional description as a 'cool' viticultural area. At the same time its afternoon relative humidities are slightly less, rainfall moderately so, and sunshine hours greater than in the Hunter. The ratio of growing season sunshine hours to effective temperature summation substantially exceeds that of the Hunter.

The sum of these differences amply explains the robustness of Mudgee wines, as opposed to the more delicate and complex, yet still fully flavoured wines of the Hunter. Berry sugar contents are lower on the Hunter; but as a coastal valley it benefits from early and regular sea breezes and much warmer nights, allowing full and rapid flavour ripening despite a lower sugar content.

North-west slopes

Murrurundi (Table 99) is not on the true north-west slopes, but together with **Coonabarabran** (Table 84), it gives some guide to the southern end of that region. Each is at about 500 m, but I have adjusted the estimates of ripening dates and conditions to 600 m and moderate slopes. At that altitude, estimated ripening temperatures are more or less identical with those at Mudgee. The summer-dominant rainfall pattern here is still more firmly established than at Mudgee, giving a growing season and ripening period rainfall very similar to that of the lower and middle Hunter Valley (Cessnock, Table 83; Jerry's Plains, Table 93).

Coonabarabran has lower relative humidities and slightly more sunshine hours than Mudgee; climatically it therefore appears less suitable for table wines, although still good for fortified wines. With generally poor soils and greater remoteness from markets there is little to recommend the Coonabarabran area as a whole. One environment which does warrant consideration, however, is the Warrumbungle Range, west of Coonabarabran. This isolated volcanic outcrop has good soils and many moderate slopes with excellent air drainage at 600–700 m or higher, which could produce good, full-bodied table wines. The east-facing slopes may still receive some benefit from afternoon sea breezes blowing up the Hunter-Goulburn Valley (Figure 14). The main danger is variability of the rainfall. There may be excess during ripening or harvest in occasional years, and drought in others. Water catchment would be desirable to enable supplementary watering in dry years as needed.

Of possibly greater interest are the slopes of the Liverpool Range along its east-west arm between Murrurundi and Coolah. Rising to about 900 m, and more or less isolated, these hills should have very good air drainage. Despite their altitude they are probably less subject to killing spring frosts than the surrounding flat countryside. Average temperatures and relative humidities are very suitable for intermediate to warm climate styles of table wine from about 600 m up. Fairly ample growing season rainfall and sunshine hours, and friable, basalt-derived clay loam soils make this – in theory – one of the more promising inland environments of New South Wales for viticulture. The dangers of excess or deficient summer rainfall are similar to those for the Warrumbungles.

A distinction can be drawn between the northern and southern faces of the Liverpool Range. The southern face benefits from cooling afternoon breezes coming up the Hunter-Goulburn Valley, as already noted. Also, it would normally expect more rain from southern rain-bearing weather systems in winter and spring than the northern face. Conversely, the northern face stands to get more rain from monsoonal and tropical cyclonic systems in summer, but has greater exposure to hot north-west winds and little to sea breezes, as is evident in Figure 14. Possibly the best combination of climatic elements is with adequate water storage on the southern slopes of the range, preferably with easterly aspects.

Viticulture is currently developing at **Quirindi**, as a commercial extension from the Hunter Valley. Quirindi (390 m, just north of the Liverpool Range and inland from the Great Dividing Range) is warmer and drier than nearby Murrurundi. Its ripening mean temperatures can be expected to resemble quite closely those of the Hunter Valley, but with a little less rain, more sunshine, and cooler nights. The ratio of growing season sunshine hours to effective temperature summation of about 1.1 at Murrurundi, and probably about the same at Quirindi (cf. Cessnock 0.9, Mudgee 1.2) indicates more robust wines on average than in the Hunter, though perhaps less so than at Mudgee. A probable disadvantage compared with the lower and middle Hunter is greater variability of the minimum temperature, leading to a greater frost risk in spring. Nevertheless, with good soils and supplementary water as needed, and summer relative humidities that are probably high enough to minimize the worst stresses on the vines, this appears to be a reasonably favourable area for producing warm climate styles of table wine on a large scale. It should have clear quality advantages over the established irrigation areas on the Murrumbidgee and Murray rivers.

The altitude of **Tamworth** (Table 105), at 378 m, places it climatically more with the plains than with slopes which might be considered for wine production. Tamworth itself is clearly hotter and drier than desirable for winegrape production, despite its summer rainfall pattern. On the other hand the nearby, largely granitic Moonbi Range could provide cooler sites at

700 m or more. Such altitudes and the lack of variability in summer maximum temperatures mean that few extremes are experienced despite quite high average temperatures during ripening. Moreover, rainfall starts to diminish by the time the grapes are ripening, especially for later-maturing varieties. This effect is accentuated as ripening is retarded still higher up the slopes (compare, for instance, Armidale, Table 77). Thus, throughout the region there are clear advantages in lateness of grape maturity.

Inverell (Table 92) illustrates typical climatic conditions on the more northerly north-west slopes at about 600 m. Ripening temperatures are a little above the postulated optimum range for naturally balanced table wines. On the other hand the average highest monthly maxima are extremely moderate for a warm to hot climate. Despite predominantly summer rainfall the sunshine hours remain very ample, with a nearly 1.2 ratio of growing season sunshine hours to effective temperature summation. The friable brown clay loam soils, derived from tertiary basalt, should be very suitable for viticulture.

The Inverell environment accords well with descriptions of the wines from early vineyards in the area. Cox (1967) recorded that 'Inverell and Bukkulla in the north, with good soils and warm climate, produced some remarkably palatable full-bodied dry and sweet reds'. Similarly Halliday (1985a) noted that the 'dark, rich and strongly flavoured Bukkulla wines' were highly regarded in the late 19th century, and were valued for blending with the lighter wines of the Hunter Valley.

But that was before the advent of downy and powdery mildew. Of a recent planting just south of Inverell, at 760 m, Halliday records that: 'Spring frost and summer humidity, the latter leading to all forms of mildew, are the main viticulture problems'. Average afternoon relative humidities are in fact not high, and even at 760 m altitude are below the postulated optimum for avoidance of vine stress. But combined with warm temperatures, high night humidities, and intermittent periods of summer rain accompanied by very high humidity, they are enough for mildews to flourish.

The Inverell climate illustrates well the dilemma of all inland warm, humid viticultural climates, as discussed in general terms in Chapter 3. Afternoon relative humidities high enough to avoid severe vine moisture stress go together with dangerously high humidities at all other times. Coastal and near-coastal valleys, by contrast, can enjoy the optimum combination of frequently dry land winds at night and in the early morning, and mild, humid sea breezes in the afternoon.

The figures for Inverell illustrate also the danger from spring frosts that characterizes all the western slopes of New South Wales. Being from a valley floor they doubtless over-estimate the risk at most potential viticultural sites. Nevertheless the north-west slopes are notorious for late frost damage in wheat crops,

suggesting that the problem for vines may be greater there than in most places. It results from the combination in early spring of a very wide diurnal temperature range with great variability of the minimum, which is typical of inland, semi-arid sub-tropical environments before the summer monsoon arrives. Selection of hills or rises with the best possible air drainage, and of grape varieties having late enough budburst, is therefore essential to any viticulture on the north-west slopes.

Southern tablelands

A small wine industry exists on the New South Wales southern tablelands, so far largely as a hobby outlet for citizens of Canberra. The data for **Canberra** (Table 82) show that it has in many respects an inhospitable climate for viticulture. Spring frosts can be severe right into November, necessitating the most rigorous site selection if viticulture is to be feasible. Rainfall is marginal, and growing season relative humidities fairly low. Being on the inland side of the Great Dividing Range, this area does not get sea breezes. Afternoon winds are predominantly from the north-west or west throughout the year. Variability of the rainfall is another problem, both for rain-fed vines and for water conservation. On the other hand ripening period mean and extreme temperatures, together with sunshine hours, are quite favourable for full-bodied table wines.

The location of existing vineyards near Canberra is notable. One group abuts Lake George, with two of the four vineyards on a high escarpment having outstanding air drainage both to Lake George on the east, and down the Yass Valley to the west. The other is at Murrumbateman, on the west-projecting ridge between Murrumbateman Creek to the south, and the Yass River to the north. The best sites here are on slopes of the numerous small, isolated hills that constitute much of the ridge. Both localities are protected in some degree from cold night air drifting north from the snowfields of the Australian Alps, because such air would tend to be intercepted and diverted west along the Murrumbidgee Valley. Lake George has benefits of its own on local air convection (when it contains water). The best existing vineyards, therefore, appear to be about as well situated for frost avoidance and wine quality as they could be within the region. But even then, spring frosts remain a problem. The region must be regarded as marginal for viticulture, although able to make good cool-climate styles of wine.

Goulburn (Table 90), **Moss Vale** (Table 97) and **Bowral** (Table 81) show a gradient of increasing coastal influence compared with Canberra. The most obvious changes along the gradient are in summer rainfall and afternoon relative humidities, of which the whole-season averages are 47, 53 and 54 per cent respectively, compared with Canberra's 40 per cent.

A very serious risk remains of killing frosts in spring, as is shown by the Moss Vale and Bowral data. The lesser apparent frost risk recorded for Goulburn should perhaps be treated with caution. It is out of line with comparable surrounding stations, and the fact that Climatic Averages Australia (1975) shows a new recording site suggests that the old one may have been unsatisfactory. At best it can be said that potential viticultural sites throughout the whole southern tablelands region would have to be selected with extreme care for reduced frost risk, as at Canberra. The likelihood of finding extensive, fully satisfactory areas for commercial viticulture seems remote.

I have included **Bombala** (Table 80), at 747 m on the extreme south-eastern tableland, as representative of some of the high-altitude New South Wales environments which were at one time suggested for cool-climate viticulture, on the basis of homoclime comparisons with cool European viticultural environments. The detailed figures show that frosts in spring preclude any possibility of commercial-scale viticulture, even with the most rigorous site selection. Bombala has the same raw temperature summation as Reims (Champagne: Chapter 16, Table 137) in France, and Geisenheim (Rheingau: Chapter 17, Table 145) and Stuttgart (Württemberg etc.: Table 146) in Germany. But even if vines were able to survive the frosts, the wide diurnal temperature range so reduces the effective temperature summation (column 11 versus column 10 in the tables), that no grape varieties could be expected to ripen reliably.

Better prospects may exist at lower altitudes in some of the coastal valleys that penetrate the eastern scarp of the southern tablelands, abutting the locations already discussed under the Sydney area and southern New South Wales coast. They might include, for instance, sites with superior air drainage projecting into the Wollondilly Valley above Lake Burragorang, or the valleys of the Shoalhaven River and its tributaries inland from Nowra. To estimate for these, I have adjusted the Bowral temperature data (but not those for relative humidity) to 500 m. Like the estimates based on coastal data they show good ripening conditions for medium to full-bodied table wines if the summer is dry enough, but at the same time the risk of too much summer-autumn rain and humidity in many seasons.

Central tablelands

Bathurst (Table 78, altitude 713 m) and **Lithgow** (Table 96, altitude 917 m) represent lower and middle altitudes respectively, on the central tablelands.

Estimated ripening conditions at Bathurst show promise for full-bodied table wines: particularly in having very moderate highest maximum temperatures, which result from the small variability of the maxima in late summer and autumn. But as on the southern tablelands, the spring frosts are very severe. They explain the disappearance of the early Bathurst vineyards, as noted at the start of this chapter. It seems unlikely that otherwise satisfactory sites could be found which are free enough from frosts for commercial viticulture.

The **Lithgow** recording site appears viticulturally more favourable than most because its minimum temperatures vary only within a fairly narrow range, but there is still an unacceptable frost risk. The biologically effective temperature summation suggests anyway that only the very earliest-maturing grape varieties could ripen regularly at this altitude. Wine quality could not compete with that of areas able to ripen the better-quality grape varieties of Maturity Group 3 and later (Chapter 7).

Oberon, at 1105 m, is another of the localities directly canvassed as possibly suitable for making cool-climate styles of wine, on the basis of homoclime matching with Europe (Smart 1977). I have not attempted a detailed estimate for it because reliable long-term climatic data are lacking; but a comparison of the short-term data in Climatic Averages Australia (1975) shows Oberon to have October minima on average 1.8°C lower than at Lithgow, and November minima 1.6°C lower. The October-April average temperature at Oberon is 1.04°C lower than at Lithgow, with a diurnal temperature range 0.88°C wider. The young shoots would be regularly obliterated by spring frosts under these conditions. If they did survive, not even the earliest-maturing varieties could be expected to ripen fully in average seasons.

Doyle and Hedberg (1990) describe viticulture at an altitude of 840 to 870 m on the slopes of Mount Canobolas, near Orange. This is an isolated hill, of which the slopes would enjoy excellent air drainage. The area is far enough north to benefit considerably from the moderating monsoonal influence in summer, but far enough inland to avoid the often excessive summer rainfall and humidity of the coast. Times and conditions of ripening depend largely on altitude, and are potentially quite good for many table wine styles. Temperature variability remains a hazard, however, as is evidenced by the reported occasional snow in the area right up to late December. This means that site selection for best frost avoidance is essential. Accepting that, the area around Orange and Mount Canoblas appears to have the best prospects of any in the central tablelands for commercial viticulture.

In summary, the central tablelands of New South Wales are broadly counter-indicated for viticulture. In most areas the frost risk in spring is too great, and the wide diurnal temperature range early in the growing season results in a greatly reduced biological effectiveness of any given average mean temperature. Even if the vines survive frost, ripening potential is much less than in the classical Northern Hemisphere viticultural areas with comparable average mean temperatures and raw temperature summations. Viticulture is possible on the slopes of a few isolated hills with outstanding air drainage, and on the main criteria I have spelled out the potential for table wine quality appears reasonably good there at appropriate altitudes. But overall, the prospects for competitive commercial viticulture on the central tablelands appear very limited.

Northern tablelands (New England region)

The climates of **Armidale** (Table 77, altitude 980 m) and **Glen Innes** (Table 89, altitude 1072 m) are typical of the northern tablelands. **Tenterfield** (Table 106, altitude 863 m) is at the northern tip, and in terms of climate, landforms and soils, belongs more with the granite belt of southern Queensland, typified by Stanthorpe (Chapter 13, Table 109).

Both Armidale and Glen Innes offer excellent potential temperatures, sunshine hours and relative humidities during ripening for making medium-bodied table wines. The highest maxima during ripening, and indeed, through the whole growing season, are remarkably moderate: the result of a firmly established summer monsoon influence in this high-altitude subtropical climate. Such equability during ripening is found in few other places in Australia. At the same time, sunshine hours and afternoon relative humidities are both very favourable. Growing season rainfall is just about optimal. Its summer dominance increases to the north, but at both Armidale and Glen Innes the average rainfall diminishes quite rapidly through the ripening period, leading into a mild, sunny, fairly dry autumn. Midseason to late grape varieties from Bordeaux, the Rhône and northern Italy should be the best adapted to this ripening environment.

Apart from altitude, as discussed above for the central tablelands, the most important viticultural weakness of the whole region is frost after budburst in spring: a weakness shared with the rest of the New South Wales tablelands and western slopes. This means that site selection is crucial, together with choosing grape varieties that have late enough budburst. The other main problem is effective disease control, because as on the north-west slopes, favourable average afternoon relative humidities are accompanied by often excessive relative humidities during wet spells and at other times of day. The difficulties of disease control may not be exceptional by world standards, but they represent an economic cost compared with most other Australian environments.

The most suitable part of the northern tablelands is probably their western edge, on projecting hills or ridges enjoying superior air drainage. Temperatures are such that some loss of altitude can be afforded without ripening conditions becoming too hot. Certain outlying peaks and high plateau areas in the region of the north-west slopes can be added, including particularly the slopes of Mount Kaputar, a small, isolated basalt range east of Narrabri. In addition to having good viticultural soils, favourable ripening temperatures and reasonable freedom from spring frosts, these inland slopes should run less risk of the excessive summer rainfall and humidities than on the northern tablelands proper.

The eastern slopes of the Great Dividing Range in this region are almost certainly too wet and humid for *vinifera* grapes.

Tenterfield, at the northern tip of the tablelands, is still more prone to spring frost than either Armidale or Glen Innes. This is due to greater values in October and November for both diurnal temperature range and, in particular, variability of the minimum temperature. The result is that average lowest minima in October and November are comparable to those at Armidale and Glen Innes, despite a clearly warmer climate overall which could be expected to result in earlier budburst. Nevertheless, the fact that Stanthorpe, nearby in south-east Queensland, has a long-established vine industry shows that some sites can probably be found which are frost-free enough for commercial viticulture.

The granite belt environment is discussed further under Stanthorpe, Queensland (Chapter 13).

TABLE 77

ARMIDALE, NEW SOUTH WALES. 30°31'S, 151°39'E. ALT. 980m.

	1	2	3	4	5	6	7	8	9	10	11	12	13	14	15	16	17	18
	Temperature °C									Summation of Day Degrees above 10° with 19° Cut-Off			Temperature Variability Index					
PERIOD	Lowest Recorded Min. (64YRS)	No.Mths with <0.3° (30YRS)	No.Mths with <2.5° (30YRS)	Av.Mthly Lowest Min. (30YRS)	Av.Min. Minus Col.4	Daily Mean (30YRS)	Daily Range /2 (30YRS)	Av.Mthly Highest Max.	Col.8 Minus Av.Max.	Raw	Adj.1 for Lat. & Daily Temp Rng	Adj.2 for Vine Sites	Raw	Adj. for Vine Sites	Av. Sunshine Hrs	Av. Number Rain Days	Av. Rainfall mm	Av. % R.H. 1500 HRS (31YRS)
OCTOBER	−3.3	14	25	0.6	6.7	14.3	7.0	28.0	6.7	133	116		41.4		268	9	69	42
NOVEMBER	−0.1	2	13	3.2	6.9	17.3	7.2	31.0	6.5	220	198		42.2		281	9	80	41
DECEMBER	2.2	0	1	5.7	6.7	19.3	6.9	33.0	6.8	279	265		41.1		285	10	88	44
JANUARY	4.4	0	0	8.5	5.1	20.4	6.8	33.9	6.7	279	279		39.0		273	10	101	44
FEBRUARY	3.3	0	0	9.2	4.0	19.8	6.6	32.1	5.7	252	252		36.1		232	10	87	46
MARCH	−0.6	1	6	5.3	5.9	17.6	6.4	30.2	6.2	236	233		37.7		235	10	67	49
APRIL	−3.9	10	22	1.1	6.5	13.9	6.3	26.4	6.2	117	120		37.9		222	8	46	51
OCT-APRIL	—	27	67	—	6.0	17.5	6.7	—	6.4	1,516	1,463		39.3		1,796	66	538	45
MAY-SEPT	—	—	—	—	—	8.5	—	—	—	—	—		—		—	—	257	—

DATA FOR MONTH TO MATURITY

MATURITY GROUP	EST. MATURITY DATE																	
1	FEBRUARY 23					20.0	6.6	32.6	5.9				36.8		259	11	96	46
2	FEBRUARY 28					19.8	6.6	32.2	5.7				36.3		255	11	95	46
3	MARCH 5					19.5	6.6	31.9	5.8				36.6		250	11	91	46
4	MARCH 11					19.2	6.5	31.6	5.9				37.0		246	11	86	47
5	MARCH 17					18.8	6.5	31.2	6.0				37.3		242	10	79	48
6	MARCH 25					18.1	6.4	30.7	6.1				37.5		238	10	72	48
7	APRIL 2					17.4	6.4	30.1	6.2				37.7		234	10	65	49
8	APRIL 12					16.3	6.4	29.0	6.2				37.8		231	9	57	50

TABLE 78

BATHURST (Agricultural Research Station), NEW SOUTH WALES. 33°26'S, 149°34'E. ALT. 713m.

	1	2	3	4	5	6	7	8	9	10	11	12	13	14	15	16	17	18
	Temperature °C									Summation of Day Degrees above 10° with 19° Cut-Off			Temperature Variability Index					
PERIOD	Lowest Recorded Min. (62YRS)	No.Mths with <0.3° (30YRS)	No.Mths with <2.5° (30YRS)	Av.Mthly Lowest Min. (30YRS)	Av.Min. Minus Col.4	Daily Mean (29YRS)	Daily Range /2 (29YRS)	Av.Mthly Highest Max. (30YRS)	Col.8 Minus Av.Max.	Raw	Adj.1 for Lat. & Daily Temp Rng	Adj.2 for Vine Sites	Raw	Adj. for Vine Sites	Av. Sunshine Hrs	Av. Number Rain Days	Av. Rainfall mm	Av. % R.H. 1500 HRS
OCTOBER	−3.9	21	28	−0.3	7.0	13.4	6.7	27.4	7.3	107	100		41.1		271	9	60	53
NOVEMBER	−1.1	4	21	2.0	7.5	16.6	7.1	31.3	7.6	197	180		43.5		292	8	59	46
DECEMBER	0.8	0	5	4.4	7.9	19.3	7.0	33.9	7.6	279	268		43.5		311	7	66	48
JANUARY	1.8	0	0	7.1	6.1	20.3	7.1	35.1	7.7	279	279		42.2		302	7	68	52
FEBRUARY	2.8	0	0	7.0	6.4	20.4	7.0	33.3	5.9	252	252		40.3		258	7	56	55
MARCH	−2.2	2	9	4.1	7.1	17.9	6.7	30.6	6.0	245	240		39.9		248	6	49	50
APRIL	−3.3	17	25	0.4	7.0	13.5	6.1	26.3	6.7	104	107		38.1		228	7	42	44
OCT-APRIL	—	44	88	—	7.0	17.3	6.8	—	7.0	1,463	1,426		41.2		1,910	51	400	50
MAY-SEPT	—	—	—	—	—	8.5	—	—	—	—	—		—		—	—	229	—

DATA FOR MONTH TO MATURITY

MATURITY GROUP	EST. MATURITY DATE																	
1	FEBRUARY 25					20.4	7.0	33.6	6.1				40.5		279	7	61	55
2	MARCH 3					20.2	7.0	33.1	5.9				40.3		273	7	59	55
3	MARCH 8					20.0	7.0	32.7	5.9				40.2		267	7	57	54
4	MARCH 13					19.7	6.9	32.2	5.9				40.1		260	7	55	53
5	MARCH 20					19.2	6.8	31.6	6.0				40.0		252	6	52	52
6	MARCH 28					18.3	6.7	30.9	6.0				39.9		243	6	49	50
7	APRIL 6					17.0	6.6	29.8	6.1				39.5		237	6	47	49
8	APRIL 19					15.2	6.4	27.8	6.4				38.8		232	7	46	47

TABLE 79

BEGA, NEW SOUTH WALES. 36°40'S, 149°50'E. ALT. 13m.

PERIOD	1 Lowest Recorded Min. (60YRS)	2 No.Mths with <0.3° (30YRS)	3 No.Mths with <2.5° (30YRS)	4 Av.Mthly Lowest Min. (30YRS)	5 Av.Min. Minus Col.4	6 Daily Mean (29YRS)	7 Daily Range /2 (29YRS)	8 Av.Mthly Highest Max. (30YRS)	9 Col.8 Minus Av.Max.	10 Raw	11 Adj.1 for Lat. & Daily Temp Rng	12 Adj.2 for Vine Sites	13 Raw	14 Adj. for Vine Sites	15 Av. Sunshine Hrs	16 Av. Number Rain Days	17 Av. Rainfall mm	18 Av. % R.H. 1500 HRS
				Temperature °C						Summation of Day Degrees above 10° with 19° Cut-Off			Temperature Variability Index					
OCTOBER	−2.2	3	14	2.4	5.7	15.7	7.6	31.2	7.9	177	158	159	44.0	41.0	253	8	66	56
NOVEMBER	−1.7	1	4	4.5	6.0	17.5	7.0	33.2	8.7	225	213	210	42.7	39.7	254	8	65	56
DECEMBER	2.2	0	1	6.7	6.4	19.7	6.6	34.8	8.5	279	279	279	41.3	38.3	256	8	81	57
JANUARY	3.9	0	0	8.2	5.9	20.7	6.6	37.2	9.9	279	279	279	42.2	39.2	251	8	86	57
FEBRUARY	3.9	0	0	8.3	5.9	20.9	6.7	35.4	7.8	252	252	252	40.5	37.5	240	7	91	60
MARCH	2.2	0	0	6.7	5.7	19.3	6.9	33.8	7.6	279	279	277	40.9	37.9	222	8	89	58
APRIL	−2.1	3	14	2.4	6.6	16.1	7.1	29.4	6.2	183	176	174	41.2	38.2	203	6	69	50
OCT-APRIL	—	7	33	—	6.0	18.6	6.9	—	8.1	1,674	1,636	1,630	41.8	38.8	1,679	53	547	56
MAY-SEPT	—	—	—	—	—	11.1	—	—	—	—	—	—	—	—	—	—	324	—

				DATA FOR MONTH TO MATURITY														
MATURITY GROUP	EST. MATURITY DATE					*	*	*						*				
1	FEBRUARY 14					20.6	5.9	35.2	8.8					38.3	255	8	92	59
2	FEBRUARY 19					20.6	5.9	34.9	8.4					37.9	257	8	95	59
3	FEBRUARY 25					20.5	5.9	34.6	7.9					37.6	259	7	97	60
4	MARCH 3					20.4	6.0	34.1	7.8					37.5	257	7	97	60
5	MARCH 8					20.2	6.0	33.9	7.8					37.6	252	8	95	60
6	MARCH 14					19.9	6.0	33.6	7.7					37.7	245	8	94	59
7	MARCH 19					19.6	6.1	33.4	7.7					37.8	237	8	93	59
8	MARCH 25					19.2	6.1	33.1	7.6					37.8	230	8	91	58

* Adjusted for vine sites.

TABLE 80

BOMBALA, NEW SOUTH WALES. 36°54'S, 149°14'E. ALT 747m.

PERIOD	1 Lowest Recorded Min. (57YRS)	2 No.Mths with <0.3° (30YRS)	3 No.Mths with <2.5° (30YRS)	4 Av.Mthly Lowest Min. (30YRS)	5 Av.Min. Minus Col.4	6 Daily Mean (29YRS)	7 Daily Range /2 (29YRS)	8 Av.Mthly Highest Max. (30YRS)	9 Col.8 Minus Av.Max.	10 Raw	11 Adj.1 for Lat. & Daily Temp Rng	12 Adj.2 for Vine Sites	13 Raw	14 Adj. for Vine Sites	15 Av. Sunshine Hrs	16 Av. Number Rain Days	17 Av. Rainfall mm	18 Av. % R.H. 1500 HRS
				Temperature °C						Summation of Day Degrees above 10° with 19° Cut-Off			Temperature Variability Index					
OCTOBER	−6.1	23	29	−0.7	5.5	11.9	7.1	26.7	7.7	58	49		41.6		252	10	54	51
NOVEMBER	−1.7	13	28	0.5	6.2	14.1	7.4	30.9	9.4	122	108		45.2		272	9	59	49
DECEMBER	−1.1	3	13	2.6	6.2	16.4	7.6	33.4	9.4	198	177		46.0		280	9	68	50
JANUARY	−2.2	2	4	3.9	5.7	17.2	7.6	34.7	9.9	224	202		46.0		302	8	67	50
FEBRUARY	−0.3	1	6	3.9	6.3	17.6	7.4	33.4	8.4	214	197		44.3		252	7	60	58
MARCH	−1.7	3	17	2.3	6.1	15.5	7.1	31.2	8.6	170	161		43.1		226	8	56	54
APRIL	−5.0	22	30	−0.8	6.0	11.7	6.5	26.3	8.1	51	51		40.1		178	7	43	51
OCT-APRIL	—	67	127	—	6.0	14.9	7.2	—	8.8	1,037	945		43.8		1,762	58	407	52
MAY-SEPT	—	—	—	—	—	6.7	—	—	—	—	—		—		—	—	233	—

				DATA FOR MONTH TO MATURITY														
MATURITY GROUP	EST. MATURITY DATE																	
1	—																	
2	—																	
3	—																	
4	—																	
5	—																	
6	—																	
7	—																	
8	—																	

TABLE 81

BOWRAL, NEW SOUTH WALES. 34°30'S, 150°24'E. ALT 661m.

	1	2	3	4	5	6	7	8	9	10	11	12	13	14	15	16	17	18
	Temperature °C									Summation of Day Degrees above 10° with 19° Cut-Off			Temperature Variability Index					
PERIOD	Lowest Recorded Min. (62YRS)	No.Mths with <0.3° (30YRS)	No.Mths with <2.5° (30YRS)	Av.Mthly Lowest Min. (30YRS)	Av.Min. Minus Col.4	Daily Mean (31YRS)	Daily Range /2 (31YRS)	Av.Mthly Highest Max. (30YRS)	Col.8 Minus Av.Max.	Raw	Adj.1 for Lat. & Daily Temp Rng	Adj.2 for Vine Sites	Raw	Adj. for Vine Sites	Av. Sunshine Hrs	Av. Number Rain Days	Av. Rainfall mm	Av. % R.H. 1500 HRS
OCTOBER	−4.4	17	26	−0.1	6.9	13.9	7.1	28.4	7.4	121	110	152	42.7	40.9	256	13	83	52
NOVEMBER	−2.2	3	19	2.0	7.2	16.7	7.5	32.0	7.8	201	180	222	45.0	43.2	258	13	102	55
DECEMBER	−1.7	2	7	4.2	6.8	18.5	7.5	34.0	8.0	264	239	279	44.8	43.0	275	13	100	53
JANUARY	1.1	0	3	5.7	6.4	19.4	7.3	35.3	8.6	279	269	279	44.2	42.4	277	14	100	54
FEBRUARY	2.2	0	1	6.2	6.1	19.3	7.0	33.9	7.6	252	247	252	41.7	39.9	223	13	80	58
MARCH	−1.1	3	12	3.6	7.1	17.1	6.4	31.3	7.8	220	219	255	40.5	38.7	224	14	80	55
APRIL	−3.9	15	26	0.6	6.3	12.9	6.0	26.5	7.6	87	88	123	37.9	36.1	206	12	76	52
OCT-APRIL	—	40	94	—	6.7	16.8	7.0	—	7.8	1,424	1,352	1,562	42.4	40.6	1,719	92	621	54
MAY-SEPT	—	—	—	—	—	8.3	—	—	—	—	—	—	—	—	—	—	304	—

	DATA FOR MONTH TO MATURITY																
MATURITY GROUP	EST. MATURITY DATE			*	*	*							*				
1	FEBRUARY 14			20.7	6.7	35.4	8.0						41.0	260	14	94	56
2	FEBRUARY 19			20.7	6.7	35.2	7.8						40.6	254	14	92	57
3	FEBRUARY 25			20.6	6.6	34.9	7.7						40.2	248	14	90	58
4	MARCH 2			20.4	6.5	34.5	7.6						39.9	242	14	88	58
5	MARCH 8			20.1	6.4	34.1	7.7						39.7	237	14	86	57
6	MARCH 14			19.7	6.3	33.6	7.7						39.5	232	14	84	57
7	MARCH 21			19.2	6.2	33.0	7.7						39.2	227	14	82	56
8	MARCH 28			18.5	6.0	32.3	7.8						38.8	222	14	81	55

* Adjusted for vine sites.

TABLE 82

CANBERRA (Forestry School, Acton), AUSTRALIAN CAPITAL TERRITORY. 35°18'S, 149°06'E. ALT. 518m.

	1	2	3	4	5	6	7	8	9	10	11	12	13	14	15	16	17	18
	Temperature °C									Summation of Day Degrees above 10° with 19° Cut-Off			Temperature Variability Index					
PERIOD	Lowest Recorded Min. (29YRS)	No.Mths with <0.3° (29YRS)	No.Mths with <2.5° (29YRS)	Av.Mthly Lowest Min. (29YRS)	Av.Min. Minus Col.4	Daily Mean (22YRS)	Daily Range /2 (22YRS)	Av.Mthly Highest Max. (29YRS)	Col.8 Minus Av.Max.	Raw	Adj.1 for Lat. & Daily Temp Rng	Adj.2 for Vine Sites	Raw	Adj. for Vine Sites	Av. Sunshine Hrs	Av. Number Rain Days	Av. Rainfall mm	Av. % R.H. 1500 HRS (16YRS)
OCTOBER	−2.8	20	28	−0.2	6.3	13.1	7.0	27.9	7.8	96	87	119	42.1	38.5	250	11	73	43
NOVEMBER	−2.2	6	12	2.5	6.5	16.4	7.4	31.9	8.1	192	174	211	44.2	40.6	268	9	61	40
DECEMBER	0.0	1	3	5.3	6.5	19.1	7.3	34.7	8.3	279	261	279	44.0	40.4	283	8	56	35
JANUARY	3.4	0	0	6.8	6.1	20.4	7.5	36.0	8.1	279	279	279	44.2	40.6	273	8	58	35
FEBRUARY	0.6	0	2	6.5	6.3	20.2	7.4	34.7	7.1	252	252	252	43.0	39.4	245	7	59	38
MARCH	−0.6	2	7	4.2	6.4	17.6	7.0	31.9	7.3	236	227	259	41.7	38.1	234	7	59	41
APRIL	−3.1	15	25	0.2	6.4	13.0	6.4	26.2	6.8	90	91	115	38.8	35.3	200	7	51	51
OCT-APRIL	—	44	77	—	6.4	17.1	7.1	—	7.6	1,424	1,371	1,514	42.6	39.0	1,753	57	417	40
MAY-SEPT	—	—	—	—	—	7.8	—	—	—	—	—	—	—	—	—	—	234	—

	DATA FOR MONTH TO MATURITY																
MATURITY GROUP	EST. MATURITY DATE			*	*	*							*				
1	FEBRUARY 18			21.2	6.6	35.1	7.5						39.9	270	8	60	36
2	FEBRUARY 24			21.1	6.6	34.8	7.3						39.7	267	8	62	37
3	MARCH 2			20.9	6.6	34.5	7.1						39.4	264	7	64	38
4	MARCH 7			20.7	6.5	34.1	7.1						39.2	260	7	63	38
5	MARCH 13			20.4	6.5	33.6	7.2						38.9	255	7	62	39
6	MARCH 18			20.0	6.4	33.0	7.2						38.6	249	7	61	39
7	MARCH 24			19.4	6.3	32.3	7.3						38.4	242	7	60	40
8	APRIL 1			18.4	6.2	31.7	7.3						38.1	234	7	59	41

* Adjusted for vine sites.

TABLE 83

CESSNOCK, NEW SOUTH WALES. 32°54'S, 151°21'E. ALT 76m. (Lower Hunter Valley)

	1	2	3	4	5	6	7	8	9	10	11	12	13	14	15	16	17	18
				Temperature °C						Summation of Day Degrees above 10° with 19° Cut-Off			Temperature Variability Index					
PERIOD	Lowest Recorded Min. (36YRS)	No.Mths with <0.3° (30YRS)	No.Mths with <2.5° (30YRS)	Av.Mthly Lowest Min. (30YRS)	Av.Min. Minus Col.4	Daily Mean (14YRS)	Daily Range /2 (14YRS)	Av.Mthly Highest Max. (30YRS)	Col.8 Minus Av.Max.	Raw	Adj.1 for Lat. & Daily Temp Rng	Adj.2 for Vine Sites	Raw	Adj. for Vine Sites	Av. Sunshine Hrs	Av. Number Rain Days	Av. Rainfall mm	Av. % R.H. 1500 HR
OCTOBER	−0.6	1	3	4.6	5.8	17.8	7.4	33.5	8.3	242	223	225	43.7	40.3	245	7	55	47
NOVEMBER	4.4	0	0	7.5	6.1	20.7	7.1	36.9	9.1	270	270	270	43.6	40.2	250	7	56	41
DECEMBER	2.2	0	0	10.0	5.4	22.6	7.2	38.8	9.0	279	279	279	43.2	39.8	250	8	81	44
JANUARY	4.4	0	0	11.8	5.3	23.8	6.7	39.7	9.2	279	279	279	41.3	37.9	265	7	78	44
FEBRUARY	5.0	0	0	12.3	4.3	23.3	6.7	37.7	7.7	252	252	252	38.8	35.4	230	8	85	47
MARCH	5.0	0	0	9.6	5.3	21.6	6.7	35.8	7.5	279	279	279	39.6	36.2	230	8	74	44
APRIL	0.0	0	1	5.9	5.4	17.8	6.5	31.8	7.5	234	238	228	38.9	35.5	190	7	64	46
OCT-APRIL	—	1	4	—	5.4	21.1	6.9	—	8.3	1,835	1,820	1,812	41.3	37.9	1,660	52	493	45
MAY-SEPT	—	—	—	—	—	12.6	—	—	—	—	—	—	—	—	—	—	253	—

DATA FOR MONTH TO MATURITY

MATURITY GROUP	EST. MATURITY DATE					*	*	*					#	*				
1	JANUARY 31					23.4	5.9	38.5	9.2				41.3	37.9	265	7	78	44
2	FEBRUARY 6					23.4	5.9	38.3	8.9				40.7	37.3	263	7	82	45
3	FEBRUARY 11					23.3	5.9	37.8	8.6				40.1	36.7	261	8	86	45
4	FEBRUARY 17					23.2	5.8	37.2	8.2				39.6	36.2	258	8	89	46
5	FEBRUARY 22					23.1	5.8	36.8	7.9				39.1	35.7	255	9	91	46
6	FEBRUARY 28					23.0	5.8	36.5	7.7				38.8	35.4	251	9	93	47
7	MARCH 6					22.8	5.8	36.2	7.7				38.9	35.5	247	9	91	47
8	MARCH 11					22.5	5.7	35.8	7.6				39.1	35.7	243	8	87	46

* Adjusted to sites on the slopes of Brokenback Range at 200m. # See Cessnock commentary.

TABLE 84

COONABARABRAN, NEW SOUTH WALES. 31°16'S, 149°18'E. ALT 510m.

	1	2	3	4	5	6	7	8	9	10	11	12	13	14	15	16	17	18
				Temperature °C						Summation of Day Degrees above 10° with 19° Cut-Off			Temperature Variability Index					
PERIOD	Lowest Recorded Min. (64YRS)	No.Mths with <0.3° (30YRS)	No.Mths with <2.5° (30YRS)	Av.Mthly Lowest Min. (30YRS)	Av.Min. Minus Col.4	Daily Mean (30YRS)	Daily Range /2 (30YRS)	Av.Mthly Highest Max. (30YRS)	Col.8 Minus Av.Max.	Raw	Adj.1 for Lat. & Daily Temp Rng	Adj.2 for Vine Sites	Raw	Adj. for Vine Sites	Av. Sunshine Hrs	Av. Number Rain Days	Av. Rainfall mm	Av. % R.H. 1500 HR (31YRS
OCTOBER	−5.0	11	25	0.6	6.7	15.7	8.4	30.4	6.3	177	143	144	46.6	43.6	285	7	58	36
NOVEMBER	0.0	2	10	3.5	7.5	19.3	8.3	34.1	6.5	270	240	241	47.2	44.2	300	7	59	35
DECEMBER	1.7	0	1	6.2	7.5	21.7	8.0	36.4	6.7	279	279	279	46.2	43.2	320	7	64	34
JANUARY	4.2	0	0	8.9	6.0	23.1	8.2	37.5	6.2	279	279	279	45.0	42.0	321	7	83	34
FEBRUARY	5.0	0	0	9.1	5.5	22.7	8.1	35.2	4.4	252	252	252	42.3	39.3	268	6	83	37
MARCH	0.0	1	3	5.7	6.2	20.0	8.1	32.9	4.8	279	279	279	43.4	40.4	270	6	64	40
APRIL	−3.3	11	26	0.9	6.4	15.3	8.0	28.5	5.2	159	140	141	43.6	40.6	237	5	53	46
OCT-APRIL	—	25	69	—	6.5	19.7	8.2	—	5.7	1,695	1,612	1,615	44.9	41.9	2,001	45	464	37
MAY-SEPT	—	—	—	—	—	10.2	—	—	—	—	—	—	—	—	—	—	266	—

DATA FOR MONTH TO MATURITY

MATURITY GROUP	EST. MATURITY DATE					*	*	*						*				
1	FEBRUARY 12					22.7	7.4	35.4	5.4					40.8	308	7	85	35
2	FEBRUARY 18					22.7	7.4	35.0	5.0					40.2	302	7	86	36
3	FEBRUARY 23					22.6	7.4	34.5	4.7					39.7	296	6	87	36
4	MARCH 1					22.4	7.4	34.0	4.4					39.3	291	6	89	37
5	MARCH 7					22.1	7.4	33.5	4.4					39.4	287	6	86	37
6	MARCH 12					21.8	7.4	33.1	4.5					39.6	283	6	81	38
7	MARCH 18					21.4	7.4	32.7	4.6					39.9	279	6	75	38
8	MARCH 23					20.8	7.4	32.3	4.7					40.1	275	6	69	39

* Adjusted for vine sites.

TABLE 85

COOTAMUNDRA, NEW SOUTH WALES. 34°39'S, 148°02'E. ALT 330m.

PERIOD	1 Lowest Recorded Min. (69YRS)	2 No.Mths with <0.3° (30YRS)	3 No.Mths with <2.5° (30YRS)	4 Av.Mthly Lowest Min. (30YRS)	5 Av.Min. Minus Col.4	6 Daily Mean (28YRS)	7 Daily Range /2 (28YRS)	8 Av.Mthly Highest Max. (30YRS)	9 Col.8 Minus Av.Max.	10 Raw	11 Adj.1 for Lat. & Daily Temp Rng	12 Adj.2 for Vine Sites	13 Raw	14 Adj. for Vine Sites	15 Av. Sunshine Hrs	16 Av. Number Rain Days	17 Av. Rainfall mm	18 Av. % R.H. 1500 HRS
						Temperature °C				Summation of Day Degrees above 10° with 19° Cut-Off			Temperature Variability Index					
OCTOBER	-2.2	10	26	0.8	6.0	14.9	8.1	30.6	7.6	152	125	134	46.0	43.6	278	9	60	45
NOVEMBER	-1.7	0	11	3.4	7.0	18.8	8.4	34.9	7.7	264	228	237	48.3	45.9	303	6	47	35
DECEMBER	1.7	0	0	6.2	7.3	21.9	8.4	37.6	7.3	279	279	279	48.2	45.8	335	5	48	31
JANUARY	4.4	0	0	8.8	6.2	23.4	8.4	39.5	7.7	279	279	279	47.5	45.1	338	5	51	32
FEBRUARY	4.3	0	0	8.3	6.8	23.3	8.2	37.7	6.2	252	252	252	45.8	43.4	280	4	40	34
MARCH	-1.1	0	1	5.8	6.6	20.3	7.9	34.5	6.3	279	279	279	44.5	42.1	271	5	49	37
APRIL	-2.2	7	21	1.6	6.5	15.2	7.1	29.4	7.1	156	149	158	42.0	39.6	235	6	47	44
OCT-APRIL	—	17	59	—	6.6	19.7	8.1	—	7.1	1,661	1,591	1,618	46.0	43.6	2,040	40	342	37
MAY-SEPT	—	—	—	—	—	9.4	—	—	—	—	—	—	—	—	—	—	274	—

DATA FOR MONTH TO MATURITY

MATURITY GROUP	EST. MATURITY DATE			*	*									*				
1	FEBRUARY 14					23.5	7.7	38.0	6.9					44.2	319	5	47	33
2	FEBRUARY 19					23.5	7.7	37.8	6.6					43.9	311	5	45	33
3	FEBRUARY 25					23.4	7.6	37.4	6.3					43.6	304	4	44	34
4	MARCH 3					23.2	7.6	36.9	6.2					43.3	298	4	44	34
5	MARCH 8					22.8	7.5	36.4	6.2					43.1	292	4	45	35
6	MARCH 14					22.3	7.5	35.8	6.2					42.9	287	5	46	35
7	MARCH 19					21.7	7.4	35.2	6.3					42.6	282	5	47	36
8	MARCH 25					21.2	7.3	34.6	6.3					42.3	277	5	48	37

* Adjusted for vine sites.

TABLE 86

COROWA, NEW SOUTH WALES. 36°00'S, 146°24'E. ALT 125m.

PERIOD	1 Lowest Recorded Min. (49YRS)	2 No.Mths with <0.3° (30YRS)	3 No.Mths with <2.5° (30YRS)	4 Av.Mthly Lowest Min. (30YRS)	5 Av.Min. Minus Col.4	6 Daily Mean (26YRS)	7 Daily Range /2 (26YRS)	8 Av.Mthly Highest Max. (30YRS)	9 Col.8 Minus Av.Max.	10 Raw	11 Adj.1 for Lat. & Daily Temp Rng	12 Adj.2 for Vine Sites	13 Raw	14 Adj. for Vine Sites	15 Av. Sunshine Hrs	16 Av. Number Rain Days	17 Av. Rainfall mm	18 Av. % R.H. 1500 HRS
						Temperature °C				Summation of Day Degrees above 10° with 19° Cut-Off			Temperature Variability Index					
OCTOBER	-0.6	3	15	2.7	5.6	15.6	7.3	31.0	8.1	174	159		42.9		275	8	52	42
NOVEMBER	-1.7	1	1	5.1	6.3	19.4	8.0	36.1	8.7	270	254		47.0		304	6	39	33
DECEMBER	1.1	0	1	6.9	7.0	22.2	8.3	39.3	8.8	279	279		49.0		340	5	39	28
JANUARY	2.8	0	0	8.9	6.3	23.3	8.1	40.6	9.2	279	279		47.9		340	4	36	27
FEBRUARY	4.4	0	0	8.8	6.3	23.2	8.1	38.8	7.5	252	252		46.2		280	4	37	29
MARCH	3.1	0	0	6.6	6.1	20.5	7.8	35.9	7.6	279	279		44.9		278	5	40	35
APRIL	-0.6	1	10	3.0	5.4	15.5	7.1	29.9	7.3	165	158		41.1		242	6	38	45
OCT-APRIL	—	5	27	—	6.1	20.0	7.8	—	8.2	1,698	1,660		45.6		2,059	38	281	34
MAY-SEPT	—	—	—	—	—	10.0	—	—	—	—	—		—		—	—	255	—

DATA FOR MONTH TO MATURITY

MATURITY GROUP	EST. MATURITY DATE																	
1	FEBRUARY 9					23.4	8.1	40.2	8.7					47.5	326	4	37	28
2	FEBRUARY 15					23.4	8.1	39.8	8.3					47.2	320	4	38	28
3	FEBRUARY 20					23.4	8.1	39.4	7.9					46.9	313	4	39	29
4	FEBRUARY 26					23.3	8.1	38.9	7.6					46.6	307	4	40	29
5	MARCH 3					23.1	8.1	38.4	7.5					46.3	301	4	40	30
6	MARCH 9					22.9	8.0	37.9	7.5					46.0	296	5	40	31
7	MARCH 15					22.5	8.0	37.4	7.6					45.7	290	5	40	32
8	MARCH 20					22.0	7.9	36.9	7.6					45.4	286	5	40	33

TABLE 87
COWRA, NEW SOUTH WALES. 33°51′S, 148°40′E. ALT 298m.

	1	2	3	4	5	6	7	8	9	10	11	12	13	14	15	16	17	18
	Temperature °C									Summation of Day Degrees above 10° with 19° Cut-Off			Temperature Variability Index					
PERIOD	Lowest Recorded Min. (60YRS)	No.Mths with <0.3° (30YRS)	No.Mths with <2.5° (30YRS)	Av.Mthly Lowest Min. (30YRS)	Av.Min. Minus Col.4	Daily Mean (27YRS)	Daily Range /2 (27YRS)	Av.Mthly Highest Max. (30YRS)	Col.8 Minus Av.Max.	Raw	Adj.1 for Lat. & Daily Temp Rng	Adj.2 for Vine Sites	Raw	Adj. for Vine Sites	Av. Sunshine Hrs	Av. Number Rain Days	Av. Rainfall mm	Av. % R.H. 1500 HRS
OCTOBER	−3.3	10	28	0.6	6.9	15.8	8.3	30.9	6.8	180	149	156	46.9	43.9	280	8	56	45
NOVEMBER	−1.1	2	12	3.2	7.8	19.6	8.6	35.5	7.3	270	248	255	49.5	46.5	305	6	48	34
DECEMBER	3.3	0	0	6.4	7.9	22.6	8.3	38.3	7.4	279	279	279	48.5	45.5	332	6	58	34
JANUARY	3.3	0	0	8.5	7.4	24.4	8.5	39.8	6.9	279	279	279	48.3	45.3	331	5	58	34
FEBRUARY	3.9	0	0	8.4	6.8	23.9	8.7	37.6	5.0	252	252	252	46.6	43.6	275	5	44	37
MARCH	−1.1	1	2	5.6	6.9	20.9	8.4	35.3	6.0	279	279	279	46.5	43.5	267	5	51	38
APRIL	−3.3	8	18	1.8	6.5	16.0	7.7	29.9	6.2	180	165	172	43.5	40.5	235	5	46	47
OCT-APRIL	—	21	60	—	7.2	20.5	8.4	—	6.5	1,719	1,651	1,672	47.1	44.1	2,025	40	361	38
MAY-SEPT	—	—	—	—	—	10.0	—	—	—	—	—	—	—	—	—	—	251	—
DATA FOR MONTH TO MATURITY																		
MATURITY GROUP	EST. MATURITY DATE					*	*	*						*				
1	FEBRUARY 9					24.2	7.8	38.2	6.3					44.1	318	5	55	35
2	FEBRUARY 15					24.1	7.8	37.8	5.8					44.0	313	5	53	36
3	FEBRUARY 21					24.0	7.9	37.4	5.4					43.8	307	5	51	36
4	FEBRUARY 26					23.9	7.9	36.9	5.1					43.7	302	5	49	37
5	MARCH 4					23.6	7.9	36.5	5.1					43.6	296	5	49	37
6	MARCH 9					23.3	7.8	36.0	5.3					43.5	290	5	50	37
7	MARCH 15					22.9	7.8	35.6	5.5					43.5	284	5	50	38
8	MARCH 20					22.4	7.7	35.2	5.7					43.5	278	5	50	38

* Adjusted for vine sites.

TABLE 88
FORBES, NEW SOUTH WALES. 33°27′S, 148°05′E. ALT 238m.

	1	2	3	4	5	6	7	8	9	10	11	12	13	14	15	16	17	18
	Temperature °C									Summation of Day Degrees above 10° with 19° Cut-Off			Temperature Variability Index					
PERIOD	Lowest Recorded Min. (64YRS)	No.Mths with <0.3° (30YRS)	No.Mths with <2.5° (30YRS)	Av.Mthly Lowest Min. (30YRS)	Av.Min. Minus Col.4	Daily Mean (29YRS)	Daily Range /2 (29YRS)	Av.Mthly Highest Max. (30YRS)	Col.8 Minus Av.Max.	Raw	Adj.1 for Lat. & Daily Temp Rng	Adj.2 for Vine Sites	Raw	Adj. for Vine Sites	Av. Sunshine Hrs	Av. Number Rain Days	Av. Rainfall mm	Av. % R.H. 1500 HRS (29YRS)
OCTOBER	−1.1	3	17	2.3	7.6	17.3	7.4	32.3	7.6	226	208		44.8		288	7	49	38
NOVEMBER	1.7	0	0	5.1	8.3	21.2	7.8	36.6	7.6	270	270		47.1		315	5	40	31
DECEMBER	5.0	0	0	7.7	8.6	24.0	7.7	38.9	7.2	279	279		46.6		340	5	47	32
JANUARY	5.6	0	0	9.2	8.5	25.5	7.8	40.4	7.1	279	279		46.8		340	5	47	32
FEBRUARY	6.1	0	0	9.7	7.9	25.3	7.7	38.1	5.1	252	252		43.8		282	5	45	34
MARCH	2.8	0	0	6.7	7.9	22.1	7.5	35.6	6.0	279	279		43.9		278	5	44	38
APRIL	−0.2	1	11	2.7	7.5	17.0	6.8	30.6	6.8	210	208		41.5		247	5	40	44
OCT-APRIL	—	4	28	—	8.0	21.8	7.5	—	6.8	1,795	1,775		44.9		2,090	37	312	36
MAY-SEPT	—	—	—	—	—	11.2	—	—	—	—	—		—		—	—	188	—
DATA FOR MONTH TO MATURITY																		
MATURITY GROUP	EST. MATURITY DATE																	
1	FEBRUARY 2					25.5	7.8	40.4	7.0				46.7		388	5	47	32
2	FEBRUARY 8				·	25.6	7.8	40.1	6.5				46.2		334	5	47	32
3	FEBRUARY 13					25.6	7.8	39.7	6.0				45.6		328	5	47	33
4	FEBRUARY 18					25.5	7.7	39.2	5.6				45.0		320	5	48	33
5	FEBRUARY 24					25.4	7.7	38.8	5.3				44.4		312	5	48	34
6	MARCH 2					25.2	7.7	38.3	5.1				43.8		305	5	48	34
7	MARCH 7					24.9	7.7	37.8	5.2				43.9		299	5	47	35
8	MARCH 13					24.6	7.6	37.2	5.4				43.9		294	5	46	36

TABLE 89

GLEN INNES, NEW SOUTH WALES. 29°42'S, 151°43'E. ALT 1072m.

	1	2	3	4	5	6	7	8	9	10	11	12	13	14	15	16	17	18
	Temperature °C									Summation of Day Degrees above 10° with 19° Cut-Off			Temperature Variability Index					
PERIOD	Lowest Recorded Min. (62YRS)	No.Mths with <0.3° (30YRS)	No.Mths with <2.5° (30YRS)	Av.Mthly Lowest Min. (30YRS)	Av.Min. Minus Col.4	Daily Mean (29YRS)	Daily Range /2 (29YRS)	Av.Mthly Highest Max. (30YRS)	Col.8 Minus Av.Max.	Raw	Adj.1 for Lat. & Daily Temp Rng	Adj.2 for Vine Sites	Raw	Adj. for Vine Sites	Av. Sunshine Hrs	Av. Number Rain Days	Av. Rainfall mm	Av. % R.H. 1500 HR
OCTOBER	−4.4	16	26	0.4	6.3	14.1	7.4	27.0	5.5	127	109		41.4		270	9	82	49
NOVEMBER	−1.1	2	13	3.3	6.1	16.8	7.4	29.8	5.6	204	180		41.3		280	9	80	48
DECEMBER	0.8	0	0	5.8	5.6	18.6	7.2	31.2	5.4	267	239		39.8		284	10	104	46
JANUARY	2.2	0	0	8.0	4.6	19.8	7.2	32.3	5.3	279	275		38.7		285	10	110	51
FEBRUARY	3.9	0	0	9.0	3.2	19.1	6.9	30.8	4.8	252	239		35.6		234	9	87	53
MARCH	0.0	0	3	6.1	4.6	17.3	6.6	28.6	4.7	226	224		35.7		236	8	70	52
APRIL	−4.2	8	21	1.2	6.0	13.8	6.6	25.8	5.4	114	117		37.8		228	6	41	48
OCT-APRIL	—	26	63	—	5.2	17.1	7.0	—	5.2	1,469	1,383		38.6		1,817	61	574	50
MAY-SEPT	—	—	—	—	—	8.4	—	—	—	—	—		—		—	—	266	—

DATA FOR MONTH TO MATURITY

MATURITY GROUP	EST. MATURITY DATE					6	7	8	9				13		15	16	17	18
1	MARCH 2					19.1	6.9	30.8	4.8				35.6		256	10	95	53
2	MARCH 7					18.9	6.8	30.5	4.8				35.6		251	10	89	53
3	MARCH 14					18.6	6.8	30.2	4.8				35.6		246	9	83	53
4	MARCH 21					18.1	6.7	29.6	4.7				35.7		241	9	78	52
5	MARCH 29					17.5	6.6	28.8	4.7				35.7		237	8	72	52
6	APRIL 6					16.7	6.6	28.1	4.8				35.9		233	8	65	51
7	APRIL 19					15.3	6.6	27.0	5.1				36.7		230	7	55	50
8	MAY 7					13.3	6.6	25.4	5.4				38.0		228	6	40	48

TABLE 90

GOULBURN, NEW SOUTH WALES. 34°45'S, 149°52'E. ALT. 639m.

	1	2	3	4	5	6	7	8	9	10	11	12	13	14	15	16	17	18
	Temperature °C									Summation of Day Degrees above 10° with 19° Cut-Off			Temperature Variability Index					
PERIOD	Lowest Recorded Min. (58YRS)	No.Mths with <0.3° (30YRS)	No.Mths with <2.5° (30YRS)	Av.Mthly Lowest Min. (30YRS)	Av.Min. Minus Col.4	Daily Mean (30YRS)	Daily Range /2 (30YRS)	Av.Mthly Highest Max. (30YRS)	Col.8 Minus Av.Max.	Raw	Adj.1 for Lat. & Daily Temp Rng	Adj.2 for Vine Sites	Raw	Adj. for Vine Sites	Av. Sunshine Hrs	Av. Number Rain Days	Av. Rainfall mm	Av. % R.H. 1500 HR (32YRS)
OCTOBER	−1.7	5	23	1.7	5.6	14.1	6.8	28.4	7.5	127	121		40.3		264	9	69	47
NOVEMBER	−0.6	0	7	4.1	5.9	16.9	6.9	32.6	8.8	207	195		42.3		270	8	63	43
DECEMBER	0.6	0	0	6.4	6.0	19.4	7.0	35.7	9.3	279	274		43.3		285	8	63	44
JANUARY	3.9	0	0	7.9	5.5	20.4	7.0	36.8	9.4	279	279		42.9		285	8	65	43
FEBRUARY	2.5	0	0	8.1	5.4	20.4	6.9	34.8	7.5	252	252		40.5		238	7	61	45
MARCH	0.0	0	2	5.7	6.0	18.1	6.4	32.9	8.4	251	250		40.0		229	7	56	50
APRIL	−4.0	5	23	1.4	6.6	13.8	5.8	28.1	8.5	114	116		38.3		205	8	51	58
OCT-APRIL	—	10	55	—	5.9	17.6	6.7	—	8.5	1,509	1,487		41.1		1,776	55	428	47
MAY-SEPT	—	—	—	—	—	8.6	—	—	—	—	—		—		—	—	284	—

DATA FOR MONTH TO MATURITY

MATURITY GROUP	EST. MATURITY DATE					6	7	8	9				13		15	16	17	18
1	FEBRUARY 21					20.6	6.9	35.2	7.9				41.0		267	8	66	44
2	FEBRUARY 26					20.5	6.9	34.9	7.7				40.6		262	8	67	45
3	MARCH 4					20.3	6.8	34.5	7.6				40.4		256	8	66	45
4	MARCH 9					20.0	6.7	34.2	7.8				40.3		249	8	64	46
5	MARCH 15					19.6	6.6	33.9	8.0				40.2		242	7	61	47
6	MARCH 22					19.1	6.5	33.5	8.2				40.1		236	7	59	48
7	MARCH 29					18.4	6.4	33.1	8.4				40.0		231	7	57	50
8	APRIL 6					17.2	6.3	32.5	8.4				39.8		226	7	55	52

TABLE 91
GRIFFITH, NEW SOUTH WALES. 34°17'S, 146°03'E. ALT 128m.

	1	2	3	4	5	6	7	8	9	10	11	12	13	14	15	16	17	18
	Temperature °C									Summation of Day Degrees above 10° with 19° Cut-Off			Temperature Variability Index					
PERIOD	Lowest Recorded Min. (55YRS)	No.Mths with <0.3° (30YRS)	No.Mths with <2.5° (30YRS)	Av.Mthly Lowest Min. (30YRS)	Av.Min. Minus Col.4	Daily Mean (22YRS)	Daily Range /2 (22YRS)	Av.Mthly Highest Max. (30YRS)	Col.8 Minus Av.Max.	Raw	Adj.1 for Lat. & Daily Temp Rng	Adj.2 for Vine Sites	Raw	Adj. for Vine Sites	Av. Sunshine Hrs	Av. Number Rain Days	Av. Rainfall mm	Av. % R.H. 1500 HR
OCTOBER	−1.0	2	16	2.2	7.3	16.6	7.1	32.6	8.9	205	192		44.6		287	7	42	35
NOVEMBER	1.7	0	0	5.0	7.6	20.1	7.5	36.9	9.3	270	270		46.9		320	5	31	28
DECEMBER	4.5	0	0	7.2	8.2	22.8	7.4	39.1	8.9	279	279		46.7		350	5	32	28
JANUARY	6.2	0	0	8.9	7.7	24.2	7.6	40.1	8.3	279	279		46.4		360	4	28	28
FEBRUARY	5.7	0	0	8.8	8.0	24.3	7.5	38.6	6.8	252	252		44.8		293	4	30	28
MARCH	2.8	0	0	6.4	7.8	21.5	7.3	35.0	6.2	279	279		43.2		285	5	35	34
APRIL	−0.9	3	14	2.3	7.7	16.4	6.4	30.2	7.4	192	195		40.7		245	6	34	39
OCT-APRIL	—	5	30	—	7.6	20.8	7.3	—	8.0	1,756	1,746		44.8		2,140	36	232	31
MAY-SEPT	—	—	—	—	—	10.9	—	—	—	—	—		—		—	—	179	—

DATA FOR MONTH TO MATURITY

MATURITY GROUP	EST. MATURITY DATE																	
1	FEBRUARY 4					24.3	7.6	40.0	8.1					46.2	343	4	28	28
2	FEBRUARY 9					24.4	7.6	39.7	7.9					45.9	338	4	28	28
3	FEBRUARY 14					24.5	7.5	39.3	7.6					45.6	332	4	29	28
4	FEBRUARY 19					24.5	7.5	39.0	7.4					45.3	325	4	29	28
5	FEBRUARY 25					24.4	7.5	38.7	7.1					45.0	318	4	30	28
6	MARCH 4					24.2	7.5	38.3	6.9					44.7	311	4	30	29
7	MARCH 9					23.9	7.5	37.7	6.7					44.3	304	4	31	30
8	MARCH 15					23.5	7.4	37.0	6.5					44.0	296	4	32	31

TABLE 92
INVERELL, NEW SOUTH WALES. 29°47'S, 151°07'E. ALT 604m.

	1	2	3	4	5	6	7	8	9	10	11	12	13	14	15	16	17	18
	Temperature °C									Summation of Day Degrees above 10° with 19° Cut-Off			Temperature Variability Index					
PERIOD	Lowest Recorded Min. (64YRS)	No.Mths with <0.3° (30YRS)	No.Mths with <2.5° (30YRS)	Av.Mthly Lowest Min. (30YRS)	Av.Min. Minus Col.4	Daily Mean (30YRS)	Daily Range /2 (30YRS)	Av.Mthly Highest Max. (30YRS)	Col.8 Minus Av.Max.	Raw	Adj.1 for Lat. & Daily Temp Rng	Adj.2 for Vine Sites	Raw	Adj. for Vine Sites	Av. Sunshine Hrs	Av. Number Rain Days	Av. Rainfall mm	Av. % R.H. 1500 HR
OCTOBER	−5.7	17	26	0.2	6.8	15.8	8.8	30.6	6.0	180	139	149	48.0	45.0	280	8	69	36
NOVEMBER	−0.2	0	12	3.4	7.5	19.3	8.4	33.8	6.1	270	237	247	47.2	44.2	291	8	73	33
DECEMBER	1.7	0	1	6.4	7.1	21.6	8.1	35.3	5.6	279	279	279	45.1	42.1	302	8	91	36
JANUARY	1.7	0	0	8.9	5.7	22.7	8.1	36.5	5.7	279	279	279	43.8	40.8	304	8	98	38
FEBRUARY	4.1	0	0	9.4	4.4	22.1	8.3	34.9	4.5	252	252	252	42.1	39.1	255	7	80	39
MARCH	0.6	0	2	6.2	5.5	19.9	8.2	32.8	4.7	279	278	279	43.0	40.0	254	7	65	40
APRIL	−4.1	8	19	1.4	6.0	15.8	8.4	29.6	5.4	174	150	160	45.0	42.0	246	5	43	38
OCT-APRIL	—	25	60	—	6.1	19.6	8.3	—	5.4	1,713	1,614	1,645	44.9	41.9	1,932	51	519	37
MAY-SEPT	—	—	—	—	—	9.8	—	—	—	—	—		—		—	—	246	—

DATA FOR MONTH TO MATURITY

MATURITY GROUP	EST. MATURITY DATE					*	*							*				
1	FEBRUARY 12					22.6	7.4	35.1	5.1					40.1	293	8	93	38
2	FEBRUARY 17					22.5	7.5	34.8	4.9					39.9	288	8	91	39
3	FEBRUARY 22					22.3	7.5	34.5	4.7					39.5	283	8	89	39
4	FEBRUARY 28					22.1	7.5	34.2	4.5					39.1	278	7	87	39
5	MARCH 5					21.8	7.5	33.8	4.5					39.2	273	7	84	39
6	MARCH 10					21.5	7.4	33.4	4.6					39.3	268	7	81	39
7	MARCH 16					21.1	7.4	33.0	4.6					39.5	264	7	77	40
8	MARCH 22					20.7	7.4	32.6	4.6					39.7	260	7	72	40

* Adjusted for vine sites.

TABLE 93
JERRY'S PLAINS, NEW SOUTH WALES. 32°30'S, 150°55'E. ALT 82m. (Middle Hunter Valley)

	1	2	3	4	5	6	7	8	9	10	11	12	13	14	15	16	17	18
	Temperature °C									Summation of Day Degrees above 10° with 19° Cut-Off			Temperature Variability Index					
PERIOD	Lowest Recorded Min. (19YRS)	No.Mths with <0.3° (19YRS)	No.Mths with <2.5° (19YRS)	Av.Mthly Lowest Min. (19YRS)	Av.Min. Minus Col.4	Daily Mean (18YRS)	Daily Range /2 (18YRS)	Av.Mthly Highest Max. (19YRS)	Col.8 Minus Av.Max.	Raw	Adj.1 for Lat. & Daily Temp Rng	Adj.2 for Vine Sites	Raw	Adj. for Vine Sites	Av. Sunshine Hrs	Av. Number Rain Days	Av. Rainfall mm	Av. % R.H. 1500 HRS
OCTOBER	1.7	0	3	4.0	6.8	17.9	7.1	33.3	8.3	245	230		43.5		256	7	51	40
NOVEMBER	4.4	0	0	7.3	6.1	20.6	7.2	37.0	9.2	270	270		44.1		264	7	57	34
DECEMBER	5.0	0	0	9.2	6.6	22.6	6.8	38.2	8.8	279	279		42.6		270	7	69	40
JANUARY	8.9	0	0	11.6	5.6	23.6	6.4	38.9	8.9	279	279		40.1		284	7	78	44
FEBRUARY	8.9	0	0	12.3	5.0	23.4	6.1	37.5	8.0	252	252		37.4		235	7	67	47
MARCH	6.1	0	0	8.7	6.5	21.6	6.4	35.0	7.0	279	279		39.1		234	7	56	45
APRIL	2.7	0	0	4.9	6.6	18.3	6.8	31.1	6.0	247	249		39.8		200	6	45	43
OCT-APRIL	—	0	3	—	6.2	21.1	6.7	—	8.0	1,851	1,838		40.9		1,743	48	423	42
MAY-SEPT	—	—	—	—	—	12.4	—	—	—	—	—		—		—	—	212	—

					DATA FOR MONTH TO MATURITY													
MATURITY GROUP	EST. MATURITY DATE																	
1	JANUARY 31					23.6	6.4	38.9	8.9				40.1		284	7	78	44
2	FEBRUARY 6					23.7	6.3	38.8	8.8				39.6		280	7	78	44
3	FEBRUARY 11					23.7	6.2	38.6	8.6				39.0		274	7	77	45
4	FEBRUARY 17					23.7	6.2	38.3	8.4				38.4		268	7	76	46
5	FEBRUARY 22					23.6	6.1	37.9	8.2				37.8		262	8	75	47
6	FEBRUARY 28					23.4	6.1	37.5	8.0				37.4		257	8	73	47
7	MARCH 6					23.2	6.2	37.1	7.8				37.7		252	7	70	47
8	MARCH 11					22.9	6.2	36.7	7.6				38.1		248	7	67	46

TABLE 94
JUNEE, NEW SOUTH WALES. 34°53'S, 147°34'E. ALT. 301m.

	1	2	3	4	5	6	7	8	9	10	11	12	13	14	15	16	17	18
	Temperature °C									Summation of Day Degrees above 10° with 19° Cut-Off			Temperature Variability Index					
PERIOD	Lowest Recorded Min. (63YRS)	No.Mths with <0.3° (30YRS)	No.Mths with <2.5° (30YRS)	Av.Mthly Lowest Min. (30YRS)	Av.Min. Minus Col.4	Daily Mean (27YRS)	Daily Range /2 (27YRS)	Av.Mthly Highest Max. (30YRS)	Col.8 Minus Av.Max.	Raw	Adj.1 for Lat. & Daily Temp Rng	Adj.2 for Vine Sites	Raw	Adj. for Vine Sites	Av. Sunshine Hrs	Av. Number Rain Days	Av. Rainfall mm	Av. % R.H. 1500 HRS
OCTOBER	-2.8	12	27	0.9	6.1	14.9	7.9	31.3	8.5	152	128	120	46.2	43.2	276	8	51	
NOVEMBER	0.0	2	11	3.3	7.0	19.0	8.7	36.8	9.1	270	230	222	50.9	47.9	303	6	44	
DECEMBER	3.3	0	0	6.1	7.5	22.2	8.6	39.3	8.5	279	279	279	50.4	47.4	335	5	41	
JANUARY	4.4	0	0	8.4	6.8	23.5	8.3	41.0	9.2	279	279	279	49.2	46.2	348	4	39	
FEBRUARY	2.8	0	0	8.2	7.4	23.7	8.1	39.2	7.4	252	252	252	47.2	44.2	281	5	38	
MARCH	0.6	0	1	6.3	6.5	20.5	7.7	35.7	7.5	279	279	279	44.8	41.8	276	4	42	
APRIL	-1.7	4	21	2.1	6.2	15.5	7.2	30.1	7.4	165	157	148	42.4	39.4	237	6	43	
OCT-APRIL	—	18	60	—	6.8	19.9	8.1	—	8.2	1,676	1,604	1,579	47.3	44.3	2,056	38	298	
MAY-SEPT	—	—	—	—	—	9.8	—	—	—	—	—	—	—	—	—	—	235	

					DATA FOR MONTH TO MATURITY													
MATURITY GROUP	EST. MATURITY DATE					*	*	*						*				
1	FEBRUARY 17					23.1	7.4	38.9	8.1					45.0	320	5	40	
2	FEBRUARY 23					23.1	7.4	38.4	7.7					44.6	312	5	41	
3	FEBRUARY 28					23.0	7.4	37.8	7.4					44.2	306	5	41	
4	MARCH 6					22.9	7.3	37.2	7.4					43.8	300	5	41	
5	MARCH 11					22.6	7.3	36.6	7.4					43.4	294	5	41	
6	MARCH 17					22.0	7.2	36.0	7.5					43.0	288	4	42	
7	MARCH 23					21.1	7.1	35.3	7.5					42.5	283	4	42	
8	MARCH 29					20.1	7.0	34.6	7.5					42.0	278	4	42	

* Adjusted for vine sites.

TABLE 95

LEETON, NEW SOUTH WALES. 34°30'S, 146°34'E. ALT 142m.

	1	2	3	4	5	6	7	8	9	10	11	12	13	14	15	16	17	18
	Temperature °C									Summation of Day Degrees above 10° with 19° Cut-Off			Temperature Variability Index					
PERIOD	Lowest Recorded Min. (67YRS)	No.Mths with <0.3° (30YRS)	No.Mths with <2.5° (30YRS)	Av.Mthly Lowest Min. (30YRS)	Av.Min. Minus Col.4	Daily Mean (28YRS)	Daily Range /2 (28YRS)	Av.Mthly Highest Max. (30YRS)	Col.8 Minus Av.Max.	Raw	Adj.1 for Lat. & Daily Temp Rng	Adj.2 for Vine Sites	Raw	Adj. for Vine Sites	Av. Sunshine Hrs	Av. Number Rain Days	Av. Rainfall mm	Av. % R.H. 1500 HRS
OCTOBER	1.1	0	9	3.2	6.6	16.5	6.7	32.1	8.9	202	195		42.3		282	8	43	38
NOVEMBER	1.9	0	0	6.0	7.1	20.2	7.1	37.2	9.9	270	270		45.4		315	5	34	30
DECEMBER	4.4	0	0	8.3	7.7	23.1	7.1	39.4	9.2	279	279		45.3		346	5	31	28
JANUARY	6.7	0	0	10.4	7.0	24.5	7.1	40.7	9.1	279	279		44.5		356	4	31	28
FEBRUARY	5.1	0	0	10.1	7.4	24.5	7.0	38.9	7.4	252	252		42.8		290	4	30	32
MARCH	4.4	0	0	7.4	7.6	21.6	6.6	35.4	7.2	279	279		41.2		285	4	34	34
APRIL	0.6	0	7	3.6	7.1	16.6	5.9	30.2	7.7	198	201		38.4		245	6	37	41
OCT-APRIL	—	0	16	—	7.2	21.0	6.8	—	8.5	1,759	1,755		42.8		2,119	36	240	33
MAY-SEPT	—	—	—	—	—	10.8	—	—	—	—	—		—		—	—	192	—

DATA FOR MONTH TO MATURITY

MATURITY GROUP	EST. MATURITY DATE																	
1	FEBRUARY 4					24.5	7.1	40.5	8.9				44.2		351	4	31	28
2	FEBRUARY 9					24.6	7.1	40.3	8.6				43.9		344	4	30	29
3	FEBRUARY 15					24.6	7.1	40.0	8.2				43.6		336	4	30	30
4	FEBRUARY 20					24.6	7.0	39.7	7.8				43.3		327	4	30	31
5	FEBRUARY 26					24.5	7.0	39.4	7.5				43.0		319	4	30	31
6	MARCH 3					24.4	7.0	39.0	7.4				42.7		312	4	30	32
7	MARCH 9					24.2	6.9	38.4	7.4				42.4		306	4	31	32
8	MARCH 15					23.8	6.8	37.3	7.3				42.0		300	4	32	33

TABLE 96

LITHGOW, NEW SOUTH WALES. 33°28'S, 150°09'E. ALT. 917m.

	1	2	3	4	5	6	7	8	9	10	11	12	13	14	15	16	17	18
	Temperature °C									Summation of Day Degrees above 10° with 19° Cut-Off			Temperature Variability Index					
PERIOD	Lowest Recorded Min. (50YRS)	No.Mths with <0.3° (30YRS)	No.Mths with <2.5° (30YRS)	Av.Mthly Lowest Min. (30YRS)	Av.Min. Minus Col.4	Daily Mean (27YRS)	Daily Range /2 (27YRS)	Av.Mthly Highest Max. (30YRS)	Col.8 Minus Av.Max.	Raw	Adj.1 for Lat. & Daily Temp Rng	Adj.2 for Vine Sites	Raw	Adj. for Vine Sites	Av. Sunshine Hrs	Av. Number Rain Days	Av. Rainfall mm	Av. % R.H. 1500 HRS
OCTOBER	-5.6	16	28	0.2	4.8	11.9	6.9	26.5	7.7	59	52		40.1		260	12	87	53
NOVEMBER	-4.3	7	17	2.0	5.4	14.7	7.3	30.2	8.2	141	124		42.8		272	14	73	47
DECEMBER	-2.4	1	6	4.1	5.7	16.9	7.1	33.1	9.1	214	196		43.2		285	12	94	50
JANUARY	0.1	0	2	6.0	5.0	18.3	7.3	34.1	8.5	257	235		42.7		278	15	141	55
FEBRUARY	-1.0	0	3	6.3	4.9	18.1	6.9	32.4	7.4	227	215		39.9		230	18	120	56
MARCH	-2.9	3	9	4.1	5.0	15.7	6.6	29.9	7.6	177	174		39.0		222	13	72	55
APRIL	-4.9	12	24	0.6	5.3	11.8	5.9	24.9	7.2	54	55		36.1		197	8	44	55
OCT-APRIL	—	39	89	—	5.2	15.3	6.9	—	8.0	1,129	1,051		40.5		1,744	92	631	53
MAY-SEPT	—	—	—	—	—	6.9	—	—	—	—	—		—		—	—	269	—

DATA FOR MONTH TO MATURITY

MATURITY GROUP	EST. MATURITY DATE																	
1	APRIL 30					11.8	5.9	24.9	7.2				36.1		197	8	44	55
2	—					—	—	—	—				—		—	—	—	—
3	—					—	—	—	—				—		—	—	—	—
4	—					—	—	—	—				—		—	—	—	—
5	—					—	—	—	—				—		—	—	—	—
6	—					—	—	—	—				—		—	—	—	—
7	—					—	—	—	—				—		—	—	—	—
8	—					—	—	—	—				—		—	—	—	—

TABLE 97

MOSS VALE, NEW SOUTH WALES. 34°33'S, 150°22'E. ALT. 672m.

	1	2	3	4	5	6	7	8	9	10	11	12	13	14	15	16	17	18
	Temperature °C									Summation of Day Degrees above 10° with 19° Cut-Off			Temperature Variability Index					
PERIOD	Lowest Recorded Min. (62YRS)	No.Mths with <0.3° (30YRS)	No.Mths with <2.5° (30YRS)	Av.Mthly Lowest Min. (30YRS)	Av.Min. Minus Col.4	Daily Mean (29YRS)	Daily Range /2 (29YRS)	Av.Mthly Highest Max. (30YRS)	Col.8 Minus Av.Max.	Raw	Adj.1 for Lat. & Daily Temp Rng	Adj.2 for Vine Sites	Raw	Adj. for Vine Sites	Av. Sunshine Hrs	Av. Number Rain Days	Av. Rainfall mm	Av. % R.H. 1500 HRS
OCTOBER	−3.9	18	28	−0.1	7.0	13.7	6.8	27.4	6.9	115	108		41.1		258	10	77	52
NOVEMBER	−2.2	5	17	2.1	7.0	16.3	7.2	31.5	8.0	189	173		43.8		260	10	71	49
DECEMBER	0.0	0	5	4.2	7.2	18.3	6.9	33.9	8.7	257	242		43.5		277	10	83	51
JANUARY	1.2	0	2	6.3	6.1	19.2	6.8	35.5	9.5	279	276		42.8		279	11	96	52
FEBRUARY	1.9	0	1	6.4	6.4	19.4	6.6	34.0	8.0	252	252		40.8		226	11	95	54
MARCH	−1.1	2	6	3.8	6.9	17.1	6.4	30.8	7.3	220	219		39.8		226	11	90	56
APRIL	−3.3	10	27	0.7	6.8	13.4	5.9	26.0	6.7	102	104		37.1		208	10	84	56
OCT-APRIL	—	35	86	—	6.8	16.8	6.7	—	7.9	1,414	1,374		41.3		1,734	73	596	53
MAY-SEPT	—	—	—	—	—	8.6	—	—	—	—	—		—		—	—	394	—
DATA FOR MONTH TO MATURITY																		
MATURITY GROUP	EST. MATURITY DATE																	
1	FEBRUARY 28					19.4	6.6	34.0	8.0				40.8		247	12	104	54
2	MARCH 6					19.1	6.6	33.6	7.8				40.5		242	12	101	54
3	MARCH 13					18.7	6.5	33.0	7.7				40.3		237	12	98	55
4	MARCH 20					18.2	6.5	32.2	7.5				40.1		233	11	95	55
5	MARCH 28					17.6	6.4	31.2	7.4				39.9		229	11	92	56
6	APRIL 7					16.2	6.3	29.8	7.2				39.4		224	11	89	56
7	APRIL 20					14.6	6.1	27.6	6.9				38.1		217	10	86	56
8	—					—	—	—	—				—		—	—	—	—

TABLE 98

MUDGEE, NEW SOUTH WALES. 32°36'S, 149°36'E. ALT. 454m.

	1	2	3	4	5	6	7	8	9	10	11	12	13	14	15	16	17	18
	Temperature °C									Summation of Day Degrees above 10° with 19° Cut-Off			Temperature Variability Index					
PERIOD	Lowest Recorded Min. (62YRS)	No.Mths with <0.3° (30YRS)	No.Mths with <2.5° (30YRS)	Av.Mthly Lowest Min. (30YRS)	Av.Min. Minus Col.4	Daily Mean (28YRS)	Daily Range /2 (28YRS)	Av.Mthly Highest Max. (30YRS)	Col.8 Minus Av.Max.	Raw	Adj.1 for Lat. & Daily Temp Rng	Adj.2 for Vine Sites	Raw	Adj. for Vine Sites	Av. Sunshine Hrs	Av. Number Rain Days	Av. Rainfall mm	Av. % R.H. 1500 HRS
OCTOBER	−5.6	9	22	1.2	6.1	15.6	8.3	30.8	6.8	174	142	140	46.2	43.2	280	7	59	43
NOVEMBER	−1.1	1	7	4.0	6.6	19.1	8.5	35.1	7.5	270	233	231	48.1	45.1	288	6	59	39
DECEMBER	0.6	0	0	7.3	6.4	21.8	8.1	37.3	7.4	279	279	279	46.2	43.2	315	6	65	38
JANUARY	3.3	0	0	9.7	5.5	23.3	8.1	38.2	6.8	279	279	279	44.7	41.7	310	6	66	42
FEBRUARY	3.9	0	0	10.3	4.8	23.1	8.0	36.4	5.4	252	252	252	42.1	39.1	260	6	64	43
MARCH	0.6	0	2	6.7	5.9	20.4	7.8	33.9	5.8	279	279	279	42.8	39.8	260	5	48	42
APRIL	−2.2	4	16	2.5	5.7	15.7	7.5	29.2	5.9	171	159	157	41.7	38.7	230	5	44	45
OCT-APRIL	—	14	47	—	5.9	19.9	8.0	—	6.5	1,704	1,623	1,617	44.6	41.6	1,943	41	405	42
MAY-SEPT	—	—	—	—	—	9.8	—	—	—	—	—	—	—	—	—	—	262	—
DATA FOR MONTH TO MATURITY																		
MATURITY GROUP	EST. MATURITY DATE					*	*	*					*					
1	FEBRUARY 15					22.9	7.3	36.0	5.7					40.0	299	6	68	43
2	FEBRUARY 20					22.9	7.3	35.8	5.5					39.6	294	6	69	43
3	FEBRUARY 25					22.8	7.2	35.5	5.4					39.2	289	6	70	43
4	MARCH 3					22.5	7.2	35.2	5.5					39.1	284	6	68	43
5	MARCH 8					22.2	7.2	34.9	5.6					39.2	279	6	65	43
6	MARCH 13					21.8	7.1	34.5	5.6					39.3	274	5	62	43
7	MARCH 18					21.3	7.1	34.0	5.7					39.5	269	5	57	42
8	MARCH 24					20.6	7.0	33.5	5.8					39.7	264	5	52	42

* Adjusted for vine sites.

TABLE 99

MURRURUNDI, NEW SOUTH WALES. 31°45′S, 150°51′E. ALT. 472m.

	1	2	3	4	5	6	7	8	9	10	11	12	13	14	15	16	17	18
					Temperature °C					Summation of Day Degrees above 10° with 19° Cut-Off			Temperature Variability Index					
PERIOD	Lowest Recorded Min. (57YRS)	No.Mths with <0.3° (30YRS)	No.Mths with <2.5° (30YRS)	Av.Mthly Lowest Min. (30YRS)	Av.Min. Minus Col.4	Daily Mean (21YRS)	Daily Range /2 (21YRS)	Av.Mthly Highest Max. (30YRS)	Col.8 Minus Av.Max.	Raw	Adj.1 for Lat. & Daily Temp Rng	Adj.2 for Vine Sites	Raw	Adj. for Vine Sites	Av. Sunshine Hrs	Av. Number Rain Days	Av. Rainfall mm	Av. % R.H. 1500 HRS
OCTOBER	−1.1	9	27	1.1	8.1	16.9	7.7	30.6	6.0	213	191	186	44.9	41.9	276	8	72	42
NOVEMBER	0.0	0	7	3.9	7.8	19.7	8.0	34.7	7.0	270	257	252	46.8	43.8	286	7	68	41
DECEMBER	2.2	0	0	6.2	7.8	22.1	8.1	36.4	6.2	279	279	279	46.4	43.4	300	7	86	47
JANUARY	2.9	0	0	8.5	7.5	23.8	7.8	37.6	6.0	279	279	279	44.7	41.7	302	7	88	47
FEBRUARY	2.8	0	0	8.5	6.4	22.5	7.6	36.2	6.1	252	252	252	42.9	39.9	247	6	76	51
MARCH	0.6	0	4	5.8	7.1	20.5	7.6	33.8	5.7	279	279	279	43.2	40.2	247	5	59	46
APRIL	−1.1	6	20	1.8	7.3	16.2	7.1	29.8	6.5	187	181	173	42.2	39.2	224	6	54	46
OCT-APRIL	—	15	58	—	7.4	20.3	7.7	—	6.2	1,759	1,718	1,700	44.4	41.4	1,882	46	503	46
MAY-SEPT	—	—	—	—	—	10.6	—	—	—						—	—	304	—
DATA FOR MONTH TO MATURITY																		
MATURITY GROUP	EST. MATURITY DATE					*	*	*						*				
1	FEBRUARY 7					23.0	7.0	36.0	6.0					41.2	293	7	86	48
2	FEBRUARY 12					22.9	7.0	35.9	6.0					40.7	287	7	86	49
3	FEBRUARY 18					22.7	6.9	35.7	6.1					40.3	281	7	86	49
4	FEBRUARY 23					22.4	6.9	35.4	6.1					40.0	275	7	85	50
5	MARCH 1					22.1	6.9	35.0	6.1					39.9	269	7	83	51
6	MARCH 6					21.7	6.8	34.6	6.0					40.0	264	6	80	50
7	MARCH 11					21.3	6.8	34.1	6.0					40.1	259	6	76	49
8	MARCH 17					20.9	6.8	33.6	5.9					40.1	254	5	71	48

* Adjusted for vine sites.

TABLE 100

NARRABRI, NEW SOUTH WALES. 30°19′S, 149°47′E. ALT. 212m.

	1	2	3	4	5	6	7	8	9	10	11	12	13	14	15	16	17	18
					Temperature °C					Summation of Day Degrees above 10° with 19° Cut-Off			Temperature Variability Index					
PERIOD	Lowest Recorded Min. (63YRS)	No.Mths with <0.3° (30YRS)	No.Mths with <2.5° (30YRS)	Av.Mthly Lowest Min. (30YRS)	Av.Min. Minus Col.4	Daily Mean (29YRS)	Daily Range /2 (29YRS)	Av.Mthly Highest Max. (30YRS)	Col.8 Minus Av.Max.	Raw	Adj.1 for Lat. & Daily Temp Rng	Adj.2 for Vine Sites	Raw	Adj. for Vine Sites	Av. Sunshine Hrs	Av. Number Rain Days	Av. Rainfall mm	Av. % R.H. 1500 HRS
OCTOBER	−0.6	0	7	4.4	7.4	19.7	7.9	34.0	6.4	279	271		45.4		288	5	53	
NOVEMBER	3.3	0	0	7.5	8.0	23.4	7.9	38.9	7.6	270	270		47.2		307	5	63	
DECEMBER	5.0	0	0	10.6	7.2	25.5	7.7	39.4	6.2	279	279		44.2		318	6	66	
JANUARY	4.4	0	0	13.2	6.0	26.9	7.7	40.3	5.7	279	279		42.5		313	6	79	
FEBRUARY	4.4	0	0	12.6	6.3	26.4	7.5	38.4	4.5	252	252		40.8		276	5	71	
MARCH	4.4	0	0	9.7	6.6	23.8	7.5	35.6	4.3	279	279		40.9		275	4	61	
APRIL	−1.2	0	2	5.3	6.5	19.2	7.4	31.8	5.2	270	270		41.3		253	4	39	
OCT-APRIL	—	0	9	—	6.9	23.6	7.7	—	5.7	1,908	1,900		43.2		2,030	35	432	
MAY-SEPT	—	—	—	—	—	12.8	—	—	—				—		—	—	230	
DATA FOR MONTH TO MATURITY																		
MATURITY GROUP	EST. MATURITY DATE																	
1	JANUARY 26					26.9	7.7	40.2	5.8				42.8		314	6	77	
2	JANUARY 31					26.9	7.7	40.3	5.7				42.5		313	6	79	
3	FEBRUARY 5					26.9	7.7	40.0	5.5				42.2		311	6	79	
4	FEBRUARY 12					26.8	7.6	39.6	5.3				41.8		309	6	78	
5	FEBRUARY 17					26.7	7.6	39.2	5.0				41.4		307	6	78	
6	FEBRUARY 22					26.6	7.5	38.8	4.7				41.1		304	5	77	
7	FEBRUARY 28					26.4	7.5	38.4	4.5				40.8		301	5	76	
8	MARCH 6					26.2	7.5	37.9	4.5				40.8		297	5	74	

TABLE 101

PARRAMATTA, NEW SOUTH WALES. 33°49′S, 151°00′E. ALT. 15m.

	1	2	3	4	5	6	7	8	9	10	11	12	13	14	15	16	17	18
	Temperature °C									Summation of Day Degrees above 10° with 19° Cut-Off			Temperature Variability Index					
PERIOD	Lowest Recorded Min. (49YRS)	No.Mths with <0.3° (30YRS)	No.Mths with <2.5° (30YRS)	Av.Mthly Lowest Min. (30YRS)	Av.Min. Minus Col.4	Daily Mean (24YRS)	Daily Range /2 (24YRS)	Av.Mthly Highest Max. (30YRS)	Col.8 Minus Av.Max.	Raw	Adj.1 for Lat. & Daily Temp Rng	Adj.2 for Vine Sites	Raw	Adj. for Vine Sites	Av. Sunshine Hrs	Av. Number Rain Days	Av. Rainfall mm	Av. % R.H. 1500 HRS
OCTOBER	2.2	0	1	5.0	5.9	17.5	6.6	32.5	8.4	233	227		40.7		240	9	63	53
NOVEMBER	4.3	0	0	7.8	5.6	19.7	6.3	36.1	10.1	270	270		40.9		242	9	63	57
DECEMBER	7.2	0	0	10.0	5.8	21.8	6.0	37.5	9.7	279	279		39.5		245	9	71	59
JANUARY	9.0	0	0	11.4	5.3	22.6	5.9	38.4	9.9	279	279		38.8		240	9	92	56
FEBRUARY	7.2	0	0	12.0	4.4	22.2	5.8	36.3	8.3	252	252		35.9		217	10	97	61
MARCH	4.9	0	0	9.8	4.8	20.6	6.0	34.5	7.9	279	279		36.7		215	10	99	61
APRIL	2.2	0	0	6.0	5.5	17.6	6.1	30.1	6.4	228	232		36.3		193	9	93	61
OCT-APRIL	—	0	1	—	5.3	20.3	6.1	—	8.7	1,820	1,818		38.4		1,592	65	578	59
MAY-SEPT	—	—	—	—	—	12.8	—	—	—	—	—		—		—	—	352	—

DATA FOR MONTH TO MATURITY

MATURITY GROUP	EST. MATURITY DATE																	
1	JANUARY 31					22.6	5.9	38.4	9.9					38.8	240	9	92	60
2	FEBRUARY 5					22.6	5.9	37.9	9.4					38.2	240	9	95	60
3	FEBRUARY 11					22.5	5.9	37.4	9.0					37.4	239	10	98	60
4	FEBRUARY 16					22.5	5.8	37.0	8.7					36.6	239	10	101	61
5	FEBRUARY 22					22.4	5.8	36.6	8.5					36.2	238	11	104	61
6	FEBRUARY 28					22.2	5.8	36.3	8.3					35.9	237	11	106	61
7	MARCH 5					22.0	5.8	36.0	8.2					36.0	234	11	105	61
8	MARCH 11					21.8	5.9	35.7	8.1					36.2	230	10	103	61

TABLE 102

PICTON, NEW SOUTH WALES. 34°11′S, 150°37′E. ALT. 171m.

	1	2	3	4	5	6	7	8	9	10	11	12	13	14	15	16	17	18
	Temperature °C									Summation of Day Degrees above 10° with 19° Cut-Off			Temperature Variability Index					
PERIOD	Lowest Recorded Min. (62YRS)	No.Mths with <0.3° (30YRS)	No.Mths with <2.5° (30YRS)	Av.Mthly Lowest Min. (30YRS)	Av.Min. Minus Col.4	Daily Mean (30YRS)	Daily Range /2 (30YRS)	Av.Mthly Highest Max. (30YRS)	Col.8 Minus Av.Max.	Raw	Adj.1 for Lat. & Daily Temp Rng	Adj.2 for Vine Sites	Raw	Adj. for Vine Sites	Av. Sunshine Hrs	Av. Number Rain Days	Av. Rainfall mm	Av. % R.H. 1500 HRS
OCTOBER	−1.2	8	23	1.8	7.1	16.7	7.8	32.6	8.1	208	184	176	46.4	43.4	253	9	62	49
NOVEMBER	1.0	0	5	4.3	7.5	19.2	7.4	35.8	9.2	270	254	246	46.3	43.3	254	8	68	48
DECEMBER	2.7	0	0	6.7	7.7	21.4	7.0	37.5	9.1	279	279	279	44.8	41.8	260	9	75	49
JANUARY	3.3	0	0	8.4	7.0	22.4	7.0	38.7	9.3	279	279	279	44.3	41.3	260	10	90	51
FEBRUARY	1.7	0	1	9.2	6.2	22.3	6.9	37.1	7.9	252	252	252	41.7	38.7	219	9	85	53
MARCH	−0.1	0	2	6.2	7.1	20.3	7.0	34.4	7.1	279	279	279	42.2	39.2	222	10	85	52
APRIL	−2.2	3	15	2.2	7.6	16.7	6.9	30.1	6.5	201	198	185	41.7	38.7	202	8	72	50
OCT-APRIL	—	11	46	—	7.2	19.9	7.1	—	8.2	1,768	1,725	1,696	43.9	40.9	1,670	63	537	50
MAY-SEPT	—	—	—	—	—	11.4	—	—	—	—	—	—	—	—	—	—	269	—

DATA FOR MONTH TO MATURITY

MATURITY GROUP	EST. MATURITY DATE					*	*	*						*				
1	FEBRUARY 9					21.9	6.2	36.4	8.9					40.4	253	10	91	51
2	FEBRUARY 14					21.9	6.2	36.3	8.7					40.0	249	10	91	52
3	FEBRUARY 20					21.8	6.2	36.1	8.4					39.4	245	10	92	52
4	FEBRUARY 25					21.7	6.2	35.9	8.1					38.9	241	10	93	53
5	MARCH 3					21.5	6.2	35.5	7.9					38.7	238	10	92	53
6	MARCH 8					21.2	6.2	35.1	7.7					38.8	235	10	91	53
7	MARCH 14					20.9	6.2	34.6	7.5					38.9	232	10	90	53
8	MARCH 19					20.6	6.2	34.0	7.3					39.0	229	10	88	52

* Adjusted for vine sites.

172

TABLE 103

RICHMOND (Hawkesbury Agricultural College), NEW SOUTH WALES. 33°37'S, 150°45'E. ALT. 18m.

	1	2	3	4	5	6	7	8	9	10	11	12	13	14	15	16	17	18
	Temperature °C									Summation of Day Degrees above 10° with 19° Cut-Off			Temperature Variability Index					
PERIOD	Lowest Recorded Min. (65YRS)	No.Mths with <0.3° (30YRS)	No.Mths with <2.5° (30YRS)	Av.Mthly Lowest Min. (30YRS)	Av.Min. Minus Col.4	Daily Mean (30YRS)	Daily Range /2 (30YRS)	Av.Mthly Highest Max. (30YRS)	Col.8 Minus Av.Max.	Raw	Adj.1 for Lat. & Daily Temp Rng	Adj.2 for Vine Sites	Raw	Adj. for Vine Sites	Av. Sunshine Hrs	Av. Number Rain Days	Av. Rainfall mm	Av. % R.H. 1500 HRS
OCTOBER	0.1	0	5	4.7	5.7	17.8	7.4	34.2	9.0	242	223		44.3		248	9	56	51
NOVEMBER	3.3	0	0	7.7	5.4	20.2	7.1	37.4	10.1	270	270		43.9		250	9	69	55
DECEMBER	5.0	0	0	9.9	5.5	22.2	6.8	39.3	10.3	279	279		43.0		248	10	78	57
JANUARY	7.5	0	0	11.2	5.5	23.1	6.4	39.7	10.2	279	279		41.3		258	10	96	58
FEBRUARY	6.2	0	0	11.6	5.1	23.1	6.4	37.9	8.4	252	252		39.1		220	11	86	59
MARCH	4.1	0	0	9.4	5.2	20.9	6.3	35.5	8.3	279	279		38.7		215	11	86	59
APRIL	0.8	0	0	4.9	6.4	17.4	6.1	30.9	7.4	222	226		38.2		190	9	67	59
OCT-APRIL	—	0	5	—	5.5	20.7	6.6	—	9.1	1,823	1,808		41.2		1,629	69	538	57
MAY-SEPT	—	—	—	—	—	12.2	—	—	—	—	—		—		—	—	256	—

DATA FOR MONTH TO MATURITY

MATURITY GROUP	EST. MATURITY DATE					6	7	8	9				13		15	16	17	18
1	FEBRUARY 1					23.1	6.4	39.7	10.2				41.3		258	10	96	58
2	FEBRUARY 6					23.2	6.4	39.4	9.9				40.7		256	10	96	58
3	FEBRUARY 12					23.3	6.4	39.0	9.5				40.1		253	11	95	58
4	FEBRUARY 17					23.3	6.4	38.6	9.0				39.6		249	11	95	59
5	FEBRUARY 23					23.2	6.4	38.2	8.6				39.3		245	12	94	59
6	MARCH 1					23.1	6.4	37.8	8.4				39.1		240	12	94	59
7	MARCH 6					22.9	6.4	37.5	8.4				39.0		236	12	93	59
8	MARCH 12					22.5	6.4	37.1	8.4				38.9		231	11	91	59

TABLE 104

SCONE, NEW SOUTH WALES. 32°00'S, 150°54'E. ALT. 207m. (Upper Hunter Valley)

	1	2	3	4	5	6	7	8	9	10	11	12	13	14	15	16	17	18
	Temperature °C									Summation of Day Degrees above 10° with 19° Cut-Off			Temperature Variability Index					
PERIOD	Lowest Recorded Min. (60YRS)	No.Mths with <0.3° (30YRS)	No.Mths with <2.5° (30YRS)	Av.Mthly Lowest Min. (30YRS)	Av.Min. Minus Col.4	Daily Mean (25YRS)	Daily Range /2 (25YRS)	Av.Mthly Highest Max. (30YRS)	Col.8 Minus Av.Max.	Raw	Adj.1 for Lat. & Daily Temp Rng	Adj.2 for Vine Sites	Raw	Adj. for Vine Sites	Av. Sunshine Hrs	Av. Number Rain Days	Av. Rainfall mm	Av. % R.H. 1500 HRS
OCTOBER	−1.1	0	10	3.5	6.0	18.0	8.5	32.5	6.0	248	212		46.0		267	7	51	42
NOVEMBER	1.1	0	0	6.6	5.7	21.0	8.7	36.9	7.2	270	270		47.7		277	7	53	36
DECEMBER	4.4	0	0	8.7	5.8	23.0	8.5	38.9	7.4	279	279		47.2		283	7	70	38
JANUARY	5.6	0	0	10.6	5.8	24.6	8.2	39.8	7.0	279	279		45.6		294	7	80	40
FEBRUARY	4.4	0	0	10.9	5.0	23.7	7.8	38.0	6.5	252	252		42.7		237	6	69	44
MARCH	1.7	0	0	8.6	5.0	21.4	7.8	35.1	5.9	279	279		42.1		237	6	53	41
APRIL	−0.6	1	6	3.8	5.7	17.1	7.6	30.7	6.0	212	201		42.1		212	6	43	40
OCT-APRIL	—	1	16	—	5.6	21.3	8.1	—	6.6	1,819	1,772		44.8		1,807	46	419	40
MAY-SEPT	—	—	—	—	—	11.7	—	—	—	—	—		—		—	—	215	—

DATA FOR MONTH TO MATURITY

MATURITY GROUP	EST. MATURITY DATE					6	7	8	9				13		15	16	17	18
1	FEBRUARY 2					24.6	8.2	39.8	7.0				45.5		291	7	80	40
2	FEBRUARY 7					24.5	8.1	39.6	6.9				45.0		284	7	79	40
3	FEBRUARY 13					24.3	8.0	39.2	6.8				44.4		277	7	78	41
4	FEBRUARY 18					24.1	7.9	38.8	6.7				43.7		270	7	77	42
5	FEBRUARY 24					23.9	7.9	38.4	6.6				43.1		264	7	76	43
6	MARCH 2					23.6	7.8	37.9	6.5				42.7		258	7	74	44
7	MARCH 7					23.3	7.8	37.4	6.4				42.5		253	6	71	43
8	MARCH 13					22.9	7.8	36.9	6.3				42.4		249	6	66	42

TABLE 105

TAMWORTH, NEW SOUTH WALES. 31°06′S, 150°56′E. ALT 378m.

	1	2	3	4	5	6	7	8	9	10	11	12	13	14	15	16	17	18
	Temperature °C									Summation of Day Degrees above 10° with 19° Cut-Off			Temperature Variability Index					
PERIOD	Lowest Recorded Min. (39YRS)	No.Mths with <0.3° (30YRS)	No.Mths with <2.5° (30YRS)	Av.Mthly Lowest Min. (30YRS)	Av.Min. Minus Col.4	Daily Mean /2 (31YRS)	Daily Range /2 (31YRS)	Av.Mthly Highest Max. (30YRS)	Col.8 Minus Av.Max.	Raw	Adj.1 for Lat. & Daily Temp Rng	Adj.2 for Vine Sites	Raw	Adj. for Vine Sites	Av. Sunshine Hrs	Av. Number Rain Days	Av. Rainfall mm	Av. % R.H. 1500 HRS
OCTOBER	−3.3	2	20	2.8	7.0	17.8	8.0	32.7	6.9	242	213		45.9		277	8	60	36
NOVEMBER	2.2	0	1	6.8	6.7	21.5	8.0	35.6	6.1	270	270		44.8		293	7	66	31
DECEMBER	2.2	0	2	8.9	7.1	23.7	7.7	37.9	6.5	279	279		44.4		303	7	73	33
JANUARY	5.0	0	0	11.0	6.4	25.1	7.7	39.1	6.3	279	279		43.5		305	7	81	35
FEBRUARY	7.2	0	0	11.0	6.2	24.6	7.4	37.3	5.3	252	252		41.1		255	6	67	35
MARCH	2.8	0	0	8.1	6.5	21.9	7.3	33.3	4.1	279	279		39.8		254	5	50	36
APRIL	−1.1	2	11	3.0	7.2	17.6	7.4	30.6	5.6	228	220		42.4		236	5	42	37
OCT-APRIL	—	4	34	—	6.7	21.7	7.6	—	5.8	1,829	1,792		43.1		1,923	45	439	35
MAY-SEPT	—	—	—	—	—	11.4	—	—	—	—	—		—		—	—	234	—

DATA FOR MONTH TO MATURITY

MATURITY GROUP	EST. MATURITY DATE			6	7	8	9				13		15	16	17	18
1	FEBRUARY 2			25.1	7.7	39.1	6.3				43.4		303	7	81	35
2	FEBRUARY 7			25.1	7.6	38.8	6.1				43.0		298	7	80	35
3	FEBRUARY 12			25.0	7.6	38.4	5.9				42.7		293	7	78	35
4	FEBRUARY 17			24.9	7.5	38.0	5.6				42.3		288	7	76	35
5	FEBRUARY 23			24.8	7.4	37.6	5.4				41.7		283	7	74	35
6	MARCH 1			24.6	7.4	37.1	5.2				41.1		278	7	72	35
7	MARCH 6			24.3	7.4	36.6	5.0				40.7		274	6	69	35
8	MARCH 12			23.8	7.4	36.0	4.8				40.3		270	6	65	35

TABLE 106

TENTERFIELD, NEW SOUTH WALES. 29°03′S, 152°01′E. ALT. 863m.

	1	2	3	4	5	6	7	8	9	10	11	12	13	14	15	16	17	18
	Temperature °C									Summation of Day Degrees above 10° with 19° Cut-Off			Temperature Variability Index					
PERIOD	Lowest Recorded Min. (63YRS)	No.Mths with <0.3° (30YRS)	No.Mths with <2.5° (30YRS)	Av.Mthly Lowest Min. (30YRS)	Av.Min. Minus Col.4	Daily Mean /2 (28YRS)	Daily Range /2 (28YRS)	Av.Mthly Highest Max. (30YRS)	Col.8 Minus Av.Max.	Raw	Adj.1 for Lat. & Daily Temp Rng	Adj.2 for Vine Sites	Raw	Adj. for Vine Sites	Av. Sunshine Hrs	Av. Number Rain Days	Av. Rainfall mm	Av. % R.H. 1500 HRS
OCTOBER	−2.8	13	24	0.6	7.6	15.5	7.3	28.9	6.1	171	153		42.9		274	8	75	44
NOVEMBER	0.0	3	10	3.7	7.3	18.2	7.2	31.6	6.2	246	222		42.3		276	8	82	45
DECEMBER	0.6	0	1	5.8	7.2	19.9	6.9	33.0	6.2	279	279		41.0		279	9	103	47
JANUARY	4.5	0	0	9.4	4.8	20.9	6.7	33.8	6.2	279	279		37.8		277	10	117	53
FEBRUARY	3.9	0	0	9.7	4.2	20.3	6.4	32.4	5.7	252	252		35.5		229	9	90	57
MARCH	−0.6	0	1	6.3	5.9	18.5	6.3	30.3	5.5	264	261		36.6		226	9	80	51
APRIL	−2.8	8	17	1.5	7.0	15.0	6.5	27.2	5.7	150	154		38.7		224	7	45	46
OCT-APRIL	—	24	53	—	6.3	18.3	6.8	—	5.9	1,641	1,600		39.3		1,785	60	592	49
MAY-SEPT	—	—	—	—	—	9.6	—	—	—	—	—		—		—	—	255	—

DATA FOR MONTH TO MATURITY

MATURITY GROUP	EST. MATURITY DATE			6	7	8	9				13		15	16	17	18
1	FEBRUARY 13			20.7	6.6	33.3	6.0				36.5		271	10	108	55
2	FEBRUARY 18			20.6	6.5	33.1	5.9				36.2		268	10	105	55
3	FEBRUARY 24			20.4	6.5	32.6	5.8				35.8		264	10	102	56
4	MARCH 2			20.2	6.4	32.3	5.7				35.5		259	10	99	57
5	MARCH 7			20.0	6.4	32.0	5.7				35.6		253	10	96	56
6	MARCH 12			19.8	6.4	31.6	5.6				35.8		246	10	92	55
7	MARCH 18			19.5	6.3	31.2	5.5				36.0		238	9	88	54
8	MARCH 25			19.0	6.3	30.7	5.5				36.3		229	9	83	52

TABLE 107

YOUNG, NEW SOUTH WALES. 34°19'S, 148°18'E. ALT. 457m.

	1	2	3	4	5	6	7	8	9	10	11	12	13	14	15	16	17	18
	Temperature °C									Summation of Day Degrees above 10° with 19° Cut-Off			Temperature Variability Index					
PERIOD	Lowest Recorded Min. (63YRS)	No.Mths with <0.3° (30YRS)	No.Mths with <2.5° (30YRS)	Av.Mthly Lowest Min. (30YRS)	Av.Min. Minus Col.4	Daily Mean (29YRS)	Daily Range /2 (29YRS)	Av.Mthly Highest Max. (30YRS)	Col.8 Minus Av.Max.	Raw	Adj.1 for Lat. & Daily Temp Rng	Adj.2 for Vine Sites	Raw	Adj. for Vine Sites	Av. Sunshine Hrs	Av. Number Rain Days	Av. Rainfall mm	Av. % R.H. 1500 HRS
OCTOBER	−3.3	17	28	0.1	7.0	14.8	7.7	30.2	7.7	149	128	135	45.5	43.7	280	8	61	48
NOVEMBER	−0.6	1	18	2.6	7.8	18.6	8.2	34.6	7.8	258	225	232	48.4	46.6	306	6	49	40
DECEMBER	0.6	0	3	5.2	7.9	21.4	8.3	37.4	7.7	279	279	279	48.8	47.0	332	5	52	34
JANUARY	4.4	0	0	8.1	6.7	22.9	8.1	39.0	8.0	279	279	279	47.1	45.3	338	5	52	33
FEBRUARY	3.3	0	0	7.8	7.3	23.1	8.0	37.4	6.3	252	252	252	45.6	43.8	278	5	44	36
MARCH	0.5	0	5	5.2	6.8	19.7	7.7	34.1	6.7	279	279	279	44.3	42.5	272	5	51	40
APRIL	−3.4	9	21	1.4	6.6	15.1	7.1	20.9	6.7	153	146	153	41.7	39.9	235	6	50	47
OCT–APRIL	—	27	75	—	7.2	19.4	7.9	—	7.3	1,649	1,588	1,609	45.9	44.1	2,041	40	359	40
MAY–SEPT	—	—	—	—	—	9.5	—	—	—	—	—	—	—	—	—	—	298	—
DATA FOR MONTH TO MATURITY																		
MATURITY GROUP	EST. MATURITY DATE					*	*							*				
1	FEBRUARY 15					23.2	7.6	37.9	7.1					44.5	318	5	49	34
2	FEBRUARY 20					23.2	7.6	37.5	6.7					44.2	311	5	48	35
3	FEBRUARY 26					23.1	7.6	37.1	6.4					43.9	304	5	48	36
4	MARCH 3					23.0	7.6	36.7	6.3					43.6	298	5	48	36
5	MARCH 9					22.6	7.5	36.2	6.3					43.4	293	5	48	37
6	MARCH 14					22.0	7.5	35.7	6.4					43.2	288	5	49	38
7	MARCH 19					21.4	7.4	35.1	6.5					43.0	283	5	49	39
8	MARCH 25					20.7	7.4	34.5	6.6					42.8	278	5	50	40

* Adjusted for vine sites.

Commentaries and vineyard site adjustments

Armidale (Table 77)

Vineyard site adjustments: None

Notes

The tablelands around Armidale are generally flat, with only limited scope for site selection. Better potential probably exists on the plateau's western edge and associated slopes.

Bathurst (Table 78)

Vineyard site adjustments: None

Bega (Table 79)

Vineyard site adjustments, °C

Min.	Max.	
+1.0	−0.5	Moderate slopes
−0.6	−0.6	Altitude 100–120 m (Bega 13 m)
+0.4	−1.1	Mean temperature −0.35°, variability index −3.0

Bombala (Table 80)

Vineyard site adjustments: None

Bowral (Table 81)

Vineyard site adjustments, °C

Min.	Max.	
+0.6	−0.3	Moderate slopes (Bowral fairly flat)
+1.0	+1.0	Altitude 500 m (Bowral 661 m)
+1.6	+0.7	Mean temperature +1.15°, variability index −1.8

Notes

The estimate is for valley slopes coastwards from Bowral.

Canberra (Forestry School, Acton) (Table 82)

Vineyard site adjustments, °C

Min.	Max.	
+0.5	–	Slopes of moderately steep or free-standing hills
+0.6	+0.2	Slopes facing sun
+0.6	–0.3	Moderately stony soils
+1.7	–0.1	Mean temperature +0.8°, variability index –3.6

Notes

The raw figures for Acton indicate reasonable air drainage, so the additional site adjustment for slopes is only partial.

Actual vineyard areas in the Canberra region are small, and of necessity highly selected to escape frost. I have therefore allowed fairly major adjustments. Such favourable site characteristics are unlikely to be available for large-scale viticulture.

Cessnock (lower Hunter Valley) (Table 83)

Vineyard site adjustments, °C

Min.	Max.	
+1.0	–0.5	Moderate slopes (Cessnock in valley)
+0.3	+0.1	Aspects predominantly north-east
–0.8	–0.8	Altitude ~ 200 m (Cessnock 76 m)
+0.5	–1.2	Mean temperature –0.35°, variability index –3.4

Coonabarabran (Table 84)

Vineyard site adjustments, °C

Min.	Max.	
+1.0	–0.5	Moderate slopes, vs valley floor
–0.6	–0.6	Slopes ~ 100 m above Coonabarabran
+0.4	–1.1	Mean temperature –0.35°, variability index –3.0

Cootamundra (Table 85)

Vineyard site adjustments, °C

Min.	Max.	
+0.8	–0.4	Moderate slopes, vs (?) slight slope
–0.2	–0.2	Vineyards 350–400 m (Cootamundra 330 m)
+0.6	–0.6	Mean temperature unchanged, variability index –2.4

Corowa (Table 86)

Vineyard site adjustments: None

Notes

The relative humidity figures for Corowa in Climatic Averages Australia (1975) are erratic and look too high. Probably they came from a short period. The recording site in that publication is also noted as being the Corowa Bowling Club, which by its nature might be atypical for relative humidity.

Cowra (Table 87)

Vineyard site adjustments, °C

Min.	Max.	
+1.0	–0.5	Moderate slopes
–0.4	–0.4	Vine sites at 350–400 m (Cowra 298 m)
+0.6	–0.9	Mean temperature –0.15°, variability index –3.0

Notes

Existing Cowra vineyards are largely in the valley, on only slightly undulating terrain at up to 350 m. Their temperatures would be probably between the raw figures for Cowra and those estimated for more optimal hillside sites.

Forbes (Table 88)

Vineyard site adjustments: None

Glen Innes (Table 89)

Vineyard site adjustments: None

Notes

Comparison of the figures for Glen Innes Post Office with those for the Agricultural Research Station, over the period covered by Climatic Averages Australia (1975), suggests that the town temperatures may now be a little above those in the open countryside at similar altitudes. The wider diurnal temperature range at the Post Office suggests, further, that it is in a flat or valley situation. In not adjusting the 1911–1940 Post Office figures for vineyard sites, because of the generally flat nature of the tableland country, I am therefore perhaps being conservative.

Goulburn (Table 90)

Vineyard site adjustments: None

Notes

The Goulburn figures show anomalously little frost compared with nearby stations, and are suspect. See main text.

Griffith (Table 91)

Vineyard site adjustments: None

Inverell (Table 92)

Vineyard site adjustments, °C

Min.	Max.	
+1.0	−0.5	Moderate slopes (Inverell on valley flats)
−0.3	−0.3	Altitude 650 m (Inverell 604 m)
+0.7	−0.8	Mean temperature −0.05°, variability index −3.0

Jerry's Plains (middle Hunter Valley) (Table 93)

Vineyard site adjustments: None

Notes

The available temperature data from Climatic Averages Australia (1975) refer to the period 1957–74. Direct comparisons with nearby other centres, for which the records came mostly from 1911–40, are therefore subject to reservation. However I can find no evidence of any consistent change in temperatures along the east coast between the two periods, so in broad terms the figures for Jerry's Plains are probably valid.

Junee (Table 94)

Vineyard site adjustments, °C

Min.	Max.	
+1.0	−0.5	Moderate slopes, vs Junee probably flat
−0.9	−0.9	Hillsides at ~ 450 m (Junee 301 m)
+0.1	−1.4	Mean temperature −0.65°, variability index −3.0

Notes

The relative humidity data for Junee in Climatic Averages Australia (1975) are very erratic. Presumably they are from a short period.

Leeton (Table 95)

Vineyard site adjustments: None

Lithgow (Table 96)

Vineyard site adjustments: None

Notes

Lithgow's fairly narrow recorded diurnal temperature range suggests that the recording site is climatically typical of where viticulture might be attempted (if attempted).

Moss Vale (Table 97)

Vineyard site adjustments: None

Notes

The raw data for Moss Vale show a slightly lesser diurnal temperature range and variability index than those of most surrounding stations. They may therefore represent the better site climates on the surrounding tablelands at comparable altitudes.

Mudgee (Table 98)

Vineyard site adjustments, °C

Min.	Max.	
+1.0	−0.5	Moderate slopes, vs Mudgee in valley
−0.7	−0.7	Vineyards at 550–600 m (Mudgee 454 m)
+0.3	−1.2	Mean temperature −0.45°, variability index −3.0

Murrurundi (Table 99)

Vineyard site adjustments, °C

Min.	Max.	
+1.0	−0.5	Moderate slopes (Murrurundi in valley)
−0.8	−0.8	Vineyard sites at 600 m (Murrurundi 472 m)
+0.2	−1.3	Mean temperature −0.55°, variability index −3.0

Narrabri (Table 100)

Vineyard site adjustments: None

Parramatta (Table 101)

Vineyard site adjustments: None

Notes

Relative humidity (column 18) is estimated from Riverview (33°50'S, 151°10'E, altitude 22 m), less 1 per cent for Parramatta's slightly greater distance inland. Climatic Averages Australia (1975) give relative humidity figures for Parramatta, but they are erratic and appear to be from a short period.

Picton (Table 102)

Vineyard site adjustments, °C

Min.	Max.	
+1.0	−0.5	Moderate slopes (Picton in valley)
−0.9	−0.9	Altitude 300–350 m (Picton 171 m)
+0.1	−1.4	Mean temperature −0.65°, variability index −3.0

Richmond (Hawkesbury Agricultural College) (Table 103)

Vineyard site adjustments: None

Notes

Relative humidity (column 18) is estimated from Riverview (33°50'S, 151°10'E, altitude 22 m), less 3 per cent to allow for Richmond's greater distance inland.

Scone (upper Hunter Valley) (Table 104)

Vineyard site adjustments: None

Notes

Scone is at a greater altitude than most plantings in the upper Hunter, but its diurnal temperature range indicates a valley climate typical of the vineyards there. Temperature estimates are made without adjustment, on the assumption that Scone's distance north approximately offsets its greater altitude.

Tamworth (Table 105)

Vineyard site adjustments: None

Notes

The unadjusted Tamworth figures illustrate the climate at the foot of the western slopes adjacent to the northern tablelands, mainly as a basis from which to estimate site climates at various altitudes up the slopes.

Tenterfield (Table 106)

Vineyard site adjustments: None

Young (Table 107)

Vineyard site adjustments, °C

Min.	Max.	
+0.6	−0.3	Vine sites on moderate slopes (Young ? slight slopes)
−0.15	−0.15	Altitude up to 500 m (Young 457 m)
+0.45	−0.45	Mean temperature unchanged, variability index −1.8

13

Viticultural Environments of Queensland

Apart from an old-established vineyard producing mainly fortified wines at Roma, in the hot western region, recent viticulture in Queensland for winemaking has been confined to around Stanthorpe, in the high-altitude 'Granite Belt' of the extreme south-east (Figure 15). The first true winegrapes, Shiraz, were not planted there until 1965 (Halliday 1985a). Since then the plantings of wine varieties have grown steadily, and the red wines, in particular, have achieved significant success. Scudamore-Smith (1988) has reviewed experimental results to that date.

The soils of the region are formed from decomposing granite, and are described as well drained and easily worked. They also support a thriving deciduous fruit industry. The vineyards are mostly at between 750 and 900 m, on rolling topography which gives a wide variety of slopes and exposures.

Stanthorpe (Table 109), at 810 m, is climatically representative. Comments on it apply equally to Tenterfield, at 860 m just across the border in New South Wales. I present the Stanthorpe data unadjusted, but the vineyard sites must have less killing frosts for a commercial grape industry to have survived. The table therefore gives a conservative measure of climatic quality for viticulture.

Summer temperatures and sunshine hours are good for producing full-bodied table wines, apart from years when too much rain falls during ripening or harvest. As in the northern tablelands of New South Wales, the highest maximum temperatures are very moderate because of the sub-tropical summer monsoon influence in minimising temperature variability. Relative humidities are high enough to eliminate any vine stress and promote grape and wine quality; but they are also high enough for there to be a considerable risk of vine diseases. Against that the ratio of growing season sunshine hours to effective temperature summation is a fairly generous 1.07, suggesting that the wines will have good body, colour and flavour intensity, but still be reasonably soft. Some of these advantages may be offset by high altitude and a consequently reduced atmospheric carbon dioxide concentration.

Shiraz grapes have already shown their adaptation to the Stanthorpe environment. So should Merlot, in view of its evident tolerance of heavy growing season and ripening period rainfall: see, for instance, Lugarno, Switzerland (Chapter 18, Table 148). Other red varieties known to be adapted to such conditions include Cabernet Sauvignon, Mondeuse and Tannat, while northern Italian varieties in general should also do well.

Killarney (Table 108), at 515 m, illustrates potential viticultural climates at lower altitudes. Grape growing is feasible, but ripening temperatures exceed the optimum range for table wines. Perhaps more importantly, the earlier ripening means a reduced chance of escaping heavy rain during ripening and harvest.

I include **Toowoomba** (Table 110) to show that north from the Granite Belt the summer rains become too heavy for successful commercial viticulture; relative humidities are also too high for *vinifera* vines at the prevailing temperatures. These adverse factors can be avoided by moving to drier areas further inland, but they in turn are too hot. Coastal areas, such as Brisbane, are much too wet and humid for *vinifera* grapes.

Three broad conclusions can be drawn. First, the Granite Belt of south-east Queensland, typified by Stanthorpe, offers generally good ripening conditions for table wines where spring frosts can be avoided, albeit with a danger of excessive ripening and harvest period rain. Second, high altitudes are needed for ripening to be late enough to escape the heaviest summer rains. Third, even there, the midseason and later grape varieties are the best adapted.

The Granite Belt appears thus to be most naturally an area for red wines of the softer Bordeaux styles, and perhaps especially those of the lower Rhône and northern Italy. Early-maturing red grape varieties such as Pinot Noir, together with most white grape varieties and wine styles, can be expected to be less successful. These conclusions accord well with initial practical and experimental results as reported by Scudamore-Smith (1988).

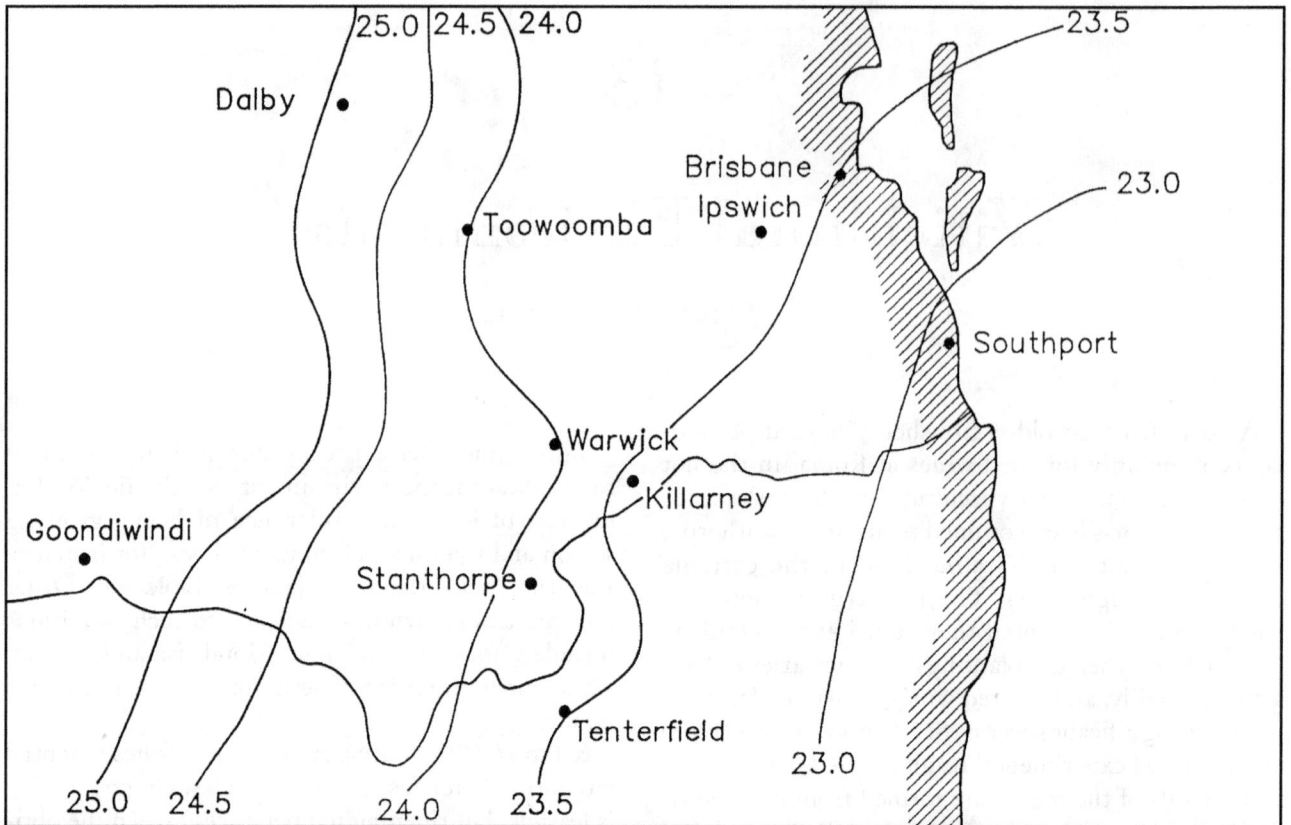

Figure 15. South-east Queensland vinegrowing region, with isotherms of October-April average mean temperatures, reduced to sea level.

TABLE 108

KILLARNEY, QUEENSLAND. 28°20'S, 152°18'E. ALT. 515m.

	1	2	3	4	5	6	7	8	9	10	11	12	13	14	15	16	17	18
	Temperature °C									Summation of Day Degrees above 10° with 19° Cut-Off			Temperature Variability Index					
PERIOD	Lowest Recorded Min. (67YRS)	No.Mths with <0.3° (30YRS)	No.Mths with <2.5° (30YRS)	Av.Mthly Lowest Min. (30YRS)	Av.Min. Minus Col.4	Daily Mean (28YRS)	Daily Range /2 (28YRS)	Av.Mthly Highest Max. (30YRS)	Col.8 Minus Av.Max.	Raw	Adj.1 for Lat. & Daily Temp Rng	Adj.2 for Vine Sites	Raw	Adj. for Vine Sites	Av. Sunshine Hrs	Av. Number Rain Days	Av. Rainfall mm	Av. % R.H. 1500 HRS
OCTOBER	−2.2	7	21	1.1	8.1	17.3	8.1	31.2	5.8	226	194		46.3		275	7	64	
NOVEMBER	1.7	0	2	4.3	8.3	20.2	7.6	34.1	6.3	270	270		45.0		275	8	70	
DECEMBER	2.2	0	1	7.4	6.9	21.7	7.4	35.3	6.2	279	279		42.7		270	9	91	
JANUARY	4.2	0	0	10.7	5.2	22.8	6.9	35.4	5.7	279	279		38.5		252	9	98	
FEBRUARY	6.4	0	0	11.1	4.6	22.2	6.5	34.6	5.9	252	252		36.5		210	9	86	
MARCH	2.2	0	0	7.9	6.1	20.5	6.5	31.9	4.9	279	279		37.0		220	9	77	
APRIL	−2.2	5	11	3.1	6.9	17.1	7.1	28.9	4.7	213	211		40.0		223	7	45	
OCT-APRIL	—	12	35	—	6.6	20.3	7.2	—	5.6	1,798	1,764		40.9		1,725	58	531	
MAY-SEPT	—	—	—	—	—	11.4	—	—	—	—	—		—		—	—	214	

	DATA FOR MONTH TO MATURITY																	
MATURITY GROUP	EST. MATURITY DATE																	
1	FEBRUARY 4					22.9	6.9	35.4	5.7				38.3		249	9	98	
2	FEBRUARY 9					22.9	6.8	35.3	5.8				37.8		244	9	97	
3	FEBRUARY 14					22.8	6.7	35.1	5.8				37.3		239	10	96	
4	FEBRUARY 19					22.6	6.6	34.9	5.8				36.9		235	10	95	
5	FEBRUARY 25					22.4	6.5	34.7	5.9				36.7		232	10	94	
6	MARCH 2					22.1	6.5	34.4	5.9				36.6		229	10	93	
7	MARCH 8					21.9	6.5	34.0	5.8				36.6		227	10	90	
8	MARCH 14					21.6	6.5	33.4	5.6				36.7		225	10	86	

TABLE 109
STANTHORPE, QUEENSLAND. 28°40'S, 151°56'E. ALT. 810m.

	1	2	3	4	5	6	7	8	9	10	11	12	13	14	15	16	17	18
	Temperature °C									Summation of Day Degrees above 10° with 19° Cut-Off			Temperature Variability Index					
PERIOD	Lowest Recorded Min. (64YRS)	No.Mths with <0.3° (30YRS)	No.Mths with <2.5° (30YRS)	Av.Mthly Lowest Min. (30YRS)	Av.Min. Minus Col.4	Daily Mean (28YRS)	Daily Range /2 (28YRS)	Av.Mthly Highest Max. (30YRS)	Col.8 Minus Av.Max.	Raw	Adj.1 for Lat. & Daily Temp Rng	Adj.2 for Vine Sites	Raw	Adj. for Vine Sites	Av. Sunshine Hrs	Av. Number Rain Days	Av. Rainfall mm	Av. % R.H. 1500 HRS
OCTOBER	-3.6	18	24	0.7	8.0	15.9	7.2	28.7	5.5	183	166		42.4		275	8	68	44
NOVEMBER	0.0	1	10	4.0	8.0	18.9	6.9	31.4	5.6	267	246		41.2		280	8	69	41
DECEMBER	1.7	0	1	6.9	6.8	20.5	6.8	33.4	6.1	279	279		40.1		280	10	92	47
JANUARY	5.2	0	0	9.6	5.5	21.6	6.5	33.4	5.3	279	279		36.8		270	10	97	49
FEBRUARY	3.3	0	0	10.2	4.5	20.9	6.2	32.3	5.2	252	252		34.5		225	9	85	53
MARCH	0.9	0	3	6.3	6.8	19.1	6.0	30.1	5.0	279	279		35.8		235	9	67	52
APRIL	-5.9	14	20	0.9	8.0	15.5	6.6	26.9	4.8	165	169		39.2		225	6	41	49
OCT-APRIL	—	33	58	—	6.8	18.9	6.6	—	5.4	1,704	1,670		38.6		1,790	60	519	48
MAY-SEPT	—	—	—	—	—	9.8	—	—	—	—	—		—		—	—	241	—

DATA FOR MONTH TO MATURITY

MATURITY GROUP	EST. MATURITY DATE																	
1	FEBRUARY 10					21.5	6.4	33.1	5.3				35.9		263	10	95	50
2	FEBRUARY 15					21.4	6.3	32.9	5.2				35.4		257	10	95	51
3	FEBRUARY 20					21.2	6.2	32.6	5.2				35.0		252	10	94	52
4	FEBRUARY 26					21.0	6.2	32.3	5.2				34.7		248	10	93	53
5	MARCH 4					20.8	6.2	32.0	5.2				34.7		245	10	90	53
6	MARCH 9					20.5	6.1	31.6	5.1				34.8		243	10	85	53
7	MARCH 14					20.2	6.1	31.3	5.1				35.0		241	10	79	53
8	MARCH 20					19.8	6.1	30.9	5.1				35.3		239	10	73	52

TABLE 110
TOOWOOMBA, QUEENSLAND. 27°35'S, 151°56'E. ALT. 585m.

	1	2	3	4	5	6	7	8	9	10	11	12	13	14	15	16	17	18
	Temperature °C									Summation of Day Degrees above 10° with 19° Cut-Off			Temperature Variability Index					
PERIOD	Lowest Recorded Min. (64YRS)	No.Mths with <0.3° (30YRS)	No.Mths with <2.5° (30YRS)	Av.Mthly Lowest Min. (30YRS)	Av.Min. Minus Col.4	Daily Mean (30YRS)	Daily Range /2 (30YRS)	Av.Mthly Highest Max. (30YRS)	Col.8 Minus Av.Max.	Raw	Adj.1 for Lat. & Daily Temp Rng	Adj.2 for Vine Sites	Raw	Adj. for Vine Sites	Av. Sunshine Hrs	Av. Number Rain Days	Av. Rainfall mm	Av. % R.H. 1500 HRS
OCTOBER	0.2	1	5	4.4	6.9	17.9	6.6	30.6	6.1	245	235		39.4		275	9	72	42
NOVEMBER	3.8	0	0	7.4	6.2	20.2	6.6	33.2	6.4	270	270		39.0		275	9	85	45
DECEMBER	4.5	0	0	10.3	5.0	21.6	6.3	33.8	5.9	279	279		36.1		274	10	118	49
JANUARY	7.5	0	0	12.2	4.0	22.2	6.0	34.4	6.2	279	279		34.2		260	12	140	53
FEBRUARY	7.8	0	0	12.6	3.5	21.7	5.6	32.9	5.6	252	252		31.5		218	11	121	57
MARCH	0.6	0	0	10.5	4.4	20.3	5.4	30.6	4.9	279	279		30.9		230	11	100	55
APRIL	-0.6	0	2	5.4	6.0	17.2	5.8	28.0	5.0	216	223		34.2		228	8	61	51
OCT-APRIL	—	1	7	—	5.1	20.2	6.0	—	5.7	1,820	1,817		35.0		1,760	70	697	50
MAY-SEPT	—	—	—	—	—	12.4	—	—	—	—	—		—		—	—	258	—

DATA FOR MONTH TO MATURITY

MATURITY GROUP	EST. MATURITY DATE																	
1	JANUARY 30					22.2	6.0	34.4	6.2				34.4		261	12	140	53
2	FEBRUARY 5					22.2	5.9	34.1	6.1				34.0		258	12	139	53
3	FEBRUARY 10					22.1	5.8	33.8	6.0				33.3		254	12	138	54
4	FEBRUARY 16					22.0	5.7	33.4	5.8				32.5		249	12	136	55
5	FEBRUARY 21					21.9	5.6	33.1	5.7				31.8		244	12	134	56
6	FEBRUARY 27					21.8	5.6	32.9	5.6				31.4		241	12	132	57
7	MARCH 4					21.6	5.6	32.7	5.5				31.3		239	12	129	57
8	MARCH 10					21.4	5.6	32.4	5.4				31.2		238	12	124	56

Commentaries and vineyard site adjustments

Killarney (Table 108)

Vineyard site adjustments: None

Stanthorpe (Table 109)

Vineyard site adjustments: None

Toowoomba (Table 110)

Vineyard site adjustments: None

14

Viticultural Environments of Tasmania

Only New South Wales has a longer history of viticulture than Tasmania. Bartholomew Broughton established Tasmania's first commercial vineyard in 1823, at New Town, now a Hobart suburb. By 1827 he was able to advertise that he had 300 gallons (1365 L) of wine for sale (Halliday 1985a).

Further plantings followed around Hobart, and also in the Tamar Valley in the north-east of the State. Vine cuttings from Tasmania were among the first to be planted in South Australia and Victoria.

But despite early promise, and attempted revivals later in the 19th century, the industry did not survive. Apparent reasons include a lack of market interest in the light wines produced, loss of labour to the gold rushes, inexperience, and a climate which was often difficult for viticulture.

A contributing reason may have been global cooling, such as became apparent in Western Europe after the 1870s (see Chapter 23). It is notable that the first Tasmanian plantings coincided with climatic recovery in Europe from the early 19th century period of 'Dickensian Winters', and the start of a mostly warm period which lasted world-wide for some 50 years.

The re-birth of Tasmanian viticulture began in 1956, when Jean Miguet, a French engineer, planted a hectare of assorted winegrape varieties at Lalla, near Lilydale, about 20 km north of Launceston. In 1958 Claudio Alcorso, not knowing of Miguet's venture, planted a half hectare of vines on his Moorilla Estate, on the bank of the Derwent Estuary just north of Hobart. These included Riesling, Traminer and, most notably, Pinot Noir. The ultimate success of Alcorso's wines gave the first modern proof of Tasmania's potential for winemaking.

A further important step towards a true Tasmanian wine industry was the establishment in 1975 of the Pipers Brook Vineyard, by the brothers Andrew and David Pirie, on a basalt rise overlooking Bass Straight some 40 km north of Launceston. Andrew Pirie had selected this site on the basis of his environmental studies at Sydney University, as I described in Chapter 2. These studies, and the practical experience that followed, became a basis for further commercial development in the area. For more extensive accounts of the Tasmanian vine industry and its history, see Halliday (1985a, 1991a), Bulleid (1987) and Pirie (1988).

All viticultural climates in Tasmania are cool. As such they are unusual in being highly maritime. That is, there is only a small temperature difference between winter and summer, in contrast to the highly 'continental' temperature patterns of the classic cool viticultural climates of north-western and central Europe (Table 1, Chapter 3).

A consequence has been variable adaptation of the traditional cool-climate winegrape varieties. As in many of the cool or mild, south-coastal environments of mainland Australia, problems have included irregular budburst in susceptible varieties; weather damage to the young shoots and bunches during the prolonged cool, and often windy, spring; consequent poor berry set; and irregular fruitfulness of the buds in the following year if the spring and early summer have been cool, wet or cloudy. To some extent these factors are offset by the excellent totals of sunshine hours received through the growing season.

Another result of the maritime temperature pattern is that vine phenology in Tasmania can be expected to be influenced more strongly by local differences in temperature or altitude than almost anywhere else. Monthly average mean temperatures remain below the 'saturation' level of 19°C through the whole growth and ripening period; therefore temperature effects on vine phenological development accumulate over the whole season.

Site differences in vine phenology are not only maximal; they are also the most readily predicted. If one assumes a seven-month season with all monthly average mean temperatures below 19°C, and that ripening is not complete until April 30, it follows that a difference of 0.24°C in average mean temperature is equivalent to 50 effective day degrees up to ripeness. That is, the temperature difference resulting from a simple altitude difference of 40 m (other topographic and soil factors remaining the same) should shift ripening dates by the equivalent of one vine maturity group (Chapter 7, Table 5). I use this relationship below, to predict vine performance at different altitudes in the Launceston – Pipers Brook area; also in Chapter 22,

Table 111. Tasmanian viticultural regions: average mean temperatures, raw temperature summations, diurnal temperature ranges/2, and average mean temperatures adjusted to sea level (0.6°C per 100 m altitude), for the October-April growing season. Calculated from Climatic Averages Australia, editions of 1956 and 1975

Location	Alt.	1956 Edition			1975 Edition			Diurnal range/2*	Sea-level temperature	
		No. of years	Mean temp.	Raw day°	No. of years	Mean temp.	Raw day°		1956 Ed.	1975 Ed.
North-east										
Launceston	81	51	15.47	1156	–	–	-	6.2	15.96	-
Elphin	8	–	–	-	< 5	15.64	1178	6.2	-	15.69
7EX	107	–	–	-	< 5	15.04	1063	5.5	-	15.68
Mt Pleasant	140	–	–	-	10	14.68	987	5.6	-	15.52
George Town	15	–	–	-	7	14.95	1043	4.6	-	15.04
Bridport	14	–	–	-	< 5	15.44	1147	4.7	-	15.52
Scottsdale	200	–	–	-	16	14.17	882	5.1	-	15.37
East coast										
St Helens	5	35	14.67	986	18	15.04	1064	5.5	14.70	15.07
Bicheno	31	–	–	-		15.34	1129	4.3	-	15.53
Swansea	8	39	14.75	1008	18	14.66	984	5.2	14.80	14.71
Orford	9	–	–	-		14.81	1015	5.4	-	14.86
South-east										
Hobart	55	30	14.62	976	27	14.85	1025	4.6	14.95	15.18
Risdon	43	23	14.73	1001	18	15.11	1080	4.8	14.99	15.37
New Norfolk	31	–	–	-		14.77	1006	5.8	-	14.95
Bushy Park	53	19	14.52	956	18	14.29	905	6.2	14.84	14.61

* From all available records.

in the commentary on Portland, Oregon, to predict vine phenology as related to site characteristics in the Willamette Valley.

Equally, vine phenology will fluctuate more from season to season than in warmer and/or more continental climates, given similar differences in temperature among seasons. However the effects of this may be offset by the slower fall in autumn temperatures and sunshine, giving a longer potential ripening period which can be exploited provided that the autumn is not too wet.

Figure 16 shows sea-level isotherms for the October-April growing season across Tasmania, based on long-term temperature records. After allowance has been made for altitude (subtract 0.6°C per 100 m), and for individual site characteristics of slope, aspect and soil type (Chapter 4, Table 3), a final, effective mean of 14.5°C represents the approximate cool limit for satisfactory ripening of the earliest (Maturity Group 1)

grape varieties; 15.0°C for Maturity Group 3 varieties such as Pinot Noir for red wines, and Chardonnay; and 15.5°C for red wines from early midseason (Maturity Group 5) varieties such as Shiraz, Cabernet Franc and Merlot. Ripening of Maturity Group 3 varieties represents about the cool limit for fully competitive commercial viticulture – other than perhaps for sparkling wines – because varieties maturing before that are of mostly inferior quality (Chapter 7).

A question to be examined, in the light of the postulated extreme responsiveness of vines to temperature in the Tasmanian climate, is how far recent viticultural developments have been made possible by what might prove to be only a short-term fluctuation in climate. Table 111 shows the average mean temperatures for October-April at locations for which data are available from the two editions of Climatic Averages Australia: firstly as actually recorded, and secondly, reduced to sea level by adding 0.6°C per 100 m of altitude. Those

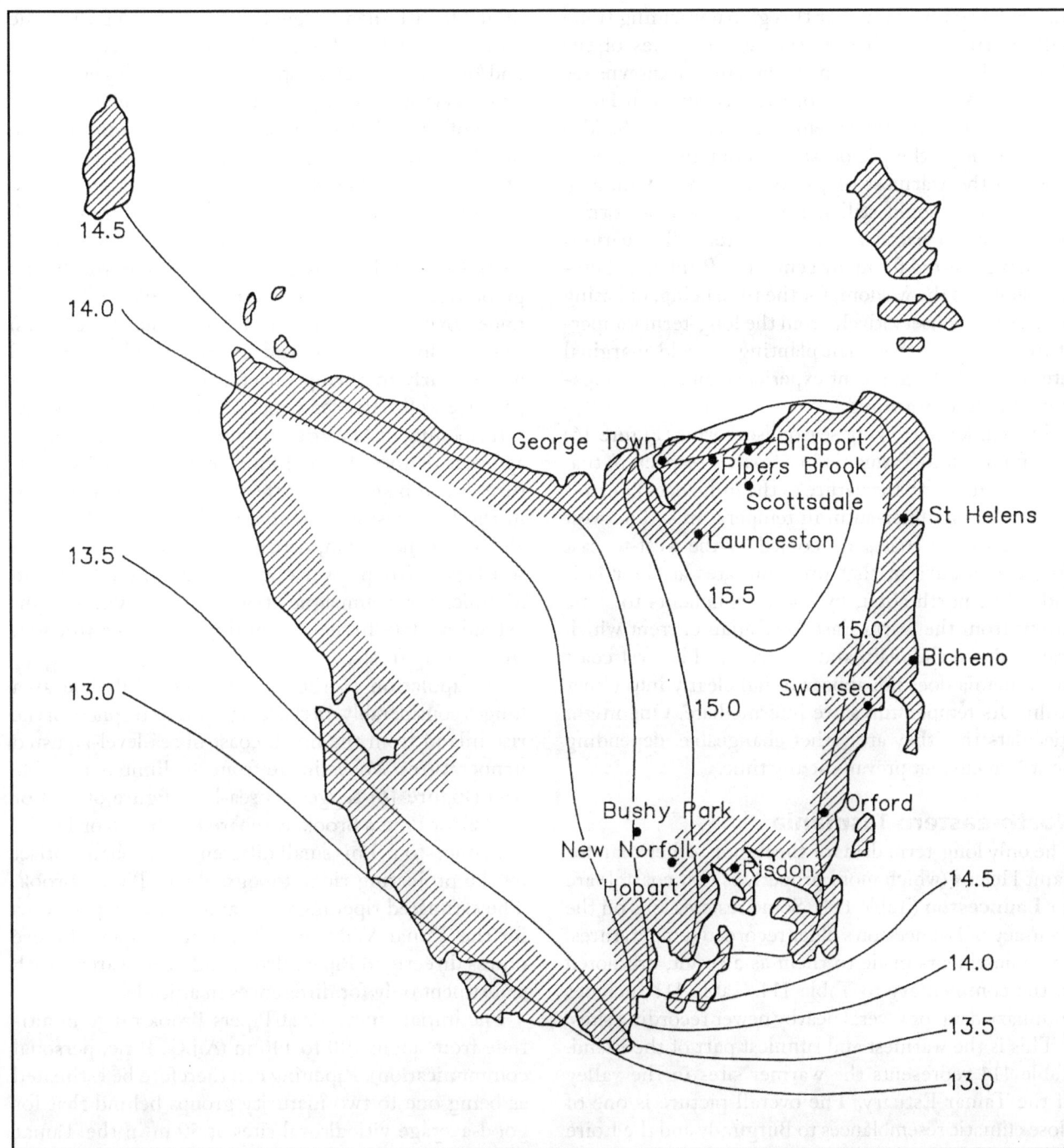

Figure 16. Tasmania, with isotherms of October-April average mean temperatures, reduced to sea level.

from the 1956 edition came predominantly from the 30-year period 1911–1940; and those from the 1975 edition, from varying (often short) periods starting in 1957 or later.

Of the six locations for which temperatures were recorded in both editions, three (Hobart, Risdon and Saint Helens) showed apparent rises in their sea level-adjusted growing season temperatures, and the remainder (Launceston, Swansea and Bushy Park) showed apparent falls. The Launceston figures are suspect because of changed recording sites, as discussed in the commentary to Table 114. Overall, however, Tasmania appears to conform to the pattern of the mainland's

east coast, rather than that of its south coast, in showing no consistent rise in recorded summer or growing season temperatures through the first three-quarters of the 20th century. Any slight rise in the average figure would be well within the range expected from urban growth and specifically urban warming of the recording sites (Kukla et al. 1986; Karl et al. 1988; this book, Chapter 23).

On the other hand it does seems likely that real, but perhaps temporary, increases in temperature and advances in ripening occurred during the globally warm decade of the 1980s, which is the period from which most present commercial experience in

185

Tasmania stems.[1] Whether this global warming trend will continue, or whether the temperatures of the 1980s will even be sustained, remains unknown: see Chapter 23. Local studies of the tree rings in Huon pines (Cook et al. 1991) seem consistent with the idea that, although the period since about 1970 has been close to the warmest of the present millennium in Tasmania, it could still merely represent a normal peak within a pattern of medium-term fluctuations that has persisted for many centuries. Prudence therefore suggests the wisdom, for the time being, of basing projections conservatively from the long-term temperature records, rather than planting on cold-marginal sites in the light of recent experience and in anticipation of further warming.

The sea-level growing season isotherms (Figure 16) are of interest to Tasmania in a further respect, in that they reveal clearly the nature of the influences controlling spring-summer-autumn temperatures. The west and south-west coasts are cooled by the west-to-east cool ocean current that predominates at their latitude. The north coast, by contrast, benefits to some extent from the warm East Australian current which travels down the mainland east coast. The east coast of Tasmania does not appear to fall clearly into either realm. Its temperatures are intermediate. One might speculate that they are rather changeable, depending on which current prevails at any time.

North-eastern Tasmania

The only long-term data for the region, other than for Saint Helens (which more properly is east coastal) are for **Launceston** (Table 114). Some reservations on the accuracy of Launceston's older recorded temperatures, and adjustments made to them as a result, are noted in the commentary to Table 114. Table 111 includes summary data for several nearby newer recording sites.

This is the warmest and sunniest part of the island. Table 114 represents the warmer sites in the valley of the Tamar Estuary. The overall picture is one of close climatic resemblances to Burgundy and the Loire Valley in France, the main growing season differences being that the Tamar has substantially more sunshine hours, slightly lower relative humidities, and a more protracted potential ripening period in autumn but significantly cooler summer months. On the good-average sites to which Table 114 applies directly, Cabernet Sauvignon (Maturity Group 6) could be expected to ripen in an average season, but late and under much lower temperatures than in Bordeaux. Probably there would not be enough warmth during average ripening for the greatest colour and flavour development. Maturity groups up to 5, however, should ripen well except in cold seasons, to produce well balanced wines comparable with those of Burgundy and the Loire.

Higher temperatures, or effective temperatures, might be attained through lower altitudes than the 50 m assumed in Table 114; through steeper slopes and/or due northerly aspects; or through very stony or otherwise specially warm soils. Various combinations of these factors could advance ripening by one or two maturity groups, with ripening temperatures up to 0.5°C higher than indicated by Table 114 for given maturity dates. Such highly selected sites, which at most would only be available in very small areas, should allow full and regular ripening of Bordeaux red grape varieties of Maturity Groups 5 and 6, albeit still under lower temperatures (but greater sunshine hours) than in Bordeaux, to produce fresh medium-bodied wines. Early-maturing varieties on such sites should produce well-balanced wines of medium to full body.

If ripening of Maturity Group 3 varieties in an average year defines the cool limit of commercial viticulture, it can be estimated that the upper altitude limit in the Launceston/Tamar estuary area lies between about 150 and 220 m, depending on slope, aspect and soil type. Group 2 varieties such as Pinot Noir and Meunier for champagne-style sparkling wines might extend a little higher, assuming sufficient freedom from spring frosts.

Extrapolation to the Pipers Brook–Lilydale area hinges considerably on the location and steepness of the rise inland from the north coast in sea level-adjusted temperatures. My estimate from the limited available data (Figure 16) suggests a sea-level figure of 15.4 or 15.5°C for Pipers Brook, compared with 15.6 or 15.7°C at Launceston. This small difference is probably offset by the projecting ridge topography of Pipers Brook. The estimated ripening dates and mean temperatures for the Tamar Valley can therefore be applied more or less directly to Pipers Brook and nearby areas, with adjustment only for differences in altitude.

The initial vineyards at Pipers Brook range in altitude from about 100 to 140 m (A.J.G. Pirie, personal communication). Ripening can therefore be estimated as being one to two maturity groups behind that for good-average viticultural sites at 50 m in the Tamar Valley. That is, given the average climatic conditions of the 20th century, maturity groups up to 4 should ripen satisfactorily throughout, to produce light to medium-bodied wines; Group 5 only in the warmer parts of the vineyards to produce nearly always light, delicate wines; and Cabernet Sauvignon (Group 6) only ripening fully in the warmest parts in warmer-than-average years, more typically producing rather thin, high-acid wines of vegetal rather than berry character. This theoretically-based projection agrees well with actual experience as described by Halliday (1991a).

North-coastal Tasmania enjoys one climatic advantage which in Australia is more or less unique: that of very low variability of the maximum temperature. As discussed in Chapter 3, this is due to the heat of northerly winds being tempered by their passage across Bass Straight, whereas cooler southerly winds

---

1 See Gladstones (1994).

are warmed during the day through having to traverse the full length of the island. Combined with rather late ripening under low mean temperatures, but very adequate sunshine hours, this means that the highest maxima experienced during ripening are much less than elsewhere in Australia. Indeed, they are less even than in the Loire Valley, and compare favourably with those in Burgundy, Champagne and the Rhine Valley. Relative humidities are also optimal. Delicacy of flavour and aroma should therefore be attainable here better than in any other Australian viticultural region. In particular, north-eastern Tasmania clearly has the best climate for early-ripening, heat-sensitive red grape varieties which nevertheless need plenty of sunshine, such as Pinot Noir and Gamay, and perhaps Bastardo (for table wines), Dolcetto and Durif.

The climatic problems of north-eastern Tasmania stem from its lack of continentality: leading, in susceptible varieties, to irregular budburst, poor setting, rank vegetative growth, and poor fruitfulness of the buds. Late frosts are a potential problem away from the immediate influence of coast or estuary. Wind can also have serious effects in exposed areas such as Pipers Brook. Grape varieties and management practices (e.g. Pirie 1982, 1988) must be such as to minimize these problems, if the potential of the region is to be fully achieved.

South-eastern Tasmania

The data for **Hobart** (Table 113) and **Risdon** (Table 115) typify the near-coastal viticultural climates of south-eastern Tasmania. **Bushy Park** (Table 112), inland from Hobart, is cooler and viticulturally very marginal.

Reduced to sea level, the average mean temperatures for October-April (Figure 16) show a broad warm strip extending north-south through east-central Tasmania. Hobart on this basis is only 0.5°C cooler than Launceston. Its warmth probably results from summer warming of the relatively sedentary waters of the Derwent Estuary, and of Pitt Water and perhaps Frederick Henry bay to the east. These are largely cut off from the cool ocean current that flows from the west around the south-west and south coasts, so that a steep fall occurs in sea-level temperatures southward and westward from Hobart (Figure 16). The Derwent and Huon estuary regions also benefit from the adiabatic warming of winds that have dropped their moisture over the mountains to the west: a rain-shadow effect which is reflected favourably in both temperatures and sunshine hours.

In contrast to Launceston, the small temperature variability indices of Hobart and Risdon result primarily from their narrow diurnal temperature ranges, combined with a fairly restricted variability of the minimum temperatures. Both can be attributed to the influence of partly surrounding waters, which also shows in the very low figures for continentality

(Chapter 3, Table 1). But variability of the maximum temperature (Column 9 in the climatic tables) is quite marked: a typical south coast feature. The more inland location of Bushy Park widens both diurnal temperature range and variability of the minimum, without reducing the variability of the maximum temperatures.

Viticulture on any scale in south-eastern Tasmania is thus effectively restricted to areas under the direct climatic influence of the Derwent Estuary and adjacent bays. Possibly some sheltered pockets are warmer than indicated by Hobart and Risdon, especially around the more or less fully enclosed Pitt Water (off Frederick Henry Bay).

Regular satisfactory ripening as indicated by the figures is confined to early-maturing grape varieties, up to Maturity Group 3; or possibly Group 4 on the warmest sites. Ripening of varieties as late as Cabernet Sauvignon (Group 6 for red wines or 5 for pink wines) has been reported (Smart et al. 1980). This is, of course, possible at selected sites in some seasons; but it seems too uncertain, and is probably too incomplete in most years, for acceptable commercial results.

Wines made in the Hobart area are predictably delicate in character, with the best from early-maturing varieties such as Pinot Noir. It could be a good area for sparkling wines, given commercially adequate yields.

East coastal Tasmania

Long-term records for **Saint Helens** (Table 116) and for **Swansea** (not presented in detail, but see Table 111) show that temperatures along the east coast of Tasmania are, at best, marginal for viticulture. Short-term figures for Orford, though not for **Bicheno** (see Table 111) agree with this. The more favourable figures for Bicheno are explained by its north-facing and unusually sheltered mesoclimate, as described elsewhere.

As also described in that reference, at least one low-altitude area exists inland from the east coast which has viticultural conditions a good deal better than those of the exposed coast itself, necessitating a revision for the uniformly adverse evaluation of the east coastal region contained in the original first printing of this book and still shown in Figure 21 (p. 274). The area lies in the valleys of the Swan and Apsley rivers, just inland between Swansea and Bicheno.

As well as being sheltered by the coastal hills from easterly sea breezes, the basin surrounds the shallow, but quite extensive, Moulting Lagoon. This should help to retain warmth into the late autumn. Within the basin are a number free-standing small hills and ridges, providing warm, stony slopes with excellent air drainage. The revised Table 116 estimates that such sites should approximately match the Tamar Valley in ripening capacity, and provide excellent conditions for wine quality, More widespread, gently sloping sites within the basin probably match more closely the conditions of Hobart (Table 113).

TABLE 112
BUSHY PARK, TASMANIA. 42°42′S, 146°54′E. ALT. 53m.

	1	2	3	4	5	6	7	8	9	10	11	12	13	14	15	16	17	18
	Temperature °C									Summation of Day Degrees above 10° with 19° Cut-Off			Temperature Variability Index					
PERIOD	Lowest Recorded Min. (36YRS)	No.Mths with <0.3° (30YRS)	No.Mths with <2.5° (30YRS)	Av.Mthly Lowest Min. (30YRS)	Av.Min. Minus Col.4	Daily Mean (38YRS)	Daily Range /2 (38YRS)	Av.Mthly Highest Max. (30YRS)	Col.8 Minus Av.Max.	Raw	Adj.1 for Lat. & Daily Temp Rng	Adj.2 for Vine Sites	Raw	Adj. for Vine Sites	Av. Sunshine Hrs	Av. Number Rain Days	Av. Rainfall mm	Av. % R.H. 1500 HRS
OCTOBER	−2.6	17	30	0.0	5.8	11.6	5.8	26.4	9.0	50	50	74	38.0	33.6	198	16	57	53
NOVEMBER	−0.6	1	20	1.9	5.8	13.3	5.6	29.6	10.7	99	101	127	38.9	34.5	224	13	54	51
DECEMBER	0.3	0	7	3.4	5.8	15.4	6.2	33.2	11.6	167	171	190	42.2	37.8	228	12	52	49
JANUARY	0.6	0	6	4.1	5.8	16.7	6.8	34.1	10.6	208	212	231	43.6	39.2	242	9	42	46
FEBRUARY	−0.6	1	9	3.5	6.5	16.8	6.8	32.2	8.6	190	193	210	42.3	37.9	207	8	36	46
MARCH	−1.4	3	21	1.9	6.8	15.1	6.4	30.9	9.4	158	159	177	41.8	37.4	201	10	39	49
APRIL	−4.4	17	28	−0.2	6.7	12.1	5.6	25.6	7.9	63	63	88	37.0	32.6	150	11	48	57
OCT-APRIL	—	39	121	—	6.2	14.4	6.2	—	9.7	935	949	1,097	40.5	36.1	1,450	79	328	50
MAY-SEPT	—	—	—	—	—	7.8	—	—	—	—	—	—	—	—	—	—	251	—
DATA FOR MONTH TO MATURITY																		
MATURITY GROUP	EST. MATURITY DATE					*	*	*						*				
1	APRIL 12					14.5	5.0	29.0	8.8					35.8	188	10	42	52
2	MAY 3					12.4	4.5	24.8	7.9					32.4	146	11	48	57
3	—					—	—	—						—	—	—	—	—
4	—					—	—	—						—	—	—	—	—
5	—					—	—	—						—	—	—	—	—
6	—					—	—	—						—	—	—	—	—
7	—					—	—	—						—	—	—	—	—
8	—					—	—	—						—	—	—	—	—

* Adjusted for vine sites.

TABLE 113
HOBART, TASMANIA. 42°53′S, 147°20′E. ALT. 55m.

	1	2	3	4	5	6	7	8	9	10	11	12	13	14	15	16	17	18
	Temperature °C									Summation of Day Degrees above 10° with 19° Cut-Off			Temperature Variability Index					
PERIOD	Lowest Recorded Min. (102YRS)	No.Mths with <0.3° (30YRS)	No.Mths with <2.5° (30YRS)	Av.Mthly Lowest Min. (30YRS)	Av.Min. Minus Col.4	Daily Mean (100YRS)	Daily Range /2 (60YRS)	Av.Mthly Highest Max. (30YRS)	Col.8 Minus Av.Max.	Raw	Adj.1 for Lat. & Daily Temp Rng	Adj.2 for Vine Sites	Raw	Adj. for Vine Sites	Av. Sunshine Hrs	Av. Number Rain Days	Av. Rainfall mm	Av. % R.H. 1500 HRS (30YRS)
OCTOBER	0.0	0	10	2.7	5.1	12.4	4.6	25.4	8.4	74	81	93	31.9	31.1	195	17	60	54
NOVEMBER	1.7	0	1	4.4	4.8	13.8	4.6	28.9	10.5	114	122	134	33.7	32.9	220	15	61	55
DECEMBER	3.3	0	0	5.8	5.1	15.5	4.6	32.0	11.9	171	181	193	35.4	34.6	222	13	56	56
JANUARY	4.5	0	0	6.8	5.0	16.7	4.9	33.2	11.6	208	214	226	36.2	35.4	240	11	45	52
FEBRUARY	3.9	0	0	6.5	5.6	16.8	4.7	31.1	9.6	190	197	209	34.0	33.2	200	9	41	56
MARCH	1.8	0	0	5.6	5.1	15.3	4.6	30.0	10.1	164	171	182	33.6	32.8	198	11	44	56
APRIL	−1.1	0	7	3.4	5.7	13.2	4.1	25.0	7.7	96	109	121	29.8	29.0	148	13	52	60
OCT-APRIL	—	0	18	—	5.2	14.8	4.6	—	10.0	1,017	1,075	1,158	33.5	32.7	1,423	89	359	56
MAY-SEPT	—	—	—	—	—	9.3	—	—	—	—	—	—	—	—	—	—	263	—
DATA FOR MONTH TO MATURITY																		
MATURITY GROUP	EST. MATURITY DATE					*	*	*						*				
1	APRIL 3					15.4	4.4	29.7	9.9					32.5	194	11	44	56
2	APRIL 13					14.7	4.2	28.2	9.3					31.2	178	12	46	58
3	APRIL 27					13.8	3.9	25.6	8.0					29.3	155	13	51	60
4	MAY 26					11.2	3.5	22.0	7.0					27.0	120	14	50	62
5	—					—	—	—						—	—	—	—	—
6	—					—	—	—						—	—	—	—	—
7	—					—	—	—						—	—	—	—	—
8	—					—	—	—						—	—	—	—	—

* Adjusted for vine sites.

TABLE 114
LAUNCESTON, TASMANIA. 41°27′S, 147°10′E. ALT. 81m.

PERIOD	Lowest Recorded Min. (78YRS)	No.Mths with <0.3° (30YRS)	No.Mths with <2.5° (30YRS)	Av.Mthly Lowest Min. (30YRS)	Av.Min. Minus Col.4	Daily Mean (51YRS)	Daily Range /2 (51YRS)	Av.Mthly Highest Max. (30YRS)	Col.8 Minus Av.Max.	Raw	Adj.1 for Lat. & Daily Temp Rng	Adj.2 for Vine Sites	Raw	Adj. for Vine Sites	Av. Sunshine Hrs	Av. Number Rain Days	Av. Rainfall mm (30YRS)	Av. %R.H. 1500 HRS (58YRS)
	1	2	3	4	5	6	7	8	9	10	11	12	13	14	15	16	17	18
OCTOBER	-3.9	8	27	1.0	5.7	12.3	5.6	23.6	5.7	71	72	93	33.8	29.8	240	14	68	57
NOVEMBER	0.0	1	11	2.9	5.6	14.6	6.1	27.0	6.3	138	139	154	36.3	32.3	245	11	46	51
DECEMBER	-0.3	0	4	4.0	6.2	16.5	6.3	30.5	7.7	202	204	219	39.1	35.1	270	10	54	49
JANUARY	1.1	0	2	4.7	6.4	17.7	6.6	31.2	6.9	239	240	257	39.7	35.7	270	8	39	46
FEBRUARY	0.9	0	2	4.9	6.6	18.2	6.7	30.2	5.3	230	228	245	38.7	34.7	235	6	38	47
MARCH	-0.6	1	13	2.9	7.0	16.1	6.2	29.4	7.1	189	189	205	38.9	34.9	215	8	43	52
APRIL	-2.8	12	22	1.0	6.3	12.9	5.6	24.2	5.7	87	87	108	34.4	30.4	185	10	60	59
OCT-APRIL	—	22	81	—	6.3	15.5	6.2	—	6.4	1,156	1,159	1,281	37.3	33.3	1,660	67	348	52
MAY-SEPT	—	—	—	—	—	8.9	—	—	—	—	—	—	—	—	—	—	377	—

DATA FOR MONTH TO MATURITY

MATURITY GROUP	EST. MATURITY DATE			*	*	*						*				*
1	MARCH 11	18.1	5.5	29.4	6.0			34.8	234	7	42	50
2	MARCH 19	17.5	5.4	29.2	6.5			34.8	233	7	42	51
3	MARCH 27	16.9	5.2	29.0	7.0			34.8	232	8	43	53
4	APRIL 6	16.0	5.0	28.0	6.8			34.5	225	8	46	55
5	APRIL 19	14.6	4.8	26.2	6.2			32.5	206	9	52	58
6	MAY 10	12.4	4.4	22.5	5.5			29.0	170	11	63	63
7	—	—	—	—	—			—	—	—	—	—
8	—	—	—	—	—			—	—	—	—	—

* Adjusted for vine sites. Relative humidities increased by 2% for greater proximity to coast.

TABLE 115
RISDON, TASMANIA. 42°50′S, 147°21′E. ALT. 43m.

PERIOD	Lowest Recorded Min. (33YRS)	No.Mths with <0.3° (30YRS)	No.Mths with <2.5° (30YRS)	Av.Mthly Lowest Min. (30YRS)	Av.Min. Minus Col.4	Daily Mean (42YRS)	Daily Range /2 (42YRS)	Av.Mthly Highest Max. (30YRS)	Col.8 Minus Av.Max.	Raw	Adj.1 for Lat. & Daily Temp Rng	Adj.2 for Vine Sites	Raw	Adj. for Vine Sites	Av. Sunshine Hrs	Av. Number Rain Days	Av. Rainfall mm	Av. %R.H. 1500 HRS (42YRS)
	1	2	3	4	5	6	7	8	9	10	11	12	13	14	15	16	17	18
OCTOBER	0.0	1	12	2.6	5.2	12.5	4.7	25.4	8.2	78	83	99	32.2	30.6	198	16	59	54
NOVEMBER	1.1	0	2	4.2	5.1	14.0	4.7	28.4	9.7	120	127	142	33.6	32.0	222	14	53	55
DECEMBER	2.2	0	1	5.6	4.9	15.4	4.9	31.6	11.3	167	173	189	35.8	34.2	227	12	57	55
JANUARY	4.4	0	0	6.4	5.2	16.8	5.2	32.7	10.7	211	215	228	36.7	35.1	243	9	42	51
FEBRUARY	1.9	0	2	6.1	5.8	16.9	5.0	31.8	9.9	193	196	210	35.7	34.1	205	9	46	54
MARCH	0.1	1	1	4.9	5.9	15.6	4.8	30.9	10.5	174	177	192	35.6	34.0	200	10	44	53
APRIL	0.6	0	11	3.1	5.9	13.2	4.2	25.1	7.7	96	107	122	30.4	28.8	150	12	55	57
OCT-APRIL	—	2	29	—	5.4	14.9	4.8	—	9.7	1,039	1,078	1,182	34.3	32.7	1,445	82	356	54
MAY-SEPT	—	—	—	—	—	9.3	—	—	—	—	—	—	—	—	—	—	240	—

DATA FOR MONTH TO MATURITY

MATURITY GROUP	EST. MATURITY DATE			*	*	*						*				*
1	MARCH 29	16.0	4.4	30.9	10.5			34.0	202	10	44	53
2	APRIL 8	15.4	4.3	30.0	9.8			33.2	190	10	46	54
3	APRIL 19	14.5	4.1	28.2	8.9			31.4	172	11	50	55
4	MAY 8	12.8	3.7	24.0	7.5			28.0	140	13	50	57
5	—	—	—	—	—			—	—	—	—	—
6	—	—	—	—	—			—	—	—	—	—
7	—	—	—	—	—			—	—	—	—	—
8	—	—	—	—	—			—	—	—	—	—

* Adjusted for vine sites.

TABLE 116

SAINT HELENS, TASMANIA. 41°20'S, 148°14'E. ALT. 5m. (East coast, adjusted for warmest sites)

	1	2	3	4	5	6	7	8	9	10	11	12	13	14	15	16	17	18
	Temperature °C									Summation of Day Degrees above 10° with 19° Cut-Off			Temperature Variability Index					
PERIOD	Lowest Recorded Min. (60YRS)	No.Mths with <0.3° (30YRS)	No.Mths with <2.5° (30YRS)	Av.Mthly Lowest Min. (30YRS)	Av.Min. Minus Col.4	Daily Mean (54YRS)	Daily Range /2 (54YRS)	Av.Mthly Highest Max. (30YRS)	Col.8 Minus Av.Max.	Raw	Adj.1 for Lat. & Daily Temp Rng	Adj.2 for Vine Sites	Raw	Adj. for Vine Sites	Av. Sunshine Hrs	Av. Number Rain Days	Av. Rainfall mm	Av. % R.H. 1500 HRS (31YRS)
OCTOBER	−2.7	13	30	0.2	6.5	12.2	5.5	24.2	6.5	68	69	105	35.0	28.6	227	13	70	60
NOVEMBER	−2.8	1	19	2.0	6.3	13.9	5.6	27.0	7.5	117	118	151	36.2	29.8	235	12	56	61
DECEMBER	−0.6	0	7	3.9	6.2	15.5	5.4	30.2	9.3	171	172	209	37.1	30.7	253	11	62	61
JANUARY	1.1	0	4	4.2	6.8	16.7	5.7	31.6	9.2	208	210	242	38.8	32.4	253	8	50	59
FEBRUARY	0.0	1	5	4.3	7.2	17.0	5.5	30.7	8.2	196	197	229	37.4	31.0	230	9	64	59
MARCH	−0.8	1	13	3.3	6.6	15.5	5.6	28.6	7.5	170	171	206	36.5	30.1	215	10	67	59
APRIL	−1.7	11	25	0.8	6.6	12.9	5.5	24.7	6.3	87	87	122	34.9	28.5	200	12	62	62
OCT-APRIL	—	27	103	—	6.6	14.8	5.5	—	7.8	1,017	1,024	1,264	36.6	30.2	1,613	75	431	60
MAY-SEPT	—	—	—	—	—	9.2	—	—	—	—	—	—		—	—	—	351	—

	DATA FOR MONTH TO MATURITY																	
MATURITY GROUP	EST. MATURITY DATE			*	*	*								*				*
1	MAR 16			16.9	4.0	28.6	7.7							30.5	222	10	66	57
2	MAR 24			16.5	4.0	28.0	7.5							30.3	218	10	67	57
3	APRIL 2			16.1	4.0	27.3	7.2							30.0	214	10	67	57
4	APRIL 12			15.1	4.0	26.0	6.9							29.4	209	11	65	58
5	APRIL 26			14.0	3.9	24.2	6.3							28.7	202	12	63	59
6	MAY 19			11.9	3.8	21.3	5.6							27.6	190	13	61	62
7	—			—	—	—	—							—	—	—	—	—
8	—			—	—	—	—							—	—	—	—	—

*Adjusted for vine sites. Relative humidity minus 2% for more inland locations.

Commentaries and vineyard site adjustments

Bushy Park (Table 112)

Vineyard site adjustments, °C

Min.	Max.	
+1.3	−0.5	Moderate-steep slopes, vs valley floor
+0.6	+0.2	Northerly aspect
−0.2	−0.2	Altitude 60–100 m, vs Bushy Park 53 m.
+1.7	−0.5	Mean +0.6°, variability index −4.4

Notes

The topographical adjustments are about as great as can be made, and would apply only to carefully picked small areas on the lower slopes of the Derwent south bank upstream from New Norfolk. Up to about one maturity group equivalent could be gained if the soil were markedly stony or rocky.

The figures indicate that only very early-maturing grape varieties could ripen regularly in the Derwent Valley beyond the estuary, and then barely. The frost risk is very high. Lacking the favourable coastal influences enjoyed by Hobart and Risdon (q.v.), this is at best a very marginal environment for vine growing.

Hobart (Table 113)

Vineyard site adjustments, °C

Min.	Max.	
+0.6	+0.2	Northerly aspects on moderate to steep slopes
−0.1	−0.1	Altitude 50–100 m, vs Hobart 55 m
+0.5	+0.1	Mean +0.3°, variability index −0.8

Notes

The area immediately around Hobart gains substantially in its calculated ripening capacity, compared with mainland Australian sites, from the adjustments for latitude and especially for diurnal temperature range. The slow rate of temperature decline in autumn, coupled with continuing reasonable sunshine hours in late autumn, also means that some extra sugar accumulation can probably continue in midseason and later varieties into May; but whether that would be accompanied by much physiological ripening under the low temperatures of May is a different question.

Launceston (Table 114)

Vineyard site adjustments, °C

Min.	Max.	
+1.3	−0.5	Moderate slopes, greater estuary influence
+0.3	+0.1	East-north-west aspects
−0.3	−0.3	Adjustment for atypical recording site: see Notes, below
+0.2	+0.2	Altitude ~ 50 m (Launceston 81 m)
+1.5	−0.5	Mean +0.5°, variability index −4.0

Relative humidities increased 2 per cent for closer proximity to coast

Notes

The 51-year data for the original Launceston recording site, as published in Climatic Averages Australia (1956) and used in Table 114, must be regarded as suspect. Adjusted to sea-level, the mean temperature for October-April of 15.96°C is 0.33°C higher than the equivalent average recorded later in Climatic Averages Australia (1975) for the three replacement sites (Table 111). Weighting the three new sites for their lengths of recording period points to a difference perhaps closer to 0.40°C. This suggests that the old site may have been abandoned because of already excessive urban warming. In estimating vineyard site climates, I have compensated by reducing the old Launceston temperatures by 0.3°C. Some doubts must obviously remain about the north-east Tasmanian temperatures.

Risdon (Table 115)

Vineyard site adjustments, °C

Min.	Max.	
+0.4	−0.2	Moderate slopes at 50 m (Risdon ? slight slopes)
+0.3	+0.1	Predominantly northern aspects
+0.7	−0.1	Mean +0.3°, variability index −1.6

Notes

The data for Risdon give an alternative estimate for the Hobart area. Both raw and adjusted figures are very similar.

Saint Helens (Table 116)

Vineyard site adjustments, °C

Min.	Max.	
+1.4	−0.5	Moderate to steep slopes of semi-isolate hills
+0.6	+0.2	Mostly northern aspect, shelter from coast
+0.6	−0.3	Moderately stony soil
−0.4	−0.4	Altitude ~ 70 m
+2.2	−1.0	Mean +0.6°, variability index −6.4

Notes

The vineyard site adjustments for Table 116 fit the topographic characteristics of the Freycinet Estate vineyard, inland from Bicheno, chosen as representing the warmest mesoclimates of the Tasmanian east coast. The resulting effective temperature summations and estimated ripening dates fit actual vineyard experience well. However, the distance south and size of the topographic adjustments relative to Saint Helens, together with uncertainties due to apparent temperature changes over time on the east coast, suggest that caution is needed. See more detailed discussion by Gladstones (1994).

Accepting this estimate for the rare warmest mesoclimates, those more generally available throughout the Swan-Apsley river basin cannot be expected to have as great a ripening capacity, even through they might be at lower altitudes. They still all enjoy a degree of shelter from early sea breezes as compared with the official weather stations, which are more or less immediately coastal; but the importance of this is hard to estimate, since it must at least be partly offset by the colder nights inland. Full ripening is probably general on such sites up to about Maturity Group 3 or 4, or perhaps of slightly later varieties on the warmest stony soils.

15

Viticultural Environments of New Zealand

The main existing wine producing areas of New Zealand are represented by **Auckland** (Table 117), in the north of the North Island; **Napier** (Table 119), on Hawke Bay on the east coast of the North Island; and **Blenheim** (Table 118), on Cloudy Bay at the north-east tip of the South Island.

Initial grape plantings in the early 19th century were entirely of *Vitis vinifera* varieties, which spread in small plantings through much of the North Island in the mid-century. But when wine production began to expand in the 1930s and 1940s, following the original industry's near-extinction under prohibitionist pressures and labour shortages during the First World War, most plantings were of direct-producing species hybrids. In 1965, *vinifera* varieties were only 32 per cent of total plantings (Thorpy 1973). The preference for hybrids was due both to Phylloxera and to fungal diseases, which flourish in the mild, humid climate of the North Island but were presumably absent from the earliest plantings, as in Australia. It was only in the mid to late 1960s that *vinifera* varieties again began to replace the hybrids. Virtually all recent commercial plantings are of *vinifera* varieties.

Commercial-scale viticulture in the South Island did not begin until 1973; but the outstanding success of dry white wines from Sauvignon Blanc and Chardonnay grapes grown in the Wairau Valley, near Blenheim in Marlborough Province, has already shown this to be perhaps the best environment of all in New Zealand for making table wines of classic European cool-climate style.

The climate recorded for **Nelson** (not presented) is a little less favourable than that of Blenheim, having an October-April raw temperature summation of 1077 day degrees and rainfall of 542 mm, but a still generous 1616 sunshine hours. However Nelson itself is hardly representative of potential vineyard areas on Tasman Bay. East and north-east facing slopes on the Motueka and Wai-iti river systems, being more in the direct rain and wind shadows of the Tasman Mountains to the west, should be warmer, drier, and sunnier. Selected sites are thus probably similar, climatically, to Blenheim.

Jackson and Cherry (1988) claim that varieties as late as Cabernet Sauvignon ripen satisfactorily at Christchurch, in Canterbury Province, because of a long growing season irrespective of temperature summation as traditionally measured. I estimate a raw October-April summation of 897, based on 52 years of records covering the first half of the 20th century (Meteorological Office, Great Britain, 1958b). That is a full 200 day degrees less than at Blenheim. On such a basis only a few outstandingly warm sites and soils, unlikely to be available in commercially significant areas, could be expected to ripen any high-quality grape varieties regularly. To ripen Cabernet Sauvignon would be exceptional, although it may well happen in occasional very warm years such as some of those experienced during the 1980s. In any case ripening, when it occurs, must be under very low temperatures. The question has to be asked: do the grapes develop enough flavour?

A final area in the South Island of New Zealand which has created much viticultural interest is Central Otago. A recent report (Anon 1988) gives a temperature summation of 1100 day degrees and annual sunshine hours of up to 2000 on the north-facing slopes of Karawau Gorge, east of Queenstown. Effective temperature summations of that order, with reasonable freedom from spring frosts, do seem feasible for selected low-altitude sites in such an area, sheltered as it is by high mountains to the west, and having the strong influence on local air convection of large water bodies such as Lake Wakatipu. Adiabatically warmed north-westerly winds off the alps bring significant local warming in spring and autumn. The temperature pattern would in any case be more continental than at Christchurch, so that for a given total temperature summation the critical flowering, bud differentiation and early ripening periods should be under higher temperatures and probably sunshine hours. The grape varieties mentioned as being successful include Chasselas and Traminer: a fact which, together with the nature of the terrain, invites analogy with Switzerland (see Geneva, Chapter 18, Table 147).

Similar warm and sunny enclaves doubtless exist to

the north through Canterbury Province, in the rain shadow of the Southern Alps: mainly on the southern sides of lakes and deep valleys. At comparable altitudes they presumably become warmer with distance north.

As in Tasmania, such areas appear to offer better prospects than the immediate east coastal region, which is more directly influenced by cool southerly winds.

TABLE 117

AUCKLAND, NEW ZEALAND. 36°51'S, 174°41'E. ALT. 49m.

	1	2	3	4	5	6	7	8	9	10	11	12	13	14	15	16	17	18
	Temperature °C									Summation of Day Degrees above 10° with 19° Cut-Off			Temperature Variability Index					
PERIOD	Lowest Recorded Min. (77YRS)	No.Mths with <0.3°	No.Mths with <2.5°	Av.Mthly Lowest Min. (85YRS)	Av.Min. Minus Col.4	Daily Mean (85YRS)	Daily Range /2 (85YRS)	Av.Mthly Highest Max. (85YRS)	Col.8 Minus Av.Max.	Raw	Adj.1 for Lat. & Daily Temp Rng	Adj.2 for Vine Sites	Raw	Adj. for Vine Sites	Av. Sunshine Hrs (41YRS)	Av. Number Rain Days	Av. Rainfall mm (92YRS)	Av. % R.H. 1500 HRS (9YRS)
OCTOBER	2.2			6.8	4.1	14.1	3.2	20.0	2.7	127	153	138	19.6	23.6	176	16	102	66
NOVEMBER	5.0			8.6	3.8	15.8	3.4	22.8	3.6	174	195	180	21.0	25.0	200	15	89	64
DECEMBER	6.1			9.9	4.2	17.6	3.5	24.7	3.6	236	254	238	21.8	25.8	222	12	79	64
JANUARY	7.2			11.4	4.2	19.1	3.5	26.1	3.5	279	279	279	21.7	25.7	231	10	79	62
FEBRUARY	8.3			11.8	4.0	19.3	3.5	26.1	3.3	252	252	252	21.3	25.3	195	10	94	61
MARCH	5.6			10.6	4.3	18.3	3.4	24.8	3.1	257	279	266	21.0	25.0	187	11	81	65
APRIL	3.9			8.2	4.9	16.3	3.2	22.7	3.2	189	218	203	20.9	24.9	151	14	97	69
OCT-APRIL	—		—		4.2	17.2	3.4	—	3.3	1,514	1,630	1,556	21.0	25.0	1,362	88	621	64
MAY-SEPT	—		—	—		12.0	—	—	—	—	—	—	—	—	—	—	627	—
DATA FOR MONTH TO MATURITY																		
MATURITY GROUP	EST. MATURITY DATE					*	*							*				
1	FEBRUARY 24					19.4	4.5	27.1	3.3					25.4	214	10	98	61
2	MARCH 2					19.3	4.5	27.1	3.3					25.3	210	10	101	61
3	MARCH 7					19.2	4.5	26.9	3.2					25.3	206	10	98	62
4	MARCH 13					19.0	4.5	26.7	3.2					25.2	202	10	94	63
5	MARCH 19					18.8	4.4	26.4	3.2					25.2	197	11	89	63
6	MARCH 25					18.6	4.4	26.1	3.1					25.1	192	11	84	64
7	MARCH 31					18.3	4.4	25.8	3.1					25.0	187	11	81	65
8	APRIL 7					17.9	4.3	25.3	3.1					25.0	180	12	84	66

* Adjusted for vine sites.

TABLE 118

BLENHEIM, NEW ZEALAND. 41°30′S, 173°58′E. ALT. 12m.

	1	2	3	4	5	6	7	8	9	10	11	12	13	14	15	16	17	18
				Temperature °C						Summation of Day Degrees above 10° with 19° Cut-Off			Temperature Variability Index					
PERIOD	Lowest Recorded Min.	No.Mths with <0.3°	No.Mths with <2.5°	Av.Mthly Lowest Min. (20YRS)	Av.Min. Minus Col.4	Daily Mean (20YRS)	Daily Range /2 (20YRS)	Av.Mthly Highest Max. (20YRS)	Col.8 Minus Av.Max.	Raw	Adj.1 for Lat. & Daily Temp Rng	Adj.2 for Vine Sites	Raw	Adj. for Vine Sites	Av. Sunshine Hrs (24YRS)	Av. Number Rain Days	Av. Rainfall mm (30YRS)	Av. % R.H. 1500 HRS
OCTOBER				−0.2	7.4	12.3	5.1	23.6	6.2	71	72	90	34.0	31.0	219		54	
NOVEMBER				1.7	6.8	14.1	5.6	27.2	7.5	123	124	134	36.7	33.7	242		43	
DECEMBER				3.7	7.0	16.3	5.6	29.8	7.9	195	197	208	37.3	34.3	257		52	
JANUARY				4.9	6.9	17.3	5.5	31.4	8.6	226	229	241	37.5	34.5	261		48	
FEBRUARY				5.0	7.0	17.4	5.4	29.9	7.1	207	209	221	35.7	32.7	227		63	
MARCH				2.5	7.5	15.8	5.8	28.2	6.6	180	180	188	37.3	34.3	222		39	
APRIL				0.2	7.8	13.3	5.3	25.8	7.2	99	99	113	36.2	33.2	176		48	
OCT–APRIL				—	7.2	15.2	5.5	—	7.3	1,101	1,110	1,195	36.4	33.4	1,604		347	
MAY–SEPT				—	—	8.6	—	—	—				—	—	—		296	

DATA FOR MONTH TO MATURITY

MATURITY GROUP	EST. MATURITY DATE				*	*	*							*				
1	MARCH 25						16.5	5.0	28.0	6.8					34.0	226		43	
2	APRIL 4						15.8	4.9	27.4	6.7					33.7	218		41	
3	APRIL 16						14.8	4.7	26.3	6.9					33.4	201		44	
4	MAY 3						13.2	4.5	25.0	7.3					33.0	170		49	
5	—						—	—	—	—					—	—		—	
6	—						—	—	—	—					—	—		—	
7	—						—	—	—	—					—	—		—	
8	—						—	—	—	—					—	—		—	

* Adjusted for vine sites.

TABLE 119

NAPIER, NEW ZEALAND. 39°29′S, 176°55′E. ALT 2m.

	1	2	3	4	5	6	7	8	9	10	11	12	13	14	15	16	17	18
				Temperature °C						Summation of Day Degrees above 10° with 19° Cut-Off			Temperature Variability Index					
PERIOD	Lowest Recorded Min. (34YRS)	No.Mths with <0.3°	No.Mths with <2.5°	Av.Mthly Lowest Min. (25YRS)	Av.Min. Minus Col.4	Daily Mean (25YRS)	Daily Range /2 (25YRS)	Av.Mthly Highest Max. (25YRS)	Col.8 Minus Av.Max.	Raw	Adj.1 for Lat. & Daily Temp Rng	Adj.2 for Vine Sites	Raw	Adj. for Vine Sites	Av. Sunshine Hrs (41YRS)	Av. Number Rain Days	Av. Rainfall mm (63YRS)	Av. % R.H. 1500 HRS (2YRS)
OCTOBER	−0.6			3.2	5.8	13.6	4.6	24.7	6.5	112	118	105	30.7	33.5	222	9	56	63
NOVEMBER	1.7			4.9	5.8	15.3	4.6	27.4	7.5	159	165	153	31.7	34.5	235	9	61	61
DECEMBER	3.3			7.3	5.5	17.3	4.5	28.7	6.9	226	233	219	30.4	33.2	263	8	58	61
JANUARY	5.0			8.7	5.2	18.6	4.7	30.4	7.1	267	270	260	31.1	33.9	261	8	74	60
FEBRUARY	3.3			8.3	5.9	18.6	4.4	29.4	6.4	241	249	235	29.9	32.7	212	8	76	61
MARCH	0.0			7.5	4.9	16.9	4.5	27.9	6.5	214	222	208	29.4	32.2	211	8	74	66
APRIL	−0.6			4.1	6.2	14.7	4.4	25.6	6.5	141	150	135	30.3	33.1	179	8	76	69
OCT–APRIL	—			—	5.6	16.4	4.5	—	6.8	1,360	1,407	1,315	30.5	33.3	1,583	58	475	63
MAY–SEPT	—			—	—	10.0	—	—	—				—	—	—	—	417	—

DATA FOR MONTH TO MATURITY

MATURITY GROUP	EST. MATURITY DATE				*	*	*							*				*
1	MARCH 11						17.9	5.1	29.7	6.4					32.5	218	8	79	61
2	MARCH 18						17.5	5.2	29.4	6.5					32.4	215	8	77	62
3	MARCH 26						17.1	5.2	28.9	6.5					32.3	211	8	75	63
4	APRIL 4						16.4	5.2	28.3	6.5					32.3	207	8	75	64
5	APRIL 14						15.7	5.2	27.6	6.5					32.6	198	8	75	65
6	APRIL 26						14.8	5.1	26.7	6.5					33.0	184	8	76	67
7	MAY 17						12.8	5.1	24.8	6.5					32.5	166	8	86	67
8	—						—	—	—	—					—	—	—	—	—

* Adjusted for vine sites. Relative humidities minus 2% for vines being further inland.

Commentaries and vineyard site adjustments

Auckland (Table 117)

Vineyard site adjustments, °C

Min.	Max.	
–1.0	+1.0	Vines further inland (Auckland highly maritime)
–	–	Altitude and topography ~ as Auckland
–	–	Sandy to clay soils
–1.0	+1.0	Mean unchanged, variability index +4.0

Notes

The main problem for *vinifera* varieties is high rainfall and humidity throughout the season, combined with sub-optimal sunshine hours. Disease control is therefore paramount, and direct-producing hybrid varieties almost certainly the better adapted. With so mild a winter and maritime a temperature pattern, lack of dormancy and irregular bud-burst is presumably also a problem with some varieties. On the other hand growing season and ripening average mean temperatures are highly favourable, similar to or slightly lower than in Bordeaux; while temperature variability is extraordinarily low. Given good disease control and trellising, the data point to Bordeaux grape varieties: Sémillon and Sauvignon Blanc for white wines and Cabernet Sauvignon, Cabernet Franc and perhaps especially (but depending on its setting) Merlot for red wines. Medium-bodied wines of considerable merit should be achievable in occasional seasons with well-above-average sunshine hours. In normal seasons, the low average ratio of seasonal sunshine hours to effective temperature summation (Chapter 3, and Figure 2) suggests that the region will typically produce light, probably soft, styles of wine for early consumption.

Blenheim (Table 118)

Vineyard site adjustments, °C

Min.	Max.	
–	–	Flat valley; altitude as Blenheim
+1.0	–0.5	Soil very gravelly: basalt pebbles
+1.0	–0.5	Mean +0.25°, variability index –3.0

Notes

Vines are grown on the river flats of the Wairau Valley. The figures for Blenheim show a fairly strong risk of spring frosts, due mainly to variability of the minimum temperature (Column 5). Gravelly soils are an offsetting factor. The heat-absorbency of the basalt pebbles that characterize the best soils, combined with the ample sunshine to heat them, possibly results in more night warming and frost protection on the better sites than is suggested by the site climate adjustment used above.

Table 118 indicates full ripening of Maturity Groups 1–4 to produce true cool-climate styles of wine. Good ripening period sunshine hours for such a cool environment, minimal ripening period rain following good rainfall in February, and on average a continuation of mean temperatures over 10°C until at least mid May suggest that ripening even of Maturity Group 4 should be reasonably reliable – if under cool conditions. The climatic data underline the critical role of soil stoniness in the full ripening of Maturity Group 3 and 4 varieties, as well as in frost avoidance.

The very mild highest maxima throughout growth and ripening are especially notable. Temperature equability fully matches that of comparable cool climates in northern Europe, and should theoretically result in outstanding delicacy and aroma retention in the fruit and wines. White wines from Sauvignon Blanc and Chardonnay (Maturity Group 3) have already shown the excellent adaptation of those varieties to the better stony soils in the area. Pinot Noir (frost allowing) and Meunier should also be well adapted for Champagne-style wines in cool seasons, and for still dry red wines in warmer seasons.

The climate of Blenheim is intermediate between those of Hobart and Launceston, Tasmania (Tables 113 and 114). Probably its closest parallel is with Launceston – Pipers Brook at altitudes of about 120–160 m.

Napier (Table 119)

Vineyard site adjustments, °C

Min.	Max.	
−0.7	+0.7	Further inland (Napier on coast)
–	–	More or less flat; fertile alluvial soils
−0.2	−0.2	Altitudes to 50 m or more (Napier 2 m)
−0.9	+0.5	Mean −0.2°, variability index +2.8

Relative humidities reduced 2 per cent because of more inland vineyard locations.

Notes

Ripening mean temperatures, temperature variability index, and seasonal and ripening sunshine hours are all excellent for light to medium-bodied wines from all varieties up to Maturity Group 4. Midseason (Maturity Groups 5 and 6) Bordeaux and similar red varieties should also ripen, but at about the lower limit of temperature and sunshine hours for making red wines. Combined with the high ripening period rainfall, this means that only light red wines can be expected on average, and those of sunless years may be very light. On the other hand the temperature equability and high relative humidities give the potential for making outstanding Bordeaux-style red wines in warm, sunny seasons.

For all varieties, high rainfall throughout the year combined with the mainly fertile soils of the Hawke Bay area mean strong vine vigour, so that body and character in the wine will depend greatly on proper attention to trellising and yield levels. The relative humidities indicate problems with disease control, although less than at Auckland.

16

Viticultural Environments of France

The table wines of France are the historical benchmark against which most others are compared. Because of that, and the availability of good climatic and topographic data and soil descriptions, I have included as many of the major wine-producing regions as possible.

Note that here, as for other non-Australian countries apart from New Zealand and the USA, I have included ripening period estimates only for the vine maturity groups that are actually grown. The reason is to highlight the identities and ripening conditions of those maturity groups that over the centuries have proved best adapted to them, for ready matching against estimated ripening conditions for all maturity groups in New World environments.

TABLE 120

ANGERS, FRANCE. 47°30′N, 0°35′W. ALT. 68m. (Anjou, Ancenis, Chinon and Bourgueil)

	1	2	3	4	5	6	7	8	9	10	11	12	13	14	15	16
	Temperature °C							Summation of Day Degrees above 10° with 19° Cut-Off			Temperature Variability Index					
PERIOD	Lowest Recorded Min.	Av.Mthly Lowest Min.	Av.Min Minus Col.2	Daily Mean (40YRS)	Daily Range /2 (40YRS)	Av.Mthly Highest Max.	Col.6 Minus Av.Max.	Raw #	Adj.1 for Lat. & Daily Temp Rng	Adj.2 for Vine Sites	Raw	Adj. for Vine Sites	Av. Sunshine Hours (10YRS)	Av. No. Rain Days (15YRS)	Av. Rainfall mm (40YRS)	Av. % R.H. 1300 HRS (10YRS)
APRIL		−0.5	6.1	10.6	5.0	22.8	7.2	25	26	42	33.3	29.7	176	15	47	63
MAY		2.7	6.3	14.5	5.5	27.2	7.2	140	147	166	35.5	31.9	215	12	53	63
JUNE		6.1	5.5	17.4	5.8	30.9	7.7	222	236	250	36.4	32.8	228	11	50	63
JULY		8.2	5.2	19.4	6.0	33.5	8.1	279	279	279	37.3	33.7	240	11	42	59
AUGUST		7.7	5.5	19.1	5.9	33.0	8.0	279	279	279	37.1	33.5	223	10	41	59
SEPTEMBER		4.1	6.9	16.6	5.6	29.9	7.7	198	200	217	37.0	33.4	180	10	41	65
OCTOBER		−0.2	8.0	12.2	4.4	23.1	6.5	68	76	101	32.1	28.5	127	15	74	72
APR-OCTOBER		—	6.2	15.7	5.5	—	7.5	1,211	1,243	1,334	35.5	31.9	1,389	84	348	63
NOV-MARCH		—	—	6.2	—	—	—	—	—	—	—	—	—	—	273	—
DATA FOR MONTH TO MATURITY																
MATURITY GROUP	EST. MATURITY DATE		*	*	*							*				
1	—		—	—	—	—						—	—	—	—	—
2	—		—	—	—	—						—	—	—	—	—
3	—		—	—	—	—						—	—	—	—	—
4	SEPTEMBER 24		17.3	4.7	30.2	7.8						33.4	191	10	41	64
5	OCTOBER 4		16.0	4.6	28.8	7.6						33.0	174	10	41	66
6	—		—	—	—	—						—	—	—	—	—
7	—		—	—	—	—						—	—	—	—	—
8	—		—	—	—	—						—	—	—	—	—

April and October day degree summations derived from temperature curve, not directly from monthly average means. * Adjusted for vine sites.

TABLE 121

ANGOULÊME, FRANCE. 45°43′N, 0°10′E. ALT. 82m. (Charente: Cognac)

	1	2	3	4	5	6	7	8	9	10	11	12	13	14	15	16
	Temperature °C							Summation of Day Degrees above 10° with 19° Cut-Off			Temperature Variability Index					
PERIOD	Lowest Recorded Min.	Av.Mthly Lowest Min.	Av.Min Minus Col.2	Daily Mean (40YRS)	Daily Range /2 (40YRS)	Av.Mthly Highest Max. (30YRS)	Col.6 Minus Av.Max.	Raw	Adj.1 for Lat. & Daily Temp Rng	Adj.2 for Vine Sites	Raw	Adj. for Vine Sites	Av. Sunshine Hours (10YRS)	Av. No. Rain Days (15YRS)	Av. Rainfall mm (40YRS)	Av. % R.H. 1300 HRS (10YRS)
APRIL		−0.1	5.9	11.5	5.7	25.3	8.1	45	46	52	36.8	34.4	180	15	64	59
MAY		3.1	5.9	15.2	6.2	29.8	8.4	161	167	173	39.1	36.7	211	14	70	59
JUNE		6.7	5.2	18.3	6.4	32.7	8.0	249	261	267	38.8	36.4	223	11	74	59
JULY		8.8	5.0	20.5	6.7	34.5	7.3	279	279	279	39.1	36.7	254	11	51	55
AUGUST		8.2	5.2	20.2	6.8	34.5	7.5	279	279	279	39.9	37.5	235	11	55	57
SEPTEMBER		5.1	6.2	17.8	6.5	31.7	7.4	234	236	242	39.6	37.2	204	10	56	61
OCTOBER		0.1	7.7	13.1	5.3	25.2	6.8	96	95	106	35.7	33.3	135	14	80	69
APR-OCTOBER		—	5.9	16.7	6.2	—	7.6	1,343	1,363	1,398	38.4	36.0	1,442	86	450	60
NOV-MARCH		—	—	6.8	—	—	—	—	—	—	—	—	—	—	354	—

DATA FOR MONTH TO MATURITY

MATURITY GROUP	EST. MATURITY DATE	*	*	*						*				
1	—	—	—	—	—					—	—	—	—	—
2	—	—	—	—	—					—	—	—	—	—
3	—	—	—	—	—					—	—	—	—	—
4	—	—	—	—	—					—	—	—	—	—
5	SEPTEMBER 25	18.7	6.0	32.2	7.4					37.3	211	10	56	60
6	OCTOBER 1	17.9	5.9	31.1	7.4					37.1	202	10	57	61
7	OCTOBER 9	16.6	5.6	29.5	7.2					36.3	188	11	61	63
8	—	—	—	—	—					—	—	—	—	—

* Adjusted for vine sites.

TABLE 122

AUXERRE, FRANCE. 47°79′N, 3°34′E. ALT. 99m. (Chablis)

	1	2	3	4	5	6	7	8	9	10	11	12	13	14	15	16
	Temperature °C							Summation of Day Degrees above 10° with 19° Cut-Off			Temperature Variability Index					
PERIOD	Lowest Recorded Min. (17YRS)	Av.Mthly Lowest Min. (17YRS)	Av.Min Minus Col.2	Daily Mean (40YRS)	Daily Range /2 (40YRS)	Av.Mthly Highest Max. (17YRS)	Col.6 Minus Av.Max.	Raw #	Adj.1 for Lat. & Daily Temp Rng	Adj.2 for Vine Sites	Raw	Adj. for Vine Sites	Av. Sunshine Hours	Av. No. Rain Days (15YRS)	Av. Rainfall mm (40YRS)	Av. % R.H. 1200 HRS (66YRS)
APRIL	−5.0	−1.7	6.2	10.2	5.7	25.0	9.1	20	21	28	38.1	32.7	175	13	51	52
MAY	−1.7	0.6	7.6	14.4	6.2	28.9	8.3	136	144	149	40.7	35.3	200	13	59	56
JUNE	2.2	5.0	6.2	17.5	6.3	31.7	7.9	225	241	245	39.3	33.9	223	11	71	61
JULY	5.0	7.2	5.6	19.2	6.4	34.4	8.8	279	279	279	40.0	34.6	236	12	63	56
AUGUST	4.4	6.7	5.5	18.8	6.6	33.3	7.9	273	279	279	39.8	34.4	228	12	58	56
SEPTEMBER	−0.6	2.2	7.5	15.9	6.2	29.4	7.3	177	179	184	39.6	34.2	184	9	53	61
OCTOBER	−6.7	−2.2	8.5	11.3	5.0	25.0	8.7	42	41	53	37.2	31.8	125	14	69	72
APR-OCTOBER	—	—	6.7	15.3	6.1	—	8.3	1,152	1,184	1,217	39.2	33.8	1,371	84	424	59
NOV-MARCH	—	—	—	4.8	—	—	—	—	—	—	—	—	—	—	253	—

DATA FOR MONTH TO MATURITY

MATURITY GROUP	EST. MATURITY DATE	*	*	*						*				
1	—	—	—	—	—					—	—	—	—	—
2	—	—	—	—	—					—	—	—	—	—
3	SEPTEMBER 27	16.4	4.9	28.6	7.4					34.2	189	9	54	60
4	—	—	—	—	—					—	—	—	—	—
5	—	—	—	—	—					—	—	—	—	—
6	—	—	—	—	—					—	—	—	—	—
7	—	—	—	—	—					—	—	—	—	—
8	—	—	—	—	—					—	—	—	—	—

April and October day degree summations derived from temperature curves, not directly from monthly average means. * Adjusted for vine sites.

TABLE 123

BEAUJOLAIS (district), FRANCE. Approx. 46°03′N, 4°38′E. ALT. approx. 300m. (All data estimated)

	1	2	3	4	5	6	7	8	9	10	11	12	13	14	15	16
	Temperature °C							Summation of Day Degrees above 10° with 19° Cut-Off			Temperature Variability Index					
PERIOD	Lowest Recorded Min.	Av.Mthly Lowest Min.	Av.Min Minus Col.2	Daily Mean	Daily Range /2	Av.Mthly Highest Max.	Col.6 Minus Av.Max.	Raw #	Adj.1 for Lat. & Daily Temp Rng	Adj.2 for Vine Sites	Raw	Adj. for Vine Sites	Av. Sunshine Hours	Av. No. Rain Days	Av. Rainfall mm	Av. % R.H. 1300 HRS
APRIL	−1.0	6.4	10.5	5.1	23.0	7.4	25	26		33.9		176	12	63	57
MAY	2.7	6.7	14.9	5.5	27.7	7.3	152	158		36.2		222	12	69	56
JUNE	6.6	5.5	18.0	5.9	31.3	7.4	240	252		36.7		236	11	79	54
JULY	9.2	5.2	20.5	6.1	33.8	7.2	279	279		36.7		269	10	82	52
AUGUST	8.5	5.2	19.7	6.0	32.9	7.2	279	279		36.5		244	11	78	52
SEPTEMBER	4.1	6.9	16.4	5.4	28.9	7.1	192	194		35.4		199	11	74	57
OCTOBER	−0.2	7.3	11.4	4.3	22.9	7.2	46	54		31.8		124	12	78	66
APR-OCTOBER	—	6.2	15.9	5.4	—	7.3	1,213	1,242		35.3		1,470	79	523	56
NOV-MARCH	—	—	4.4	—	—	—	—	—		—		—	—	280	—

DATA FOR MONTH TO MATURITY

MATURITY GROUP	EST. MATURITY DATE															
1	—		—	—	—	—				—		—	—	—	—
2	—		—	—	—	—				—		—	—	—	—
3	SEPTEMBER 22	17.5	5.6	30.1	7.1				35.7		210	11	75	56
4	—		—	—	—	—				—		—	—	—	—
5	—		—	—	—	—				—		—	—	—	—
6	—		—	—	—	—				—		—	—	—	—
7	—		—	—	—	—				—		—	—	—	—
8	—		—	—	—	—				—		—	—	—	—

April and October day degree summations derived from temperature curve, not directly from monthly average means.

TABLE 124

BERGERAC, FRANCE. 44°51′N, 0°29′E. ALT. 33m. (Bergerac, Monbazillac)

	1	2	3	4	5	6	7	8	9	10	11	12	13	14	15	16
	Temperature °C							Summation of Day Degrees above 10° with 19° Cut-Off			Temperature Variability Index					
PERIOD	Lowest Recorded Min.	Av.Mthly Lowest Min.	Av.Min Minus Col.2	Daily Mean (40YRS)	Daily Range /2 (40YRS)	Av.Mthly Highest Max.	Col.6 Minus Av.Max.	Raw	Adj.1 for Lat. & Daily Temp Rng	Adj.2 for Vine Sites	Raw	Adj. for Vine Sites	Av. Sunshine Hours (10YRS)	Av. No. Rain Days (15YRS)	Av. Rainfall mm (40YRS)	Av. % R.H. 1300 HRS
APRIL	−0.3	6.3	11.7	5.7	25.4	8.0	51	52	52	37.1	33.5	164	15	63	58
MAY	2.8	6.3	15.6	6.5	30.2	8.1	174	179	176	40.4	36.8	185	13	64	58
JUNE	7.3	5.4	19.0	6.3	33.1	7.8	270	270	270	38.4	34.8	210	10	59	58
JULY	9.1	5.4	21.1	6.6	34.9	7.2	279	279	279	39.0	35.4	248	10	49	56
AUGUST	8.3	5.8	20.9	6.8	35.7	8.0	279	279	279	41.0	37.4	241	10	60	54
SEPTEMBER	4.2	7.0	18.0	6.8	32.2	7.4	240	237	234	41.6	38.0	198	10	44	59
OCTOBER	−0.3	8.0	13.3	5.6	26.7	7.8	102	101	103	38.2	34.6	136	13	64	68
APR-OCTOBER	—	6.3	17.1	6.3	—	7.8	1,395	1,397	1,393	39.4	35.8	1,382	81	403	59
NOV-MARCH	—	—	6.8	—	—	—	—	—	—	—	—	—	—	327	—

DATA FOR MONTH TO MATURITY

MATURITY GROUP	EST. MATURITY DATE	*	*	*							*					
1	—		—	—	—	—				—		—	—	—	—
2	—		—	—	—	—				—		—	—	—	—
3	SEPTEMBER 12	19.9	5.9	33.7	7.8					37.6	228	10	55	56
4	SEPTEMBER 18	19.4	5.9	32.9	7.7					37.7	219	10	51	57
5	SEPTEMBER 25	18.7	5.9	32.0	7.5					37.9	208	10	47	58
6	—		—	—	—	—				—		—	—	—	—
7	—		—	—	—	—				—		—	—	—	—
8	—		—	—	—	—				—		—	—	—	—

* Adjusted for vine sites.

199

TABLE 125

BORDEAUX, FRANCE. 44°50′N, 0°43′W. ALT. 47m. (a: Graves, Sauternes)

	1	2	3	4	5	6	7	8	9	10	11	12	13	14	15	16
			Temperature °C					Summation of Day Degrees above 10° with 19° Cut-Off			Temperature Variability Index					
PERIOD	Lowest Recorded Min. (41YRS)	Av.Mthly Lowest Min. (41YRS)	Av.Min Minus Col.2	Daily Mean (40YRS)	Daily Range /2 (40YRS)	Av.Mthly Highest Max. (41YRS)	Col.6 Minus Av.Max.	Raw	Adj.1 for Lat. & Daily Temp Rng	Adj.2 for Vine Sites	Raw	Adj. for Vine Sites	Av. Sunshine Hours (10YRS)	Av. No. Rain Days (15YRS)	Av. Rainfall mm (40YRS)	Av. % R.H. 1300 HRS (10YRS)
APRIL	−5.6	0.6	5.9	11.7	5.2	25.0	8.1	51	52	62	34.8	33.6	185	16	67	58
MAY	0.6	3.9	5.9	15.4	5.6	29.4	8.4	167	173	182	36.7	35.5	205	15	65	58
JUNE	2.8	7.2	5.2	18.3	5.9	32.2	8.0	249	259	268	36.8	35.6	227	14	60	58
JULY	5.6	9.4	5.0	20.5	6.1	33.9	7.3	279	279	279	36.7	35.5	257	12	52	56
AUGUST	0.6	8.9	5.2	20.5	6.4	34.4	7.5	279	279	279	38.3	37.1	250	10	47	54
SEPTEMBER	−1.7	6.1	6.2	18.3	6.0	31.7	7.4	249	251	260	37.6	36.4	205	12	55	59
OCTOBER	−5.6	1.1	7.7	13.8	5.0	25.6	6.8	118	116	130	34.5	33.3	143	14	81	68
APR-OCTOBER	—	—	5.9	16.9	5.7	—	7.6	1,392	1,409	1,460	36.5	35.3	1,472	93	427	59
NOV-MARCH	—	—	—	7.4	—	—	—	—	—	—	—	—	—	—	406	—

DATA FOR MONTH TO MATURITY

MATURITY GROUP	EST. MATURITY DATE	*	*	*					*				
1	—		—	—	—				—		—	—	—
2	—		—	—	—				—		—	—	—
3	SEPTEMBER 9	20.4	6.0	33.4	7.5				36.9	241	10	49	55
4	SEPTEMBER 14	20.0	5.9	33.1	7.5				36.8	232	11	50	56
5	SEPTEMBER 20	19.5	5.9	32.6	7.4				36.6	222	11	52	57
6	SEPTEMBER 27	18.9	5.8	32.0	7.4				36.4	210	12	54	58
7	OCTOBER 4	18.0	5.6	31.0	7.3				36.0	195	12	58	60
8	—		—	—	—				—	—	—	—	—

* Adjusted for vine sites.

TABLE 126

BORDEAUX, FRANCE. 44°50′N, 0°43′W. ALT. 47m. (b: Médoc)

	1	2	3	4	5	6	7	8	9	10	11	12	13	14	15	16
			Temperature °C					Summation of Day Degrees above 10° with 19° Cut-Off			Temperature Variability Index					
PERIOD	Lowest Recorded Min. (41YRS)	Av.Mthly Lowest Min. (41YRS)	Av.Min Minus Col.2	Daily Mean (40YRS)	Daily Range /2 (40YRS)	Av.Mthly Highest Max. (41YRS)	Col.6 Minus Av.Max.	Raw	Adj.1 for Lat. & Daily Temp Rng	Adj.2 for Vine Sites	Raw	Adj. for Vine Sites	Av. Sunshine Hours (10YRS)	Av. No. Rain Days (15YRS)	Av. Rainfall mm (40YRS)	Av. % R.H. 1300 HRS (10YRS)
APRIL	−5.6	0.6	5.9	11.7	5.2	25.0	8.1	51	52	66	34.8	30.4	185	16	67	58
MAY	0.6	3.9	5.9	15.4	5.6	29.4	8.4	167	173	181	36.7	32.3	205	15	65	58
JUNE	2.8	7.2	5.2	18.3	5.9	32.2	8.0	249	259	262	36.8	32.4	227	14	60	58
JULY	5.6	9.4	5.0	20.5	6.1	33.9	7.3	279	279	279	36.7	32.3	257	12	52	56
AUGUST	0.6	8.9	5.2	20.5	6.4	34.4	7.5	279	279	279	38.3	33.9	250	10	47	54
SEPTEMBER	−1.7	6.1	6.2	18.3	6.0	31.7	7.4	249	251	253	37.6	33.2	205	12	55	59
OCTOBER	−5.6	1.1	7.7	13.8	5.0	25.6	6.8	118	116	133	34.5	30.1	143	14	81	68
APR-OCTOBER	—	—	5.9	16.9	5.7	—	7.6	1,392	1,409	1,453	36.5	32.1	1,472	93	427	59
NOV-MARCH	—	—	—	7.4	—	—	—	—	—	—	—	—	—	—	406	—

DATA FOR MONTH TO MATURITY

MATURITY GROUP	EST. MATURITY DATE	*	*	*					*				*
1	—	—	—	—	—				—	—	—	—	—
2	—	—	—	—	—				—	—	—	—	—
3	—	—	—	—	—				—	—	—	—	—
4	SEPTEMBER 15	19.7	5.1	32.1	7.5				33.5	232	11	51	59
5	SEPTEMBER 21	19.3	5.0	31.6	7.4				33.4	221	11	52	60
6	SEPTEMBER 28	18.6	4.9	31.0	7.4				33.2	209	12	54	61
7	OCTOBER 6	17.5	4.7	29.5	7.3				32.7	194	12	58	63
8	—	—	—	—	—				—	—	—	—	—

* Adjusted for vine sites.

TABLE 127
CAHORS, FRANCE. 44°28'N, 0°26'E. ALT. 122m.

	1	2	3	4	5	6	7	8	9	10	11	12	13	14	15	16
	Temperature °C							Summation of Day Degrees above 10° with 19° Cut-Off			Temperature Variability Index					
PERIOD	Lowest Recorded Min.	Av.Mthly Lowest Min.	Av.Min Minus Col.2	Daily Mean (40YRS)	Daily Range /2 (40YRS)	Av.Mthly Highest Max.	Col.6 Minus Av.Max.	Raw	Adj.1 for Lat. & Daily Temp Rng	Adj.2 for Vine Sites	Raw	Adj. for Vine Sites	Av. Sunshine Hours	Av. No. Rain Days (15YRS)	Av. Rainfall mm (40YRS)	Av. % R.H. 1300 HRS
APRIL	0.1	6.1	11.8	5.6	25.1	7.7	54	55	64	36.2	31.4	164	14	77	59	
MAY	3.5	5.9	15.6	6.2	29.6	7.8	174	179	179	38.5	33.7	185	13	78	58	
JUNE	7.5	5.1	18.9	6.3	32.8	7.6	267	270	270	37.9	33.1	210	11	77	57	
JULY	9.6	5.0	21.2	6.6	34.9	7.1	279	279	279	38.5	33.7	248	9	54	53	
AUGUST	8.9	5.4	21.1	6.8	35.7	7.8	279	279	279	40.4	35.6	241	8	51	51	
SEPTEMBER	5.2	6.6	18.2	6.4	32.0	7.4	246	247	247	39.6	34.8	198	9	64	57	
OCTOBER	−0.1	7.9	13.3	5.5	26.4	7.6	102	101	112	37.5	32.7	136	11	66	66	
APR-OCTOBER	—	6.0	17.2	6.2	—	7.6	1,401	1,410	1,430	38.4	33.6	1,382	75	467	57	
NOV-MARCH	—	—	6.8	—	—	—	—	—	—	—	—	—	—	334	—	

								DATA FOR MONTH TO MATURITY								
MATURITY GROUP	EST. MATURITY DATE		*	*	*							*				
1	—		—	—	—	—						—	—	—	—	—
2	—		—	—	—	—						—	—	—	—	—
3	—		—	—	—	—						—	—	—	—	—
4	SEPTEMBER 15		20.0	5.5	33.3	7.6						35.2	226	9	55	54
5	—		—	—	—	—						—	—	—	—	—
6	—		—	—	—	—						—	—	—	—	—
7	—		—	—	—	—						—	—	—	—	—
8	—		—	—	—	—						—	—	—	—	—

* Adjusted for vine sites.

TABLE 128
COLMAR, FRANCE. 48°04'N, 7°20'E. ALT. 188m. (Southern Alsace)

	1	2	3	4	5	6	7	8	9	10	11	12	13	14	15	16
	Temperature °C							Summation of Day Degrees above 10° with 19° Cut-Off			Temperature Variability Index					
PERIOD	Lowest Recorded Min.	Av.Mthly Lowest Min.	Av.Min Minus Col.2	Daily Mean (40YRS)	Daily Range /2 (40YRS)	Av.Mthly Highest Max.	Col.6 Minus Av.Max.	Raw #	Adj.1 for Lat. & Daily Temp Rng	Adj.2 for Vine Sites	Raw	Adj. for Vine Sites	Av. Sunshine Hours	Av. No. Rain Days (15YRS)	Av. Rainfall mm (40YRS)	Av. % R.H. 1300 HRS
APRIL	0.2	5.5	10.4	4.7	24.2	9.1	24	28	28	33.4	29.8	168	13	39	59	
MAY	3.8	6.0	15.0	5.2	29.1	8.9	155	164	162	35.7	32.1	206	14	52	58	
JUNE	8.2	4.7	18.2	5.3	31.2	7.7	246	263	260	33.6	30.0	230	13	53	58	
JULY	10.3	4.4	20.1	5.4	34.2	8.7	279	279	279	34.7	31.1	242	12	60	57	
AUGUST	8.9	5.0	19.2	5.3	32.7	8.2	279	279	279	34.4	30.8	226	12	53	57	
SEPTEMBER	5.1	5.9	15.7	4.7	28.7	8.3	171	177	179	33.0	29.4	179	10	50	62	
OCTOBER	−0.1	7.0	10.8	3.9	23.7	9.0	32	41	41	31.6	28.0	118	15	48	70	
APR-OCTOBER	—	5.5	15.6	4.9	—	8.6	1,186	1,231	1,228	33.8	30.2	1,369	89	355	60	
NOV-MARCH	—	—	3.9	—	—	—	—	—	—	—	—	—	—	147	—	

								DATA FOR MONTH TO MATURITY								
MATURITY GROUP	EST. MATURITY DATE		*	*	*							*				
1	—		—	—	—	—						—	—	—	—	—
2	SEPTEMBER 21+ ⎫		—	—	—	—						—	—	—	—	—
3	SEPTEMBER 21 ⎭		16.3	4.0	29.3	8.3						29.8	194	11	51	60
4	OCTOBER 6		14.3	3.7	26.8	8.4						29.1	168	11	50	63
5	—		—	—	—	—						—	—	—	—	—
6	—		—	—	—	—						—	—	—	—	—
7	—		—	—	—	—						—	—	—	—	—
8	—		—	—	—	—						—	—	—	—	—

April and October from temperature curve, not from monthly average means. * Adjusted for vine sites. + On cooler sites: see commentary.

VITICULTURE & ENVIRONMENT

TABLE 129

COSNE, FRANCE. 47°25′N, 2°55′E. ALT. 173m. (Sancerre, Pouilly-sur-Loire)

PERIOD	Lowest Recorded Min.	Av.Mthly Lowest Min.	Av.Min Minus Col.2	Daily Mean (40YRS)	Daily Range /2	Av.Mthly Highest Max.	Col.6 Minus Av.Max.	Raw #	Adj.1 for Lat. & Daily Temp Rng	Adj.2 for Vine Sites	Raw	Adj. for Vine Sites	Av. Sunshine Hours	Av. No. Rain Days (15YRS)	Av. Rainfall mm (40YRS)	Av. % R.H. 1200 HRS
APRIL		−1.8	6.2	10.0	5.6	24.7	9.1	15	15	25	37.7	32.9	172	15	39	52
MAY		0.4	7.6	14.0	6.0	28.3	8.3	124	130	142	39.9	35.3	204	14	43	56
JUNE		4.6	6.2	17.0	6.2	31.1	7.9	210	224	233	38.9	34.1	227	11	57	61
JULY		7.4	5.6	19.0	6.0	33.8	8.8	279	279	279	38.4	33.6	239	10	58	56
AUGUST		6.9	5.5	18.6	6.2	32.7	7.9	267	277	279	38.2	33.4	229	10	55	56
SEPTEMBER		2.5	7.5	15.7	5.7	28.7	7.3	171	173	189	37.6	32.8	184	9	46	61
OCTOBER		−2.4	8.5	10.8	4.7	24.2	8.7	33	35	55	36.0	31.2	124	14	63	72
APR-OCTOBER	—	6.7	15.0	5.8	—	8.3		1,099	1,133	1,202	38.1	33.3	1,379	83	361	59
NOV-MARCH	—	—	6.8	—	—	—		—	—	—	—	—	—	—	227	—

DATA FOR MONTH TO MATURITY

MATURITY GROUP	EST. MATURITY DATE	*	*	*					*				
1	—	—	—	—	—				—	—	—	—	—
2	OCTOBER 1+ }												
3	OCTOBER 1	15.9	4.5	27.7	7.3				32.8	182	9	46	61
4	—	—	—	—	—				—	—	—	—	—
5	—	—	—	—	—				—	—	—	—	—
6	—	—	—	—	—				—	—	—	—	—
7	—	—	—	—	—				—	—	—	—	—
8	—	—	—	—	—				—	—	—	—	—

April and October from temperature curve, not directly from monthly average means. * Adjusted for vine sites. + See commentary.

TABLE 130

DIJON, FRANCE. 47°16′N, 5°05′E. ALT. 220m. (Burgundy)

PERIOD	Lowest Recorded Min. (20YRS)	Av.Mthly Lowest Min. (20YRS)	Av.Min Minus Col.2	Daily Mean (40YRS)	Daily Range /2 (40YRS)	Av.Mthly Highest Max. (20YRS)	Col.6 Minus Av.Max.	Raw #	Adj.1 for Lat. & Daily Temp Rng	Adj.2 for Vine Sites	Raw	Adj. for Vine Sites	Av. Sunshine Hours (10YRS)	Av. No. Rain Days (15YRS)	Av. Rainfall mm (40YRS)	Av. % R.H. 1300 HRS (10YRS)
APRIL	−9.4	−1.7	6.5	10.3	5.5	23.9	8.1	20	21	26	36.6	33.0	175	15	50	58
MAY	−0.6	2.2	6.6	14.5	5.7	27.2	7.0	140	146	153	36.4	32.8	212	14	55	58
JUNE	0.6	6.1	5.7	17.6	5.8	31.1	7.7	228	242	247	36.6	33.0	241	12	69	58
JULY	2.8	8.3	5.3	19.6	6.0	33.3	7.7	279	279	279	37.0	33.4	258	11	62	57
AUGUST	4.4	8.3	4.5	19.0	6.2	33.3	8.1	279	279	279	37.4	33.8	242	12	61	53
SEPTEMBER	−0.6	3.9	6.5	16.1	5.7	28.9	7.1	183	185	191	36.4	32.8	192	11	54	59
OCTOBER	−3.9	−0.6	6.8	10.9	4.7	23.9	8.3	35	37	48	33.9	30.3	129	15	78	70
APR-OCTOBER	—	—	6.0	15.4	5.7	—	7.7	1,164	1,189	1,223	36.3	32.7	1,449	90	429	59
NOV-MARCH	—	—	—	4.1	—	—	—	—	—	—	—	—	—	—	267	—

DATA FOR MONTH TO MATURITY

MATURITY GROUP	EST. MATURITY DATE	*	*	*					*				
1	—	—	—	—	—				—	—	—	—	—
2	—								—				
3	SEPTEMBER 24	17.0	4.9	28.8	7.3				33.0	202	11	55	58
4	—	—	—	—	—				—	—	—	—	—
5	—	—	—	—	—				—	—	—	—	—
6	—	—	—	—	—				—	—	—	—	—
7	—	—	—	—	—				—	—	—	—	—
8	—	—	—	—	—				—	—	—	—	—

April and October day degree summations derived from temperature curve, not directly from monthly average means. * Adjusted for vine sites.

TABLE 131

LYON (ST. GENIS-LAVAL), FRANCE. 45°42′N, 4°42′E. ALT. 299m. (Ain, Isère, Lower Savoie)

	1	2	3	4	5	6	7	8	9	10	11	12	13	14	15	16
			Temperature °C					Summation of Day Degrees above 10° with 19° Cut-Off			Temperature Variability Index					
PERIOD	Lowest Recorded Min. (70YRS)	Av.Mthly Lowest Min. (70YRS)	Av.Min Minus Col.2	Daily Mean (40YRS)	Daily Range /2 (40YRS)	Av.Mthly Highest Max. (70YRS)	Col.6 Minus Av.Max.	Raw #	Adj.1 for Lat. & Daily Temp Rng	Adj.2 for Vine Sites	Raw	Adj. for Vine Sites	Av. Sunshine Hours (10YRS)	Av. No. Rain Days (15YRS)	Av. Rainfall mm (40YRS)	Av. % R.H. 1300 HRS (11YRS)
APRIL	−3.3	−0.6	6.0	10.5	5.1	23.3	7.7	25	26	33	34.1	32.3	176	11	53	57
MAY	−3.9	2.8	6.4	14.8	5.6	28.3	7.9	149	155	160	36.7	34.9	225	11	69	55
JUNE	−3.3	7.2	5.2	18.2	5.8	31.7	7.7	246	258	262	36.1	34.3	235	10	71	53
JULY	5.6	9.4	5.0	20.5	6.1	33.9	7.3	279	279	279	36.7	34.9	273	10	75	51
AUGUST	4.4	8.9	5.0	20.0	6.1	33.3	7.2	279	279	279	36.6	34.8	244	10	82	52
SEPTEMBER	0.0	4.4	6.8	16.6	5.4	29.4	7.4	198	200	206	35.8	34.0	201	11	71	56
OCTOBER	−6.7	0.0	7.2	11.7	4.5	23.3	7.1	53	59	64	32.3	30.5	122	12	84	64
APR-OCTOBER	—	—	5.9	16.0	5.5	—	7.5	1,229	1,256	1,283	35.5	33.7	1,476	75	505	55
NOV-MARCH	—	—	—	4.5	—	—	—	—	—	—	—	—	—	—	232	—

DATA FOR MONTH TO MATURITY

MATURITY GROUP	EST. MATURITY DATE	*	*	*						*				
1	—	—	—	—	—					—		—	—	—
2	—	—	—	—	—					—		—	—	—
3	SEPTEMBER 18	18.3	5.2	30.9	7.3					34.3		221	11	75
4	SEPTEMBER 26	17.3	4.9	29.9	7.4					34.1		209	11	72
5	OCTOBER 9 ⎫	15.3	4.6	27.7	7.3					33.2		183	11	75
6	OCTOBER 9+ ⎭	—	—	—	—					—		—	—	—
7	—	—	—	—	—					—		—	—	—
8	—	—	—	—	—					—		—	—	—

April and October day degree summations derived from temperature curve, not directly from monthly average means. +See commentary.

TABLE 132

MONTPELLIER, FRANCE. 43°36′N, 3°53′E. ALT. 80m. (Frontignan, Mireval, Lunel, Clairette du Languedoc, etc.)

	1	2	3	4	5	6	7	8	9	10	11	12	13	14	15	16
			Temperature °C					Summation of Day Degrees above 10° with 19° Cut-Off			Temperature Variability Index					
PERIOD	Lowest Recorded Min.	Av.Mthly Lowest Min.	Av.Min Minus Col.2	Daily Mean (40YRS)	Daily Range /2 (40YRS)	Av.Mthly Highest Max.	Col.6 Minus Av.Max.	Raw #	Adj.1 for Lat. & Daily Temp Rng	Adj.2 for Vine Sites	Raw	Adj. for Vine Sites	Av. Sunshine Hours	Av. No. Rain Days (15YRS)	Av. Rainfall mm (40YRS)	Av. % R.H.
APRIL		1.3	5.5	12.8	6.0	25.0	6.2	84	85	94	35.7	30.9	223	10	58	
MAY		4.7	5.4	16.6	6.5	30.2	7.1	205	210	216	38.5	33.7	277	8	61	
JUNE		8.5	4.9	20.3	6.9	34.6	7.4	270	270	270	39.9	35.1	298	6	40	
JULY		10.8	4.7	22.9	7.4	37.0	6.7	279	279	279	41.0	36.2	342	4	27	
AUGUST		10.3	4.9	22.4	7.2	35.3	5.7	279	279	279	39.4	34.6	294	5	48	
SEPTEMBER		6.6	5.9	19.3	6.8	31.6	5.5	270	270	270	38.6	33.8	240	7	72	
OCTOBER		2.5	6.2	14.5	5.8	25.6	5.3	139	138	150	34.7	29.9	190	11	108	
APR-OCTOBER	—	—	5.4	18.4	6.7	—	6.3	1,526	1,531	1,558	38.3	33.5	1,864	51	414	
NOV-MARCH	—	—	—	8.0	—	—	—	—	—	—	—	—	—	—	318	

DATA FOR MONTH TO MATURITY

MATURITY GROUP	EST. MATURITY DATE	*	*	*						*				
1	—	—	—	—	—					—		—	—	—
2	—	—	—	—	—					—		—	—	—
3	SEPTEMBER 2	22.4	6.0	33.9	5.7					34.5		291	5	50
4	SEPTEMBER 7	22.1	6.0	33.4	5.7					34.4		281	5	53
5	SEPTEMBER 12	21.7	5.9	32.8	5.6					34.3		271	6	57
6	SEPTEMBER 17	21.2	5.8	32.1	5.6					34.2		261	6	61
7	SEPTEMBER 23	20.5	5.7	31.4	5.5					34.0		251	7	66
8	SEPTEMBER 30	19.5	5.6	30.6	5.5					33.8		240	7	72

* Adjusted for vine sites.

203

TABLE 133

NANTES, FRANCE. 47°10′N, 1°35′W. ALT. 37m. (Loire-Inférieure: Muscadet)

	1	2	3	4	5	6	7	8	9	10	11	12	13	14	15	16
	Temperature °C							Summation of Day Degrees above 10° with 19° Cut-Off			Temperature Variability Index					
PERIOD	Lowest Recorded Min. (25YRS)	Av.Mthly Lowest Min. (25YRS)	Av.Min Minus Col.2	Daily Mean (50YRS)	Daily Range /2 (50YRS)	Av.Mthly Highest Max. (25YRS)	Col.6 Minus Av.Max.	Raw #	Adj.1 for Lat. & Daily Temp Rng	Adj.2 for Vine Sites	Raw	Adj. for Vine Sites	Av. Sunshine Hours (10YRS)	Av. No. Rain Days (15YRS)	Av. Rainfall mm (35YRS)	Av. % R.H. 1200 HRS (12YRS)
APRIL	−2.8	−0.6	5.9	10.6	5.3	22.2	6.3	24	25	29	33.4	32.2	200	9	56	61
MAY	−1.1	2.2	6.1	14.0	5.7	26.1	6.4	124	130	136	35.3	34.1	226	10	56	61
JUNE	2.8	5.6	5.5	16.8	5.7	30.0	7.5	204	216	222	35.8	34.7	250	9	50	61
JULY	5.6	7.8	5.1	18.8	5.9	31.7	7.0	273	279	279	35.7	34.6	277	8	53	59
AUGUST	3.9	7.2	5.4	18.6	6.0	32.2	7.6	267	276	279	37.0	35.8	259	7	50	59
SEPTEMBER	−0.6	3.9	7.1	16.7	5.7	29.4	7.0	201	203	209	36.9	35.7	201	7	50	62
OCTOBER	−5.0	0.0	7.6	12.3	4.7	22.2	5.2	71	74	85	31.6	30.4	147	13	89	73
APR-OCTOBER	—	—	6.1	15.4	5.6	—	6.7	1,164	1,203	1,239	35.1	33.9	1,560	63	404	62
NOV-MARCH	—	—	—	6.2	—	—	—	—	—	—	—	—	—	—	391	—
DATA FOR MONTH TO MATURITY																
MATURITY GROUP	EST. MATURITY DATE		*	*	*							*				
1	—	•	—	—	—							—	—	—	—	—
2	—		—	—	—							—	—	—	—	—
3	SEPTEMBER 25		17.0	5.4	29.4							35.7	203	7	50	62
4	OCTOBER 12		15.2	5.1	26.9							34.0	178	9	62	65
5	—		—	—	—							—	—	—	—	—
6	—		—	—	—							—	—	—	—	—
7	—		—	—	—							—	—	—	—	—
8	—		—	—	—							—	—	—	—	—

\# April day degree summation derived from temperature curve, not directly from monthly average mean. * Adjusted for vine sites.

TABLE 134

ORANGE, FRANCE. 44°07′N, 4°48′E. ALT. 48m. (Chateauneuf du Pape, Tavel, etc.)

	1	2	3	4	5	6	7	8	9	10	11	12	13	14	15	16
	Temperature °C							Summation of Day Degrees above 10° with 19° Cut-Off			Temperature Variability Index					
PERIOD	Lowest Recorded Min.	Av.Mthly Lowest Min.	Av.Min Minus Col.2	Daily Mean (40YRS)	Daily Range /2 (40YRS)	Av.Mthly Highest Max.	Col.6 Minus Av.Max.	Raw	Adj.1 for Lat. & Daily Temp Rng	Adj.2 for Vine Sites	Raw	Adj. for Vine Sites	Av. Sunshine Hours	Av. No. Rain Days (15YRS)	Av. Rainfall mm (40YRS)	Av. % R.H. 1300 HRS
APRIL				13.1	5.9			93	94	100			221	9	62	52
MAY				17.0	6.4			217	223	229			277	8	73	51
JUNE				20.5	6.5			270	270	270			296	5	58	48
JULY				22.9	6.8			279	279	279			341	4	39	41
AUGUST				22.1	6.6			279	279	279			303	5	78	44
SEPTEMBER				18.9	6.0			267	268	270			246	6	88	52
OCTOBER				14.1	5.6			127	126	137			185	7	120	59
APR-OCTOBER				18.4	6.3			1,532	1,539	1,564			1,869	44	518	50
NOV-MARCH				7.4	—			—	—	—		—	—	—	294	—
DATA FOR MONTH TO MATURITY																
MATURITY GROUP	EST. MATURITY DATE		*	*												
1	—		—	—									—	—	—	—
2	—		—	—									—	—	—	—
3	—		—	—									—	—	—	—
4	—		—	—									—	—	—	—
5	SEPTEMBER 11		21.4	5.4									284	5	82	47
6	SEPTEMBER 16		20.9	5.2									274	6	84	48
7	SEPTEMBER 22		20.2	5.1									263	6	86	49
8	SEPTEMBER 28		19.5	5.0									250	6	87	51

* Adjusted for vine sites.

TABLE 135

ORLÉANS, FRANCE. 47°56′N, 1°53′E. ALT. 125m. (Loiret)

	1	2	3	4	5	6	7	8	9	10	11	12	13	14	15	16
	Temperature °C							Summation of Day Degrees above 10° with 19° Cut-Off			Temperature Variability Index					
PERIOD	Lowest Recorded Min. (21YRS)	Av.Mthly Lowest Min. (21YRS)	Av.Min Minus Col.2	Daily Mean (40YRS)	Daily Range /2 (40YRS)	Av.Mthly Highest Max. (21YRS)	Col.6 Minus Av.Max.	Raw #	Adj.1 for Lat. & Daily Temp Rng	Adj.2 for Vine Sites	Raw	Adj. for Vine Sites	Av. Sunshine Hours (10YRS)	Av. No. Rain Days (15YRS)	Av. Rainfall mm (40YRS)	Av. % R.H. 1300 HRS (10YRS)
APRIL	−4.4	−1.7	6.3	9.9	5.3	23.3	8.1	13	13	21	35.6	32.8	175	15	51	60
MAY	−0.6	1.7	6.6	13.8	5.5	27.2	7.9	118	124	133	36.5	33.7	197	15	56	59
JUNE	1.1	5.6	5.4	16.5	5.5	30.0	8.0	195	209	218	35.4	32.6	219	12	58	58
JULY	4.4	7.7	5.3	18.6	5.6	33.3	9.1	267	279	279	36.8	34.0	230	13	56	55
AUGUST	4.4	7.2	5.6	18.3	5.5	32.2	8.4	257	268	277	36.0	33.2	224	12	45	56
SEPTEMBER	0.6	3.9	6.7	15.8	5.2	29.4	8.4	174	176	190	35.9	33.1	182	11	42	60
OCTOBER	−5.6	−1.7	8.4	11.1	4.4	23.3	7.8	38	44	57	33.8	31.0	124	16	70	69
APR-OCTOBER	—	—	6.3	14.9	5.3	—	8.2	1,062	1,113	1,175	35.7	32.9	1,351	94	378	60
NOV-MARCH	—	—	—	5.0	—	—	—	—	—	—	—	—	—	—	258	—
DATA FOR MONTH TO MATURITY																
MATURITY GROUP	EST. MATURITY DATE			*	*	*						*				
1	—			—	—	—						—	—	—	—	—
2	SEPTEMBER 26			16.5	4.6	29.5	8.4					33.1	188	11	42	59
3	—			—	—	—						—	—	—	—	—
4	—			—	—	—						—	—	—	—	—
5	—			—	—	—						—	—	—	—	—
6	—			—	—	—						—	—	—	—	—
7	—			—	—	—						—	—	—	—	—
8	—			—	—	—						—	—	—	—	—

April and October day degree summations derived from temperature curve, not directly from monthly average means. * Adjusted for vine sites.

TABLE 136

POMEROL (district), FRANCE. Approx. 44°56′N, 0°12′W. ALT. approx. 25m. (All data estimated)

	1	2	3	4	5	6	7	8	9	10	11	12	13	14	15	16
	Temperature °C							Summation of Day Degrees above 10° with 19° Cut-Off			Temperature Variability Index					
PERIOD	Lowest Recorded Min.	Av.Mthly Lowest Min.	Av.Min Minus Col.2	Daily Mean	Daily Range /2	Av.Mthly Highest Max.	Col.6 Minus Av.Max.	Raw	Adj.1 for Lat. & Daily Temp Rng	Adj.2 for Vine Sites	Raw	Adj. for Vine Sites	Av. Sunshine Hours	Av. No. Rain Days	Av. Rainfall mm	Av. % R.H. 1300 HRS
APRIL		0.9	5.9	12.1	5.3	25.5	8.1	63	64	67	35.2	34.0	175	16	64	58
MAY		4.1	5.9	15.8	5.8	30.0	8.4	180	186	189	37.5	36.3	195	15	60	58
JUNE		8.1	5.2	19.3	6.0	33.3	8.0	270	270	270	37.2	36.0	218	13	55	58
JULY		10.0	5.0	21.3	6.3	34.9	7.3	279	279	279	37.5	36.3	253	11	50	56
AUGUST		9.8	5.2	21.3	6.3	35.1	7.5	279	279	279	37.9	36.7	245	10	54	54
SEPTEMBER		6.4	6.2	18.6	6.0	32.0	7.4	258	260	263	37.6	36.4	201	11	51	59
OCTOBER		1.0	7.7	13.8	5.1	25.7	6.8	118	116	122	34.9	33.7	140	14	71	68
APR-OCTOBER		—	5.9	17.5	5.8	—	7.6	1,447	1,454	1,469	36.8	35.6	1,427	90	405	59
NOV-MARCH		—	—	7.1	—	—	—	—	—	—	—	—	—	—	346	—
DATA FOR MONTH TO MATURITY																
MATURITY GROUP	EST. MATURITY DATE			*	*	*						*				
1	—			—	—	—						—	—	—	—	—
2	—			—	—	—						—	—	—	—	—
3	—			—	—	—						—	—	—	—	—
4	—			—	—	—						—	—	—	—	—
5	SEPTEMBER 20			19.9	5.8	33.0						36.5	218	11	52	57
6	—			—	—	—						—	—	—	—	—
7	—			—	—	—						—	—	—	—	—
8	—			—	—	—						—	—	—	—	—

* Adjusted for vine sites.

TABLE 137

REIMS, FRANCE. 49°18'N, 4°02'E. ALT. 83m. (Champagne)

	1	2	3	4	5	6	7	8	9	10	11	12	13	14	15	16
	Temperature °C							Summation of Day Degrees above 10° with 19° Cut-Off			Temperature Variability Index					
PERIOD	Lowest Recorded Min. (10YRS)	Av.Mthly Lowest Min. (10YRS)	Av.Min Minus Col.2	Daily Mean (40YRS)	Daily Range /2 (40YRS)	Av.Mthly Highest Max. (10YRS)	Col.6 Minus Av.Max.	Raw #	Adj.1 for Lat. & Daily Temp Rng	Adj.2 for Vine Sites	Raw	Adj. for Vine Sites	Av. Sunshine Hours (10YRS)	Av. No. Rain Days (15YRS)	Av. Rainfall mm (40YRS)	Av. % R.H. 1300 HRS (10YRS)
APRIL	-3.9	-1.7	6.4	10.1	5.4	24.4	8.9	14	14	27	36.9	30.9	175	12	47	58
MAY	-2.2	0.6	7.3	13.4	5.5	27.8	8.9	105	112	139	38.2	32.2	217	11	54	55
JUNE	2.8	4.4	6.8	16.9	5.7	30.6	8.0	207	224	247	37.6	31.6	224	13	53	56
JULY	3.3	6.7	6.4	18.8	5.7	32.8	8.3	273	279	279	37.5	31.5	228	14	67	58
AUGUST	3.9	6.7	5.5	18.2	6.0	31.1	6.9	254	267	279	36.4	30.4	197	10	58	58
SEPTEMBER	0.6	2.8	6.7	15.2	5.7	27.8	6.9	156	158	179	36.4	30.4	178	12	42	62
OCTOBER	-6.7	-2.8	8.9	10.4	4.3	22.8	8.1	22	28	41	34.2	28.2	118	15	67	71
APR-OCTOBER	—	—	6.9	14.7	5.5	—	8.0	1,031	1,082	1,191	36.7	30.7	1,337	87	388	60
NOV-MARCH	—	—	—	4.0	—	—	—				—	—		—	251	—
DATA FOR MONTH TO MATURITY																
MATURITY GROUP	EST. MATURITY DATE			*	*	*										
1	—						—					—	—	—	—	—
2	SEPTEMBER 30			15.5	4.2	26.6	6.9					30.4	178	12	42	62
3	SEPTEMBER 30 +			—	—	—	—					—	—	—	—	—
4	—			—	—	—	—					—	—	—	—	—
5	—			—	—	—	—					—	—	—	—	—
6	—			—	—	—	—					—	—	—	—	—
7	—			—	—	—	—					—	—	—	—	—
8	—			—	—	—	—					—	—	—	—	—

\# April and October from temperature curve, not directly from monthly average means. * Adjusted for vine sites. + See commentary.

TABLE 138

SAINT EMILION (district), FRANCE. 44°54'N, 0°09'W. ALT. approx. 70m. (All data estimated)

	1	2	3	4	5	6	7	8	9	10	11	12	13	14	15	16
	Temperature °C							Summation of Day Degrees above 10° with 19° Cut-Off			Temperature Variability Index					
PERIOD	Lowest Recorded Min.	Av.Mthly Lowest Min.	Av.Min Minus Col.2	Daily Mean	Daily Range /2	Av.Mthly Highest Max.	Col.6 Minus Av.Max.	Raw #	Adj.1 for Lat. & Daily Temp Rng	Adj.2 for Vine Sites	Raw	Adj. for Vine Sites	Av. Sunshine Hours	Av. No. Rain Days	Av. Rainfall mm	Av. % R.H. 1300 HRS
APRIL		0.6	5.9	11.8	5.3	25.2	8.1	54	55	81	35.2	30.4	175	16	64	58
MAY		3.8	5.9	15.5	5.8	29.7	8.4	170	176	195	37.5	32.7	195	15	60	58
JUNE		7.8	5.2	19.0	6.0	33.0	8.0	270	270	270	37.2	32.4	218	13	55	58
JULY		9.7	5.0	21.0	6.3	34.6	7.3	279	279	279	37.5	32.7	253	11	50	56
AUGUST		9.5	5.2	21.0	6.3	34.8	7.5	279	279	279	37.9	33.1	245	10	54	54
SEPTEMBER		6.1	6.2	18.3	6.0	31.7	7.4	249	251	266	37.6	32.8	201	11	51	59
OCTOBER		0.7	7.7	13.5	5.1	25.4	6.8	109	108	136	34.9	30.1	140	14	71	68
APR-OCTOBER	—	—	5.9	17.2	5.8	—	7.6	1,410	1,417	1,506	36.8	32.0	1,427	90	405	59
NOV-MARCH	—	—	—	6.8	—	—	—	—			—	—		—	346	—
DATA FOR MONTH TO MATURITY																
MATURITY GROUP	EST. MATURITY DATE			*	*	*						*				
1	—			—	—	—	—					—	—	—	—	—
2	—			—	—	—	—					—	—	—	—	—
3	—			—	—	—	—					—	—	—	—	—
4	—			—	—	—	—					—	—	—	—	—
5	SEPTEMBER 17			20.2	4.9	32.5	7.4					32.9	223	11	52	57
6	SEPTEMBER 23			19.6	4.9	31.9	7.4					32.9	213	11	52	57
7	—			—	—	—	—					—	—	—	—	—
8	—			—	—	—	—					—	—	—	—	—

* Adjusted for vine sites.

TABLE 139
SERRIÈRES, FRANCE. 46°18′N, 4°46′E. ALT. 314m. (Mâconnais)

	1	2	3	4	5	6	7	8	9	10	11	12	13	14	15	16
	Temperature °C							Summation of Day Degrees above 10° with 19° Cut-Off			Temperature Variability Index					
PERIOD	Lowest Recorded Min.	Av.Mthly Lowest Min.	Av.Min Minus Col.2	Daily Mean (40YRS)	Daily Range /2 (40YRS)	Av.Mthly Highest Max.	Col.6 Minus Av.Max.	Raw #	Adj.1 for Lat. & Daily Temp Rng	Adj.2 for Vine Sites	Raw	Adj. for Vine Sites	Av. Sunshine Hours	Av. No. Rain Days (15YRS)	Av. Rainfall mm (40YRS)	Av. % R.H. 1300 HRS
APRIL	−1.4	6.8	10.4	5.0	22.3	6.9	24	25	34	33.7	31.3	176	14	72	57	
MAY	2.4	7.0	14.9	5.5	27.1	6.7	152	158	166	35.7	33.3	218	12	70	56	
JUNE	6.1	5.7	17.9	6.1	31.2	7.2	237	249	255	37.3	34.9	238	11	88	55	
JULY	9.0	5.5	20.5	6.0	33.6	7.1	279	279	279	36.6	34.2	265	11	88	53	
AUGUST	8.0	5.3	19.3	6.0	32.5	7.2	279	279	279	36.5	34.1	243	12	73	53	
SEPTEMBER	3.9	7.0	16.2	5.3	28.3	6.8	186	188	198	35.0	32.6	197	10	77	57	
OCTOBER	−0.4	7.5	11.2	4.1	22.6	7.3	41	50	57	31.2	28.8	126	13	72	68	
APR-OCTOBER	—	6.4	15.8	5.4	—	7.0	1,198	1,228	1,268	35.1	32.7	1,463	83	540	57	
NOV-MARCH	—	—	4.3	—	—	—	—	—	—	—	—	—	—	328	—	

DATA FOR MONTH TO MATURITY																
MATURITY GROUP	EST. MATURITY DATE		*	*	*							*				
1	—		—	—	—	—						—	—	—	—	—
2	—		—	—	—	—						—	—	—	—	—
3	SEPTEMBER 18		17.9	5.0	29.6	7.0						33.3	218	11	76	55
4	—		—	—	—	—						—	—	—	—	—
5	—		—	—	—	—						—	—	—	—	—
6	—		—	—	—	—						—	—	—	—	—
7	—		—	—	—	—						—	—	—	—	—
8	—		—	—	—	—						—	—	—	—	—

April and October day degree summations derived from temperature curve, not directly from monthly average means. * Adjusted for vine sites.

TABLE 140
TOULON, FRANCE. 43°06′N, 5°55′E. ALT. 24m. (Provence: Bandol, Cassis, etc.)

	1	2	3	4	5	6	7	8	9	10	11	12	13	14	15	16
	Temperature °C							Summation of Day Degrees above 10° with 19° Cut-Off			Temperature Variability Index					
PERIOD	Lowest Recorded Min. (49YRS)	Av.Mthly Lowest Min. (49YRS)	Av.Min Minus Col.2	Daily Mean (49YRS)	Daily Range /2 (49YRS)	Av.Mthly Highest Max. (49YRS)	Col.6 Minus Av.Max.	Raw	Adj.1 for Lat. & Daily Temp Rng	Adj.2 for Vine Sites	Raw	Adj. for Vine Sites	Av. Sunshine Hours	Av. No. Rain Days (49YRS)	Av. Rainfall mm (49YRS)	Av. % R.H. 1230 HRS (8YRS)
APRIL	2.2	4.8	4.6	13.3	3.9	22.2	5.0	99	117	100	25.2	28.4	273	6	61	63
MAY	4.4	8.5	4.2	16.9	4.2	25.8	4.7	214	231	218	25.7	28.9	322	6	48	64
JUNE	9.4	12.8	3.5	20.6	4.3	29.1	4.2	270	270	270	24.9	28.1	363	3	25	62
JULY	13.3	15.5	3.3	23.1	4.3	31.6	4.2	279	279	279	24.7	27.9	384	2	8	58
AUGUST	12.2	15.1	3.8	23.1	4.2	31.5	4.2	279	279	279	24.8	28.0	332	2	20	62
SEPTEMBER	7.8	11.7	4.8	20.4	3.9	28.5	4.2	270	270	270	24.6	27.8	276	4	61	63
OCTOBER	−3.3	7.2	5.6	16.4	3.6	24.6	4.6	198	218	196	24.6	27.8	229	8	112	66
APR-OCTOBER	—	—	4.3	19.1	4.1	—	4.4	1,609	1,664	1,612	24.9	28.1	2,179	31	335	63
NOV-MARCH	—	—	—	9.8	—	—	—	—	—	—	—	—	—	—	389	—

DATA FOR MONTH TO MATURITY																
MATURITY GROUP	EST. MATURITY DATE		*	*	*							*				*
1	—		—	—	—	—						—	—	—	—	—
2	—		—	—	—	—						—	—	—	—	—
3	—		—	—	—	—						—	—	—	—	—
4	—		—	—	—	—						—	—	—	—	—
5	SEPTEMBER 12		22.3	4.9	31.5	4.2						27.9	310	3	32	59
6	SEPTEMBER 17		21.8	4.8	31.0	4.2						27.9	300	3	39	59
7	SEPTEMBER 22		21.3	4.8	30.4	4.2						27.8	291	4	47	60
8	SEPTEMBER 27		20.8	4.7	29.7	4.2						27.8	282	4	56	60

* Adjusted for vine sites. Relative humidities estimated as Toulon minus 3%, due to vineyards being more inland.

TABLE 141

TOURS, FRANCE. 47°25′N, 0°41′E. ALT. 96m. (Vouvray)

	1	2	3	4	5	6	7	8	9	10	11	12	13	14	15	16
	Temperature °C							Summation of Day Degrees above 10° with 19° Cut-Off			Temperature Variability Index					
PERIOD	Lowest Recorded Min.	Av.Mthly Lowest Min.	Av.Min Minus Col.2	Daily Mean (40YRS)	Daily Range /2 (40YRS)	Av.Mthly Highest Max.	Col.6 Minus Av.Max.	Raw #	Adj.1 for Lat. & Daily Temp Rng	Adj.2 for Vine Sites	Raw	Adj. for Vine Sites	Av. Sunshine Hours (10YRS)	Av. No. Rain Days (15YRS)	Av. Rainfall mm (40YRS)	Av. % R.H. 1300 HRS (10YRS)
APRIL	−0.6	6.3	10.7	5.0	23.5	7.8	27	28	42	34.1	31.1	162	15	51	59
MAY	2.5	6.6	14.6	5.5	27.6	7.5	143	150	162	36.1	33.1	196	13	59	59
JUNE	6.0	5.7	17.5	5.8	31.2	7.9	225	239	247	36.8	33.8	217	11	56	59
JULY	8.2	5.5	19.7	6.0	33.7	8.0	279	279	279	37.5	34.5	244	11	49	54
AUGUST	7.2	5.9	19.1	6.0	33.0	7.9	279	279	279	37.8	34.8	232	12	46	55
SEPTEMBER	3.9	7.2	16.4	5.3	29.2	7.5	192	194	208	35.9	32.9	176	10	44	60
OCTOBER	−0.5	8.3	11.9	4.1	23.1	7.1	59	71	90	31.8	28.8	122	14	71	68
APR-OCTOBER	—	6.5	15.7	5.4	—	7.7	1,204	1,240	1,307	35.7	32.7	1,349	86	376	59
NOV-MARCH	—	—	5.7	—	—	—	—	—	—	—	—	—	—	294	—

	DATA FOR MONTH TO MATURITY															
MATURITY GROUP	EST. MATURITY DATE		*	*	*							*				
1	—		—	—	—							—	—	—	—	—
2	—		—	—	—							—	—	—	—	—
3	—		—	—	—							—	—	—	—	—
4	—		—	—	—							—	—	—	—	—
5	OCTOBER 9		15.2	4.2	27.0	7.4						31.9	160	11	51	62
6	—		—	—	—							—	—	—	—	—
7	—		—	—	—							—	—	—	—	—
8	—		—	—	—							—	—	—	—	—

\# April day degree summation derived from temperature curve, not directly from monthly average mean. * Adjusted for vine sites.

TABLE 142

VALENCE, FRANCE. 44°58′N, 4°51′E. ALT. 26m. (Côtes du Rhône, Hermitage)

	1	2	3	4	5	6	7	8	9	10	11	12	13	14	15	16
	Temperature °C							Summation of Day Degrees above 10° with 19° Cut-Off			Temperature Variability Index					
PERIOD	Lowest Recorded Min.	Av.Mthly Lowest Min.	Av.Min Minus Col.2	Daily Mean (40YRS)	Daily Range /2 (40YRS)	Av.Mthly Highest Max.	Col.6 Minus Av.Max.	Raw #	Adj.1 for Lat. & Daily Temp Rng	Adj.2 for Vine Sites	Raw	Adj. for Vine Sites	Av. Sunshine Hours	Av. No. Rain Days (15YRS)	Av. Rainfall mm (40YRS)	Av. % R.H. 1300 HRS
APRIL	−1.6	7.1	11.3	5.8	22.9	5.8	39	42	65	36.1	28.9	196	12	71	55
MAY	1.5	7.4	15.5	6.6	28.4	6.3	171	176	191	40.1	32.9	245	11	86	55
JUNE	6.6	5.6	18.8	6.6	32.1	6.7	264	270	270	38.7	31.5	263	9	77	52
JULY	8.4	5.8	21.5	7.3	35.3	6.5	279	279	279	41.5	34.3	304	7	43	48
AUGUST	8.0	6.0	21.0	7.0	34.2	6.2	279	279	279	40.2	33.0	273	8	101	50
SEPTEMBER	3.8	7.6	17.7	6.3	30.6	6.6	231	233	252	39.4	32.2	220	8	105	55
OCTOBER	−0.7	8.1	12.6	5.2	24.0	6.2	81	79	115	35.1	27.9	147	11	142	63
APR-OCTOBER	—	6.8	16.9	6.4	—	6.3	1,344	1,358	1,451	38.7	31.5	1,648	66	625	54
NOV-MARCH	—	—	5.9	—	—	—	—	—	—	—	—	—	—	279	—

	DATA FOR MONTH TO MATURITY															
MATURITY GROUP	EST. MATURITY DATE		*	*	*							*				
1	—		—	—	—	—						—	—	—	—	—
2	—		—	—	—	—						—	—	—	—	—
3	—		—	—	—	—						—	—	—	—	—
4	—		—	—	—	—						—	—	—	—	—
5	SEPTEMBER 19		19.6	4.7	30.6	6.4						32.5	242	8	103	53
6	SEPTEMBER 25		18.8	4.4	29.7	6.5						32.3	230	8	104	54
7	—		—	—	—	—						—	—	—	—	—
8	—		—	—	—	—						—	—	—	—	—

\# April day degree summation derived from temperature curve, not directly from monthly average mean. * Adjusted for vine sites.

Commentaries and vineyard site adjustments

Angers (Anjou, Ancenis, Chinon and Bourgueil) (Table 120)

Vineyard site adjustments, °C

Min.	Max.	
+0.4	−0.2	Undulating topography, some site selection
–	–	All aspects (perhaps slight bias to southerly)
+0.8	−0.4	Calcareous soils
–	–	Altitudes ~ same as Angers
+0.1	+0.1	Latitude adjustment (vineyards S of Angers)
+1.3	−0.5	Mean +0.4°, variability index −3.6

Notes

Dry to sweet white wines from Chenin Blanc (Maturity Group 5), usually left on the vines until mid to late October to attain maximum ripeness and some degree of noble rot; and light red and rosé wines, mainly from Cabernet Franc (5 for red and 4 for rosé). The fewer sunshine hours at Angers compared with Nantes (Table 133), despite slightly higher summer temperatures, reflect the transition from a sub-mediterranean climate close to the coast to a sub-continental climate further up the Loire Valley.

Angoulême (Charente: Cognac) (Table 121)

Vineyard site adjustments, °C

Min.	Max.	
–	–	Mainly flat (Angoulême typical)
–	−0.3	Closer to coast than Angoulême
+0.6	−0.3	Variably calcareous soils
+0.2	+0.2	Vineyard alt. ~ 50 m (Angoulême 82 m)
+0.8	−0.4	Mean +0.2°, variability index −2.4

Notes

Folle Blanche (Maturity Group 5), Colombard (6) and latterly Trebbiano, syn. St Emilion (7), as base wines for brandy distillation. Until recently the main varieties were Folle Blanche and Colombard, which were used also to make medium-quality table wines. Both retain high acidity to maturity, even under warm ripening conditions. Folle Blanche had also, in most cases, to be picked before full maturity because of its susceptibility to bunch rot. Light, acid wines low in alcohol resulted, suitable for making very high-quality brandy. Both varieties are now largely superseded by Trebbiano, which is late maturing but has better viticultural qualities and very high yield. Ripening late in October, it likewise gives light, acid wines highly suitable for making premium-quality brandy.

Auxerre (Chablis) (Table 122)

Vineyard site adjustments, °C

Min.	Max.	
+0.8	−0.4	Moderate to steep slopes (Auxerre semi-valley)
+0.45	+0.15	Predominantly SE to SW aspects
+0.8	−0.4	Soils variably calcareous
−0.6	−0.6	Average vineyard alt. 200 m (Auxerre 99 m)
+1.45	−1.25	Mean +0.1°, variability index −5.4

Notes

Light to medium-bodied dry white wines from Chardonnay (Maturity Group 3). Ripening of the Chardonnay here is later and more precarious than on the Côte d'Or (Dijon, Table 130), and under conditions of lower temperature and less sunshine hours. The wines are on average more austere and acid, with fullness of body and character being highly dependent on season and site favourability. Spring frosts are also a serious hazard, consistent with the rather high temperature variability index.

Beaujolais (Table 123). Estimated as means of Lyon (Table 131) and Serrières (Table 139)

Vineyard site adjustments: None

Notes

Fruity light red wines from Gamay (Maturity Group 3). Beaujolais is almost exactly half way between Lyon and Serrières in both latitude and altitude, and has similar topography. The soils, derived from granite and metamorphic rocks, do not call for special adjustment, and on the lower slopes all aspects are used. The best vineyards tend to be on the steeper slopes at greater altitudes, and to have a south-easterly aspect (see Johnson 1971). Site factors there would approximately cancel out the effect of altitude on mean temperature, but result in lower temperature variability indices than in Table 123.

Bergerac (Table 124)

Vineyard site adjustments, °C

Min.	Max.	
+0.8	0.4	Undulating low hills (Bergerac in valley)
–	–	All aspects
+0.4	–0.2	Fertile soils, somewhat calcareous
–0.4	–0.4	Average vineyard alt. ~ 100 m (Bergerac 33 m)
+0.8	–1.0	Mean –0.1°, variability index –3.6

Notes

Dry to sweet white wines from Sémillon and Muscadelle (Maturity Group 4) and Sauvignon Blanc (3); medium-bodied red wines from Cabernet Franc and Merlot (5), and Malbec (4). Sunshine hours and effective temperature summations are less than in Bordeaux, showing (as in the Loire Valley) the attenuation of coastal sub-Mediterranean climatic influences and increasing continentality with distance inland. A result is that generally earlier-maturing red varieties are grown than in Bordeaux. White varieties, as in Sauternes, are commonly left on the vine well beyond their indicated maturity dates; but the wines do not, as a rule, achieve the same richness.

Bordeaux (a. Graves, Sauternes) (Table 125)

Vineyard site adjustments, °C

Min.	Max.	
–	–	Flat to slightly undulating
+0.8	–0.4	Stony, partly calcareous soils
–	–	Altitudes ~ as Bordeaux
–0.4	+0.2	Greater distance than Bordeaux from sea and estuary
+0.2	+0.2	Latitude adjustment: vineyards S of Bordeaux
+0.6	0.0	Mean +0.3°, variability index –1.2

Notes

Medium-bodied red wines from Merlot and Cabernet Franc (Maturity Group 5) and Cabernet Sauvignon (6); dry to very sweet full-bodied white wines from Sémillon and a little Muscadelle (4) together with Sauvignon Blanc (3). The white varieties are normally mature for making dry wines by about mid-September, but for sweet wines are left until October or later, to attain partial dehydration through late Botrytis (noble rot) infection of the berries. As in the Médoc (Table 126), Merlot (5) is grown mainly on the cooler soils and exposures, so that ripening tends to be at a time and under conditions perhaps closer to those shown here for Maturity Group 6.

Bordeaux (b. Médoc) (Table 126)

Vineyard site adjustments, °C

Min.	Max.	
+0.2	–0.1	Vineyards on slight rises
+0.5	–0.5	Peninsular location
+0.6	–0.3	Variably stony soils
+0.1	+0.1	Average altitude ~ 30 m (Bordeaux 47 m)
–0.3	–0.3	Latitude adjustment: vineyards N of Bordeaux
+1.1	–1.1	Mean unchanged, variability index –4.4

Relative humidities of vineyards estimated as Bordeaux plus 2 per cent.

Notes

Medium-bodied red wines from Merlot and Cabernet Franc (Maturity Group 5), Cabernet Sauvignon (6) and a little Petit Verdot (7) and occasionally Malbec (4). Of the two main varieties, Cabernet Sauvignon tends to be planted on the warmer, and Merlot on the cooler, soils and exposures.

The Médoc is notable for the local climatic advantage resulting from its peninsular location between the Atlantic to the West and the Gironde Estuary to the east. The fact that virtually all land winds must cross 5 km or more of tidal water reduces its diurnal temperature range and moderates the incidence of both frost in spring and high temperatures in summer and autumn. The combined equable climate and stony soils of the Médoc help to explain its pre-eminence (with Burgundy) as a producer of red wines.

Cahors (Table 127)

Vineyard site adjustments, °C

Min.	Max.	
+1.0	–0.5	Moderate slopes (Cahors in valley)
–	–	All aspects
+0.6	–0.3	Variably calcareous soils
–0.4	–0.4	Average altitude 180–200 m (Cahors 122 m)
+1.2	–1.2	Mean unchanged, variability index –4.8

Notes

Cahors is noted for its fine 'black' wines made principally from Malbec (Maturity Group 4). As traditionally made these were deep red, full-bodied wines which develop very well with prolonged ageing in wood and bottle.

Colmar (Southern Alsace) (Table 128)

Vineyard site adjustments, °C

Min.	Max.	
+0.8	−0.4	Moderate slopes (Colmar + on flats)
–	–	Aspects predominantly east
+0.4	−0.2	Mixed soils, some calcareous, gravelly or rocky
−0.7	−0.7	Average altitude ~ 300 m (Colmar 188 m)
+0.5	−1.3	Mean −0.4°, variability index −3.6

Notes

Southern Alsace is in the rain-shadow of the highest part of the Vosges Mountains. Vines benefit from the resulting adiabatic warming, with reduced cloudiness, rainfall, and relative humidity as compared with elsewhere in the Rhine Valley. Grapes attain full natural ripeness, and produce dry wines of often substantial body.

Table 128 represents average to good exposures in the vicinity of Colmar. Riesling (Maturity Group 4) is grown on the most favoured south-easterly exposures, and therefore may ripen slightly earlier and under warmer conditions than indicated in the Table for average sites. Group 2 varieties such as Chasselas, on the other hand, tend to be grown on less favoured and generally cooler sites, so that ripening time and conditions are perhaps best represented by those for Group 3 varieties on average sites. Besides Riesling and Chasselas, important varieties for white wines are Pinot Gris and Muscat Ottonel (Maturity Group 2) and Sylvaner and Traminer (3). A little light red or rosé wine is made from Pinot Noir (3 for red, 2 for rosé).

Cosne (Sancerre, Pouilly-sur-Loire) (Table 129)

Vineyard site adjustments, °C

Min.	Max.	
+0.6	−0.3	Moderate to steep slopes (Cosne semi-valley)
+1.0	−0.5	Chalky and/or gravelly soils
–	–	All aspects
−0.2	−0.2	Vineyard altitudes ~ 200 m (Cosne 173 m)
+0.1	+0.1	Latitude adjustment (vineyards S of Cosne)
+1.5	−0.9	Mean +0.3°, variability index −4.8

Notes

Dry white wine from Chasselas (Maturity Group 2) or Sauvignon Blanc (3); light red or rosé wines from Pinot Noir (3 for red, 2 for rosé). According to Debuigne (1976), the Sauvignon Blanc gets most of the best exposures, with Chasselas together with Pinot Noir for rosé

relegated to the less favourable. Ripening times and mean temperatures for all varieties therefore probably approach those indicated in Table 129 for Group 3 varieties on the average to good sites.

Dijon (Burgundy) (Table 130)

Vineyard site adjustments, °C

Min.	Max.	
+0.4	−0.2	Slight to moderate slopes (Dijon semi-valley)
–	–	Predominantly eastern aspects
+0.8	−0.4	Calcareous, rubbly, but clayey soils
−0.4	−0.4	Vineyards typically at 250–300 m (Dijon 220 m)
+0.2	+0.2	Latitude adjustment (all vineyards south of Dijon)
+1.0	−0.8	Mean +0.1°, variability index −3.6

Notes

Medium-bodied, fruity dry red wines – the classic red burgundies – from Pinot Noir (Maturity Group 3 for red wines); some light red wines from Gamay (3); medium to full-bodied dry white wines, the classic white burgundies, from Chardonnay (3); lesser white wines from Aligoté (3).

The Côtes of Burgundy owe their warmth to the sheltering ranges of hills to the west, in whose rain shadow they lie. The calcareous, rubbly nature of the soils appears also to play an important part, perhaps not only by influencing soil drainage and temperature, and hence immediate above-ground microclimate, but also through their calcium saturation in the specific adaptations of the Pinot Noir and Chardonnay grapes. Reflection of red light wavelengths from the reddish soil may be a further factor allowing Pinot Noir to attain sufficient colour for making red wines: see Chapter 3.

In parts of the Côte d'Or, aspects tend towards the south-east rather than the east (Johnson 1971), but in general the slopes are gentle enough that this should not greatly influence the effective temperature summation. Where slopes are steeper, the warming effects of slope and aspect on mean temperature are mostly offset by greater altitude; but the temperature variability index may be reduced, as on the Hill of Corton. Such sites produce some of the greatest red and white burgundies.

Lyon (Ain, Isère, Lower Savoie) (Table 131)

Vineyard site adjustments, °C

Min.	Max.	
–	–	Hilly topography as for recording station; all aspects
+0.6	–0.3	Variably stony and calcareous soils
–	–	Altitude and latitude as for Lyon (299 m)
+0.6	–0.3	Mean +0.15°, variability index –1.8

Notes

Red wines from Gamay (Maturity Group 3), Durif (4) and Mondeuse (6); at greater altitudes, dry white wines from Chardonnay, Aligoté and Melon (3), together with light red wines from various early-maturing varieties; at still greater altitudes, white wines from Chasselas (2). Ripening temperatures at these higher altitudes probably correspond roughly to those for Maturity Groups 4 or 5 as calculated for Lyon.

The adaptation of the relatively late-maturing Mondeuse, to at least the lower and warmer parts of this area, is notable. Viala and Vermorel (1901–1904) state that Mondeuse is highly resistant to rain during ripening, and that in fact it is very dependent on good August rains to ripen the heavy crops of which it is capable.

Montpellier (Frontignan, Mireval, Lunel, Clairette du Languedoc etc.) (Table 132)

Vineyard site adjustments, °C

Min.	Max.	
+0.6	–0.3	Moderate slopes (Montpellier nearly flat)
+1.0	–0.5	Stony, and/or calcareous soils (better growths only)
–	–	All aspects, east predominantly
–0.2	–0.2	Vineyards mainly at ~ 120 m (Montpellier 80 m)
+1.4	–1.0	Mean +0.2°, variability index –4.8

Notes

Sweet, fortified wines from part-raisined Frontignac (Maturity Group 3); dry red wines from Cinsaut and Cabernet Franc (5), Cabernet Sauvignon (6), and on the irrigated plains from Aramon, Mataro and Carignan (7), Terret Noir (8) etc.; dry white wines from Trebbiano (syn. Saint Emilion, Ugni Blanc) (7), Clairette (syn. Blanquette) (8) etc.

The most famous areas producing sweet Muscat wines (Frontignan, Mireval) are very close to the coast, and almost certainly enjoy less temperature variability than shown in Table 132: probably approaching that calculated for the vineyards around Toulon (Table 140). Equability of the maximum temperature is characteristic of south coastal areas in the Northern Hemisphere, as it is of north coastal areas in the Southern Hemisphere. The south coast of France has quite moderate highest maxima during ripening, despite fairly high average mean temperatures. Apart from the risk of heavy rain towards the end of ripening, the climatic characteristics of this area appear from Australian experience to be fully compatible with high quality in table wines, provided that careful attention is paid to site selection, grape varieties, vineyard management and winemaking technology. Very good table wines were reportedly produced in the past, before the post-Phylloxera mass move from the stony hillsides to the irrigated plains. Arguably early to midseason grape varieties would be better for wine quality than the now traditional, but intrinsically inferior, late ones.

Nantes (Loire Inférieure: Muscadet) (Table 133)

Vineyard site adjustments, °C

Min.	Max.	
+0.4	–0.2	Gently undulating (Nantes semi-valley)
–	–	Non-calcareous clay loams; all aspects
–0.1	–0.1	Average altitude 50–60 m (Nantes 37 m)
+0.2	+0.2	Latitude adjustment (vineyards S of Nantes)
+0.5	–0.1	Mean +0.2°, variability index –1.2

Notes

Light to medium-bodied dry white wines from Melon, syn. Muscadet (Maturity Group 3) and Folle Blanche, syn. Gros Plant (5). The latter variety, in particular, is normally picked ahead of full maturity to avoid the bunch rot to which both varieties are prone; its harvest time probably approximates more to that of a Maturity Group 4 variety, as shown in Table 133. The common result is light, high-acid wines.

Note that the temperature data of the Meteorological Office, Great Britain (1967) are used in preference to those of Sanson (1945). Sanson's temperatures for Nantes seem anomalously low compared with those from nearby stations, and do not agree with the observed vine phenology.

Orange (Châteauneuf du Pape, Tavel, Lirac) (Table 134)

Vineyard site adjustments, °C

Min.	Max.	
+0.2	−0.1	Vineyards mostly on gentle slopes (Orange flat?)
+1.0	−0.5	Soils very rocky
−	−	All aspects
−0.1	−0.1	Average altitude ~ 70 m (Orange 48 m)
+1.1	−0.7	Mean +0.2°, variability index −3.6

Notes

Full-bodied wines from a mixture of varieties of maturity groups 5–8, with Grenache (7) predominant (Châteauneuf du Pape etc.). Dry rosé wines from Grenache (Tavel, Lirac etc.).

Orléans (Loiret) (Table 135)

Vineyard site adjustments, °C

Min.	Max.	
+0.2	−0.1	Moderate slopes (recording station slight slope?)
+0.3	+0.1	Predominantly southern aspects
+0.6	−0.3	Moderately stony and/or calcareous soils
−0.2	−0.2	Altitude ~ 150 m (Orléans 125 m)
+0.9	−0.5	Mean +0.2°, variability index −2.8

Notes

Light red and rosé wines from Meunier (Maturity Group 3 for red wines, 2 for rosé). This is now a marginal wine-producing area, of which the vine area and production have fallen greatly in the last century. In the Middle Ages, under warmer climatic conditions (see Chapter 23), it was reportedly a major supplier of highly esteemed red wines to the Paris market. These were made from Pinot Noir, which matures at a similar time to Meunier but has earlier budburst, and is therefore more liable to damage by spring frosts.

Pomerol (Table 136)

Vineyard site adjustments, °C

Min.	Max.	
−	−	Flat to slightly undulating; all aspects
+0.4	+0.2	Moderately stony soils
+0.4	−0.2	Mean +0.1°, variability index −1.2

Notes

Climatic data estimated from St André de Cubzac, Bergerac and Bordeaux. Medium to full-bodied red wines from Merlot and Cabernet Franc (Maturity Group 5). The wines of greatest repute come mainly from the higher and more undulating eastern section, where effective climatic conditions perhaps approach those of St Emilion (Table 138).

Reims (Champagne) (Table 137)

Vineyard site adjustments, °C

Min.	Max.	
+1.3	−0.5	Slopes of free-standing hill range (Reims on valley floor)
−	−	All aspects
+0.8	−0.4	Chalky, stony soils
−0.5	−0.5	Average altitude ~ 160 m (Reims 83 m)
+0.2	+0.2	Latitude adjustment (all vineyards S of Reims)
+1.8	−1.2	Mean +0.3°, variability index −6.0

Notes

Champagne, mainly from Pinot Noir (Maturity Group 2 for white wines); also from Meunier (2 for white wines), chiefly on less favourable sites for its later budburst and hence less vulnerability to spring frosts. Some Chardonnay (3) on warm exposures in southern parts of the region. With allowance for site and latitude variation it is probable that all three ripen under similar conditions, roughly as estimated here for Maturity Group 2 on good-average sites.

The actual average date for start of vintage is probably a little earlier than September 30, the date shown in Table 137. Vintage dates throughout the tables are for grapes at maturities best suited to making still wines. Grapes for champagne-style wines are normally picked when less fully mature.

Pinot Noir is still used for making red wine on south-facing slopes at Bouzy. Ripening is probably earlier and warmer than estimated here for Champagne, though later and cooler than in Burgundy. Consistent with this the wines of Bouzy are reported to be lighter than typical burgundy, and very delicate.

It is surprising to see vineyards on north-facing slopes (as just south of Reims) in so northerly and cool a region. The reason, as with the Kaiser Stuhl in the Rhine Valley and many similar locations, lies in the free-standing hill formation of the Montagne de Reims. This allows chilled night air to slip away freely and completely to the surrounding valley and plain, to be replaced by convection with warmer air from above. See Chapter 4.

Saint Emilion (Table 138)

Vineyard site adjustments, °C

Min.	Max.	
+0.8	–0.4	Moderate slopes at edge of free-standing small plateau
–	–	All aspects
+0.8	–0.4	Calcareous soils
+1.6	–0.8	Mean +0.4°, variability index –4.8

Notes

Medium to full-bodied red wines from Merlot and Cabernet Franc (Maturity Group 5) and some Cabernet Sauvignon (6). Note that superior site climatic factors offset higher altitude as compared with Pomerol (Table 136), to give a similar effective temperature summation and predicted ripening time, but less temperature variability. This is consistent with the high incidence of outstanding vineyards in the area (see Johnson 1971).

Ripening in St Emilion, as in Pomerol, is a little earlier and at appreciably higher temperatures than in the Médoc and Graves. The combination of warmer ripening and less growing season sunshine hours, together with a higher proportion of Merlot, results in wines that are in general softer and more suited to early drinking than the firm styles of the Médoc and Graves.

Serrières (Mâconnais) (Table 139)

Vineyard site adjustments, °C

Min.	Max.	
–	–	Undulating terrain, as Serrières; all aspects
+0.8	–0.4	Calcareous soils
+0.8	–0.4	Mean +0.2°, variability index –2.4

Notes

Dry white wines from Chardonnay (Maturity Group 3) and lesser white wines from Aligoté (3). Ripening is under appreciably warmer and sunnier, but also wetter, conditions than on the Côte d'Or as represented by Dijon (Table 130). Substantially more rain through the growing season presumably results in greater vine vigour.

With site factors less critical to successful ripening than on the Côte d'Or, less rigorous site selection is possible. This perhaps results in the inclusion of many site climates inferior for quality to that calculated as representative, and helps to explain the reputation of the Mâconnais as a large producer of sound wines: which, however, generally lack the distinction of the great white burgundies.

Toulon (Provence: Bandol, Cassis etc.) (Table 140)

Vineyard site adjustments, °C

Min.	Max.	
–1.0	+2.0	Vineyards on slopes, but mostly inland
+0.3	+0.1	Better vineyards with predominantly southern aspects
+0.8	–0.4	Soils rocky or calcareous
–0.9	–0.9	Typical vineyards ~ 150–200 m (Toulon 24 m)
–0.8	+0.8	Mean unchanged, variability index +3.2

Relative humidities estimated as Toulon minus 3 per cent.

Notes

Dry white wines mainly from Trebbiano (syn. Ugni Blanc) (Maturity Group 7) and Clairette (syn. Blanquette) (8); rosé and dry red wines from Cinsaut (5), and Grenache, Mataro and Carignan (7).

The site climate adjustments are only approximate, depending on distance from the sea, altitude etc. The best-known appellation d'origine contrôlée (AOC) areas, Bandol and Cassis, are nearest the coast, on slopes which predominantly face south. The larger vins délimités de qualité supérieure (VDQS) and common wine areas further inland would often have temperature variability indices higher than in Table 140.

A feature of this south-coast climate, even more than at Montpellier (Table 132), is the narrow variation of the maximum temperature (Column 7). Vines and grapes are seldom exposed to very high temperatures, despite fairly high means. The region was once a famous producer and exporter of dry red wines. Its climate could allow it to become so again.

Tours (Vouvray) (Table 141)

Vineyard site adjustments, °C

Min.	Max.	
–	–	Undulating topography (Tours typical); all aspects
+1.0	–0.5	Very calcareous soils
+1.0	–0.5	Mean +0.25°, variability index –3.0

Notes

Temperatures and sunshine hours are marginally less than in Anjou (Table 120), and ripening of the predominant variety Chenin Blanc (Maturity Group 5) is late and uncertain. The wines in good years are reported to have delicacy but strength, and great longevity. With their varietally- determined high acid content, especially in cool years, they are used very largely for making sparkling wines.

Valence (Côtes du Rhône, Hermitage) (Table 142)

Vineyard site adjustments, °C

Min.	Max.	
+1.5	–0.5	Steep slopes of projecting hills (Valence on valley floor)
+0.6	+0.2	Predominantly south-east and southerly aspects
+0.8	–0.4	Gritty skeletal soils, stone terracing
–0.4	–0.4	Average altitude ~ 200 m (Valence 126 m)
–0.3	–0.3	Latitude adjustment (all vineyards N of Valence)
+2.2	–1.4	Mean +0.4°, variability index –7.2

Notes

Dry red wines from Shiraz (Maturity Group 5) with some Viognier (5); full-bodied white wines from Roussanne and especially Marsanne (5).

The estimated temperatures broadly cover a north-south distance of 60 km, and therefore probably overestimate those for the north and under-estimate those for the south. Especially, perhaps, they under-estimate temperatures for the vineyards of the Hermitage appellation, which are to the south and have a steep southerly aspect. Ripening of the grapes which make the most famous of the Hermitage wines therefore probably takes place a few days earlier than suggested by Table 142, at an average mean temperature for the ripening month of 20°C or more, and ripening month sunshine hours of perhaps 250. A feature of the whole region is its considerable rainfall during ripening.

17

Viticultural Environments of Germany

The vineyards of Germany are at the cold limit of commercial viticulture, both in growing season temperatures and the winter survival of *Vitis vinifera*. Prescott (1965) and Becker (1977) defined the practical limit for unprotected winter survival of *vinifera* varieties as an average mean temperature of –1°C for the coldest month. Frankfurt on Main and Stuttgart, with January average mean temperatures between 0 and +1°C, are close to the limit. All German wine-producing areas depend to varying extents on special site characteristics to ensure winter survival and ripening to acceptable winemaking quality, while the addition of sugar to the fermenting musts (chaptalization) is standard practice for the commoner grades of wine.

Considering the latitude range of the German vineyards, there is a surprising uniformity in the styles of wine produced. That is because diminishing latitude towards the south is almost exactly offset by greater altitude. More important is the contrast between the west and east banks of the Rhine Valley. The east-facing slopes of the west bank are in the direct rain shadow of the mountains to their west, and as a result are the warmest, sunniest and driest part of Germany. The climate of Colmar, France (Chapter 16, Table 128), shows this in contrast to that of Freiburg, Germany (Table 144), which is on the east bank at the same latitude.

TABLE 143

BAD DURKHEIM, GERMANY. 49°28′N, 8°12′E. ALT. 130m. (Rheinpfalz, = Palatinate)

	1	2	3	4	5	6	7	8	9	10	11	12	13	14	15	16
	Temperature °C							Summation of Day Degrees above 10° with 19° Cut-Off			Temperature Variability Index					
PERIOD	Lowest Recorded Min.	Av.Mthly Lowest Min.	Av.Min Minus Col.2	Daily Mean (30YRS)	Daily Range /2	Av.Mthly Highest Max.	Col.6 Minus Av.Max.	Raw #	Adj.1 for Lat. & Daily Temp Rng	Adj.2 for Vine Sites	Raw	Adj. for Vine Sites	Av. Sunshine Hours	Av. No. Rain Days	Av. Rainfall mm (30YRS)	Av. % R.H. 1300 HRS
APRIL		–0.8	5.9	10.1	5.0	23.8	8.7	18	19	23	34.6	32.2	172	11	45	56
MAY		2.6	6.4	14.5	5.5	29.1	9.1	140	149	151	37.5	35.1	216	11	52	54
JUNE		6.9	5.1	17.7	5.7	31.2	7.8	231	251	251	35.7	33.3	229	11	69	54
JULY		9.4	4.5	19.5	5.6	33.6	8.5	279	279	279	35.4	33.0	238	11	53	55
AUGUST		7.9	5.1	18.6	5.6	31.9	7.7	267	279	279	35.2	32.8	220	11	56	56
SEPTEMBER		3.8	6.1	15.1	5.2	28.1	7.8	153	155	161	34.7	32.3	169	10	50	61
OCTOBER		–1.2	6.9	9.7	4.0	21.8	8.1	16	23	28	31.0	28.6	107	12	45	69
APR-OCTOBER		—	5.7	15.0	5.2	—	8.2	1,104	1,155	1,172	34.9	32.5	1,351	77	370	58
NOV-MARCH		—	—	3.1	—	—	—	—	—	—	—	—	—	—	202	—

DATA FOR MONTH TO MATURITY																
MATURITY GROUP	EST. MATURITY DATE		*	*	*							*				
1	—		—	—	—	—						—	—	—	—	—
2	OCTOBER 2+ ⎫		—	—	—	—						—	—	—	—	—
3	OCTOBER 2 ⎬		14.8	4.5	27.2	7.8						32.0	165	10	49	62
4	OCTOBER 2+ ⎭		—	—	—	—						—	—	—	—	—
5	—		—	—	—	—						—	—	—	—	—
6	—		—	—	—	—						—	—	—	—	—
7	—		—	—	—	—						—	—	—	—	—
8	—		—	—	—	—						—	—	—	—	—

April and October from temperature curve, not directly from monthly average means. * Adjusted for vine sites. + See commentary.

TABLE 144
FREIBURG, GERMANY. 47°59'N, 7°51'E. ALT. 285m. (South Baden)

	1	2	3	4	5	6	7	8	9	10	11	12	13	14	15	16
	Temperature °C							Summation of Day Degrees above 10° with 19° Cut-Off			Temperature Variability Index					
PERIOD	Lowest Recorded Min. (50YRS)	Av.Mthly Lowest Min. (50YRS)	Av.Min Minus Col.2	Daily Mean (50YRS)	Daily Range /2 (50YRS)	Av.Mthly Highest Max. (50YRS)	Col.6 Minus Av.Max.	Raw #	Adj.1 for Lat. & Daily Temp Rng	Adj.2 for Vine Sites	Raw	Adj. for Vine Sites	Av. Sunshine Hours	Av. No. Rain Days (40YRS)	Av. Rainfall mm (40YRS)	Av. % R.H. 1330HRS (48YRS)
APRIL	−5.6	−1.1	6.1	9.7	4.7	22.8	8.4	13	15	19	33.3		129	13	74	57
MAY	−1.7	2.8	6.1	14.2	5.3	28.3	8.8	130	137	143	36.1		180	12	89	56
JUNE	2.8	6.7	4.9	16.9	5.3	30.6	8.4	207	221	227	34.5		201	12	96	57
JULY	6.1	8.9	5.0	19.2	5.3	32.2	7.7	279	279	279	33.9		233	12	104	56
AUGUST	5.0	8.3	5.0	18.3	5.0	30.6	7.3	257	268	274	32.3		211	12	94	57
SEPTEMBER	−1.1	4.4	6.2	15.0	4.4	26.7	7.3	150	161	167	31.1		159	10	84	64
OCTOBER	−5.0	−0.6	7.3	10.0	3.3	20.6	7.3	18	29	32	27.8		127	10	79	72
APR-OCTOBER	—	—	5.8	14.8	4.8	—	7.9	1,054	1,110	1,141	32.7		1,240	81	620	60
NOV-MARCH	—	—	—	3.3	—	—	—	—	—	—			—	—	267	—

DATA FOR MONTH TO MATURITY

MATURITY GROUP	EST. MATURITY DATE	*	*	*												
1	—		—	—	—						—		—	—	—	—
2	SEPTEMBER 27	15.7	4.5	27.5							31.3		164	10	85	63
3	—		—	—	—						—		—	—	—	—
4	—		—	—	—						—		—	—	—	—
5	—		—	—	—						—		—	—	—	—
6	—		—	—	—						—		—	—	—	—
7	—		—	—	—						—		—	—	—	—
8	—		—	—	—						—		—	—	—	—

April and October day degree summations derived from temperature curve, not directly from monthly average means. * Adjusted for vine sites.

TABLE 145
GEISENHEIM, GERMANY. 50°00'N, 8°00'E. ALT. approx. 100m. (Rheingau)

	1	2	3	4	5	6	7	8	9	10	11	12	13	14	15	16
	Temperature °C							Summation of Day Degrees above 10° with 19° Cut-Off			Temperature Variability Index					
PERIOD	Lowest Recorded Min.	Av.Mthly Lowest Min.	Av.Min Minus Col.2	Daily Mean (74YRS)	Daily Range /2 (74YRS)	Av.Mthly Highest Max.	Col.6 Minus Av.Max.	Raw #	Adj.1 for Lat. & Daily Temp Rng	Adj.2 for Vine Sites	Raw	Adj. for Vine Sites	Av. Sunshine Hours (74YRS)	Av. No. Rain Days	Av. Rainfall mm (74YRS)	Av. % R.H. 1300 HRS
APRIL		−1.9	6.3	9.7	5.3	23.3	8.3	13	13	23	35.8	30.2	177	9	36	52
MAY		1.6	6.8	14.2	5.8	29.4	9.4	129	140	171	39.4	33.8	226	9	41	50
JUNE		5.5	5.6	17.2	6.1	31.1	7.8	217	235	265	37.8	32.2	228	9	53	51
JULY		8.2	4.6	18.6	5.8	32.7	8.3	267	279	279	36.1	30.5	233	10	53	53
AUGUST		7.1	5.2	18.1	5.8	31.1	7.2	250	264	279	35.6	30.0	214	10	53	54
SEPTEMBER		3.1	6.3	15.0	5.6	27.8	7.2	150	152	182	35.9	30.3	159	9	46	60
OCTOBER		−1.3	6.8	9.7	4.2	21.2	7.2	16	22	26	30.8	25.2	96	9	51	69
APR-OCTOBER		—	5.9	14.6	5.5	—	7.9	1,042	1,105	1,225	35.9	30.3	1,333	65	333	56
NOV-MARCH		—	—	2.8	—	—	—	—	—	—			—	—	185	—

DATA FOR MONTH TO MATURITY

MATURITY GROUP	EST. MATURITY DATE	*	*	*								*				
1	—		—	—	—							—	—	—	—	—
2	—		—	—	—							—	—	—	—	—
3	—		—	—	—							—	—	—	—	—
4	OCTOBER 1	15.5	4.1	26.5	7.2							30.1	157	9	46	60
5	—		—	—	—							—	—	—	—	—
6	—		—	—	—							—	—	—	—	—
7	—		—	—	—							—	—	—	—	—
8	—		—	—	—							—	—	—	—	—

April and October day degree summations derived from temperature curve, not directly from monthly average means. * Adjusted for vine sites.

217

TABLE 146

STUTTGART, GERMANY. 48°47′N, 9°10′E. ALT. 267m. (Württemberg, Neckar, etc.)

PERIOD	1 Lowest Recorded Min. (50YRS)	2 Av.Mthly Lowest Min. (50YRS)	3 Av.Min Minus Col.2	4 Daily Mean (50YRS)	5 Daily Range /2 (50YRS)	6 Av.Mthly Highest Max. (50YRS)	7 Col.6 Minus Av.Max.	8 Raw #	9 Adj.1 for Lat. & Daily Temp Rng	10 Adj.2 for Vine Sites	11 Raw	12 Adj. for Vine Sites	13 Av. Sunshine Hours	14 Av. No. Rain Days (40YRS)	15 Av. Rainfall mm (80YRS)	16 Av. % R.H. 1330 HRS (50YRS)
	Temperature °C							Summation of Day Degrees above 10° with 19° Cut-Off			Temperature Variability Index					
APRIL	−5.0	−0.6	5.1	9.2	4.7	22.8	8.9	8	10	14	32.8	31.2		11	56	57
MAY	0.0	3.3	5.6	13.9	5.0	28.3	9.4	121	127	140	35.0	33.4		11	69	56
JUNE	4.4	7.8	4.4	17.2	5.0	30.0	7.8	216	233	245	32.2	30.6		11	84	57
JULY	7.8	10.0	3.9	18.9	5.0	32.2	8.3	276	279	279	32.2	30.6		11	79	57
AUGUST	6.1	8.9	3.9	18.1	5.3	30.6	7.2	251	263	279	32.3	30.7		11	66	57
SEPTEMBER	1.1	5.0	5.6	15.0	4.4	27.2	7.8	150	161	173	31.0	29.4		10	63	63
OCTOBER	−5.6	0.0	6.1	9.7	3.6	21.7	8.4	15	25	30	28.9	27.3		9	48	69
APR-OCTOBER	—	—	4.9	14.6	4.7	—	8.3	1,037	1,098	1,160	32.1	30.5		74	465	59
NOV-MARCH	—	—	—	2.9	—	—	—	—	—	—	—	—		—	208	—

DATA FOR MONTH TO MATURITY																
MATURITY GROUP	EST. MATURITY DATE		*	*	*							*				
1	—		—	—	—	—						—		—	—	—
2	OCTOBER 8+		—	—	—	—						—		—	—	—
3	OCTOBER 8		13.9	3.8	26.0	7.9						29.1		10	60	64
4	OCTOBER 8+		—	—	—	—						—		—	—	—
5	—		—	—	—	—						—		—	—	—
6	—		—	—	—	—						—		—	—	—
7	—		—	—	—	—						—		—	—	—
8	—		—	—	—	—						—		—	—	—

April and October from temperature curve, not directly from monthly average means. * Adjusted for vine sites. + See commentary.

Commentaries and vineyard site adjustments

Bad Durkheim (Rheinpfalz, = Palatinate) (Table 143)

Vineyard site adjustments, °C

Min.	Max.	
+ 0.4	− 0.2	Mostly gentle slopes
+ 0.4	− 0.2	Sandy, some calcareous and stony soils
–	–	Most aspects
− 0.2	− 0.2	Average altitude 160 m (Bad Durkheim 130 m)
+ 0.6	− 0.6	Mean unchanged, variability index − 2.4

Notes

Being on the west bank of the Rhine, this is the warmest and sunniest part of Germany and extensive vine growing is possible without specially rigorous site selection. The Rheinpfalz produces much of Germany's commercial-grade wine.

White wines are mainly from Müller-Thurgau, Pinot Gris, Muscat Ottonel and Bacchus (Maturity Group 2), Sylvaner, Traminer, Kerner, Scheurebe and Morio-Muskat (3), and on specially warm soils and exposures, Riesling (4). Some light red wines are made, mainly from Blue Portuguese (2). Ripening conditions are estimated primarily for Group 3 varieties on average-to-good sites. It is assumed that Group 2 varieties are generally on below-average sites, and that they and Riesling on its superior sites will all ripen about as estimated for Maturity Group 3 on average sites.

Freiburg (South Baden) (Table 144)

Vineyard site adjustments, °C

Min.	Max.	
–	–	Moderate slopes (similar to Freiburg); most aspects
–	–	'Soils deep and fertile' (Hallgarten 1965)
+0.2	+0.2	Average altitude ~ 250 m (Freiburg 285 m)
+0.2	+0.2	Mean +0.2°, variability index unchanged.

Notes

Light white wines from Müller-Thurgau and Chasselas (Maturity Group 2). Note the inferior climatic conditions, particularly high rainfall, as compared with southern Alsace on the west bank of the Rhine (see Colmar, Chapter 16, Table 128).

The Kaiser Stuhl, a free-standing volcanic hill adjacent to Freiburg in the Rhine Valley, has sufficiently better site climates to allow the production of good white wines from Sylvaner and Traminer (3) and even some Riesling (4), in addition to Pinot Gris (2); also some light red wines from Pinot Noir (3). Patches of calcareous and/or rocky soils on the lower eastern slopes of the Rhine Valley to the north and south of Freiburg likewise produce some superior wines.

Geisenheim (Rheingau) (Table 145)

Vineyard site adjustments, °C

Min.	Max.	
+1.2	–0.5	Moderate to steep slopes (Geisenheim valley floor)
+0.75	+0.25	Exclusively southern aspects
+0.4	–0.2	Partly calcareous soils, plus some terracing
–0.4	–0.4	Average altitude ~ 170 m (Geisenheim ~ 100 m)
+1.95	–0.85	Mean +0.55°, variability index –5.6

Notes

Mostly semi-sweet to sweet white wines from late-picked and often noble-rotted Riesling (Maturity Group 4). The ability of Riesling to reach full ripeness so far north depends greatly on site factors: slope, southerly aspects, and to a smaller extent soil type. Close proximity to the Rhine is doubtless an additional factor, reflected in both the raw and the adjusted data for Geisenheim. Low altitude plays a part both directly and by contributing to a rain-shadow effect, resulting in anomalously low rainfall and long sunshine hours for this latitude.

Stuttgart (Württemberg, Neckar etc.) (Table 146)

Vineyard site adjustments, °C

Min.	Max.	
–	–	Valley slopes, similar to Stuttgart
+0.3	+0.1	Predominance of southerly aspects
+0.4	–0.2	Calcareous, red or bituminous marl soils
+0.2	+0.2	Average altitude 240 m (Stuttgart 267 m)
–0.3	–0.3	Latitude adjustment: vineyards mostly N of Stuttgart
+0.6	–0.2	Mean +0.2°, variability index –1.6

Notes

Sites suitable for viticulture are rare in this cold climate, and success depends greatly on favourable combinations of site factors and the choice of early-maturing vine varieties. White wines are from Chasselas, Pinot Gris and Müller-Thurgau (Maturity Group 2) and Sylvaner (3); light red wines from Blue Portuguese (2) and Trollinger (4, the wines reportedly acid and thin). Riesling (4) is grown on the very best sites, for Rhine-style wines. The average maturation data are estimated to be as for Group 3 varieties on good-average sites, with Riesling on superior sites and Group 2 varieties on inferior sites ripening at about the same time.

18

Viticultural Environments of Central and Eastern Europe

The distribution of viticulture across central and eastern Europe is limited almost entirely by climate, with winter killing of the dormant vines increasingly the main factor towards the east.

The western Swiss vineyards, typified by Geneva (Table 147), have climates closely resembling those of the Rhine Valley in Germany (Chapter 17), except that their more southern latitude gives them substantially more sunshine. Raw average mean temperatures for the coldest month (Geneva +1°C, Lugarno +2°C) are still not much above the defined threshold of –1°C for winter killing of *vinifera* vines. Corresponding raw figures for the other locations detailed in this chapter are: Austria: Vienna –1°C; Croatia/Slovenia: Zagreb –1°C; Hungary: Eger –2°C, Kesthely –1°C,

Mosonmagyarovar –2°C; and Bulgaria: Pleven –2°C, Plovdiv 0°C and Sliven +1°C. Clearly these areas approach the extreme limit for unprotected viticulture, and regular winter survival depends on the careful selection of sites.

Their growing season and grape ripening temperatures, on the other hand, and to a lesser extent sunshine hours, increase progressively to the east and southeast. Thus winter killing indirectly defines the growing season and ripening climates that can be exploited for vine growing, and therefore potential wine styles. Eastern Europe as a result only produces relatively warm climate styles of wine, just as is dictated by spring frosts in much of inland Australia.

TABLE 147

GENEVA, SWITZERLAND. 46°12′N, 6°09′E. ALT. 450m. (Lake Geneva, Chablais, Valais; also Haute Savoie, France)

	1	2	3	4	5	6	7	8	9	10	11	12	13	14	15	16
	Temperature °C							Summation of Day Degrees above 10° with 19° Cut-Off			Temperature Variability Index					
PERIOD	Lowest Recorded Min. (67YRS)	Av.Mthly Lowest Min. (67YRS)	Av.Min Minus Col.2	Daily Mean (30YRS)	Daily Range /2 (30YRS)	Av.Mthly Highest Max. (67YRS)	Col.6 Minus Av.Max.	Raw #	Adj.1 for Lat. & Daily Temp Rng	Adj.2 for Vine Sites	Raw	Adj. for Vine Sites	Av. Sunshine Hours	Av. No. Rain Days (125YRS)	Av. Rainfall mm (125YRS)	Av. % R.H. 1300 HRS (67YRS)
APRIL	–5.0	–0.6	5.6	9.7	4.7	22.2	7.8	12	14	14	32.2	27.4	195	11	64	55
MAY	–1.7	2.8	6.1	13.9	5.0	26.1	7.2	121	126	126	33.3	28.5	242	12	76	57
JUNE	2.2	6.7	6.1	17.8	5.0	29.4	6.7	234	246	246	32.7	27.9	279	11	79	55
JULY	5.6	8.9	5.5	19.7	5.3	32.2	7.2	279	279	279	33.9	29.1	301	9	74	55
AUGUST	5.0	8.3	5.6	19.2	5.3	31.1	6.7	279	279	279	33.4	28.6	276	10	91	57
SEPTEMBER	0.0	5.0	6.1	15.8	4.7	27.2	6.7	174	181	181	31.6	26.8	207	10	91	63
OCTOBER	–6.7	0.0	6.7	10.6	3.9	21.7	7.2	30	41	41	29.5	24.7	133	11	97	69
APR-OCTOBER	—	—	6.0	15.2	4.8	—	7.1	1,129	1,166	1,166	32.4	27.6	1,633	74	572	59
NOV-MARCH	—	—	—	3.4	—	—	—	—	—	—	—	—	—	—	290	---
DATA FOR MONTH TO MATURITY																
MATURITY GROUP	EST. MATURITY DATE		*	*	*							*				
1	—		—	—	—	—						—	—	—	—	—
2	SEPTEMBER 24		16.7	3.6	26.6	6.7						27.2	221	10	91	62
3	OCTOBER 8		14.5	3.3	24.1	6.8						26.1	188	10	91	64
4	—		—	—	—	—						—	—	—	—	—
5	—		—	—	—	—						—	—	—	—	—
6	—		—	—	—	—						—	—	—	—	—
7	—		—	—	—	—						—	—	—	—	—
8	—		—	—	—	—						—	—	—	—	—

April and October day degree summations derived from temperature curve, not directly from monthly average means. * Adjusted for vine sites.

TABLE 148
LUGARNO, SWITZERLAND. 46°00′N, 8°57′E. ALT. 276m. (Ticino)

	1	2	3	4	5	6	7	8	9	10	11	12	13	14	15	16
	Temperature °C							Summation of Day Degrees above 10° with 19° Cut-Off			Temperature Variability Index					
PERIOD	Lowest Recorded Min. (30YRS)	Av.Mthly Lowest Min. (30YRS)	Av.Min Minus Col.2	Daily Mean (30YRS)	Daily Range /2 (30YRS)	Av.Mthly Highest Max. (30YRS)	Col.6 Minus Av.Max.	Raw #	Adj.1 for Lat. & Daily Temp Rng	Adj.2 for Vine Sites	Raw	Adj. for Vine Sites	Av. Sunshine Hours	Av. No. Rain Days (87YRS)	Av. Rainfall mm (77YRS)	Av. % R.H. 1300 HRS (67YRS)
APRIL	−2.2	1.1	5.0	11.7	5.6	24.4	7.1	51	52	46	34.5	31.7	177	11	163	53
MAY	0.6	4.4	5.6	15.6	5.6	27.8	6.6	174	181	176	34.6	31.8	208	14	193	57
JUNE	4.4	8.9	4.4	19.4	6.1	31.1	5.6	270	270	270	34.4	31.6	243	12	185	55
JULY	7.8	10.6	4.9	21.9	6.4	32.8	4.5	279	279	279	35.0	32.2	273	11	175	53
AUGUST	6.7	10.6	4.4	21.4	6.4	32.3	4.4	279	279	279	34.4	31.6	248	10	188	55
SEPTEMBER	2.2	6.7	5.6	18.1	5.8	28.9	5.0	243	245	239	33.8	31.0	192	9	175	60
OCTOBER	−2.2	2.2	5.6	12.5	4.7	23.9	6.7	78	81	85	31.1	28.3	149	11	198	64
APR-OCTOBER	—	—	5.1	17.2	5.8	—	5.7	1,374	1,387	1,374	34.0	31.2	1,490	78	1,277	57
NOV-MARCH	—	—	—	4.9	—	—	—	—	—	—	—	—	—	—	452	—
DATA FOR MONTH TO MATURITY																
MATURITY GROUP	EST. MATURITY DATE			*	*	*						*				
1	—			—	—	—						—	—	—	—	—
2	—			—	—	—						—	—	—	—	—
3	—			—	—	—						—	—	—	—	—
4	—			—	—	—						—	—	—	—	—
5	SEPTEMBER 24			18.6	5.2	28.8	4.9					31.1	202	9	177	59
6	—			—	—	—	—					—	—	—	—	—
7	—			—	—	—	—					—	—	—	—	—
8	—			—	—	—	—					—	—	—	—	—

*Adjusted for vine sites.

TABLE 149
VIENNA, AUSTRIA. 48°14′N, 16°25′E. ALT. 202m. (Vienna, Weinviertel, Vöslau, Baden, Langenlois, Wachau)

	1	2	3	4	5	6	7	8	9	10	11	12	13	14	15	16
	Temperature °C							Summation of Day Degrees above 10° with 19° Cut-Off			Temperature Variability Index					
PERIOD	Lowest Recorded Min. (100YRS)	Av.Mthly Lowest Min. (100YRS)	Av.Min Minus Col.2	Daily Mean (50YRS)	Daily Range /2 (50YRS)	Av.Mthly Highest Max. (100YRS)	Col.6 Minus Av.Max.	Raw #	Adj.1 for Lat. & Daily Temp Rng	Adj.2 for Vine Sites	Raw	Adj. for Vine Sites	Av. Sunshine Hours	Av. No. Rain Days (50YRS)	Av. Rainfall mm (100YRS)	Av. % R.H. 1300 HRS (40YRS)
APRIL	−7.8	−0.6	5.6	9.4	4.4	22.2	8.4	11	14	16	31.6	30.0	180	9	51	51
MAY	−2.8	3.3	6.7	14.4	4.4	26.7	7.9	136	154	160	32.2	30.6	236	9	71	53
JUNE	3.9	8.9	4.4	17.5	4.2	28.9	7.2	225	254	260	28.4	26.8	261	9	69	54
JULY	7.2	11.1	3.9	19.4	4.4	31.1	7.3	279	279	279	28.8	27.2	273	9	76	54
AUGUST	5.6	10.0	4.4	18.6	4.2	30.6	7.8	267	279	279	29.0	27.4	248	10	69	55
SEPTEMBER	−0.6	5.6	5.5	15.0	3.9	26.7	7.8	150	169	175	28.9	27.3	198	7	51	58
OCTOBER	−8.9	0.0	6.6	9.7	3.1	21.1	8.3	15	30	33	27.3	25.7	124	8	51	67
APR-OCTOBER	—	—	5.3	14.9	4.1	—	7.8	1,083	1,179	1,202	29.5	27.9	1,520	61	438	56
NOV-MARCH	—	—	—	1.9	—	—	—	—	—	—	—	—	—	—	213	—
DATA FOR MONTH TO MATURITY																
MATURITY GROUP	EST. MATURITY DATE															
1	—			—	—	—						—	—	—	—	—
2	SEPTEMBER 25 + ⎱			—	—	—						—	—	—	—	—
3	SEPTEMBER 25 ⎰			16.5	3.4	27.1	7.8					27.3	207	8	54	57
4	OCTOBER 13			13.2	3.2	24.4	8.0					26.7	168	8	51	60
5	—			—	—	—	—					—	—	—	—	—
6	—			—	—	—	—					—	—	—	—	—
7	—			—	—	—	—					—	—	—	—	—
8	—			—	—	—	—					—	—	—	—	—

April and October from temperature curve, not directly from monthly average means.   * Adjusted for vine sites.   + See commentary.

221

TABLE 150

ZAGREB, CROATIA. 45°49'N, 16°00E, ALT. 163m. (Slovenia: Ljutomer, etc.)

	1	2	3	4	5	6	7	8	9	10	11	12	13	14	15	16
	Temperature °C							Summation of Day Degrees above 10° with 19° Cut-Off			Temperature Variability Index					
PERIOD	Lowest Recorded Min. (17YRS)	Av.Mthly Lowest Min. (17YRS)	Av.Min Minus Col.2	Daily Mean (14YRS)	Daily Range /2 (14YRS)	Av.Mthly Highest Max. (17YRS)	Col.6 Minus Av.Max.	Raw #	Adj.1 for Lat. & Daily Temp Rng	Adj.2 for Vine Sites	Raw	Adj. for Vine Sites	Av. Sunshine Hours	Av. No. Rain Days (17YRS)	Av. Rainfall mm (64YRS)	Av. % R.H. 1400 HRS (17YRS)
APRIL	−0.6	1.7	5.0	11.4	4.7	22.8	6.7	45	50	38	30.5			13	71	56
MAY	2.2	7.2	3.9	16.1	5.0	27.2	6.1	189	196	181	30.0			14	79	56
JUNE	8.3	12.2	2.2	19.4	5.0	30.0	5.6	270	270	270	27.8			13	99	55
JULY	11.7	13.9	2.2	21.4	5.3	31.7	5.0	279	279	279	28.4			12	81	54
AUGUST	9.4	11.7	3.8	20.8	5.3	31.7	5.6	279	279	279	30.6			10	81	55
SEPTEMBER	3.3	7.8	4.2	16.4	4.4	27.2	6.4	192	203	188	28.2			12	86	62
OCTOBER	−3.3	2.8	4.7	11.4	3.9	22.2	6.9	46	57	45	27.2			13	99	70
APR-OCTOBER	—	—	3.7	16.7	4.8	—	6.0	1,300	1,334	1,280	29.0			87	596	58
NOV-MARCH	—	—	—	3.6	—	—	—				—			—	290	—

DATA FOR MONTH TO MATURITY

MATURITY GROUP	EST. MATURITY DATE	*	*	*												
1	—			—		—	—				—			—	—	—
2	SEPTEMBER 8	19.1	5.1	30.4	5.8						30.0			11	82	56
3	SEPTEMBER 15	18.1	4.9	29.3	6.0						29.4			11	83	58
4	SEPTEMBER 23	17.0	4.6	27.9	6.2						28.7			12	85	60
5	—			—	—						—			—	—	—
6	—			—	—						—			—	—	—
7	—			—	—						—			—	—	—
8	—			—	—						—			—	—	—

April and October day degree summations derived from temperature curve, not directly from monthly average means. * Adjusted for vine sites.

TABLE 151

EGER, HUNGARY. 47°53'N, 20°23'E. ALT. 175m. (Gyöngyös, Eger, Tokay)

	1	2	3	4	5	6	7	8	9	10	11	12	13	14	15	16
	Temperature °C							Summation of Day Degrees above 10° with 19° Cut-Off			Temperature Variability Index					
PERIOD	Lowest Recorded Min. (50YRS)	Av.Mthly Lowest Min. (50YRS)	Av.Min Minus Col.2	Daily Mean (50YRS)	Daily Range /2 (50YRS)	Av.Mthly Highest Max. (50YRS)	Col.6 Minus Av.Max.	Raw #	Adj.1 for Lat. & Daily Temp Rng	Adj.2 for Vine Sites	Raw	Adj. for Vine Sites	Av. Sunshine Hours	Av. No. Rain Days (70YRS)	Av. Rainfall mm (50YRS)	Av. % R.H. 1430 HRS (40YRS)
APRIL	−6.1	−2.2	7.2	10.3	5.3	23.9	8.3	24	25	46	36.7	31.1	180	8	46	57
MAY	−2.8	2.8	6.7	15.6	6.1	28.3	6.6	174	179	200	37.7	32.1	264	9	66	55
JUNE	1.1	5.6	7.2	18.9	6.1	30.6	5.6	267	270	270	37.2	31.6	273	10	74	55
JULY	3.9	8.9	5.0	20.6	6.7	32.8	5.5	279	279	279	37.3	31.7	295	8	58	53
AUGUST	3.9	7.8	5.5	20.0	6.7	32.8	6.1	279	279	279	38.4	32.8	273	7	58	54
SEPTEMBER	−2.2	3.3	6.1	15.8	6.4	28.3	6.1	174	176	191	37.8	32.2	189	6	48	61
OCTOBER	−12.8	−2.2	6.6	10.0	5.6	22.2	6.6	21	20	35	35.6	30.0	139	8	51	70
APR-OCTOBER	—	—	6.3	15.9	6.1	—	6.4	1,218	1,228	1,300	37.2	31.6	1,613	56	401	58
NOV-MARCH	—	—	—	1.4	—	—	—				—	—		—	185	—

DATA FOR MONTH TO MATURITY

MATURITY GROUP	EST. MATURITY DATE	*	*	*								*				
1	—			—	—							—		—	—	—
2	—			—	—							—		—	—	—
3	—			—	—							—		—	—	—
4	SEPTEMBER 17	18.7	5.1	29.6	6.1							32.5	222	6	53	58
5	SEPTEMBER 27	17.0	5.0	28.0	6.1							32.3	195	6	49	60
6	OCTOBER 15	13.5	4.6	24.8	6.3							31.2	164	7	50	65
7	—			—	—							—		—	—	—
8	—			—	—							—		—	—	—

April and October day degree summations derived from temperature curve, not directly from monthly average means. * Adjusted for vine sites.

TABLE 152
KESZTHELY, HUNGARY. 46°46′N, 17°14′E. ALT. 133m. (Lake Balaton)

	1	2	3	4	5	6	7	8	9	10	11	12	13	14	15	16
	Temperature °C							Summation of Day Degrees above 10° with 19° Cut-Off			Temperature Variability Index					
PERIOD	Lowest Recorded Min. (50YRS)	Av.Mthly Lowest Min. (50YRS)	Av.Min Minus Col.2	Daily Mean (50YRS)	Daily Range /2 (50YRS)	Av.Mthly Highest Max. (50YRS)	Col.6 Minus Av.Max.	Raw #	Adj.1 for Lat. & Daily Temp Rng	Adj.2 for Vine Sites	Raw	Adj. for Vine Sites	Av. Sunshine Hours	Av. No. Rain Days (70YRS)	Av. Rainfall mm (50YRS)	Av. % R.H. 1400 HRS (40YRS)
APRIL	−5.0	−0.6	6.7	10.8	4.7	23.3	7.8	31	34	46	33.3	29.3	180	9	56	60
MAY	−1.7	4.4	6.2	15.6	5.0	27.8	7.2	174	182	201	33.4	29.4	264	10	74	59
JUNE	3.3	8.3	5.6	19.2	5.3	30.6	6.1	270	270	270	32.9	28.9	273	9	74	59
JULY	6.1	10.6	4.9	21.1	5.6	32.8	6.1	279	279	279	33.4	29.4	295	9	71	56
AUGUST	5.0	10.0	5.0	20.3	5.3	32.2	6.6	279	279	279	32.8	28.8	273	8	76	58
SEPTEMBER	0.6	5.6	6.1	16.7	5.0	28.9	7.2	201	203	221	33.3	29.3	189	8	61	64
OCTOBER	−7.2	0.6	6.1	11.1	4.4	23.3	7.8	41	46	58	31.5	27.5	139	8	66	71
APR-OCTOBER	—	—	5.8	16.4	5.0	—	7.0	1,275	1,293	1,354	32.9	28.9	1,613	61	478	61
NOV-MARCH	—	—	—	2.4	—	—	—	—	—	—	—	—	—	—	206	—
DATA FOR MONTH TO MATURITY																
MATURITY GROUP	EST. MATURITY DATE		*	*	*							*				
1	—		—	—	—							—	—	—	—	—
2	—		—	—	—							—	—	—	—	—
3	—		—	—	—							—	—	—	—	—
4	SEPTEMBER 14		19.1	4.2	30.0	6.9						29.6	231	8	69	61
5	SEPTEMBER 21		18.3	4.1	29.0	7.0						29.5	211	8	65	62
6	—		—	—	—							—	—	—	—	—
7	—		—	—	—							—	—	—	—	—
8	—		—	—	—							—	—	—	—	—

April and October day degree summations derived from temperature curve, not directly from monthly average means. * Adjusted for vine sites.

TABLE 153
MOSONMAGYAROVÁR, HUNGARY. 47°53′N, 17°16′E. ALT. 123m. (Sopron, Hungary; Burgenland, E. Austria)

	1	2	3	4	5	6	7	8	9	10	11	12	13	14	15	16
	Temperature °C							Summation of Day Degrees above 10° with 19° Cut-Off			Temperature Variability Index					
PERIOD	Lowest Recorded Min. (50YRS)	Av.Mthly Lowest Min. (50YRS)	Av.Min Minus Col.2	Daily Mean (50YRS)	Daily Range /2 (50YRS)	Av.Mthly Highest Max. (50YRS)	Col.6 Minus Av.Max.	Raw #	Adj.1 for Lat. & Daily Temp Rng	Adj.2 for Vine Sites	Raw	Adj. for Vine Sites	Av. Sunshine Hours	Av. No. Rain Days (70YRS)	Av. Rainfall mm (50YRS)	Av. % R.H. 1400 HRS (40YRS)
APRIL	−5.0	−1.7	6.7	10.3	5.3	23.3	7.7	23	24	29	35.6	33.2	180	8	43	57
MAY	−1.7	2.2	7.2	15.0	5.6	27.2	6.6	155	163	169	36.2	33.8	248	8	66	56
JUNE	2.2	6.1	6.7	18.1	5.3	30.6	7.2	243	260	270	35.1	32.7	267	8	58	56
JULY	7.2	10.0	4.5	20.3	5.8	32.8	6.7	279	279	279	34.4	32.0	285	8	66	55
AUGUST	6.7	8.3	5.6	19.7	5.8	32.2	6.7	279	279	279	35.5	33.1	260	7	58	56
SEPTEMBER	−2.2	3.3	7.2	15.8	5.3	28.3	7.2	174	176	187	35.6	33.2	192	7	51	62
OCTOBER	−9.4	−1.1	6.7	10.0	4.4	22.8	8.4	22	25	33	32.7	30.3	133	8	48	69
APR-OCTOBER	—	—	6.4	15.6	5.4	—	7.2	1,175	1,206	1,246	35.0	32.6	1,565	54	390	59
NOV-MARCH	—	—	—	1.7	—	—	—	—	—	—	—	—	—	—	208	—
DATA FOR MONTH TO MATURITY																
MATURITY GROUP	EST. MATURITY DATE		*	*	*							*				
1	—		—	—	—							—	—	—	—	—
2	—		—	—	—							—	—	—	—	—
3	SEPTEMBER 18		18.0	4.9	29.1							33.2	221	7	54	60
4	SEPTEMBER 28		16.5	4.7	28.4							33.2	196	7	51	62
5	—		—	—	—							—	—	—	—	—
6	—		—	—	—							—	—	—	—	—
7	—		—	—	—							—	—	—	—	—
8	—		—	—	—							—	—	—	—	—

April and October day degree summations derived from temperature curve, not directly from monthly average means. * Adjusted for vine sites.

TABLE 154

PLEVEN, BULGARIA. 43°25'N, 24°37'E. ALT. 109m.

	1	2	3	4	5	6	7	8	9	10	11	12	13	14	15	16
	Temperature °C							Summation of Day Degrees above 10° with 19° Cut-Off			Temperature Variability Index					
PERIOD	Lowest Recorded Min. (30YRS)	Av.Mthly Lowest Min. (30YRS)	Av.Min Minus Col.2	Daily Mean (30YRS)	Daily Range /2 (30YRS)	Av.Mthly Highest Max. (30YRS)	Col.6 Minus Av.Max.	Raw	Adj.1 for Lat. & Daily Temp Rng	Adj.2 for Vine Sites	Raw	Adj. for Vine Sites	Av. Sunshine Hours	Av. No. Rain Days (26YRS)	Av. Rainfall mm (25YRS)	Av. % R.H. 1330 HRS (25YRS)
APRIL	−4.4	−0.5	7.2	12.5	5.8	27.2	8.9	75	76	64	39.3	38.1	189	7	48	49
MAY	0.6	6.7	5.0	17.5	5.8	30.6	7.3	232	238	225	35.5	34.3	239	10	71	51
JUNE	5.6	10.0	4.5	20.6	6.1	33.3	6.6	270	270	270	35.5	34.3	269	9	84	50
JULY	7.8	12.8	3.9	23.1	6.4	35.6	6.1	279	279	279	35.6	34.4	314	7	84	45
AUGUST	8.9	11.7	4.4	22.8	6.7	35.6	6.1	279	279	279	37.3	36.1	305	5	43	42
SEPTEMBER	−0.6	7.2	5.0	18.9	6.7	33.3	7.7	267	265	256	39.5	38.3	224	6	51	48
OCTOBER	−3.3	1.1	6.1	12.8	5.6	27.8	9.4	87	86	74	37.9	38.7	152	7	48	60
APR-OCTOBER	—	—	5.2	18.3	6.2	—	7.4	1,489	1,493	1,447	37.2	36.0	1,692	51	429	49
NOV-MARCH	—	—	—	2.1	—	—	—	—	—	—	—	—	—	—	196	—

DATA FOR MONTH TO MATURITY

MATURITY GROUP	EST. MATURITY DATE	*	*	*							*				
1	—		—	—	—	—					—	—	—	—	—
2	—		—	—	—	—					—	—	—	—	—
3	SEPTEMBER 4	21.9	6.4	34.7	6.3					36.3	300	5	44	43
4	SEPTEMBER 10	21.2	6.4	34.4	6.5					36.8	285	5	45	44
5	SEPTEMBER 15	20.5	6.4	34.0	6.8					37.3	265	5	47	45
6	SEPTEMBER 22	19.6	6.4	33.4	7.2					37.8	248	6	49	47
7	—		—	—	—	—					—	—	—	—	—
8	—		—	—	—	—					—	—	—	—	—

* Adjusted for vine sites.

TABLE 155

PLOVDIV, BULGARIA. 42°09'N, 24°45'E. ALT. 161m.

	1	2	3	4	5	6	7	8	9	10	11	12	13	14	15	16
	Temperature °C							Summation of Day Degrees above 10° with 19° Cut-Off			Temperature Variability Index					
PERIOD	Lowest Recorded Min. (30YRS)	Av.Mthly Lowest Min. (30YRS)	Av.Min Minus Col.2	Daily Mean (44YRS)	Daily Range /2 (44YRS)	Av.Mthly Highest Max. (30YRS)	Col.6 Minus Av.Max.	Raw	Adj.1 for Lat. & Daily Temp Rng	Adj.2 for Vine Sites	Raw	Adj. for Vine Sites	Av. Sunshine Hours	Av. No. Rain Days (26YRS)	Av. Rainfall mm (24YRS)	Av. % R.H. 1330 HRS (23YRS)
APRIL	−3.9	−0.5	7.2	12.5	5.8	27.2	8.9	75	76	59	39.3	35.7	189	6	38	50
MAY	−0.6	5.6	5.5	17.8	6.7	31.1	6.6	242	242	226	38.9	35.3	239	9	58	51
JUNE	6.1	9.4	5.1	21.4	6.9	34.4	6.1	270	270	270	38.8	35.2	288	7	58	49
JULY	8.3	12.2	4.5	23.6	6.9	36.7	6.2	279	279	279	38.3	34.7	332	5	46	42
AUGUST	5.6	11.1	4.5	23.1	7.5	36.1	5.5	279	279	279	40.0	36.4	316	4	36	44
SEPTEMBER	1.1	5.6	6.6	19.4	7.2	33.9	7.3	270	270	265	42.7	39.1	240	4	43	47
OCTOBER	−5.6	0.0	7.2	13.6	6.4	28.3	8.3	112	111	92	41.1	37.5	170	5	38	58
APR-OCTOBER	—	—	5.8	18.8	6.8	—	7.0	1,527	1,527	1,470	39.9	36.3	1,774	40	317	49
NOV-MARCH	—	—	—	3.7	—	—	—	—	—	—	—	—	—	—	193	—

DATA FOR MONTH TO MATURITY

MATURITY GROUP	EST. MATURITY DATE	*	*	*							*				
1	—		—	—	—	—					—	—	—	—	—
2	—		—	—	—	—					—	—	—	—	—
3	SEPTEMBER 5	22.2	6.8	34.4	5.7					36.7	306	4	37	44
4	SEPTEMBER 10	21.7	6.8	34.2	6.1					37.2	295	4	39	44
5	SEPTEMBER 16	21.1	6.7	33.8	6.5					37.8	282	4	40	45
6	SEPTEMBER 21	20.4	6.7	33.3	6.8					38.4	263	4	41	46
7	SEPTEMBER 27	19.4	6.6	32.7	7.2					39.0	248	4	43	47
8	—		—	—	—	—					—	—	—	—	—

* Adjusted for vine sites.

TABLE 156
SLIVEN, BULGARIA. 42°41′N, 26°19′E. ALT. 265m.

PERIOD	1 Lowest Recorded Min. (30YRS)	2 Av.Mthly Lowest Min. (30YRS)	3 Av.Min Minus Col.2	4 Daily Mean (30YRS)	5 Daily Range /2 (30YRS)	6 Av.Mthly Highest Max. (30YRS)	7 Col.6 Minus Av.Max.	8 Raw	9 Adj.1 for Lat. & Daily Temp Rng	10 Adj.2 for Vine Sites	11 Raw	12 Adj. for Vine Sites	13 Av. Sunshine Hours	14 Av. No. Rain Days (27YRS)	15 Av. Rainfall mm (27YRS)	16 Av. % R.H. 1400 HRS (26YRS)
	Temperature °C							Summation of Day Degrees above 10° with 19° Cut-Off			Temperature Variability Index					
APRIL	−5.0	0.0	6.6	11.9	5.3	25.0	7.8	57	58		35.6		186	7	51	50
MAY	0.6	5.6	6.0	16.9	5.3	28.9	6.7	214	218		33.9		229	10	79	51
JUNE	5.6	9.4	5.6	20.6	5.6	32.2	6.0	270	270		34.0		261	9	84	50
JULY	6.7	12.8	4.5	23.1	5.8	35.0	6.1	279	279		33.8		316	6	48	44
AUGUST	7.8	11.7	5.0	22.8	6.1	34.4	5.5	279	279		34.9		313	4	33	43
SEPTEMBER	2.2	7.8	5.6	19.2	5.8	31.7	6.7	270	270		35.5		243	4	36	45
OCTOBER	−6.1	2.2	6.7	13.9	5.0	27.2	8.3	121	120		35.0		174	5	43	57
APR-OCTOBER	—	—	5.7	18.3	5.6	—	6.7	1,490	1,494		34.7		1,722	45	374	49
NOV-MARCH	—	—	—	4.3	—	—	—	—	—		—		—	—	229	—
DATA FOR MONTH TO MATURITY																
MATURITY GROUP	EST. MATURITY DATE															
1	—		—	—	—	—					—		—	—	—	—
2	AUGUST 31		22.8	6.1	34.4	5.5					34.9		313	4	33	43
3	SEPTEMBER 5		22.4	6.0	34.0	5.7					35.0		300	4	34	43
4	SEPTEMBER 11		21.8	5.9	33.5	5.9					35.1		290	4	34	44
5	SEPTEMBER 17		21.2	5.9	33.0	6.1					35.2		275	4	35	44
6	SEPTEMBER 23		20.4	5.8	32.5	6.4					35.3		260	4	35	45
7	—		—	—	—	—					—		—	—	—	—
8	—		—	—	—	—					—		—	—	—	—

Commentaries and vineyard site adjustments

1.   Switzerland

Geneva (Lake Geneva, Chablais, Valais; also Haute Savoie, France) (Table 147)

Vineyard site adjustments, °C

Min.	Max.	
+0.8	−0.3	Steep slopes, but mostly further from the lake
+0.6	−0.3	Some soils rocky and/or terraced, some calcareous
+0.6	+0.2	Predominantly southern aspects
−1.4	−1.4	Average altitude 600–700 m (Geneva 405 m)
+0.6	−1.8	Mean −0.6°, variability index −4.8

Notes

Dry white wines mainly from Chasselas (Maturity Group 2), with some Pinot Gris (2) and Sylvaner (3); light red wines from Pinot Noir and Gamay (3).

No one table can characterize fully an environment where so much depends on altitude, slope and aspect. In general the early-maturing Chasselas is grown in the higher and/or cooler parts; conditions during its maturation probably range from those tabulated for average conditions down to those indicated for Group 3 varieties. By contrast Pinot Noir and Gamay (3 for red wines, 2 for rosé) must be grown for red wines largely on warmer-than-average sites. Ripening there might be as early and warm as indicated for Group 2 varieties under 'average' conditions. Such conditions nearly match those experienced by the same varieties in Burgundy, the Beaujolais, and the upper and lower Loire.

The delicacy of Swiss wines can be related to the notable equability of conditions during ripening, including especially a complete absence of damaging high temperatures. Delicacy is clearly not compromised by having more sunshine than other comparably cool European regions.

Lugarno (Ticino) (Table 148)

Vineyard site adjustments, °C

Min.	Max.	
–	–	Moderate to steep slopes but greater distance from lake
+0.8	–0.4	Stony soils and terracing
+0.3	+0.1	Southerly aspects predominating
–0.6	–0.6	Average altitude 350–400 m (Lugarno 276 m)
+0.5	–0.9	Mean –0.2°, variability index –2.8

Notes

Principally red wines from Merlot (Maturity Group 5). A notable feature of the environment is its very high rainfall, both through the growing season and during ripening. Nevertheless the average number of rain days for the ripening month is only nine, and the average early afternoon relative humidity is not excessive. The data can be taken as demonstrating the tolerance of Merlot to high-rainfall conditions, if not necessarily to high relative humidity or lack of sunshine. Temperature equability is another feature of the climate, and is attributable to the joint influences of Lake Lugarno and constant convectional air mixing due to the mountainous terrain.

2.   Austria

Vienna (also Veinviertel, Vöslau, Baden, Langenlois, Wachau) (Table 149)

Vineyard site adjustments, °C

Min.	Max.	
–	–	Mostly moderate slopes, as Vienna
+0.3	+0.1	Moderate predominance of southerly aspects
+0.4	–0.2	Some soils rocky and/or calcareous
–0.3	–0.3	Average altitude ~ 250 m (Vienna 202 m)
+0.4	–0.4	Mean unchanged, variability index –1.6

Notes

Mainly light, dry white wines from Müller-Thurgau and Red Veltliner (Maturity Group 2), Green Veltliner, Traminer and Sylvaner (3), and Welschriesling and some Riesling (4); light red wines from Blue Portuguese (2) and Pinot Noir (3).

The Wachau, on steep slopes facing south across the Donau (Danube) west of Krems, enjoys particularly favoured exposures which should allow Riesling to ripen significantly earlier and more fully than indicated in Table 149 for Maturity Group 4, at least at the lower altitudes.

3.   Croatia, Slovenia

Zagreb (Northern Croatia and Southern Slovenia: Ljutomer etc.) (Table 150)

Vineyard site adjustments, °C

Min.	Max.	
–	–	Flat to moderate slopes, as Zagreb. All aspects
–	–	Mainly alluvial soils?
–0.2	–0.2	Average altitude ~ 200 m (Zagreb 163 m)
–0.3	–0.3	Latitude adjustment (vineyards N of Zagreb)
–0.5	–0.5	Mean –0.5°, variability index unchanged

Notes

Mostly semi-sweet white wines from Pinot Gris (Maturity Group 2), Traminer and Sauvignon Blanc (3), and Riesling and Welschriesling (Olaszrizling) (4); some red wine from Blue Portuguese (2) and other red varieties. The early (Maturity Group 2) varieties are probably often grown in cool situations, to ripen at times and under conditions roughly equivalent to those estimated here for Group 3.

The growing season is very equable. High rainfall and (probably) ample sunshine hours should be conducive to good vine growth and yields.

4.   Hungary

Eger (Gyöngyös, Eger, Tokay) (Table 151)

Vineyard site adjustments, °C

Min.	Max.	
+1.0	–0.5	Slopes of semi-isolated hills (Eger in valley)
+0.6	–0.3	Volcanic or calcareous soils
+0.6	+0.2	Southerly aspects
–0.3	–0.3	Average altitude ~ 225 m (Eger 175 m)
+1.9	–0.9	Mean +0.5°, variability index –5.6

Notes

Sunshine hours are assumed to be as for Budapest. Full-bodied sweet white wines (Tokay) from late-picked, noble-rotted Furmint and Harslevelu (Maturity Group 4); medium to full-bodied dry to sweet white wines from Welschriesling (4); red wines from Merlot (5) and other red varieties of unknown maturity; light red to rosé wines from Kadarka (6 for red wine, 5 for rosé).

Keszthely (Lake Balaton) (Table 152)

Vineyard site adjustments, °C

Min.	Max.	
+0.4	−0.2	Slopes of isolated hills; Keszthely flatter but on lake
+0.3	+0.1	Predominantly southern aspects
+0.8	−0.4	Rubbly basalt soils
−0.4	−0.4	Average altitude ~ 200 m (Keszthely 133 m)
+1.1	−0.9	Mean +0.1, variability index −4.0

Notes

Sunshine hours are assumed to be as for Budapest. Full-bodied, dry to sweet white wines from Welschriesling and Furmint (Maturity Group 4); also from Szürkebarat and Keknyelü (maturities unknown: possibly 4 or 5?). The noted strength of the wines reflects the relatively high mean temperatures and (probably) sunshine hours during ripening in this low-elevation, sheltered area.

Mosonmagyaróvár (Sopron; also Burgenland, E. Austria) (Table 153)

Vineyard site adjustments, °C

Min.	Max.	
–	–	Flat; sandy soils
+0.8	−0.4	Proximity of vineyards to Neusiedler See
+0.8	−0.4	Mean +0.2°, variability index −2.4

Notes

Sunshine hours are estimated as the averages of Vienna and Budapest. Mainly sweet white wines from late-picked Traminer (Maturity Group 3), and Riesling, Welschriesling and Furmint (4). Some red wines from Pinot Noir (3 for red wines). The most famous centres of Rust and Apetlon lie close to the west and east banks respectively of the Neusiedler See, and benefit greatly from the tempering effect of that large water body. Their traditional sweet wines are reported to bear comparison with those of the Rhine Valley and Tokay (Johnson 1971).

5. Bulgaria

Pleven (Table 154)

Vineyard site adjustments, °C

Min.	Max.	
+0.4	−0.2	Slight slopes (Pleven in valley)
–	–	Loamy soils
−0.5	−0.5	Average altitude ~ 200 m (Pleven 109 m)
−0.1	−0.7	Mean −0.4°, variability index −1.2

Notes

The Pleven and nearby areas in northern Bulgaria produce mainly red wines. Varieties grown include Cabernet Franc (Maturity Group 5), Granza, syn. Kadarka, and Cabernet Sauvignon (6) and Mavrud (maturity unknown but probably similar). Ripening-month conditions are calculated for Maturity Groups 3 and 4 as well, to cover any white and early-maturing red varieties grown.

The estimated average ripening conditions are very suitable for medium to full-bodied table wines from midseason (Groups 5–6) varieties. The temperature variability index is perhaps above optimum, but the monthly highest maxima are quite moderate. Humidities are at the low end of the optimum range and should be favourable for vine health. Rainfall is ample early in the growing season, and diminishes conveniently to a moderate amount during ripening.

One problem suggested by the figures is frost after budburst: note the variability of the minimum in April. However this is associated with a very rapid rise in temperature over the spring period, so that most frosts will be early in April, and therefore probably before general budburst.

Plovdiv (Table 155)

Vineyard site adjustments, °C

Min.	Max.	
+0.8	−0.4	Slight to moderate slopes (Plovdiv in valley/basin)
–	–	Loamy soils
−1.1	−1.1	Average altitude ~ 350 m (Plovdiv 161 m)
−0.3	−1.5	Mean −0.9°, variability index −2.4

Notes

Compared with Pleven, the Plovdiv vineyards are slightly warmer and drier and have a less pronounced summer rainfall pattern: as might be expected from their more southerly location. The wider diurnal temperature range recorded for Plovdiv may be partly related to this, but is probably due also to the city's position at the bottom of a basin-like valley. Because of this I have made a fairly large site climate adjustment, despite the vines being at most on slight to moderate slopes.

Johnson (1971), describes this as a red wine area, with the main vine varieties Karabunar, Pamid and Mavrud. I have no information on their maturities, but they are most probably midseason or late midseason. The calculations for Maturity Groups 5–7 are on this assumption. For such varieties the ripening conditions are close to optimal for producing medium to full-bodied table wines. Moisture could be a limiting factor during ripening if irrigation is not practised.

Ripening conditions for Frontignac and similar early (Groups 3 and 4) varieties appear optimal to make sweet dessert wines, and correspond closely to those experienced in southern France and at Rutherglen, Victoria.

Sliven (Table 156)

Vineyard site adjustments: None

Notes

Sliven itself is on a favoured south-facing lower foothill slope, and this is reflected in its narrow diurnal temperature range. Vineyards to the south-west towards Stara Zagora appear to have similar topography. Those to the east are less hilly but closer to the Black Sea: two factors which could roughly cancel each other out. Altitudes are mainly similar to that of Sliven.

This region produces largely white wines, from varieties including Dimiat (? = Chasselas, Maturity Group 2), Chardonnay and Sylvaner (3) and Welschriesling (4), together with Rkatsiteli, Sungurlare and local Muscat types of unknown maturity. Ripening month estimates for up to Maturity Group 6 cover these and any red varieties likely to be grown.

Ripening average mean temperatures are probably above optimum for many early varieties, but this is partly offset by the moderate highest maxima. The ample rainfall in spring and early summer is followed by quite low rainfall in August-September. The climate appears very suitable to produce both full-bodied table wines and particularly, as in the Plovdiv region, sweet, fortified wines from early-maturing varieties such as Frontignac.

19

Viticultural Environments of Italy

Italy has a very wide range of viticultural climates. I have not attempted to document the hot climates of the south, of which the wines are themselves not well documented and are used mostly for everyday local consumption or blending. Moderately cool climates in the hills of Sicily have recently started to produce superior table wines, which have been highly praised, but have not yet built up a style tradition which can readily be related to particular environments. The most widely known and reputed table wines of Italy still come from the centre and north.

Two general points can be made about the climate of northern Italy. First, it is continental rather than mediterranean. Second, sunshine hours are less than might be imagined. The ratios of growing season sunshine hours to effective day degrees, where I have them, are clearly below 1. Temperature variability indices

and extreme maxima are also low, for mild to warm climates. The expected result is soft wines, having only moderate acid, body and colour. However this seems to be offset by the use of local grape varieties, some of which, elsewhere, give wines that are highly acid and astringent (see discussion under 'Sunshine hours', Chapter 3).

Of interest to Australia is the broad resemblance of the growing and ripening conditions in central and northern Italy to those of eastern New South Wales, for instance the Hunter Valley and to a lesser extent Mudgee, and of south-east Queensland. The Italian grape varieties Barbera, Nebbiolo and Sangiovese are already grown commercially at Mudgee, with interesting and promising results. Might such varieties have a wider role to play in warm-temperate eastern Australia?

TABLE 157 BOLZANO, ITALY. 46°28'N, 11°28'E. ALT. 290m. (Alto Adige)

	1	2	3	4	5	6	7	8	9	10	11	12	13	14	15	16
	Temperature °C							Summation of Day Degrees above 10° with 19° Cut-Off			Temperature Variability Index					
PERIOD	Lowest Recorded Min.	Av.Mthly Lowest Min.	Av.Min Minus Col.2	Daily Mean	Daily Range /2	Av.Mthly Highest Max.	Col.6 Minus Av.Max.	Raw	Adj.1 for Lat. & Daily Temp Rng	Adj.2 for Vine Sites	Raw	Adj. for Vine Sites	Av. Sunshine Hours	Av. No. Rain Days	Av. Rainfall mm	Av. % R.H. 1245 HRS
APRIL		1.1	7.2	13.1	4.8	24.6	6.7	92	98	89	33.1	30.7	192	7	60	49
MAY		5.3	6.1	16.6	5.2	28.5	6.7	203	214	201	33.6	31.2	177	8	76	50
JUNE		8.9	6.1	20.9	5.9	32.4	5.6	270	270	270	35.3	32.9	213	10	68	53
JULY		10.7	5.0	22.4	6.7	34.1	5.0	279	279	279	36.8	34.4	236	9	84	53
AUGUST		9.0	6.1	21.4	6.3	33.3	5.6	279	279	279	36.9	34.5	211	9	77	54
SEPTEMBER		6.1	7.2	18.6	5.3	28.9	5.0	258	260	244	33.4	31.0	183	6	72	55
OCTOBER		0.7	7.8	12.8	4.3	23.2	6.1	87	96	87	31.1	28.7	149	6	65	55
APR-OCTOBER		—	6.5	18.0	5.5	—	5.8	1,468	1,496	1,449	34.3	31.9	1,361	55	502	53
NOV-MARCH		—	—	4.2	—	—	—	—	—	—	—	—	—	—	185	—
DATA FOR MONTH TO MATURITY																
MATURITY GROUP	EST. MATURITY DATE		*	*	*							*				
1	—		—	—	—	—						—	—	—	—	—
2	SEPTEMBER 4 + }		—	—	—	—						—	—	—	—	—
3	SEPTEMBER 4 }		20.6	5.6	31.8	5.5						34.0	208	9	76	54
4	SEPTEMBER 9		20.3	5.4	31.0	5.4						33.5	204	8	75	54
5	SEPTEMBER 15		19.7	5.2	30.1	5.3						32.9	198	8	74	55
6	SEPTEMBER 22		19.0	4.9	29.0	5.2						32.3	191	7	73	55
7	—		—	—	—	—						—	—	—	—	—
8	—		—	—	—	—						—	—	—	—	—

* Adjusted for vine sites. + Group 2 varieties on cooler sites or exposures only: see commentary.

TABLE 158
SIENA, ITALY. 43°18′N, 11°20′E. ALT. 348m. (Chianti)

	1	2	3	4	5	6	7	8	9	10	11	12	13	14	15	16
	Temperature °C							Summation of Day Degrees above 10° with 19° Cut-Off			Temperature Variability Index					
PERIOD	Lowest Recorded Min.	Av.Mthly Lowest Min.	Av.Min Minus Col.2	Daily Mean	Daily Range /2	Av.Mthly Highest Max.	Col.6 Minus Av.Max.	Raw	Adj.1 for Lat. & Daily Temp Rng	Adj.2 for Vine Sites	Raw	Adj. for Vine Sites	Av. Sunshine Hours	Av. No. Rain Days	Av. Rainfall mm	Av. % R.H. 1245 HRS
APRIL	2.1	5.8	12.2	4.3	22.9	6.4	66	78	87	29.4	27.0	129	7	62	57
MAY	5.3	5.8	15.8	4.7	26.6	6.1	180	189	198	30.7	28.3	171	9	87	56
JUNE	9.5	5.8	20.4	5.1	31.1	5.6	270	270	270	31.8	29.4	228	5	60	53
JULY	13.7	4.5	23.6	5.4	34.2	5.2	279	279	279	31.3	28.9	295	4	28	45
AUGUST	13.3	4.7	23.3	5.3	33.6	5.0	279	279	279	30.9	28.5	233	4	33	47
SEPTEMBER	10.5	4.8	19.9	4.6	30.6	6.1	270	270	270	29.3	26.9	183	6	78	54
OCTOBER	5.0	5.6	14.3	3.7	24.9	6.9	133	152	162	27.3	24.9	136	9	114	64
APR-OCTOBER	—	5.3	18.5	4.7	—	5.9	1,477	1,517	1,545	30.1	27.7	1,375	44	462	54
NOV-MARCH	—	—	6.6	—	—	—				—	—	—	—	401	—

DATA FOR MONTH TO MATURITY

MATURITY GROUP	EST. MATURITY DATE		*	*								*				
1	—	—	—	—	—						—		—	—	—	—
2	—	—	—	—	—						—		—	—	—	—
3	—	—	—	—	—						—		—	—	—	—
4	—	—	—	—	—						—		—	—	—	—
5	SEPTEMBER 15	21.9	4.4	31.7	5.4						27.7		202	5	52	51
6	SEPTEMBER 21	21.3	4.3	31.2	5.6						27.5		194	6	64	52
7	SEPTEMBER 27	20.5	4.1	30.5	5.9						27.2		188	6	74	53
8	OCTOBER 4	19.5	3.9	29.4	6.0						26.8		178	7	85	55

* Adjusted for vine sites.

TABLE 159
TURIN, ITALY. 45°00′N, 7°41′E. ALT. 241m. (Piedmont: Barolo, Asti, Monferrato, etc.)

	1	2	3	4	5	6	7	8	9	10	11	12	13	14	15	16
	Temperature °C							Summation of Day Degrees above 10° with 19° Cut-Off			Temperature Variability Index					
PERIOD	Lowest Recorded Min. (8YRS)	Av.Mthly Lowest Min. (8YRS)	Av.Min Minus Col.2	Daily Mean (8YRS)	Daily Range /2 (8YRS)	Av.Mthly Highest Max. (8YRS)	Col.6 Minus Av.Max.	Raw	Adj.1 for Lat. & Daily Temp Rng	Adj.2 for Vine Sites	Raw	Adj. for Vine Sites	Av. Sunshine Hours	Av. No. Rain Days (8YRS)	Av. Rainfall mm (8YRS)	Av. % R.H.
APRIL	0.0	2.8	4.4	11.4	4.2	22.8	7.2	42	55	61	28.4	26.0		9	79	
MAY	2.8	6.7	4.5	16.5	5.3	29.6	7.8	202	209	206	33.5	31.1		10	86	
JUNE	7.2	10.8	4.7	21.1	5.6	32.8	6.1	270	270	270	33.2	30.8		7	69	
JULY	9.6	12.2	5.6	23.6	5.8	34.7	5.3	279	279	279	34.1	31.7		5	53	
AUGUST	10.8	12.3	5.4	23.3	5.6	34.5	5.6	279	279	279	33.4	31.0		5	41	
SEPTEMBER	5.0	9.0	5.4	19.4	5.0	30.0	5.6	270	270	270	31.0	28.6		6	51	
OCTOBER	0.0	1.9	6.5	12.8	4.4	22.8	5.6	87	95	101	29.7	27.3		5	38	
APR-OCTOBER	—	—	5.2	18.3	5.1	—	6.2	1,429	1,457	1,466	31.9	29.5		47	417	
NOV-MARCH	—	—	—	4.0	—	—	—	—	—	—	—	—		—	267	

DATA FOR MONTH TO MATURITY

MATURITY GROUP	EST. MATURITY DATE		*	*	*							*				
1	—		—	—	—	—					—			—	—	
2	—		—	—	—	—					—			—	—	
3	SEPTEMBER 7	22.5	4.8	33.1	5.6						31.4			5	43	
4	—		—	—	—	—					—			—	—	
5	SEPTEMBER 18	21.4	4.6	31.4	5.6						30.2			6	47	
6	SEPTEMBER 24	20.5	4.5	30.4	5.6						29.5			6	49	
7	SEPTEMBER 30	19.3	4.4	29.3	5.6						28.7			6	51	
8	—		—	—	—	—					—			—	—	

* Adjusted for vine sites.

TABLE 160
VERONA, ITALY. 45°24′N, 10°53′E. ALT. 73m. (Soave, Valpolicella, Bardolino, Chiaretto del Garda)

	1	2	3	4	5	6	7	8	9	10	11	12	13	14	15	16
	Temperature °C							Summation of Day Degrees above 10° with 19° Cut-Off			Temperature Variability Index					
PERIOD	Lowest Recorded Min. (10YRS)	Av.Mthly Lowest Min. (10YRS)	Av.Min Minus Col.2	Daily Mean (10YRS)	Daily Range /2 (10YRS)	Av.Mthly Highest Max. (10YRS)	Col.6 Minus Av.Max.	Raw	Adj.1 for Lat. & Daily Temp Rng	Adj.2 for Vine Sites	Raw	Adj. for Vine Sites	Av. Sunshine Hours	Av. No. Rain Days (10YRS)	Av. Rainfall mm (10YRS)	Av. % R.H. 1230 HRS (10YRS)
APRIL	−6.1	0.6	6.1	13.1	6.4	25.0	5.5	93	95	85	37.2	33.8		6	51	55
MAY	4.4	6.1	5.0	16.9	5.8	28.9	6.2	214	222	213	34.4	31.0		9	64	57
JUNE	3.9	9.4	5.0	20.8	6.4	32.8	5.6	270	270	270	36.2	32.8		7	46	51
JULY	7.2	12.2	5.0	23.3	6.1	34.4	5.0	279	279	279	34.4	31.0		6	79	48
AUGUST	9.4	10.6	5.5	22.5	6.4	33.3	4.4	279	279	279	35.5	32.1		5	58	50
SEPTEMBER	6.1	7.8	5.5	19.4	6.1	30.6	5.1	270	270	270	35.0	31.6		6	71	54
OCTOBER	−4.4	1.1	7.2	13.6	5.3	25.0	6.1	112	110	108	34.5	31.1		7	69	63
APR-OCTOBER	—	—	5.6	18.5	6.1	—	5.4	1,517	1,525	1,504	35.3	31.9		46	438	54
NOV-MARCH	—	—	—	5.1	—	—	—	—	—	—	—	—		—	257	—
DATA FOR MONTH TO MATURITY																
MATURITY GROUP	EST. MATURITY DATE		*	*	*							*				
1	—		—	—	—	—					—		—	—	—	
2	—		—	—	—	—					—		—	—	—	
3	—		—	—	—	—					—		—	—	—	
4	—		—	—	—	—					—		—	—	—	
5	SEPTEMBER 12		21.3	5.4	31.1	4.7					31.9		5	63	52	
6	SEPTEMBER 18		20.8	5.3	30.6	4.8					31.8		6	66	52	
7	SEPTEMBER 25		19.8	5.3	30.0	4.9					31.7		6	69	53	
8	OCTOBER 2		18.6	5.2	29.2	5.1					31.6		6	71	54	

* Adjusted for vine sites.

Commentaries and vineyard site adjustments

Bolzano (Alto Adige) (Table 157)

Vineyard site adjustments, °C

Min.	Max.	
+0.4	−0.2	Moderate-steep slopes (Bolzano semi-valley)
–	–	Aspects largely E or W
+0.4	−0.2	Some soils stony, some terracing
−0.8	−0.8	Average altitude 400–450 m (Bolzano 290 m)
0.0	−1.2	Mean −0.6°, variability index −2.4

Notes

Dry white wines from Pinot Blanc and Müller-Thurgau (Maturity Group 2), Green Veltliner, Sylvaner and Sauvignon Blanc (3) and Riesling and Welschriesling (Italian Riesling) (4); red wines from Pinot Noir (3), Schiava (4), Merlot (5), Cabernet Sauvignon (6) and other varieties of unknown maturities.

Vine growing extends to above 600 m altitude. Presumably the earliest-maturing varieties are grown mainly under higher and cooler conditions than those tabulated.

Siena (Chianti) (Table 158)

Vineyard site adjustments, °C

Min.	Max.	
–	–	Hilly, as Siena. No particular aspects
+0.8	−0.4	Rocky, gritty soils
−0.2	−0.2	Latitude adjustment: most of Chianti N of Siena
+0.6	−0.6	Mean unchanged, variability index −2.4

Notes

Dry red wines, mainly from Sangiovese (5) with the addition of some Trebbiano (7) and other varieties. Trebbiano is also used for white wines of varying styles, including 'vino santo' from bunches hung after harvest to become partially dried. The very low temperature variability index is notable, allowing grapes to ripen under quite high mean temperatures but without exposure to damaging extremes.

Turin (Piedmont: Barolo, Asti, Monferrato etc.) (Table 159)

Vineyard site adjustments, °C

Min.	Max.	
+0.8	–0.4	Vines on rolling hills (Turin in valley)
–	–	Rich, loamy soils; all aspects
–0.4	–0.4	Average altitude + 300 m (Turin 241 m)
+0.1	+0.1	Latitude adjustment (most vineyards S of Turin)
+0.5	–0.7	Mean –0.1°, variability index –2.4

Notes

Sweet sparkling wines (Asti spumante etc.) from very ripe Frontignac (Maturity Group 3); red wines from Barbera (5), Nebbiolo (6) and Grignolino and Freisa (7). Some red wines from Dolcetto (3) in higher and cooler locations than estimated for in Table 159.

Verona (Soave, Valpolicella, Bardolino, Chiaretto del Garda) (Table 160)

Vineyard site adjustments, °C

Min.	Max.	
+1.0	–0.5	Moderate slopes (Verona on flats)
+0.3	+0.1	Moderate predominance of southerly aspects
–	–	Rich, loamy soils
–0.8	–0.8	Average altitude ~ 200 m (Verona 73 m)
+0.5	–1.2	Mean –0.35°, variability index –3.4

Notes

Dry white wine (Soave) from Garganega (maturity group unknown) and Trebbiano (7); light to medium-bodied dry red wines (Valpolicella, Bardolino, Chiaretto del Garda etc.) largely from Barbera and Merlot (5), Corvina and Cabernet Sauvignon (6) and Negrara (7).

The areas closest to Lake Garda (Bardolino Classico and Chiaretto del Garda) probably enjoy less variable temperatures than indicated in Table 160. Even elsewhere, the highest maxima during ripening (column 6) remain moderate despite warm average temperatures.

20

Viticultural Environments of Portugal, Madeira and Spain

The climate of the Iberian Peninsula is of direct interest to Australian viticulture. Many of the environments are similar, and, in the case of Portugal, climatically well documented.

The Portuguese climatic data come from the period 1931–1960, which is a little later than for most of the regions surveyed. Because this period straddled the mid-century global peak in temperatures (see Chapter 23, and Appendix 1) the Portuguese temperatures could include an upward bias of perhaps 0.1 or 0.2°C compared with other regions. A bias of this size should not greatly affect comparisons of ripening conditions at the viticulturally warm to hot end of the temperature range, so I have not attempted to adjust for it.

TABLE 161

BRAGA, PORTUGAL. 41°33′N, 8°24′W. ALT. 190m. (Douro Littoral, Minho)

	1	2	3	4	5	6	7	8	9	10	11	12	13	14	15	16
	Temperature °C							Summation of Day Degrees above 10° with 19° Cut-Off			Temperature Variability Index					
PERIOD	Lowest Recorded Min. (30YRS)	Av.Mthly Lowest Min.	Av.Min Minus Col.2	Daily Mean (30YRS)	Daily Range /2 (30YRS)	Av.Mthly Highest Max.	Col.6 Minus Av.Max.	Raw	Adj.1 for Lat. & Daily Temp Rng	Adj.2 for Vine Sites	Raw	Adj. for Vine Sites	Av. Sunshine Hours	Av. No. Rain Days (30YRS)	Av. Rainfall mm (30YRS)	Av. % R.H. 1500 HRS
APRIL	−1.3	2.4	5.7	13.6	5.5	27.1	8.0	108	109		35.7		223	11	114	57
MAY	1.5	4.5	5.4	15.4	5.5	29.3	8.4	167	169		35.8		243	12	108	59
JUNE	3.0	7.1	5.3	18.7	6.3	34.2	9.2	261	264		39.7		273	7	56	55
JULY	5.0	8.2	5.0	20.3	7.1	36.7	9.3	279	279		42.7		314	5	24	50
AUGUST	5.7	8.2	5.0	20.4	7.2	36.3	8.7	279	279		42.5		286	6	33	49
SEPTEMBER	2.5	5.8	5.7	18.4	6.9	33.3	8.0	252	247		41.3		209	10	81	54
OCTOBER	−0.8	2.6	6.6	15.4	5.9	29.4	8.1	167	167		38.6		179	13	129	60
APR–OCTOBER	—	—	5.5	17.5	6.3	—	8.5	1,513	1,514		39.5		1,727	64	545	55
NOV–MARCH	—	—	—	10.0	—	—	—	—	—		—		—	—	994	—
DATA FOR MONTH TO MATURITY																
MATURITY GROUP	EST. MATURITY DATE															
1	—		—	—	—	—					—		—	—	—	—
2	—		—	—	—	—					—		—	—	—	—
3	SEPTEMBER 6		20.1	7.2	35.9	8.6					42.4		278	6	39	50
4	SEPTEMBER 11		19.8	7.1	35.4	8.4					42.2		264	7	46	51
5	SEPTEMBER 17		19.4	7.1	34.8	8.2					42.0		244	8	55	52
6	SEPTEMBER 23		19.0	7.0	34.1	8.1					41.7		224	9	66	53
7	SEPTEMBER 30		18.4	6.9	33.3	8.0					41.3		209	10	81	54
8	—		—	—	—	—					—		—	—	—	—

233

TABLE 162

CALDAS da RAINHA, PORTUGAL. 39°24′N, 9°08′W. ALT. 61m. (Estremadura : Colares, Torres Vedras)

	1	2	3	4	5	6	7	8	9	10	11	12	13	14	15	16
	Temperature °C							Summation of Day Degrees above 10° with 19° Cut-Off			Temperature Variability Index					
PERIOD	Lowest Recorded Min. (30YRS)	Av.Mthly Lowest Min.	Av.Min Minus Col.2	Daily Mean (30YRS)	Daily Range /2 (30YRS)	Av.Mthly Highest Max.	Col.6 Minus Av.Max.	Raw	Adj.1 for Lat. & Daily Temp Rng	Adj.2 for Vine Sites	Raw	Adj. for Vine Sites	Av. Sunshine Hours (30YRS)	Av. No. Rain Days (30YRS)	Av. Rainfall mm (30YRS)	Av. % R.H. 1500 HRS (30YRS)
APRIL	3.5	5.8	4.3	14.4	4.3	25.6	6.9	132	142	149	28.4	27.2	224	9	53	67
MAY	4.2	7.8	3.9	15.8	4.1	27.7	7.8	180	193	201	28.1	26.9	256	9	44	68
JUNE	9.0	10.7	3.5	18.2	4.0	31.4	9.2	246	260	268	28.7	27.5	264	4	16	68
JULY	10.5	12.2	3.2	19.4	4.0	32.6	9.2	279	279	279	28.4	27.2	290	2	5	67
AUGUST	10.8	12.2	3.4	19.7	4.1	32.5	8.7	279	279	279	28.5	27.3	271	2	5	63
SEPTEMBER	8.0	11.0	3.9	19.2	4.3	31.1	7.6	270	270	270	28.7	27.5	227	5	35	65
OCTOBER	3.8	7.9	4.9	17.2	4.4	29.2	7.6	223	233	241	30.1	28.9	203	7	45	67
APR-OCTOBER	—	—	3.9	17.7	4.2	—	8.1	1,609	1,656	1,687	28.7	27.5	1,735	38	203	66
NOV-MARCH	—	—	—	11.8	—	—	—							—	404	—
DATA FOR MONTH TO MATURITY																
MATURITY GROUP	EST. MATURITY DATE			*	*							*				
1	—		—	—	—	—						—	—	—	—	—
2	—		—	—	—	—	—					—				
3	—		—	—	—	—						—	—	—	—	—
4	SEPTEMBER 3		19.7	3.9	32.0	8.4						27.3	265	2	7	63
5	SEPTEMBER 8		19.6	3.9	31.9	8.3						27.3	259	3	11	63
6	SEPTEMBER 14		19.5	3.9	31.7	8.1						27.4	251	4	16	64
7	SEPTEMBER 20		19.4	4.0	31.4	7.9						27.4	241	4	25	64
8	—		—	—	—	—						—	—	—	—	—

* Adjusted for vine sites.

TABLE 163

DAO (region), PORTUGAL. Approx. 40°30′N, 8°00′W. ALT. approx. 300-350m. (All data estimated)

	1	2	3	4	5	6	7	8	9	10	11	12	13	14	15	16
	Temperature °C							Summation of Day Degrees above 10° with 19° Cut-Off			Temperature Variability Index					
PERIOD	Lowest Recorded Min.	Av.Mthly Lowest Min.	Av.Min Minus Col.2	Daily Mean	Daily Range /2	Av.Mthly Highest Max.	Col.6 Minus Av.Max.	Raw	Adj.1 for Lat. & Daily Temp Rng	Adj.2 for Vine Sites	Raw	Adj. for Vine Sites	Av. Sunshine Hours	Av. No. Rain Days	Av. Rainfall mm	Av. % R.H. 1500 HRS
APRIL		3.1	5.1	13.9	5.7	26.5	6.9	117	117		34.8		228	12	77	53
MAY		5.3	4.8	15.9	5.8	29.3	7.6	183	184		35.6		258	12	68	54
JUNE		8.3	4.5	19.4	6.6	34.7	8.7	270	270		39.6		295	7	35	49
JULY		10.1	4.2	21.7	7.4	37.8	8.7	279	279		42.5		347	4	12	43
AUGUST		10.3	4.2	21.9	7.4	37.5	8.2	279	279		42.0		322	5	17	41
SEPTEMBER		8.4	4.9	19.9	6.6	33.8	7.3	270	270		38.6		238	7	43	47
OCTOBER		4.1	6.1	15.9	5.7	29.0	7.4	183	183		36.3		195	11	85	55
APR-OCTOBER	—	4.8	18.4	6.5		—	7.8	1,581	1,582		38.5		1,883	58	337	49
NOV-MARCH	—	—	9.7	—	—	—		—	—		—		—	—	654	—
DATA FOR MONTH TO MATURITY																
MATURITY GROUP	EST. MATURITY DATE															
1	—		—	—	—	—					—		—	—	—	—
2	—		—	—	—	—					—		—	—	—	—
3	SEPTEMBER 3		21.8	7.5	37.3	8.2					41.7		316	5	18	41
4	SEPTEMBER 8		21.6	7.4	36.9	8.0					41.2		304	5	21	42
5	SEPTEMBER 13		21.3	7.2	36.4	7.8					40.6		288	6	25	43
6	SEPTEMBER 18		21.0	7.0	35.7	7.6					39.9		270	6	30	44
7	SEPTEMBER 24		20.5	6.8	34.8	7.4					39.2		252	7	36	45
8	SEPTEMBER 30		19.9	6.6	33.8	7.3					38.6		238	7	43	47

234

TABLE 164

LISBON, PORTUGAL. 38°43′N, 9°09′W. ALT. 77m. (Burcelas, Carcavelos, Setubal)

	1	2	3	4	5	6	7	8	9	10	11	12	13	14	15	16
	Temperature °C							Summation of Day Degrees above 10° with 19° Cut-Off			Temperature Variability Index					
PERIOD	Lowest Recorded Min. (75YRS)	Av.Mthly Lowest Min. (75YRS)	Av.Min Minus Col.2	Daily Mean (30YRS)	Daily Range /2 (30YRS)	Av.Mthly Highest Max. (75YRS)	Col.6 Minus Av.Max.	Raw	Adj.1 for Lat. & Daily Temp Rng	Adj.2 for Vine Sites	Raw	Adj. for Vine Sites	Av. Sunshine Hours (30YRS)	Av. No. Rain Days (30YRS)	Av. Rainfall mm (30YRS)	Av. % R.H. 1500 HRS (30YRS)
APRIL	4.4	8.1	3.5	15.6	4.0	25.7	6.1	168	182	167	25.6		265	10	54	56
MAY	6.4	9.9	3.0	17.2	4.3	29.0	7.5	223	233	217	27.7		301	10	44	57
JUNE	9.8	12.6	2.8	20.1	4.7	32.8	8.0	270	270	270	29.6		330	5	16	54
JULY	12.1	14.3	2.7	22.2	5.2	35.2	7.8	279	279	279	31.3		378	2	3	48
AUGUST	13.3	14.5	2.8	22.5	5.2	35.5	7.8	279	279	279	31.4		357	2	4	49
SEPTEMBER	10.3	13.3	3.2	21.2	4.7	33.1	7.2	270	270	270	29.2		279	6	33	54
OCTOBER	6.7	10.4	3.7	18.2	4.1	27.9	5.6	254	267	252	25.7		231	9	62	59
APR-OCTOBER	—	—	3.1	19.6	4.6	—	7.1	1,743	1,780	1,734	28.6		2,141	44	216	54
NOV-MARCH	—	—	—	12.4	—	—	—	—	—	—	—		—	—	491	—
DATA FOR MONTH TO MATURITY																
MATURITY GROUP	EST. MATURITY DATE		*		*											
1	—		—	—	—	—					—		—	—	—	—
2	—		—	—	—	—					—		—	—	—	—
3	AUGUST 24		22.1	5.2	35.1	7.8					31.4		364	2	4	49
4	AUGUST 29		22.1	5.2	35.0	7.8					31.4		359	2	4	49
5	SEPTEMBER 4		21.9	5.1	34.7	7.7					31.2		351	2	6	50
6	SEPTEMBER 9		21.7	5.0	34.4	7.6					31.0		340	3	8	51
7	SEPTEMBER 15		21.5	4.9	34.0	7.5					30.7		326	4	12	52
8	SEPTEMBER 21		21.2	4.8	33.5	7.4					30.3		308	5	19	53

* Adjusted for vine sites.

TABLE 165

PINHAO, PORTUGAL. 41°10′N, 7°33′W. ALT. 130m. (Middle Douro Valley)

	1	2	3	4	5	6	7	8	9	10	11	12	13	14	15	16
	Temperature °C							Summation of Day Degrees above 10° with 19° Cut-Off			Temperature Variability Index					
PERIOD	Lowest Recorded Min. (30YRS)	Av.Mthly Lowest Min.	Av.Min Minus Col.2	Daily Mean (30YRS)	Daily Range /2 (30YRS)	Av.Mthly Highest Max.	Col.6 Minus Av.Max.	Raw	Adj.1 for Lat. & Daily Temp Rng	Adj.2 for Vine Sites	Raw	Adj. for Vine Sites	Av. Sunshine Hours (30YRS)	Av. No. Rain Days (30YRS)	Av. Rainfall mm (30YRS)	Av. % R.H. 1500 HRS
APRIL	2.0	3.7	5.2	15.2	6.3	29.0	7.5	156	157	142	37.9	33.1	206	8	47	48
MAY	3.0	6.5	4.9	17.9	6.5	32.3	7.9	245	247	231	38.8	34.0	234	8	46	46
JUNE	7.0	10.1	4.8	22.4	7.5	38.6	8.7	270	270	270	43.5	38.7	273	4	28	38
JULY	10.0	12.5	4.5	25.2	8.2	42.2	8.8	279	279	279	46.1	41.3	316	2	12	32
AUGUST	7.0	12.4	4.5	24.8	7.9	40.9	8.2	279	279	279	44.5	39.7	300	2	12	31
SEPTEMBER	7.0	9.4	5.2	22.2	7.6	37.3	7.5	270	270	270	43.1	38.3	219	4	34	40
OCTOBER	1.5	4.5	6.1	17.0	6.4	31.0	7.6	217	216	201	39.3	34.5	178	7	55	54
APR-OCTOBER	—	—	5.0	20.7	7.2	—	8.0	1,716	1,718	1,672	41.9	37.1	1,726	35	234	41
NOV-MARCH	—	—	—	10.0	—	—	—	—	—	—	—	—	—	—	425	—
DATA FOR MONTH TO MATURITY																
MATURITY GROUP	EST. MATURITY DATE		*	*	*							*				
1	—		—	—	—							—	—	—	—	—
2	—		—	—	—							—	—	—	—	—
3	AUGUST 26		24.5	6.7	39.4							39.9	305	2	12	31
4	AUGUST 31		24.2	6.7	39.2							39.7	300	2	12	31
5	SEPTEMBER 6		23.9	6.6	38.6							39.4	290	2	14	32
6	SEPTEMBER 12		23.6	6.6	37.9							39.2	275	3	19	34
7	SEPTEMBER 17		23.2	6.6	36.9							38.9	255	3	24	36
8	—		—	—	—							—	—	—	—	—

* Adjusted for vine sites.

TABLE 166
REGUA, PORTUGAL. 41°10′N, 7°48′W. ALT. 65m. (Lower Douro Valley)

	1	2	3	4	5	6	7	8	9	10	11	12	13	14	15	16
	Temperature °C							Summation of Day Degrees above 10° with 19° Cut-Off			Temperature Variability Index					
PERIOD	Lowest Recorded Min. (30YRS)	Av.Mthly Lowest Min.	Av.Min Minus Col.2	Daily Mean (30YRS)	Daily Range /2 (30YRS)	Av.Mthly Highest Max.	Col.6 Minus Av.Max.	Raw	Adj.1 for Lat. & Daily Temp Rng	Adj.2 for Vine Sites	Raw	Adj. for Vine Sites	Av. Sunshine Hours (30YRS)	Av. No. Rain Days (30YRS)	Av. Rainfall mm (30YRS)	Av. % R.H. 1500 HRS
APRIL	0.5	2.4	6.2	14.9	6.3	28.7	7.5	147	148	133	38.9	32.9	225	9	61	50
MAY	2.3	5.0	5.9	17.3	6.4	31.6	7.9	226	228	210	39.4	33.4	248	9	54	48
JUNE	2.3	7.6	5.8	20.8	7.4	36.9	8.7	270	270	270	44.1	38.1	303	5	26	40
JULY	6.8	9.5	5.5	23.2	8.2	40.2	8.8	279	279	279	47.1	41.1	349	2	12	34
AUGUST	2.3	9.3	5.5	23.2	8.4	39.8	8.2	279	279	279	47.3	41.3	322	3	14	33
SEPTEMBER	4.0	6.8	6.2	20.7	7.7	35.9	7.5	270	270	270	44.5	38.5	231	6	37	42
OCTOBER	−0.7	2.3	7.1	16.2	6.8	30.6	7.6	192	187	173	41.9	35.9	177	9	70	56
APR-OCTOBER	—	—	6.0	19.5	7.3	—	8.0	1,663	1,661	1,614	43.3	37.3	1,855	43	274	43
NOV-MARCH	—	—	—	9.9	—	—	—	—	—	—	—	—	—	—	581	—

DATA FOR MONTH TO MATURITY

MATURITY GROUP	EST. MATURITY DATE	*	*	*							*				
1	—		—	—	—	—					—	—	—	—	—
2	—		—	—	—	—					—	—	—	—	—
3	AUGUST 30	22.6	6.9	37.7	8.2						41.3	323	3	14	33
4	SEPTEMBER 4	22.3	6.8	37.4	8.1						41.0	315	3	16	34
5	SEPTEMBER 10	22.0	6.7	36.9	7.9						40.6	302	4	19	35
6	SEPTEMBER 15	21.7	6.6	36.3	7.8						40.1	284	4	23	37
7	SEPTEMBER 21	21.3	6.5	35.6	7.7						39.5	262	5	28	39
8	—		—	—	—	—					—	—	—	—	—

* Adjusted for vine sites.

TABLE 167
VILA REAL, PORTUGAL. 41°19′N, 7°44′W. ALT. 479m. (Tras-Os-Montes, Mateus, etc.)

	1	2	3	4	5	6	7	8	9	10	11	12	13	14	15	16
	Temperature °C							Summation of Day Degrees above 10° with 19° Cut-Off			Temperature Variability Index					
PERIOD	Lowest Recorded Min. (30YRS)	Av.Mthly Lowest Min.	Av.Min Minus Col.2	Daily Mean (30YRS)	Daily Range /2 (30YRS)	Av.Mthly Highest Max.	Col.6 Minus Av.Max.	Raw	Adj.1 for Lat. & Daily Temp Rng	Adj.2 for Vine Sites	Raw	Adj. for Vine Sites	Av. Sunshine Hours (30YRS)	Av. No. Rain Days (30YRS)	Av. Rainfall mm (30YRS)	Av. % R.H. 1500 HRS (30YRS)
APRIL	−0.1	1.8	5.2	12.6	5.6	25.2	7.0	78	78	69	34.6		223	10	77	53
MAY	1.0	4.2	4.9	14.9	5.8	28.1	7.4	152	153	144	35.5		264	11	61	53
JUNE	4.0	7.4	4.8	19.0	6.8	34.0	8.2	270	268	259	40.2		315	6	32	46
JULY	6.6	9.2	4.5	21.4	7.7	37.4	8.3	279	279	279	43.6		370	3	10	39
AUGUST	6.1	9.5	4.5	21.6	7.6	36.9	7.7	279	279	279	42.6		338	4	16	37
SEPTEMBER	2.6	7.3	5.2	19.1	6.6	32.7	7.0	270	270	264	38.6		239	7	38	45
OCTOBER	−1.7	2.5	6.1	14.4	5.8	27.3	7.1	136	136	127	36.4		191	10	84	55
APR-OCTOBER	—	—	5.0	17.6	6.6	—	7.5	1,464	1,463	1,421	38.8		1,940	51	318	47
NOV-MARCH	—	—	—	8.0	—	—	—	—	—	—	—		—	—	701	—

DATA FOR MONTH TO MATURITY

MATURITY GROUP	EST. MATURITY DATE	*		*											
1	—		—	—	—	—					—	—	—	—	—
2	—		—	—	—	—					—	—	—	—	—
3	SEPTEMBER 13	20.6	7.3	35.3	7.4						41.5	304	5	22	40
4	SEPTEMBER 18	20.2	7.1	34.5	7.2						40.6	282	6	26	41
5	SEPTEMBER 24	19.6	6.9	33.4	7.1						39.8	255	6	31	43
6	OCTOBER 1	18.7	6.6	32.2	7.0						38.6	235	7	39	45
7	—		—	—	—	—					—	—	—	—	—
8	—		—	—	—	—					—	—	—	—	—

* Adjusted for vine sites.

TABLE 168
FUNCHAL, MADEIRA. 32°38′N, 16°55′W. ALT. 25m.

	1	2	3	4	5	6	7	8	9	10	11	12	13	14	15	16
	Temperature °C							Summation of Day Degrees above 10° with 19° Cut-Off			Temperature Variability Index					
PERIOD	Lowest Recorded Min. (50YRS)	Av.Mthly Lowest Min. (30YRS)	Av.Min Minus Col.2	Daily Mean (30YRS)	Daily Range /2 (30YRS)	Av.Mthly Highest Max. (30YRS)	Col.6 Minus Av.Max.	Raw	Adj.1 for Lat. & Daily Temp Rng	Adj.2 for Vine Sites	Raw	Adj. for Vine Sites	Av. Sunshine Hours	Av. No. Rain Days (30YRS)	Av. Rainfall mm (30YRS)	Av. % R.H. 1500 HRS (30YRS)
APRIL	6.7	10.0	4.4	16.9	2.5	22.2	2.8	207	240	213	22.2			4	33	65
MAY	8.9	11.7	3.9	18.1	2.5	23.3	2.8	251	279	253	21.6			2	18	65
JUNE	8.9	13.3	3.9	19.7	2.5	25.0	2.8	270	270	270	21.7			1	5	68
JULY	12.8	15.6	3.3	21.4	2.5	26.1	2.2	279	279	279	20.5			0	1	67
AUGUST	11.1	16.7	2.8	21.9	2.5	28.9	4.4	279	279	279	22.2			0	1	67
SEPTEMBER	12.8	16.1	3.3	21.9	2.5	27.8	3.3	270	270	270	21.7			3	25	67
OCTOBER	8.3	13.9	4.4	20.8	2.5	26.7	3.3	279	279	279	22.8			7	76	66
APR-OCTOBER	—	—	3.7	20.1	2.5	—	3.1	1,835	1,896	1,843	21.8			17	159	66
NOV-MARCH	—	—	—	16.8	—	—	—	—	—	—	—			—	387	—
DATA FOR MONTH TO MATURITY																
MATURITY GROUP	EST. MATURITY DATE			*		*										
1	—			—	—	—	—				—			—	—	—
2	—			—	—	—	—				—			—	—	—
3	AUGUST 16			20.7	2.5	27.1	3.4				21.5			0	1	67
4	AUGUST 21			20.9	2.5	27.4	3.8				21.8			0	1	67
5	—			—	—	—	—				—			—	—	—
6	—			—	—	—	—				—			—	—	—
7	—			—	—	—	—				—			—	—	—
8	—			—	—	—	—				—			—	—	—

* Adjusted for vine sites.

TABLE 169
CIUDAD REAL, SPAIN. 38°59′N, 3°55′W. ALT. 628m. (Valdepenas)

	1	2	3	4	5	6	7	8	9	10	11	12	13	14	15	16
	Temperature °C							Summation of Day Degrees above 10° with 19° Cut-Off			Temperature Variability Index					
PERIOD	Lowest Recorded Min. (30YRS)	Av.Mthly Lowest Min. (30YRS)	Av.Min Minus Col.2	Daily Mean (30YRS)	Daily Range /2 (30YRS)	Av.Mthly Highest Max. (29YRS)	Col.6 Minus Av.Max.	Raw	Adj.1 for Lat. & Daily Temp Rng	Adj.2 for Vine Sites	Raw	Adj. for Vine Sites	Av. Sunshine Hours	Av. No. Rain Days (21YRS)	Av. Rainfall mm (30YRS)	Av. % R.H.
APRIL	-0.6	1.7	4.4	12.5	6.4	26.1	7.2	75	75	67	37.2	35.4		7	46	
MAY	1.1	4.4	5.1	16.4	6.9	30.6	7.3	198	191	190	40.0	38.2		6	33	
JUNE	5.0	8.3	5.0	21.1	7.8	35.6	6.7	270	270	270	42.9	41.1		4	23	
JULY	6.1	11.7	5.0	25.0	8.3	38.9	5.6	279	279	279	43.8	42.0		1	3	
AUGUST	7.8	12.2	4.5	25.0	8.3	38.3	5.0	279	279	279	42.7	40.9		1	3	
SEPTEMBER	3.9	7.2	5.6	20.3	7.5	35.0	7.2	270	270	270	42.8	41.0		4	25	
OCTOBER	-1.1	2.8	5.5	15.0	6.7	28.3	6.6	155	152	148	38.9	37.1		6	41	
APR-OCTOBER	—	—	5.0	19.3	7.4	—	6.5	1,526	1,516	1,503	41.2	39.4		29	174	
NOV-MARCH	—	—	—	7.5	—	—	—	—	—	—	—	—		—	206	
DATA FOR MONTH TO MATURITY																
MATURITY GROUP	EST. MATURITY DATE			*	*	*						*				
1	—			—	—	—	—					—		—	—	
2	—			—	—	—	—					—		—	—	
3	—			—	—	—	—					—		—	—	
4	SEPTEMBER 12			22.8	7.7	36.7	5.8					41.0		2	9	
5	SEPTEMBER 18			22.0	7.5	36.0	6.3					41.0		3	15	
6	SEPTEMBER 23			21.2	7.3	35.3	6.9					41.0		3	20	
7	SEPTEMBER 30			20.2	7.1	34.3	7.2					41.0		4	25	
8	OCTOBER 8			18.5	6.9	32.7	7.0					40.2		4	30	

* Adjusted for vine sites.

TABLE 170

LOGRONO (Observatory), SPAIN. 42°28′N, 2°30′W. ALT. approx. 420m. (Rioja)

PERIOD	Temperature °C							Summation of Day Degrees above 10° with 19° Cut-Off			Temperature Variability Index					
	Lowest Recorded Min. (60YRS)	Av.Mthly Lowest Min.	Av.Min Minus Col.2	Daily Mean (60YRS)	Daily Range /2 (30YRS)	Av.Mthly Highest Max.	Col.6 Minus Av.Max.	Raw	Adj.1 for Lat. & Daily Temp Rng	Adj.2 for Vine Sites	Raw	Adj. for Vine Sites	Av. Sunshine Hours	Av. No. Rain Days	Av. Rainfall mm (30YRS)	Av. % R.H. 1300 HRS
	1	2	3	4	5	6	7	8	9	10	11	12	13	14	15	16
APRIL	−2.3	0.1	5.0	11.3	5.6	25.2	8.9	39	39	55	36.3	30.9		10	37	53
MAY	−0.6	4.0	5.2	15.0	5.9	30.2	9.2	155	158	169	38.0	32.6		11	47	52
JUNE	4.2	6.8	4.7	18.6	6.6	33.0	8.3	258	261	267	39.4	34.0		8	40	49
JULY	7.2	9.1	4.7	21.3	7.1	36.3	8.3	279	279	279	41.4	36.0		4	24	46
AUGUST	6.8	9.7	4.9	21.5	6.9	36.9	8.5	279	279	279	41.0	35.6		3	12	47
SEPTEMBER	2.0	6.3	5.8	18.7	6.2	32.8	8.3	261	262	269	38.9	33.5		7	30	54
OCTOBER	−2.0	2.5	5.6	13.7	5.4	26.4	7.5	115	114	133	34.7	29.3		9	40	61
APR-OCTOBER	—	—	5.1	17.2	6.2	—	8.3	1,386	1,392	1,451	38.5	33.1		52	230	52
NOV-MARCH	—	—	—	7.1	—	—	—	—	—	—	—	—		—	162	—
DATA FOR MONTH TO MATURITY																
MATURITY GROUP	EST. MATURITY DATE	*	*	*							*					
1	—	—	—	—	—						—			—	—	—
2	—	—	—	—	—						—			—	—	—
3	—	—	—	—	—						—			—	—	—
4	SEPTEMBER 16	20.5	5.2	34.3	8.4						34.6			5	21	50
5	SEPTEMBER 22	19.9	5.1	33.4	8.4						34.2			6	24	51
6	SEPTEMBER 28	19.1	4.9	32.2	8.3						33.7			7	28	53
7	OCTOBER 5	18.0	4.7	30.5	8.2						33.0			7	31	55
8	—	—	—	—	—						—			—	—	—

* Adjusted for vine sites.

Commentaries and vineyard site adjustments

1.   Portugal

Braga (Douro Littoral, Minho) (Table 161)

Vineyard site adjustments: None

Notes

Vinho verde (both white and red) from grapes picked before full maturity. Principal white varieties include Alvarinho (maturity group not known), and red varieties the Sousão (7).

Braga is representative of the area, which is flat and has non-calcareous, mainly sandy soils. This is one of the wettest grape-growing areas of the world, with an annual average rainfall of 1540 mm at Braga. One reason for picking early, to make the distinctively high-acid wines of the region, is presumably the risk of heavy rain in autumn. Most other features of the climate – growing and ripening season mean temperatures, sunshine hours and relative humidities – are favourable for table wines, and the area is reasonably frost-free in spring; but in the absence of ameliorating topographic features, the temperature variability during ripening is excessive by other European standards.

Caldas da Rainha (Estremadura: Colares, Torres Vedras) (Table 162)

Vineyard site adjustments, °C

Min.	Max.	
–	–	Topography as for Caldas da Rainha
+0.4	−0.2	Soils tending to calcareous
+0.4	−0.2	Mean +0.1°, variability index −1.2

Notes

Colares produces a rare but highly esteemed red wine from Ramisco (Maturity Group 6). The soils are sandy rather than calcareous, but for site-climate adjustment this is compensated for by closeness to the sea. The Torres Vedras region is a bulk producer of generally good-quality table wines, both white and red, from a range of varieties of which one of the more important among the whites is Dona Branca (6).

The west coastal strip contrasts with Lisbon (Table 164) in being cooler and more cloudy, because of its proximity to the Atlantic Ocean and the cool

Canaries Current. The data suggest good climatic conditions for making premium-quality table wines, especially dry reds and perhaps full-bodied whites. There is a marked absence of frosts, and of extreme summer high temperatures, summer rain and hail. The temperature variability index is exceptionally low for a moderately warm climate. Exposure to strong winds from the ocean is a probable limitation.

Dão region (Table 163)

Vineyard site adjustments: None

Notes

Table 163 is estimated entirely from data for surrounding localities. The vineyards are probably on greater slopes than the source sites, but this is offset by the Dão region's mostly greater distance inland.

The region produces full-bodied dry red and white table wines. The reds have considerable tannin, which is reputed to take many years to soften. This seems to fit the environmental data, which resemble those of some typical South Australian vine-growing areas such as Clare. Ripening mean temperatures are in the upper range for good-quality table wines, and are combined with higher-than-desirable temperature variability and fairly low relative humidities.

Lisbon (Burcelas, Carcavelos, Setubal) (Table 164)

Vineyard site adjustments, °C

Min.	Max.	
−0.6	+0.3	Vineyards inland from Lisbon, but topography similar
–	–	All aspects
+0.6	−0.3	Soils volcanic or (at Setubal) calcareous
−0.5	−0.5	Altitude ~ 160 m (Lisbon 77 m)
−0.5	−0.5	Mean −0.5°, variability index unchanged.

Notes

Historically, Burcelas, Carcavelos and Setubal all produced natural or lightly fortified sweet wines. Those of Setubal, the remaining significant producer of such wines, are from lightly raisined Frontignac (Maturity Group 3) and Muscat Gordo Blanco (7), and enjoy a high reputation. Carcavelos has largely disappeared under urban development, and Burcelas now produces mainly dry white wines.

The notable features of the climate are its equable warmth and exceptional sunshine hours. All climatic factors point to fruity fortified wines in the lighter style. Lacking extreme high temperatures (cf. the Douro, Tables 165 and 166), it is also suitable for making very good, full-bodied table wines.

Pinhão (middle Douro Valley) (Table 165)

Vineyard site adjustments, °C

Min.	Max.	
+0.6	−0.3	Steep slopes vs valley bottom but close to river
–	–	No particular aspects
+1.0	−0.5	Rocky soils and terraces
−0.9	−0.9	Average altitude ~ 280 m (Pinhão 130 m)
+0.7	−1.7	Mean −0.5°, variability index −4.8

Notes

Sweet fortified wines (port) from very ripe Bastardo and Tinta Carvalha (Maturity Group 3), Touriga (5), Alvarelhão and Mourisco Tinto (6), and Souzão (7); some dry red wine from the same varieties and Tinta Amarella (3), mainly in higher and cooler areas.

Viala and Vermorel (1901–1904) noted that Bastardo was regarded by many as the quality foundation of the Douro wines, but that in post-Phylloxera plantings it was largely replaced because of its earliness and liability to berry scorching. Climatic changes may have played a role in the evolution of grape variety preferences in this region, as discussed in Chapter 23. The modern Douro varieties are more typically of Maturity Group 6.

The fewer sunshine hours of Pinhão compared with Régua (Table 166) are probably due to shadowing by surrounding hills. This would be a feature of many of the Douro vineyards.

Régua (lower Douro Valley) (Table 166)

Vineyard site adjustments, °C

Min.	Max.	
+1.0	−0.5	Steep slopes vs valley floor but close to river
–	–	No particular aspects
+1.0	−0.5	Rocky soils and terraces
−1.1	−1.1	Average altitude ~ 250 m (Régua 65 m)
+0.9	−2.1	Mean −0.6°, variability index −6.0

Notes

Varieties etc. as for Pinhão (Table 165). Régua represents the cooler (coastward) limit of the true port-producing region of the Douro Valley, whereas Pinhão is typical of the main central parts (Johnson 1971).

Vila Real (Tras-Os-Montes, Mateus etc.) (Table 167)

Vineyard site adjustments, °C

Min.	Max.	
–	–	Hilly topography, as Vila Real; all aspects
–	–	Sandy to loamy soils from granite and schist
–0.3	–0.3	Average altitude ~ 530 m (Vila Real 479 m)
–0.3	–0.3	Mean –0.3°, variability index unchanged

Notes

White wines from varieties including Dona Branca and Rabigato (Maturity Group 6); red and especially rosé wines from varieties including Bastardo, Tinta Carvalha and Tinta Amarella (3), Touriga (5), Alvarelhão (6) and Souzão (7).

Climatic conditions in the hilly upland country of northern Portugal are a little closer to the postulated ideal for table wines than those of the Dão region, having lower ripening-month average and extreme temperatures. The region is also perhaps a little better than the Minho in temperature equability and freedom from rain during ripening, but these factors are offset by lack of summer sea breezes compared with the coastal strip. Relative humidities are somewhat below optimum, similar to those of the Dão region. Taken as a whole, however, this appears to be an area of some climatic merit for table wines.

2.   Madeira

Funchal (Table 168)

Vineyard site adjustments, °C

Min.	Max.	
–	–	Sloping topography but further inland
–	–	All aspects
–0.9	–0.9	Average altitude 175 m, vs Funchal 25 m
–0.9	–0.9	Mean –0.9°, variability index unchanged

Notes

The above adjustments are for average altitudes in the coastal vineyards that surround the island. The high-altitude 'inland' vineyards are considered separately below.

Semi-sweet to sweet fortified wines (madeira) from very ripe Verdelho and Bastardo (Maturity Group 3), Malvasia bianca and Tinta Madeira (4) and Bual (maturity not known). Note that harvesting will be later than the maturity dates for table wines as indicated in Table 168. It is questionable also whether the full adjustment to temperature summation for the very narrow diurnal temperature range, equivalent to an increase in mean temperature of 1.25°C, is justified. If not, ripening may be anything up to a week later than calculated. However this will make little difference to the conditions of ripening, which in either case takes place during the warmest period of mid summer.

On the steep, south-facing slopes of the inland vineyards around Romeiro (at roughly 500–800 m, see Johnson 1971, pp. 172–3), the main variety grown is Sercial (Maturity Group 4). According to my calculations it should ripen there to table wine maturity about the end of September or beginning of October. This implies a ripening month mean temperature of a little over 18°C. It makes a delicate, dry madeira style.

Madeira's mild temperatures and their extreme lack of diurnal and seasonal variability are unique among the Old World's major viticultural areas, and reflect the dominant maritime influence on the island's climate. The climate's continentality is nevertheless hardly less than that of coastal south-western Australia and Tasmania (Chapter 3, Table 1). Indeed, the viticultural climates of Madeira have much in common with those of coastal locations such as Bunbury and Margaret River in Western Australia (Chapter 9, Tables 9 and 22). The notable success of Verdelho for dry table wines in Western Australia is thus perhaps no accident, and suggests that the other Madeira grape varieties would be worth trying there as well.

3. Spain

Ciudad Real (Valdepeñas) (Table 169)

Vineyard site adjustments, °C

Min.	Max.	
–	–	Topography flat
+0.6	−0.3	Loamy soils, many calcareous and/or stony
−0.4	−0.4	Average altitude 700 m (Ciudad Real 628 m)
+0.2	−0.7	Mean −0.25°, variability index −1.8

Notes

Relative humidity data given by the Meteorological Office, Great Britain (1967) for Ciudad Real show an average for the season of 58 per cent at 1230 hours. This seems far too high for the environment, and relative to nearby stations. A figure of about 45 per cent seems more probable, i.e. similar to that of Madrid.

La Mancha is Spain's biggest producer of bulk (mainly red) wines. It is fairly arid, and vine yields are not high. Although considered a hot environment, its ripening conditions as estimated here are quite moderate for midseason to late grape varieties. That is because in this quite continental climate the temperatures fall rapidly after mid summer, while the cool early part of spring can be expected to delay phenological development in all varieties. Important red grape varieties include Tempranillo (Maturity Group 4) and Grenache and Monastrell (7). The main white is Airén (maturity unknown).

Logroño (Rioja) (Table 170)

Vineyard site adjustments, °C

Min.	Max.	
+1.0	−0.5	Moderate slopes (Logroño in valley)
–	–	All aspects
+0.8	−0.4	Rocky, moderately calcareous soils
−0.3	−0.3	Average altitude ~ 470 m (Logroño ~ 420 m)
+1.5	−1.2	Mean +0.15°, variability index −5.4

Notes

I have assumed for calculation purposes that the meteorological station is in a valley-bottom situation. If not, the site climate adjustments will be less than indicated.

Table wines are made from a range of grape varieties. Red varieties include Tempranillo (Maturity Group 4), Cabernet Sauvignon (6) and Grenache (syn. Garnacha), Monastrell and Graciano (7). Grenache Blanc (8) is dominant for white wines in the lower Rioja.

Logroño is roughly on the borderline between the lower Rioja, which is hotter than estimated in Table 170, and the upper Rioja, which may be slightly cooler and has more rainfall (see Johnson 1971, pp. 168–9). The latter area produces the more highly reputed wines.

21

Viticultural Environments of England

English viticulture has a long history, of which Desmond Seward describes some of the early and more colourful aspects in his book 'Monks and Wine' (1979).

Roman settlers introduced vines to England in the late 3rd century AD, after the Emperor Probus had in 280 relaxed the earlier Edict of Domitian (AD 92) which prohibited vine planting in Spain, Gaul and Britain: a measure clearly aimed at reducing competition with a home industry that was already in a state of over-production.

English viticulture probably ceased with the end of Roman occupation, but was revived by the early Anglo-Saxon monks. By the time of the Venerable Bede (c. 673–735), it was again well established. The monastic vineyards appear to have received a big impetus when the Islamic conquests of the 8th century disrupted Mediterranean shipping and the internal trade of Gaul, so that little wine reached England by sea. The monks recorded that they pruned their vines in February and harvested in October, which they called Wyn Moneth.

Viking raids disrupted the monastries through the 9th century, but with the return of relative peace under Alfred the Great (reigned 871–900), the monastries and their vineyards again began to flourish. Secular commercial viticulture also started to become established. Of 38 vineyards listed in the Domesday Book (1084), only 12 were monastic. The Norman Conquest of 1066 gave further impetus to the monastic revival and, perhaps even more, to English viticulture. The industry reached its zenith in the late 11th and early 12th centuries.

The reasons for this were not only social and political. The climate of northern and western Europe had become markedly warmer after about AD 900, and was at the height of its 'Little Optimum' through the 11th and 12th centuries. This was the time the Vikings colonized the Arctic island to which they gave the name 'Greenland'. Summer temperatures in north-western Europe are thought to have been at least 1°C higher than now (Gribbin 1978; see also Chapter 23). English viticulture extended as far north as Leicester, with a scattering of vineyards through Suffolk, Essex, Kent, Sussex and Hampshire, along the whole of the Thames Valley, and in Wiltshire and Somerset. But most of all, vineyards were concentrated in the Severn Valley, in the present counties of Gloucester and Worcester.

A series of bad summers appears to have blighted English viticulture towards the middle of the 13th century. About the beginning of the 14th century the climate began to deteriorate catastrophically, culminating in the disastrously cold, wet seasons and famines of the 1340s which heralded the arrival of the Black Death (1348–52). Shortage of labour thereafter, followed by continued falls in temperature and (later) the decline and dissolution of the monastries, killed English viticulture more or less completely by the beginning of the 16th century.

But well before that another factor had come into play. The accession of King Henry II, in 1154, brought Britain into political union with the French counties of Anjou and Touraine, and with the duchies of Aquitaine and Gascony (which included Bordeaux). The Norman overlords preferred their French wines to the local product, which, even at the height of the Little Optimum, must often have been thin and acid. Before long, imports had largely displaced the produce of the English commercial vineyards, and viticulture again become confined to the self-sufficient (and, dare we say, ascetic) monastries.

The 'Little Ice Age' which followed the Little Optimum is generally reckoned to have lasted, with remissions in the mid 15th, early 16th, early to mid 17th and much of the 18th centuries, from about 1430 until 1830. Various records suggest that European temperatures fell to between 0.5°C and nearly 1.5°C lower than now during the coldest periods (see Gribbin 1978, and this book, Chapter 23). Since the mid 19th century there has been a progressive, if irregular, rise in the world's recorded temperatures up to the present. The reality or otherwise of this rise is canvassed in Chapter 23.

The modern resurgence of viticulture in England began soon after the end of World War II, at a time when recorded global temperatures were at a peak not exceeded again until the 1980s. Even the 1980s, however, very probably had cooler summers than those experienced during the mediaeval Little Optimum.

The temperature data used for Greenwich, Oxford and Southampton are for the period 1921–50, and should reasonably represent the 20th century average up to the 1970s. They show a very marginal climate for vine growing. Ripening grapes at all depends strongly on site factors to enhance effective temperature summation; also on favourability of season, and on growing only very early-ripening grape varieties and chaptalizing the musts. I estimate that a rise in average growing season temperatures of at least 1°C compared with the early and mid–20th century would be needed to ripen the earliest classic wine varieties

fully and regularly on the warmest sites, and perhaps 2°C to enable them to succeed fairly generally in the south of England.

Whether the relative warmth of the 1980s was a harbinger of such changes, or whether on the other hand it was merely part of the historical medium-term fluctuations of climate, is obviously a crucial question for English viticulture. I discuss it in some detail in Chapter 23.

TABLE 171

GREENWICH (LONDON), ENGLAND. 51°29′N, 0°00′. ALT. 45m.

	1	2	3	4	5	6	7	8	9	10	11	12	13	14	15	16
	Temperature °C							Summation of Day Degrees above 10° with 19° Cut-Off			Temperature Variability Index					
PERIOD	Lowest Recorded Min. (30YRS)	Av.Mthly Lowest Min. (30YRS)	Av.Min Minus Col.2	Daily Mean (30YRS)	Daily Range /2 (30YRS)	Av.Mthly Highest Max. (30YRS)	Col.6 Minus Av.Max.	Raw #	Adj.1 for Lat. & Daily Temp Rng	Adj.2 for Vine Sites	Raw	Adj. for Vine Sites	Av. Sunshine Hours (30YRS)	Av. No. Rain Days (30YRS)	Av. Rainfall mm (30YRS)	Av. % R.H. 1500 HRS (29YRS)
APRIL	−3.9	−1.1	5.6	8.9	4.4	21.7	8.3	2	3	13	31.1	26.5	159	14	46	58
MAY	−2.2	1.7	5.6	12.2	5.0	26.1	8.9	69	75	100	34.4	29.8	211	13	46	57
JUNE	1.7	5.6	5.0	15.6	5.0	28.9	8.3	167	185	209	33.3	28.7	210	11	41	57
JULY	6.7	8.3	4.4	17.8	5.0	30.6	7.8	241	265	279	32.3	27.7	205	13	51	55
AUGUST	3.9	7.2	5.0	17.2	5.0	29.4	7.2	224	238	263	32.2	27.6	189	13	56	58
SEPTEMBER	1.1	3.9	6.7	15.0	4.4	26.1	6.7	150	161	186	31.0	26.4	156	13	46	63
OCTOBER	−5.0	−0.6	7.2	10.6	3.9	20.6	6.1	27	35	57	29.0	24.4	109	14	58	70
APR-OCTOBER	—	—	5.6	13.9	4.7	—	7.6	880	962	1,107	31.9	27.3	1,239	91	344	60
NOV-MARCH	—	—	—	5.3	—	—	—	—	—	—	—	—	—	—	239	—
DATA FOR MONTH TO MATURITY																
MATURITY GROUP	EST. MATURITY DATE	*	*	*							*					
1	SEPTEMBER 30	15.3	3.3	25.2	6.7						26.4	156	13	46	63
2	OCTOBER 20	12.5	2.9	22.5	6.3						25.2	126	14	54	67
3	—	—	—	—	—						—	—	—	—	—
4	—	—	—	—	—						—	—	—	—	—
5	—	—	—	—	—						—	—	—	—	—
6	—	—	—	—	—						—	—	—	—	—
7	—	—	—	—	—						—	—	—	—	—
8	—	—	—	—	—						—	—	—	—	—

April and October day degree summations derived from temperature curve, not directly from monthly average means. * Adjusted for vine sites.

TABLE 172

OXFORD, ENGLAND. 51°46′N, 1°16′W. ALT. 63m.

	1	2	3	4	5	6	7	8	9	10	11	12	13	14	15	16
	Temperature °C							Summation of Day Degrees above 10° with 19° Cut-Off			Temperature Variability Index					
PERIOD	Lowest Recorded Min. (30YRS)	Av.Mthly Lowest Min. (30YRS)	Av.Min Minus Col.2	Daily Mean (30YRS)	Daily Range /2 (30YRS)	Av.Mthly Highest Max. (30YRS)	Col.6 Minus Av.Max.	Raw #	Adj.1 for Lat. & Daily Temp Rng	Adj.2 for Vine Sites	Raw	Adj. for Vine Sites	Av. Sunshine Hours (30YRS)	Av. No. Rain Days (30YRS)	Av. Rainfall mm (30YRS)	Av. % R.H.
APRIL	−4.4	−1.1	5.6	8.9	4.4	20.6	7.3	1	2	13	30.5	26.5	156	14	46	
MAY	−1.7	0.6	6.6	11.9	4.7	24.4	7.8	59	69	98	33.2	29.2	202	14	53	
JUNE	2.2	5.0	5.0	15.0	5.0	27.2	7.2	150	167	195	32.2	28.2	204	11	43	
JULY	5.6	7.8	4.4	16.9	4.7	28.3	6.7	214	240	269	29.9	25.9	195	14	58	
AUGUST	5.0	7.2	4.5	16.7	5.0	27.2	5.5	208	221	250	30.0	26.0	180	13	56	
SEPTEMBER	1.1	3.3	6.7	14.4	4.4	25.0	6.2	132	143	172	30.5	26.5	150	14	56	
OCTOBER	−3.9	−1.1	7.8	10.6	3.9	19.4	4.9	24	35	64	28.3	24.3	102	15	64	
APR-OCTOBER	—	—	5.8	13.5	4.6	—	6.5	788	877	1,061	30.7	26.7	1,189	95	376	
NOV-MARCH	—	—	—	5.3	—	—	—	—	—	—	—	—	—		269	

DATA FOR MONTH TO MATURITY

MATURITY GROUP	EST. MATURITY DATE	*	*	*								*				
1	OCTOBER 21	12.4	3.0	20.9	5.3							25.1	118	15	61	
2	—	—	—	—	—							—	—	—	—	
3	—	—	—	—	—							—	—	—	—	
4	—	—	—	—	—							—	—	—	—	
5	—	—	—	—	—							—	—	—	—	
6	—	—	—	—	—							—	—	—	—	
7	—	—	—	—	—							—	—	—	—	
8	—	—	—	—	—							—	—	—	—	

April and October day degree summations derived from temperature curve, not directly from monthly average means. * Adjusted for vine sites.

TABLE 173

SOUTHAMPTON, ENGLAND. 50°55′N, 1°24′W. ALT. 20m.

	1	2	3	4	5	6	7	8	9	10	11	12	13	14	15	16
	Temperature °C							Summation of Day Degrees above 10° with 19° Cut-Off			Temperature Variability Index					
PERIOD	Lowest Recorded Min. (30YRS)	Av.Mthly Lowest Min. (30YRS)	Av.Min Minus Col.2	Daily Mean (30YRS)	Daily Range /2 (30YRS)	Av.Mthly Highest Max. (67YRS)	Col.6 Minus Av.Max.	Raw #	Adj.1 for Lat. & Daily Temp Rng	Adj.2 for Vine Sites	Raw	Adj. for Vine Sites	Av. Sunshine Hours (30YRS)	Av. No. Rain Days (30YRS)	Av. Rainfall mm (30YRS)	Av. % R.H. 1300 HRS (15YRS)
APRIL	−2.8	0.0	5.0	9.4	4.4	20.6	6.8	5	8	23	29.4	24.8	171	13	53	69
MAY	−1.7	2.2	5.6	12.5	4.7	24.4	7.2	78	88	114	31.6	27.0	223	13	53	69
JUNE	3.3	6.1	4.5	15.3	4.7	26.1	6.1	158	179	203	29.4	24.8	225	11	46	68
JULY	6.7	8.3	3.9	16.9	4.7	27.2	5.6	215	239	265	28.3	23.7	217	13	58	69
AUGUST	5.0	8.9	3.3	16.9	4.7	26.7	5.1	215	233	258	27.2	22.6	198	13	64	71
SEPTEMBER	1.7	4.4	6.1	14.7	4.2	24.4	5.5	142	156	181	28.4	23.8	168	13	64	71
OCTOBER	−3.3	0.0	7.2	11.1	3.9	20.0	5.0	37	48	76	27.8	23.2	115	15	84	74
APR-OCTOBER	—	—	5.1	13.8	4.5	—	5.9	850	951	1,120	28.9	24.3	1,317	91	422	70
NOV-MARCH	—	—	—	5.9	—	—	—	—	—	—	—	—	—	—	378	—

DATA FOR MONTH TO MATURITY

MATURITY GROUP	EST. MATURITY DATE	*	*	*								*				
1	OCTOBER 2	14.9	3.0	23.3	5.5							23.8	165	13	65	71
2	OCTOBER 17	13.2	2.8	21.1	5.3							23.5	140	14	74	72
3	—	—	—	—	—							—	—	—	—	—
4	—	—	—	—	—							—	—	—	—	—
5	—	—	—	—	—							—	—	—	—	—
6	—	—	—	—	—							—	—	—	—	—
7	—	—	—	—	—							—	—	—	—	—
8	—	—	—	—	—							—	—	—	—	—

April and October day degree summations derived from temperature curve, not directly from monthly average means. * Adjusted for vine sites.

Commentaries and vineyard site adjustments

Greenwich (London) (Table 171)

Vineyard site adjustments, °C

Min.	Max.	
+0.6	−0.3	Moderate slopes, but mostly further from water
+0.3	+0.1	Predominantly southern aspects
+0.8	−0.4	Chalky, stony or other warm soils
−0.3	−0.3	Discount for urban warming at Greenwich
+1.4	−0.9	Mean +0.25°, variability index −4.6

Notes

A fairly substantial urban warming seems likely in the London area. The discount of 0.3°C brings the Greenwich data more into line with general temperature trends across southern England.

The predicted ripening-month data are probably typical enough of what can be attained with careful site selection in south-eastern England. In many areas warm soils, such as allowed for in the table, are rare; but their lack might be offset by steeper slopes on isolated or projecting hills, and by more strictly southern aspects.

Oxford (Table 172)

Vineyard site adjustments, °C

Min.	Max.	
+0.4	−0.2	Moderately steeper slopes than weather recording site
+0.3	+0.1	Predominantly southern aspects
+0.8	−0.4	Calcareous, stony or other warm soils
+1.5	−0.5	Mean +0.5°, variability index −4.0

Notes

Oxford is both cooler and less sunny than Greenwich and Southampton. The figures suggest it to be at about the extreme limit for ripening the earliest (Maturity Group 1) vinifera grape varieties anywhere in the open, under average 20th century climatic conditions.

Southampton (Table 173)

Vineyard site adjustments, °C

Min.	Max.	
+1.0	−0.5	Moderate slopes (Southampton flat)
+0.3	+0.1	Predominantly southern aspects
+0.4	−0.2	Some chalky or stony soils
−0.3	−0.3	Altitude 60–80 m (Southampton 20 m)
+1.4	−0.9	Mean +0.25°, variability index −4.4

Notes

The temperature gain from the slope and aspect of estimated vineyard sites is largely offset by their greater altitude compared with Southampton. Lower, but less sloping, sites could achieve the same effective temperature summation.

Compared with Greenwich, the south coast has similar growing season temperatures but is sunnier. However at these temperatures sunshine hours are probably not the primary limitation (Chapter 3).

22

Viticultural Environments of the United States of America

I confine treatment of viticulture in the United States of America to the west coast, and to *vinifera* grapes. Native grape species are adapted to a different type of climate, and in any case hardly come into direct comparison for commercial winemaking.

The estimates for the ripening month encompass all grape maturity groups, because, as in Australia, the matching of grape varieties to environments is still evolving in the USA. Ripening conditions for all maturity groups are therefore still of practical interest.

Although the North American stations included are few, their data permit several generalizations.

1. All west coastal locations in the USA have very high ratios of growing season sunshine hours to effective temperature summations (see Chapter 3, Figure 2). This is due to the coldness of the adjacent northern Pacific waters, and is in contrast to Australian coastal and near-coastal vine-growing areas, which generally have lower ratios and approximate more to the European pattern.
2. The summers are very dry, particularly in California. In warm to hot areas it is possible to ripen grapes with very little risk of rain during ripening. This has obvious advantages for varieties that are prone to berry splitting, such as Zinfandel.
3. Temperature variability indices tend to be high, except very near the coast. Nevertheless the variability of the minimum temperature is generally moderate, making it possible to grow vines in quite cool environments without too great a risk of frost after budburst.
4. The coastal and near-coastal areas have an unusually slow temperature decline in autumn, owing to the strong summer-cooling ocean influence. This gives them a unique capacity to ripen a wide range of maturity types under near-optimum mean temperatures.

These climatic factors, combined with an ample capacity for full irrigation or supplementary watering as needed, gives California and (to a lesser extent) Washington and Oregon an excellent environment for commercial viticulture. The climate is, however, clearly different from that in most European viticultural areas, and to a lesser degree from that in Australia. Contrasts in wine style therefore seem inevitable, regardless of winemaking techniques.

If my analysis is correct, temperature variability and low relative humidities (or high saturation deficits) constitute the main limitations to potential grape and wine quality in the non-coastal areas. As in Australia, the best wines as a rule will be from near the coast. But even there, topographic and soil factors which moderate local temperature variability and improve soil drainage can be expected to play a big role. In all regions the best wines will come from well-drained stony or (in cool areas) calcareous soils, and from hillsides with the best air drainage.

Nothing coming out of the present study contradicts the broad conclusions of Amerine and Winkler (1944), as they apply to west coastal USA. But it must be noted that the success of their classification of Californian wine-producing regions, on the basis of raw temperature summation alone, in fact hinges largely on the close correlation that exists in California between temperature and atmospheric saturation deficit, and almost equally, between average mean temperature and temperature variability index.

These correlations are weaker, or sometimes nonexistent, elsewhere. As a result the Californian experience, though valuable and valid at home, has only partial application in other parts of the world.

246

TABLE 174
FRESNO, CALIFORNIA. 36°46′N, 119°43′W. ALT. 101m. (San Joaquin Valley)

PERIOD	1 Lowest Recorded Min. (43YRS)	2 Av.Mthly Lowest Min. (20YRS)	3 Av.Min Minus Col.2	4 Daily Mean (43YRS)	5 Daily Range /2 (43YRS)	6 Av.Mthly Highest Max. (20YRS)	7 Col.6 Minus Av.Max.	8 Raw	9 Adj.1 for Lat. & Daily Temp Rng	10 Adj.2 for Vine Sites	11 Raw	12 Adj. for Vine Sites	13 Av. Sunshine Hours	14 Av. No. Rain Days (43YRS)	15 Av. Rainfall mm (53YRS)	16 Av. % R.H. 1200 HRS (18YRS)
				Temperature °C				Summation of Day Degrees above 10° with 19° Cut-Off			Temperature Variability Index					
APRIL	1.1	3.9	5.0	16.1	7.2	31.3	8.0	189	171		41.8		318	4	23	41
MAY	3.3	6.7	4.9	19.4	7.8	35.4	8.2	279	266		44.3		372	2	10	33
JUNE	5.6	10.0	5.0	23.9	8.9	40.5	7.7	270	270		48.3		405	1	2	27
JULY	10.0	13.9	4.5	27.8	9.4	42.2	5.0	279	279		47.1		425	0	0	24
AUGUST	10.6	12.8	4.5	26.7	9.4	40.9	4.8	279	279		46.9		409	0	0	26
SEPTEMBER	5.6	9.4	5.1	23.1	8.6	37.6	5.9	270	270		45.4		336	1	5	31
OCTOBER	2.2	5.6	5.0	18.1	7.5	32.8	7.2	251	238		42.2		291	3	13	38
APR-OCTOBER	—	—	4.9	22.2	8.4	—	6.7	1,817	1,773		45.1		2,556	11	53	31
NOV-MARCH	—	—	—	10.4	—	—	—	—	—		—		—	—	178	—

DATA FOR MONTH TO MATURITY

MATURITY GROUP	EST. MATURITY DATE														
1	AUGUST 8			27.7	9.4	42.0	5.0				47.1		421	0	0	24
2	AUGUST 14			27.5	9.4	41.8	4.9				47.0		420	0	0	25
3	AUGUST 19			27.2	9.4	41.6	4.9				47.0		418	0	0	25
4	AUGUST 25			26.9	9.4	41.3	4.8				46.9		415	0	0	26
5	AUGUST 30			26.6	9.4	40.9	4.8				46.9		411	0	0	26
6	SEPTEMBER 5			26.2	9.3	40.5	4.9				46.8		404	0	0	27
7	SEPTEMBER 10			25.8	9.2	40.0	5.1				46.6		392	0	1	27
8	SEPTEMBER 16			25.3	9.0	39.4	5.3				46.2		377	0	2	28

TABLE 175
NAPA, CALIFORNIA. 38°19′N, 122°18′W. ALT. approx. 40m. (Napa Valley)

PERIOD	1 Lowest Recorded Min.	2 Av.Mthly Lowest Min.	3 Av.Min Minus Col.2	4 Daily Mean (30YRS)	5 Daily Range /2 (30YRS)	6 Av.Mthly Highest Max.	7 Col.6 Minus Av.Max.	8 Raw	9 Adj.1 for Lat. & Daily Temp Rng	10 Adj.2 for Vine Sites	11 Raw	12 Adj. for Vine Sites	13 Av. Sunshine Hours	14 Av. No. Rain Days	15 Av. Rainfall mm	16 Av. % R.H.
				Temperature °C				Summation of Day Degrees above 10° with 19° Cut-Off			Temperature Variability Index					
APRIL		1.9	4.5	13.1	6.7	27.3	7.5	93	90		38.8		270			
MAY		3.5	4.6	15.2	7.1	31.3	9.0	161	150		42.0		310			
JUNE		6.4	3.9	18.1	7.8	34.8	8.9	243	221		44.0		339			
JULY		7.9	3.4	19.1	7.8	34.6	7.7	279	259		42.3		347			
AUGUST		7.6	3.7	18.9	7.6	33.7	7.2	276	257		41.3		319			
SEPTEMBER		5.9	4.2	18.4	8.3	34.3	7.6	252	224		45.0		270			
OCTOBER		3.9	4.7	16.3	7.1	31.8	7.8	195	178		43.3		263			
APR-OCTOBER	—	4.1	17.0	7.6	—	8.0		1,499	1,379		42.4		2,118			
NOV-MARCH	—	—	10.0	—	—	—		—	—		—		—			

DATA FOR MONTH TO MATURITY

MATURITY GROUP	EST. MATURITY DATE														
1	SEPTEMBER 10			18.9	7.8	34.0	7.3				42.2		305			
2	SEPTEMBER 16			18.8	8.0	34.0	7.4				43.0		290			
3	SEPTEMBER 23			18.6	8.2	34.0	7.5				44.0		280			
4	SEPTEMBER 30			18.4	8.3	34.0	7.6				45.0		270			
5	OCTOBER 8			18.0	8.2	33.7	7.6				44.6		268			
6	OCTOBER 16			17.6	8.0	33.1	7.7				44.1		266			
7	OCTOBER 26			16.8	7.8	32.3	7.8				43.5		264			
8	NOVEMBER 7			15.3	7.4	31.0	7.6				43.0		252			

TABLE 176

SALINAS, CALIFORNIA. 36°39'N, 121°40'W. ALT. approx. 20m.

	1	2	3	4	5	6	7	8	9	10	11	12	13	14	15	16
	Temperature °C							Summation of Day Degrees above 10° with 19° Cut-Off			Temperature Variability Index					
PERIOD	Lowest Recorded Min.	Av.Mthly Lowest Min.	Av.Min Minus Col.2	Daily Mean	Daily Range /2	Av.Mthly Highest Max.	Col.6 Minus Av.Max.	Raw	Adj.1 for Lat. & Daily Temp Rng	Adj.2 for Vine Sites	Raw	Adj. for Vine Sites	Av. Sunshine Hours	Av. No. Rain Days	Av. Rainfall mm	Av. % R.H.
APRIL		2.8	4.4	13.3	6.1	27.7	8.3	99	98		37.1		273			
MAY		3.7	4.8	14.2	5.7	29.4	9.5	130	128		37.1		288			
JUNE		5.9	4.0	15.5	5.6	30.6	9.5	165	161		35.9		324			
JULY		7.9	3.4	16.4	5.1	29.6	8.1	198	194		31.9		326			
AUGUST		7.8	3.9	16.6	4.9	29.0	7.5	205	203		31.0		301			
SEPTEMBER		6.4	4.2	16.7	6.1	30.8	8.0	201	200		36.6		273			
OCTOBER		3.2	5.0	15.0	6.8	30.2	8.4	155	152		40.6		260			
APR-OCTOBER	—	4.2	15.4	5.8	—	8.5	1,153	1,136		35.7		2,045			
NOV-MARCH	—	—	11.0	—	—	—	—	—		—		—			
	DATA FOR MONTH TO MATURITY															
MATURITY GROUP	EST. MATURITY DATE															
1	OCTOBER 12	16.2	6.4	30.5	8.2						33.0		269			
2	OCTOBER 22	15.6	6.6	30.3	8.3						34.9		264			
3	NOVEMBER 5	14.6	6.8	29.7	8.2						37.2		256			
4	NOVEMBER 26	12.7	6.9	26.6	7.0						39.5		230			
5	—		—	—	—	—					—		—			
6	—		—	—	—	—					—		—			
7	—		—	—	—	—					—		—			
8	—		—	—	—	—					—		—			

TABLE 177

SAN JOSE, CALIFORNIA. 37°20'N, 121°54'W. ALT. 43m. (Santa Clara Valley)

	1	2	3	4	5	6	7	8	9	10	11	12	13	14	15	16
	Temperature °C							Summation of Day Degrees above 10° with 19° Cut-Off			Temperature Variability Index					
PERIOD	Lowest Recorded Min. (24YRS)	Av.Mthly Lowest Min. (24YRS)	Av.Min Minus Col.2	Daily Mean (24YRS)	Daily Range /2 (24YRS)	Av.Mthly Highest Max. (24YRS)	Col.6 Minus Av.Max.	Raw	Adj.1 for Lat. & Daily Temp Rng	Adj.2 for Vine Sites	Raw	Adj. for Vine Sites	Av. Sunshine Hours (24YRS)	Av. No. Rain Days (57YRS)	Av. Rainfall mm (13YRS)	Av. % R.H. 1200 HRS
APRIL	0.0	2.2	4.4	13.3	6.7	28.3	8.3	99	95		39.5		270	5	28	51
MAY	1.7	3.6	4.8	15.3	6.9	31.7	9.5	164	156		41.9		298	3	13	49
JUNE	3.3	6.0	4.0	17.5	7.5	34.5	9.5	225	206		43.5		333	1	3	45
JULY	6.1	8.3	3.4	19.2	7.5	34.8	8.1	279	265		41.5		341	0	0	49
AUGUST	5.6	7.8	3.9	19.2	7.5	34.2	7.5	279	266		41.4		316	0	0	50
SEPTEMBER	3.9	5.8	4.2	17.8	7.8	33.6	8.0	234	214		43.4		270	1	8	48
OCTOBER	−0.6	3.3	.5.0	15.8	7.5	31.7	8.4	180	166		43.4		263	4	18	50
APR-OCTOBER	—	—	4.2	16.9	7.3	—	8.5	1,460	1,368		42.1		2,091	14	70	49
NOV-MARCH	—	—	—	10.4	—	—	—	—	—		—		—	—	300	—
	DATA FOR MONTH TO MATURITY															
MATURITY GROUP	EST. MATURITY DATE															
1	SEPTEMBER 8	18.9	7.6	34.0	7.6						42.0		305	0	1	49
2	SEPTEMBER 15	18.6	7.6	33.9	7.7						42.5		295	1	2	49
3	SEPTEMBER 22	18.2	7.7	33.8	7.8						43.0		282	1	4	48
4	SEPTEMBER 30	17.8	7.8	33.6	8.0						43.4		272	1	8	48
5	OCTOBER 9	17.3	7.7	33.1	8.1						43.4		268	2	10	49
6	OCTOBER 18	16.7	7.6	32.5	8.2						43.4		267	3	13	49
7	OCTOBER 29	15.9	7.5	31.9	8.4						43.4		265	4	17	50
8	NOVEMBER 12	14.3	7.2	30.0	8.0						43.0		250	6	24	51

TABLE 178

PORTLAND, OREGON. 45°32′N, 122°40′W. ALT. 47m. (NW Oregon, SW Washington)

	1	2	3	4	5	6	7	8	9	10	11	12	13	14	15	16
	Temperature °C							Summation of Day Degrees above 10° with 19° Cut-Off			Temperature Variability Index					
PERIOD	Lowest Recorded Min. (72YRS)	Av.Mthly Lowest Min. (38YRS)	Av.Min Minus Col.2	Daily Mean (72YRS)	Daily Range /2 (72YRS)	Av.Mthly Highest Max. (38YRS)	Col.6 Minus Av.Max.	Raw	Adj.1 for Lat. & Daily Temp Rng	Adj.2 for Vine Sites	Raw	Adj. for Vine Sites	Av. Sunshine Hours	Av. No. Rain Days (75YRS)	Av. Rainfall mm (77YRS)	Av. % R.H. 1630 HRS (59YRS)
APRIL	−2.2	1.1	5.0	11.1	5.0	25.6	9.5	33	34	38	34.5	33.9	207	14	71	54
MAY	0.0	3.3	5.0	13.6	5.3	29.4	10.5	112	116	116	36.7	36.1	229	13	53	51
JUNE	3.9	6.7	4.9	16.9	5.3	32.2	10.0	207	216	216	36.1	35.5	267	10	41	51
JULY	6.1	8.9	4.5	19.2	5.8	33.9	8.9	279	279	279	36.6	36.0	316	3	13	45
AUGUST	6.1	8.9	4.5	19.2	5.8	32.4	7.4	279	279	279	35.1	34.5	282	4	15	46
SEPTEMBER	1.7	5.6	5.5	16.4	5.3	29.5	7.8	192	193	193	34.5	33.9	210	8	46	53
OCTOBER	−1.7	2.8	5.5	12.5	4.2	23.1	6.4	77	88	93	28.7	28.1	149	12	84	66
APR-OCTOBER	—	—	5.0	15.6	5.2	—	8.6	1,179	1,205	1,214	34.6	34.0	1,660	64	323	52
NOV-MARCH	—	—	—	6.2	—	—	—	—	—	—	—	—	—	—	739	—

	DATA FOR MONTH TO MATURITY															
MATURITY GROUP	EST. MATURITY DATE	*	*	*								*				
1	SEPTEMBER 17	17.8	5.4	30.9	7.6							34.1	245	6	28	49
2	SEPTEMBER 26	16.9	5.2	29.7	7.7							33.9	222	7	40	52
3	OCTOBER 8	15.4	4.9	28.0	7.4							32.8	196	9	54	56
4	OCTOBER 22	13.5	4.3	24.9	6.7							29.8	165	11	72	63
5	—	—	—	—	—							—	—	—	—	—
6	—	—	—	—	—							—	—	—	—	—
7	—	—	—	—	—							—	—	—	—	—
8	—	—	—	—	—							—	—	—	—	—

* Adjusted for vine sites.

TABLE 179

ROSEBURG, OREGON. 43°13′N, 123°20′W. ALT. 146m.

	1	2	3	4	5	6	7	8	9	10	11	12	13	14	15	16
	Temperature °C							Summation of Day Degrees above 10° with 19° Cut-Off			Temperature Variability Index					
PERIOD	Lowest Recorded Min. (53YRS)	Av.Mthly Lowest Min. (20YRS)	Av.Min Minus Col.2	Daily Mean (53YRS)	Daily Range /2 (53YRS)	Av.Mthly Highest Max. (20YRS)	Col.6 Minus Av.Max.	Raw #	Adj.1 for Lat. & Daily Temp Rng	Adj.2 for Vine Sites	Raw	Adj. for Vine Sites	Av. Sunshine Hours	Av. No. Rain Days (53YRS)	Av. Rainfall mm (53YRS)	Av. % R.H. 1200 HRS (12YRS)
APRIL	−3.9	0.0	5.0	10.8	5.8	27.2	10.6	26	26	30	38.8	35.8	210	13	58	56
MAY	−0.6	1.7	5.5	13.6	6.4	30.8	10.8	112	114	120	41.9	38.9	229	11	46	50
JUNE	2.2	5.0	4.5	16.4	6.9	34.5	11.2	192	191	203	43.3	40.3	270	7	28	47
JULY	4.4	7.8	3.8	19.7	8.1	37.4	9.6	279	279	279	45.8	42.8	320	2	8	43
AUGUST	3.9	7.2	3.9	19.4	8.3	36.4	8.7	279	268	279	45.8	42.8	285	2	8	45
SEPTEMBER	−1.7	3.9	5.0	16.1	7.2	33.0	9.7	183	173	190	43.5	40.5	216	7	30	52
OCTOBER	−5.6	0.6	6.0	12.2	5.6	26.8	9.0	68	68	76	37.4	34.4	195	11	64	65
APR-OCTOBER	—	—	4.8	15.5	6.9	—	9.9	1,139	1,119	1,177	42.4	39.4	1,725	53	242	51
NOV-MARCH	—	—	—	6.7	—	—	—	—	—	—	—	—	—	—	577	—

	DATA FOR MONTH TO MATURITY															
MATURITY GROUP	EST. MATURITY DATE	*	*	*								*				
1	SEPTEMBER 20	17.5	7.2	33.7	9.4							41.5	240	5	21	49
2	SEPTEMBER 30	16.3	6.5	32.5	9.7							40.5	220	7	30	52
3	OCTOBER 15	14.3	5.7	29.6	9.4							37.8	200	9	45	58
4	—	—	—	—	—							—	—	—	—	—
5	—	—	—	—	—							—	—	—	—	—
6	—	—	—	—	—							—	—	—	—	—
7	—	—	—	—	—							—	—	—	—	—
8	—	—	—	—	—							—	—	—	—	—

April day degree summation derived from temperature curve, not directly from monthly average mean. * Adjusted for vine sites.

TABLE 180

PROSSER, WASHINGTON. 46°15′N, 119°45′W. ALT. approx. 150m.

	1	2	3	4	5	6	7	8	9	10	11	12	13	14	15	16
	Temperature °C							Summation of Day Degrees above 10° with 19° Cut-Off			Temperature Variability Index					
PERIOD	Lowest Recorded Min. (44YRS)	Av.Mthly Lowest Min. (44YRS)	Av.Min Minus Col.2	Daily Mean (44YRS)	Daily Range /2 (44YRS)	Av.Mthly Highest Max. (44YRS)	Col.6 Minus Av.Max.	Raw #	Adj.1 for Lat. & Daily Temp Rng	Adj.2 for Vine Sites	Raw	Adj. for Vine Sites	Av. Sunshine Hours	Av. No. Rain Days	Av. Rainfall mm (44YRS)	Av. % R.H.
APRIL	−10.0	−3.9	6.7	10.7	7.9	26.9	8.3	27	15	28	46.6	43.0	240		14	
MAY	−3.3	−0.5	6.9	14.8	8.4	31.9	8.7	149	126	151	49.2	45.6	285		15	
JUNE	1.1	3.6	6.0	18.2	8.6	35.3	8.5	254	227	251	48.9	45.3	303		17	
JULY	3.3	6.1	5.7	21.6	9.8	38.3	6.9	279	279	279	51.8	48.2	384		4	
AUGUST	2.8	5.9	5.2	20.7	9.6	36.5	6.2	279	279	279	49.8	46.2	356		5	
SEPTEMBER	−6.7	1.7	6.5	16.9	8.7	33.2	7.6	207	176	200	48.9	45.3	258		9	
OCTOBER	−11.7	−2.7	6.8	11.2	7.1	25.8	7.5	46	36	47	42.7	39.1	195		19	
APR-OCTOBER	—	—	6.3	16.3	8.6	—	7.7	1,241	1,138	1,235	48.3	44.7	2,021		83	
NOV-MARCH	—	—	—	2.7	—	—	—	—	—	—	—	—	—	.	107	
DATA FOR MONTH TO MATURITY																
MATURITY GROUP	EST. MATURITY DATE		*	*	*						*					
1	SEPTEMBER 8		20.5	8.3	35.1	6.7					45.9	329		6		
2	SEPTEMBER 14		19.9	8.2	34.5	7.0					45.7	310		7		
3	SEPTEMBER 23		18.9	8.0	33.5	7.4					45.5	282		8		
4	OCTOBER 4		16.5	7.7	32.0	7.6					44.8	252		10		
5	—		—	—	—	—					—	—		—		
6	—		—	—	—	—					—	—		—		
7	—		—	—	—	—					—	—		—		
8	—		—	—	—	—					—	—		—		

April and October day degree summations derived from temperature curve, not directly from monthly average means. * Adjusted for vine sites.

TABLE 181

SEATTLE, WASHINGTON. 47°36′N, 122°20′W. ALT. 38m.

	1	2	3	4	5	6	7	8	9	10	11	12	13	14	15	16
	Temperature °C							Summation of Day Degrees above 10° with 19° Cut-Off			Temperature Variability Index					
PERIOD	Lowest Recorded Min. (57YRS)	Av.Mthly Lowest Min. (57YRS)	Av.Min Minus Col.2	Daily Mean (57YRS)	Daily Range /2 (57YRS)	Av.Mthly Highest Max. (57YRS)	Col.6 Minus Av.Max.	Raw #	Adj.1 for Lat. & Daily Temp Rng	Adj.2 for Vine Sites	Raw	Adj. for Vine Sites	Av. Sunshine Hours	Av. No. Rain Days (57YRS)	Av. Rainfall mm (70YRS)	Av. % R.H. 1630 HRS (47YRS)
APRIL	−1.1	2.2	3.9	10.3	4.2	23.3	8.8	15	24	26	29.5	31.1	204	13	58	58
MAY	2.2	4.8	3.6	13.1	4.7	26.7	8.9	96	106	111	31.3	32.9	236	12	46	56
JUNE	4.4	7.8	3.3	15.8	4.7	29.4	8.9	174	190	194	31.0	32.6	255	9	36	54
JULY	7.8	9.6	2.6	17.2	5.0	30.9	8.7	223	236	245	31.3	32.9	298	4	15	51
AUGUST	7.8	10.0	2.8	17.8	5.0	30.2	7.4	242	251	260	30.2	31.8	267	5	18	54
SEPTEMBER	2.2	7.1	4.0	15.3	4.2	27.2	7.7	159	173	176	28.5	30.1	177	8	43	61
OCTOBER	−1.7	3.5	4.9	11.7	3.3	21.7	6.7	53	77	80	24.8	26.4	121	13	74	73
APR-OCTOBER	—	—	3.6	14.5	4.4	—	8.2	962	1,057	1,092	29.5	31.1	1,558	64	290	58
NOV-MARCH	—	—	—	6.3	—	—	—	—	—	—	—	—	—	—	559	—
DATA FOR MONTH TO MATURITY																
MATURITY GROUP	EST. MATURITY DATE		*	*	*						*					
1	OCTOBER 13		14.0	4.2	25.7	7.3					28.8	145		10	56	66
2	—		—	—	—	—					—	—		—	—	—
3	—		—	—	—	—					—	—		—	—	—
4	—		—	—	—	—					—	—		—	—	—
5	—		—	—	—	—					—	—		—	—	—
6	—		—	—	—	—					—	—		—	—	—
7	—		—	—	—	—					—	—		—	—	—
8	—		—	—	—	—					—	—		—	—	—

April day degree summation derived from temperature curve, not directly from monthly average mean. * Adjusted for vine sites.

TABLE 182
WALLA WALLA, WASHINGTON. 46°02'N, 118°20'W. ALT. 289m.

	1	2	3	4	5	6	7	8	9	10	11	12	13	14	15	16
	Temperature °C							Summation of Day Degrees above 10° with 19° Cut-Off			Temperature Variability Index					
PERIOD	Lowest Recorded Min. (59YRS)	Av.Mthly Lowest Min. (20YRS)	Av.Min Minus Col.2	Daily Mean (45YRS)	Daily Range /2 (45YRS)	Av.Mthly Highest Max. (20YRS)	Col.6 Minus Av.Max.	Raw #	Adj.1 for Lat. & Daily Temp Rng	Adj.2 for Vine Sites	Raw	Adj. for Vine Sites	Av. Sunshine Hours	Av. No. Rain Days (45YRS)	Av. Rainfall mm (58YRS)	Av. % R.H. 1200 HRS (15YRS)
APRIL	−6.7	0.6	5.5	11.7	5.6	26.1	8.8	51	52		36.7		246	9	38	43
MAY	0.0	3.8	5.7	15.6	6.1	30.6	8.9	174	181		39.0		304	9	43	40
JUNE	3.9	7.8	4.5	19.2	6.9	36.3	10.3	270	270		42.3		309	7	28	36
JULY	7.2	11.1	5.0	23.9	7.8	40.0	8.3	279	279		44.5		403	3	10	29
AUGUST	5.0	10.6	5.5	23.3	7.2	38.1	7.6	279	279		41.9		360	3	13	30
SEPTEMBER	−3.3	5.6	5.5	17.8	6.7	32.8	8.3	234	233		40.6		267	6	23	41
OCTOBER	−5.6	0.6	6.0	12.2	5.6	26.7	8.9	68	67		37.3		205	8	38	49
APR-OCTOBER	—	—	5.4	17.7	6.6	—	8.7	1,355	1,361		40.3		2,094	45	193	38
NOV-MARCH	—	—	—	3.8	—	—	—	—			—		—	—	231	—

DATA FOR MONTH TO MATURITY																
MATURITY GROUP	EST. MATURITY DATE															
1	AUGUST 30	23.3	7.2	38.2	7.6						41.9		362	3	13	30
2	SEPTEMBER 5	22.4	7.1	37.6	7.7						41.5		348	3	14	31
3	SEPTEMBER 10	21.5	7.0	36.7	7.8						41.2		330	4	16	33
4	SEPTEMBER 15	20.6	6.9	35.5	7.9						41.0		310	4	18	35
5	SEPTEMBER 22	19.4	6.8	34.1	8.1						40.8		288	5	21	38
6	OCTOBER 1	17.8	6.7	32.7	8.3						40.5		265	6	24	41
7	OCTOBER 15	15.0	6.3	29.8	8.6						39.2		232	7	30	44
8	—	—	—	—	—						—		—	—	—	—

Commentaries and vineyard site adjustments

1. California

Fresno (San Joaquin Valley) (Table 174)

Vineyard site adjustments: None

Notes

This is a typical flat, inland valley environment. The extremely high temperatures and sunshine hours, and lack of rainfall throughout the whole summer and period following ripening, indicate an area best suited for table and especially drying grapes. The temperatures and their variability, and the saturation deficit, are all far too high for wine-grape quality.

Napa (Napa Valley) (Table 175)

Vineyard site adjustments: None

Notes

The data refer primarily to the valley floor. However effective day degrees, ripening times and ripening mean temperatures similar to those at Napa should also extend to much greater altitudes on the valley sides, because effective day degrees are regained which on the valley floor are lost due to wide diurnal temperature range (column 9 vs column 8 in the Table). Indeed, the lowest slopes will be quite appreciably warmer than the valley floor, due to their higher minima.

Hill-side locations thus have a very marked potential quality advantage over the flats in this area through their large reductions in temperature variability, together with some increase in afternoon relative humidities with greater altitude. Vine vigour differences add further, unless the more vigorous vines on the fertile flats can be extensively trellised so as to give fully comparable canopy light relations.

Napa lies low in Amerine and Winkler's Region II. Temperatures rise northwards up the valley with increasing attenuation of the cooling breezes from San Francisco Bay, to reach typical Region III levels around Calistoga. This means that there is an ample tolerance of increased altitude above the valley floor, and scope to grow a wide range of grape varieties throughout to produce a comprehensive range of table wine styles.

Salinas (Table 176)

Vineyard site adjustments: None

Notes

Table 176 represents the flat country around Salinas itself. The raw data for Salinas show it to be towards the

251

cool limit for viticulture, with ripening-month average mean temperatures similar to those of the Rhine Valley, Champagne and the cooler parts of the Loire Valley, albeit with many more sunshine hours. Diurnal temperature range and temperature variability index are kept fairly low, especially in mid-summer, by close proximity to the sea, together with shelter from inland weather influences by the coastal ranges. Nevertheless the variability index during ripening remains higher than in the cool European vine-growing areas.

Temperatures rise markedly inland as the cold oceanic influence diminishes. Presumably temperature variability indices also rise, so that compensating site factors become increasingly desirable. Lower hillsides, with if possible more stony soils than on the valley floor, should provide excellent conditions for producing light to medium-bodied table wines further up the Salinas Valley.

San Jose (Santa Clara Valley) (Table 177)

Vineyard site adjustments: None

Notes

The estimates are for valley-floor situations comparable to San Jose, and show a climatic environment similar to that of the Napa Valley (Table 175): very suitable for making table wines, apart from a wide diurnal temperature range and therefore a larger than desirable temperature variability index.

Many vineyards extend into the foothills, and the comments on hillside sites under Napa (Table 175) and Salinas (Table 176) apply here also.

2.   Oregon

Portland (NW Oregon, SW Washington) (Table 178)

Vineyard site adjustments, °C

Min.	Max.	
–	–	Moderate foothill slopes but further from coast
+0.45	+0.15	Mostly southern aspects
−0.3	−0.3	Average altitude ~ 100 m (Portland 47 m)
+0.15	−0.15	Mean unchanged, variability index −0.6

Notes

The calculated data are for areas fairly close to Portland, with reasonably favourable southern aspects on the lower foothills but no special soil-type advantages.

Because this is a cool, fairly maritime climate with little of the growing season above 19°C average mean temperature, it is easy to estimate differences among environments in their grape ripening capacities directly from their locations, altitudes, and other site factors in the same way as for Tasmania (Chapter 14). Using the Portland data as a basis, predictions for other nearby environments can be made approximately as follows.

1. For the same altitude, add 0.2°C to the average mean temperatures per degree of latitude southwards along the Willamette Valley as far as Eugene; or subtract 0.2°C per degree north to about Centralia. In the southernmost section, from Albany to Eugene, ripening times and conditions at similar altitudes and topographies will be advanced by the equivalent of about one maturity group (50 day degrees) compared with Portland, whereas at Centralia they will be retarded by the same amount. From Centralia north, as judged by Seattle (Washington, Table 181), temperatures fall rapidly: presumably influenced by the cold waters of Puget Sound. Centralia is about the furthest north that the earliest-maturing grape varieties of reasonable winemaking quality (Maturity Group 2) can be expected to ripen regularly and satisfactorily, other than on very limited sites.

2. Steep south-facing slopes can be expected to gain by up to one maturity group compared with the calculated figures, other things being equal. Conversely, neutral (east or west) slopes would lose the equivalent of about one maturity group, and north-facing slopes the equivalent of two to three maturity groups, depending on their steepness.

3. Stony, rocky or calcareous soils could advance ripening by at least one maturity group. This is additive to any effect of slope and aspect.

4. Other things being equal, maturity will be retarded by about one maturity group for every 50 m increase in altitude above 100 m. It is doubtful whether there would be much gain through reduced altitude below 100 m, because of flatter topography; while vines on the true valley flats will almost certainly have seriously retarded maturity, as well as suffering badly from spring frosts.

Taken as a whole, the inland valley system from Eugene and Springfield, Oregon, northwards to about Centralia, Washington, bears a striking resemblance in geography and climate (other than in rainfall pattern) to the Rhine Valley of Germany. With protection from the west by the coast ranges and long summer sunshine hours, ripening of early varieties up to Maturity Group 3 should take place satisfactorily on suitable exposures in most areas, provided that they are sufficiently frost-free. Light, fresh white and rosé styles of wine should result. Group 4 varieties like Riesling for white wines, or Group 3 varieties such as Pinot Noir or Gamay for red wines, would probably need carefully selected warm exposures and soils in the central and southern Willamette Valley for regular full ripening. Further possibilities on outstanding southern exposures should be available along the lower Willamette and especially lower Columbia valleys. The slopes of isolated hill ranges within the Willamette Valley, for instance just south of Salem, are of special interest because of their very free air drainage, with no external sources of cold air drift. The warmest such sites appear to offer ripening conditions strikingly similar to those of Burgundy, France, confirming the earlier analysis of Aney (1974) and more recent practical experience (Adelsheim 1988). Such choice sites are a natural environment for Pinot Noir and Chardonnay. The data clearly suggest that midseason and later varieties are unlikely to be successful anywhere in the near-coastal valleys.

A point to be remembered is that October in this area is a time of rapidly increasing rainfall, in contrast to central Europe where rainfall declines in autumn. This is likely to create difficulties in producing late-picked wine styles, and with any marginally ripening varieties, regardless of the fact that temperatures, as such, may still be adequate. A fortiori, the region appears strictly to be one for early-maturing grape varieties, with even these requiring the warmest exposures and soils for safe ripening. Quality potential is then excellent, however.

The central Columbia Valley inland from the Cascade Range has areas with substantially warmer summers than the coastal valleys, reaching into Amerine and Winkler's Regions II and III (Aney 1974). They are covered by the data and commentary for Walla Walla, Washington State (Table 182).

Roseburg (Table 179)

Vineyard site adjustments, °C

Min.	Max.	
+0.8	−0.4	Moderate slopes of valley side (Roseburg in valley)
+0.45	+0.15	Southerly aspects predominating
−0.3	−0.3	Average altitude ~ 200 m (Roseburg 146 m)
+0.95	−0.55	Mean +0.2°, variability index −3.0

Notes

As for Portland, site climate adjustments are made to match the more favourable sites that might be expected to exist in significant areas. Stony or rocky soils, not assumed in the calculations, could improve ripening further by the equivalent of about one maturity group.

Conditions are a little less favourable, according to these figures, than those available at lower altitudes further north, in the Willamette and lower Columbia valleys. The frost risk appears to be greater. Even on the best sites only early to very early-maturing varieties could be expected to ripen reliably. Suitable areas for viticulture are therefore probably more restricted than in the Willamette Valley.

3. Washington

Prosser (Table 180)

Vineyard site adjustments, °C

Min.	Max.	
+1.0	−0.5	Moderate lower slopes, vs (?)valley floor
+0.45	+0.15	Predominantly southern aspects
−0.2	−0.2	Average vine altitude 30 m above recording station
+1.25	−0.55	Mean +0.35°, variability index −3.6

Notes

Two assumptions are implicit in the calculations for Prosser. The first is that the temperature recording site is on the valley floor, which is suggested by the wide diurnal temperature range and the fact that it is for the Irrigated Agriculture Research Centre (Nelson 1968). The second assumption is that Nelson's cited altitude of 256 m is incorrect for the recording site; about 150 m looks more likely on the topographical map. I further assume that vines would be grown on the lower valley-side slopes rather than the valley floor, if only to avoid killing spring frosts.

Ripening mean temperatures are suitable, and sunshine hours more than ample, for medium-bodied white wines from the main white grape varieties up to Maturity Group 4, or possibly 5 on warm sites; or for Burgundy and Loire – style red wines from red varieties of Maturity Groups 3–4. However the temperature variability index is greater than desirable, and relative humidities are probably quite low. It seems doubtful whether the same delicacy and refinement would be obtainable as in the coastal valleys such as the Willamette (see Portland, Oregon, Table 178). Moreover, there is an evident danger of spring frosts after budburst, and (for *vinifera* grape varieties), of winter killing. See also the data for Walla Walla (Table 182), and the analysis of Aney (1974) for the neighbouring State of Oregon.

Seattle (Table 181)

Vineyard site adjustments, °C

Min.	Max.	
–	+1.0	Moderate slopes, further from Puget Sound
+0.3	+0.1	Mainly southern aspects
−0.4	−0.4	Altitude ~ 100 m (Seattle 38 m)
−0.1	+0.7	Mean +0.3°, variability index −1.6

Notes

Temperatures in the Puget Sound area around Seattle are too low for true commercial-scale cultivation of vinifera grapes in the open for winemaking. Adequate ripening of very early (Maturity Group 1) varieties may be attained on favourable south-facing slopes of the lowest foothills. A few specially favoured sites (e.g. sheltered south-facing slopes with stony soils) may just ripen Group 2 varieties in average or warmer seasons. The conclusions based on these calculations agree well with local experience as described by Norton (1979).

Walla Walla (Table 182)

Vineyard site adjustments: None

Notes

The narrower diurnal temperature range and higher mean temperatures at Walla Walla compared with Prosser (Table 180) are due more or less entirely to its higher minima. These can probably be ascribed to Walla Walla's position in direct line with oceanic winds funnelling up the Columbia River. Prosser is occluded from this influence. Nevertheless periodic very low temperatures in winter can still result in a serious risk of winter killing in *vinifera* vines (Aney 1974).

Predicted climatic conditions for the ripening month of early-maturing grapes at Walla Walla show higher than desirable mean and extreme maximum temperatures, and lower than desirable relative humidities. The rapid fall in temperature and some rise in relative humidities through September, however, give ripening conditions that are quite favourable for midseason (Maturity Groups 5 and 6) varieties; these would on average ripen under close to optimum mean temperatures and ample sunshine for medium to full-bodied Bordeaux-style wines.

23

Changes in Climate

The history of European vineyards shows clearly that they must have experienced wide swings of climate in the past, sometimes well within the life-span of a vine. Such swings can only have been of natural origin, and can therefore be expected to continue. Superimposed on them now is the theoretical possibility of progressive climatic warming due to an enhanced 'greenhouse effect', from the carbon dioxide and other gases added to the atmosphere by modern human activities.

Any planning of new vineyards needs to envision a life span for them of at least 50–100 years, if the unavoidably great expenses of vineyard and winery development are to be fully justified. Regions, and sites within them, need therefore to be chosen so as to be sure that they will remain within a commercially optimum range of viticultural climates over the whole period. In the case of natural climatic variation that means effectively the full range of climates experienced locally in recent historical times; and in the case of the greenhouse effect, over what seems foreseeable, on top of natural variation, at least up to the mid to late 21st century.

Greenhouse warming is not necessarily the main problem, unless extreme. Vineyards can be adapted to clearly progressive changes in climate, for instance by grafting over to varieties of earlier or later maturity, and through changes in viticultural and oenological methods, and wine styles. Moreover, increased carbon dioxide may itself offset higher temperatures in determining optimum conditions and localities for viticulture, as will be discussed later in this chapter. Short and medium-term natural changes in climate, on the other hand, have a potential for disruption at both climatic margins of viticulture which seems to be insufficiently realized. They deserve at least as much attention as greenhouse-type warming.

The first part of this chapter looks at historical variation in climate, and its effects on viticulture. The second tackles the still-controversial greenhouse effect. Finally, I will discuss the important question of the direct effects of increased atmospheric carbon dioxide, and its interactions with temperature, sunlight and vine management in determining the best future climatic conditions for grape yield and quality for

winemaking. The results of these discussions, superimposed on the survey of existing viticultural conditions (Chapters 9–22), form the basis for Chapter 24, where I predict the best areas in Australia for viticulture and winemaking up to the mid 21st century.

Past climatic fluctuations, and their effects on viticulture

Chapter 21 described putative effects of climatic change on mediaeval English viticulture. These, and general shifts in Western European agriculture over latitudes and altitudes, led Gribbin (1978) to conclude that average summer temperatures in north-western Europe during the mediaeval 'Little Optimum' of climate were about 1°C above those of the 20th century. The warmth lasted from the 8th or 9th to the early 14th centuries, and was at its peak between about AD 1100 and 1300. After a catastrophic temperature fall in the early 14th century, and temporary partial recovery later in that century and early in the next, there followed a second and greater fall in the early 15th century. This ushered in a period known as the 'Little Ice Age', which lasted from about AD 1430 to 1830. Glaciers advanced everywhere in Europe over this time, but in an irregular manner. Very cold spells were interspersed with long periods which must have been at least as warm as the present.

Temperature appears to have fluctuated wildly in the 15th century, but to have regained some stability at a moderate level through the first half of the 16th. The first really prolonged cold spell of the Little Ice Age then spanned the second half of the 16th century (AD 1550–1600). A second, and perhaps the coldest, developed through the second half of the 17th century and lasted until about the turn of the century. Many records exist of extreme cold in the 1680s and 1690s, when the Thames River froze over repeatedly in winter. The worst time of all in Britain appears to have been 1697–1702, which Trevelyan (1944) describes as 'the dear years of Scottish memory, six consecutive years of disastrous weather when the harvests would not ripen'. Famine and severe depopulation of many of the Scottish upland parishes resulted. On the Continent, Johnson (1989) notes a succession of four particularly disastrous vintages in Bordeaux from 1692 to 1695. Cold returned once more, though less severely, through the second decade of the 19th century: the period of the 'Dickensian Winters'. Temperatures in the coldest decades of the Little Ice Age averaged perhaps 1.0°C below those of the early to mid 20th century, and 2.0°C or more below those in the warmest parts of the Middle Ages. Gribbin (1978) and Lamb (1982, 1984) give more detail.

Reasonably extensive thermometer records of land and marine temperatures exist from 1861 onwards (Folland et al. 1984; Jones et al. 1986a, 1986b, 1986c). Prevailing opinion among climatologists now holds that global temperatures rose by about 0.5°C between 1861 and the mid 1980s (Jones et al. 1986a; Schneider 1989a, 1989b; Jones and Wigley 1990; Kerr 1990). I return to their evidence later in this chapter.

Pfister (1984, 1988) studied the direct effects of climate on viticulture in Central Europe from the High Middle Ages (12th and 13th centuries) to 1860, as evidenced by recorded changes in the dates of vintage and other vine developmental stages. He concluded that harvest in the High Middle Ages on average started about the beginning of September, compared with late September now, and that the growing seasons must have averaged 1.7°C warmer than now.

My calculation method points to a somewhat smaller temperature difference, because it assumes vines to be temperature-saturated when the average daily mean exceeds 19°C. Also, picking in mediaeval times was probably at less complete grape maturities than now, to make the light wines favoured at the time and – perhaps

more importantly – to ensure low enough pH for sound fermentation with primitive winemaking. But even allowing for these factors, Pfister's evidence confirms that the difference in growing season temperatures could hardly have been less than 1.0°C. Maturity differences among grape varieties are unlikely to have contributed, for reasons I touch on later.

Pfister's grape harvest records through the centuries showed generally good medium-term correlation with other records related to temperature, such as the advance and retreat of glaciers, and tree growth ring measurements. One apparent exception is that average vintage dates got later through the 18th and early 19th centuries, despite temperatures which were apparently comparable with those of the previous two centuries. This can, however, be reasonably explained. Bottling of red wines for prolonged cellaring and maturation started in France towards the end of the 18th century (Shand 1964; Johnson 1989), and soon became standard for those of the higher qualities. This required maturer grapes than for making the old styles of pink and light red wines for current consumption. Historically, it may have started with the run of predominantly warm seasons in the late 18th century (see below). At the same time advances in both vine husbandry and winemaking technology were making it feasible to produce and use grapes of greater ripeness and pH, seasonal factors being equal. White wines also needed to be made from other than the earliest-maturing grape varieties, if they were to have enough body and character to benefit from bottle-maturation: see Chapter 7.

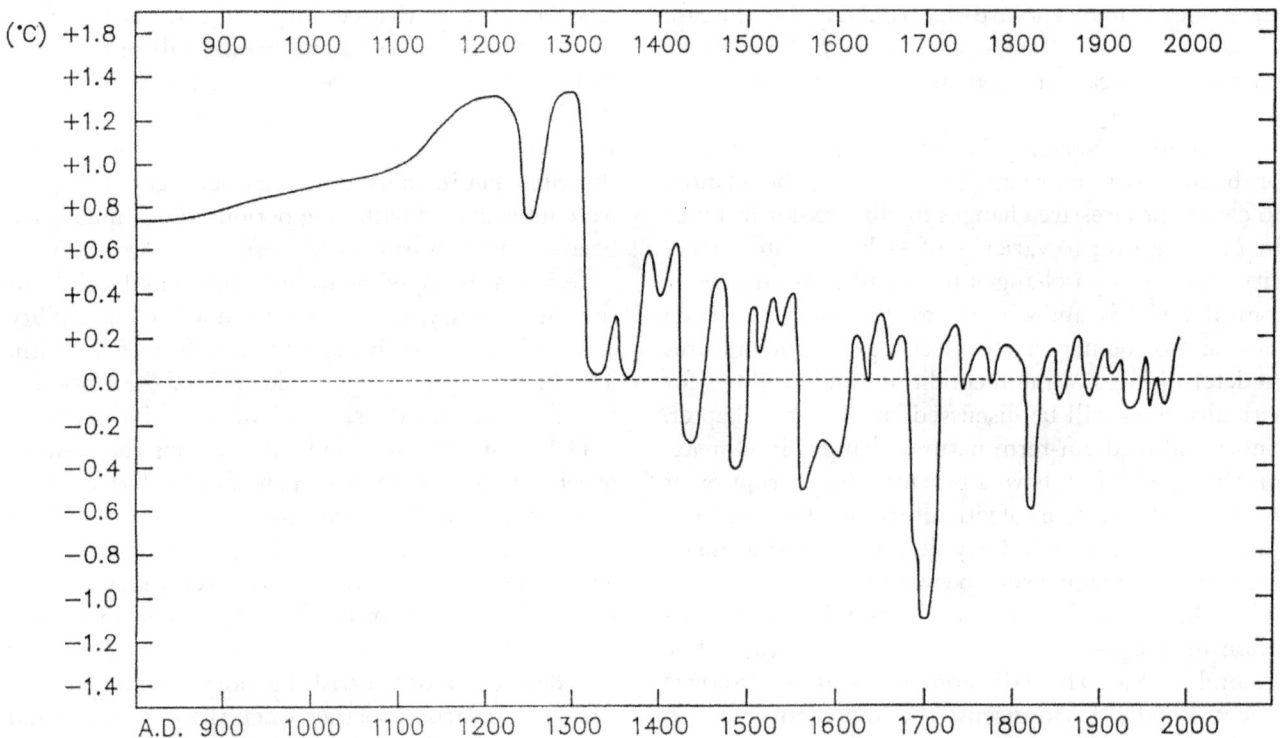

Figure 17. Estimated average mean temperature for the vine growing season in Western Europe AD 800–2000, expressed as deviations in °C from the 1900–1950 average.

These changes could only have affected vintage dates generally after about the mid 18th century, when agricultural improvement started to gain pace.

Figure 17 graphs my estimate of average temperatures for the vine growing season (April-October) in northern and western Europe from the early Middle Ages until now, against a base-line of the early to mid 20th century average. The estimate is based partly on Pfister's data, with allowance for changed winemaking practices after the mid 18th century, and partly on other sources of information as cited above and in the discussion to follow.

We are now in a position to trace in greater detail how the changes in climate influenced European viticulture and the wine trade.

England was not the only region to have had mediaeval vine industries well beyond their present cold limit. Debuigne (1976) states that the Romans established vineyards in what is now Belgium, chiefly along the Meuse and Escaut rivers. These vineyards presumably disappeared during the Dark Ages that followed. Renewed planting started there in the 9th century, around Liège and Huy in the Meuse Valley. The climate of Western Europe was then warming again, after its deterioration from the relative warmth of late Roman times (Ladurie 1972; Lamb 1982). Debuigne notes further mention of Belgian viticulture in the 10th and 11th centuries; while according to Hyams (1987), the period from AD 1150 to 1300 saw increased vineyard planting not only in England, but also in Flanders and Brabant (now parts of Belgium) and Luxembourg. Most of these vineyards disappeared again in the times of famine and general abandonment of northern viticulture which occurred in the early 14th century (Ladurie 1972; Lamb 1982, 1984).

Some of the early northward migrations of viticulture could have been made possible by the progressive selection of earlier-maturing grape varieties and clones, as Hyams (1987) emphasized. This would enable grapes to ripen in cooler and shorter growing seasons than previously, and could well have been a factor in Roman times. However ample other evidence is available – from agriculture and tree rings, and from the recorded advance and retreat of glaciers – to show that mediaeval temperature fluctuations resembling those shown in Figure 17 did in fact occur. Nor is there any evidence, except perhaps very recently in Germany, that present grape varieties are any earlier-maturing than those grown in the Middle Ages. Indeed many of them, such as Riesling, Traminer, the Pinots and Gamay, themselves date from the Middle Ages or even earlier, as Jancis Robinson documents in her admirably detailed book 'Vines, Grapes and Wines' (1986). This tends to confirm that forced viticultural shifts during the Middle Ages and later must have been caused mainly by climatic changes.

The Paris vineyards, having largely survived the climatic collapse of the early 14th century, were ravaged by the still more severe temperature downturn of the early to mid 15th century. Hyams (1987) cites contemporary French records for the period 1405–50, which stated that

> in very many of these forty-five years (around Paris) there was no vintage at all, the burgeoning of the vines being destroyed by sharp frosts in May; or the grapes failing to ripen in summers of continuous rains and low temperatures; or the ripe grapes being destroyed by wet weather in August and September.

The Paris region continued to produce some of its own wines up to the 17th century (Simon 1971; Penning-Rowsell 1976; Hyams 1987), but it was the wines of Orléans, a little to the south, which came to enjoy the greatest reputation towards the end of this period. Francis (1972) recorded that King Henry VIII of England, in the early 16th century, 'used to send his agents to the Rouen Fair, where they bought Orléans wines, famous then and until the 17th century for their strength'. Similarly Debuigne (1976) noted that at the time of King Louis XIII of France (reigned 1610–43), the Orléans wines were compared to those of Bordeaux for their richness and reputation.

Little viticulture remains around Paris, apart from some cultivation of Chasselas for tablegrapes. Orléans, too, retains only a shadow of its former fame, producing a limited amount of light red, or mainly rosé, wines from the early-maturing and frost-avoiding Meunier (see Chapter 16, Table 135). These references seem proof enough that the French climate in the early 16th, and even early 17th, centuries remained warmer than it was to become through most of the 19th and 20th centuries.

André Simon's *The History of Champagne* (1971) throws further light on apparent climatic changes in northern France through the 17th and early 18th centuries.

In the mid 17th century, champagne was a highly esteemed red wine made from Pinot Noir: a little lighter, perhaps, but comparable to the red wines of Burgundy. The first half of the 17th century was in fact a high point for Champagne, with expansion of the plantings and a growing export trade in red wines, competitive with those of Burgundy.

But after about 1660 we hear increasingly of pale pink, and even white, wines from Champagne. This would be consistent with a growing incidence of cold seasons, as might be expected from the general collapse of temperatures in north-western Europe about the time. In the coldest seasons the grapes would doubtless have failed to ripen and colour fully. I suggest that a long run of such seasons, corresponding with the coldest part of the Little Ice Age towards the end of the 17th century, helps to explain how modern 'champagne' came to be invented.

Simon records that in the 1660s the practice started to develop in London of importing new wines from

Champagne in barrel, and bottling them before the end of winter. The Londoners used strong glass bottles with tapered corks, tied down by string, as had already been employed in England for ale for some decades. Enforced late picking of the grapes in cold seasons to get the greatest possible ripeness must often have meant that fermentation was incomplete before it was stopped by the cold of winter. The resulting sweetness would in any case have been desirable to balance the acidity of wines made from only semi-ripe grapes, so there was good reason for early bottling and drinking. But when temperatures rose again in the spring, fermentation re-commenced in the bottles that remained. The result, provided that bottle and string were strong enough, was a frothy, acid but variably sweet (if cloudy) wine, which quickly became the rage of Restoration London.

Neither bottles nor corks were in general use at that time in France. Champagne winemakers were nonetheless quick to recognize and exploit the commercial potential of the new product. Perfection of the (old) champagne process by the French monk Dom Perignon, in the last two decades of the 17th century, consisted in part of finding ways to clarify the base wine of its yeast and yeast nutrients (other than sugar), so that the secondary fermentation after bottling was limited and not too many bottles exploded. It also ensured that the finished product remained somewhat sweet and reasonably clear. The modern champagne process, in which the secondary fermentation depends on a controlled addition of sugar, yeast and yeast nutrients to an already fully fermented base wine, was not perfected until the mid 19th century.

Although sparkling wine production continued in Champagne after the climate warmed again in the early 18th century, Simon (1971) emphasized that non-sparkling red wines remained the region's dominant product throughout that century. Summer conditions stayed relatively warm to the end of the century and for a few years beyond, apart from brief reversions to cold in the 1740s and 1770s.

The return to Little Ice Age conditions during the second and perhaps partly third decades of the 19th century saw a renewed preoccupation with sparkling wines, which now came to be made much more widely. According to Simon, 'sparkling burgundy' was first marketed in 1820, followed quickly by sparkling wines from the Loire, Germany and Switzerland. The industry in Europe's cooler viticultural regions thus again resorted to making sparkling wines towards the end of an abnormally cold period, when many grapes would not have ripened fully. The same pattern remains today in Germany, where failed vintages are used for making 'sekt', and in France on the Loire.

Problems of grape ripening in northern France during the 30 or 40 years up to 1710 help also to account for the quickening of viticultural development about that time in warmer southern regions, such as Bordeaux. Ultimately it was the severe killing frosts

of the 1708/9 winter throughout Europe, just as the climate had started to warm again, which led to the final abandonment of many northern vineyards (Lamb 1982). Their loss explains the 'fury of planting' (Penning-Rowsell 1976) which began at Bordeaux in or about 1709, and continued until 1725 when the authorities started placing restrictions on new plantings there.

The traditional midseason-maturing grape varieties of Bordeaux, and of other parts of southwestern France, nevertheless appear still to have needed individually warm seasons to reach optimum maturity, even after the climatic recovery. This is consistent with a climate cooler than it had been in the Middle Ages (or earlier), when the varieties originally become established there. Penning-Rowsell mentions an unusual number of fine vintages from the 1770s up to the great 'comet' year of 1811, a period which appears to have been among the warmest in late historical times. Occasional very fine vintages still occurred during the predominantly cold period that followed; but they became much more frequent again with the return to warmth of the second quarter of the 19th century, particularly the 1840s, when the estimated average growing season temperatures were once more at a peak (Figure 17). Detailed records of both vintages and temperatures are available from the 1850s onwards, and I take up their story later when discussing the greenhouse effect.

The Mediterranean coastal region of France is still warmer than Bordeaux. It was famed for its red wines in the early 19th century, with those of Bandol being exported on a large scale (Asher 1986). Whether any of this fame was related to low temperatures early in the century is unclear. In any case, it failed to survive the phylloxera devastation of the 1860s, which were a time of renewed high temperatures. The old hillside vineyards were replaced then by mass plantings of inferior high-yielding, late-maturing varieties, grafted on to phylloxera-resistant rootstocks and located on the much more fertile flats (doubtless on economic advice). But as the present study shows, France's Mediterranean coast in the 20th century remains climatically favourable for high-quality table wines, given appropriate grape varieties, wise choice of sites and soils, and modern winemaking technology.

Great Britain's wine trade with France and Portugal further illustrates the southward shift of the industry in western Europe after the Middle Ages. Changes in the wine trade can, of course, often be explained in terms of the political, commercial and dynastic rivalries of given periods, and usually are. Johnson (1989) describes some of them, largely in such terms. Nevertheless an examination of the concurrent climatic changes suggests that these could equally have contributed.

Discriminatory measures against French, and in favour of Portuguese, wines began in England under King Charles II in 1667, and culminated under Queen Anne with the signing of the second Treaty of Methuen in 1703 (see Francis 1972; Robertson 1987; Johnson 1989). The measures span, more or less exactly, what was probably the coldest period of the Little Ice Age (Figure 17).

Portuguese wine at that time was unfortified. Assuming the estimates for the European temperatures in late 17th century to be somewhere near accurate, much of Portugal must then have been climatically close to ideal for making table wines. The contemporary wines of north-western Europe, on the other hand, can at best have been meagre and acid. Indeed, the fact that the Treaty of Methuen followed immediately on what may well have been the climatically most disastrous decade of all in north-western Europe suggests that the problem could have been one of wine availability, as much as of its quality.

The modern styles of sweet, brandy-fortified Portuguese wines did not evolve until the middle of the 18th century, by which time the growing season temperatures had returned to the equivalent of 20th century levels or higher. Coinciding with that development, the English trade in table wines returned more or less exclusively to France and Germany.

The cold second half of the 16th century witnessed a parallel earlier prominence of wines from warm regions: most notably 'sherry sack' from Jerez in southern Spain, and various other sweet and/or fortified wines from the Mediterranean and Canary Islands (Francis 1972). Jeffs (1961) maintains that sack by that time was already a fortified wine, similar to modern sherry. In any case it was high in alcohol. Perhaps Shakespeare was merely reflecting the deficiencies of the contemporary French and German wines, when he had Sir John Falstaff (a notorious expert in such matters) proclaim: 'If I had a thousand sons, the first human principle I would teach them should be, to forswear thin potations, and addict themselves to sack'. Johnson (1989) likewise maintained that Sir John was comparing sack with the wines of Germany and France; but it may now be surmised further that these were more than usually lacking in Shakespeare's time, due to the cold weather that spanned most of his life. The play King Henry IV, Part 2, from which the quotation comes, was written in 1597/98, almost at the end of the cold period.[1]

---

1 The fundamental nature, and far-reaching effects, of this prolonged period of low temperature in north-western Europe is illustrated also by the history of the North Sea herring fishery in Elizabethan times. Trevelyan (1944) cites the antiquary William Camden (1551–1623) as follows. 'The herring had recently moved from the Baltic into the North Sea, and our herring fishery had sprung to importance as a result. 'These herrings', wrote Camden, 'which in the times of our grandfathers swarmed only around Norway, now in our times by the bounty of Providence swim in great shoals round our coasts every year'. Trevelyan goes on to suggest that the southward shift in the herring shoals played an important part in the build-up of England's maritime fleet during Tudor times, with momentous results for later history. Notably, the herrings stayed south, which tends to confirm the continuing downward trend of north-west European temperatures around this time.

The history of German vineyards largely parallels that of Europe's western seaboard. The Romans established vineyards in the Rhine and Mosel Valleys in the 3rd and 4th centuries AD, a period coinciding with the peak of warmth in their era. The German industry expanded still further northwards through the Middle Ages. Evidently surviving the temperature downturn of the early 14th century, it attained its greatest extent in the 15th and early 16th centuries (Hallgarten 1965; Weinhold 1978). Viticulture reached the southern limits of the Baltic States Holstein, Mecklenburg, Pomerania and West and East Prussia. Thuringia and the Electorate of Saxony were flourishing wine provinces. Vineyards ringed towns such as Erfurt and Dresden (Weinhold 1978). Both winter and growing season temperatures must clearly have been warmer than now. It was also a time of unprecedented viticultural prosperity on the Rhine and Mosel (Johnson 1989).

The end of viticulture in far northern Germany came in the second half of the 16th century (Weinhold 1978). Thereafter, the northern limit became very much as at present.

The mostly warm 18th century saw a renewed prosperity in German viticulture. Warmth, increasing removal of vineyards from flat country to the slopes, and a greater concentration on the superior (but relatively late-maturing) Riesling made possible a major improvement in wine quality, and laid the foundation of the German wine industry as we now know it. Riesling was especially appropriate because of its resistance to winter cold. Eighteenth century Europe, despite its generally warm summers, still had many cold winters (Lamb 1982).

A further reconstruction and expansion of the German vineyards started about 1830 (Ray 1979), and coincided with renewed warmth after the years of the Dickensian Winters. This warmth lasted, off and on, until about 1880. The new plantings again emphasized the Riesling, which by mid-century was dominant in all the better vineyards. The reported excellence and foothold gained by German wines from the 1840s onwards in the London market show that Rhine Valley temperatures must have been at least as high as in the 20th century, if not higher.

Temperature changes up to the early 19th century can only have had natural origins, since the global effects of human activities prior to the industrial revolution were at most very small. One of the lessons to be derived from this history is that quite large natural changes – of the order of 1°C, averaged over decades – can become established quickly, and then persist for up to several decades. Such fluctuations are additional to the normal year-to-year variability of seasons, and can have a profound influence on viticulture, particularly at its cold margin. We therefore need now to look briefly at their causes, to give a basis for projecting into future decades.

Causes of natural decadal temperature fluctuations, and their projection to future decades

The literature suggests three main causes of natural temperature changes in geologically recent times.

1. The Milankovich Model

The 'Milankovich Model', named after its proponent Milutin Milankovich, is now generally accepted as explaining the major swings of climate over the last million or so years, such as have been responsible for the comings and goings of the ice ages. The model is based on complementary effects from: 1) changes in the earth's elliptical orbit around the sun, varying over a 100,000 year cycle and causing the heat received by the earth to vary by as much as 30 per cent within the year in times of most elliptical orbit, compared with only 7 per cent at the present time of relatively circular orbit; 2) pendulum-like swings in the tilt of the earth's axis from the vertical to its plane of orbit round the sun, ranging between 21.8 and 24.4° (at present 23.4°) over a 40,000 year cycle, and likewise affecting the climatic contrasts between seasons of the year; and 3) a wobble in the earth's axis of spin, as seen in a child's spinning top, which traces a circle every 22,000 years and controls the timing of the equinoxes and solstices.

The last ice age ended abruptly some 12,000 years ago, and at its coldest had temperatures in the mid latitudes about 5°C cooler than now. The present inter-glacial reached its peak 5000–7000 years ago, with mid-latitude temperatures about 2°C higher than now. Temperatures since then have fallen gradually but irregularly, as shown for the last 1000 years in Figure 17. No writers have confidently predicted an immediate or early return to full ice age conditions, but the record does suggest a strong possibility of periodic returns to 'Little Ice Age' conditions through the coming centuries. For fuller accounts see Gribbin (1978, 1989), and Daly (1989).

2. Fluctuations in the sun's energy output

Short and medium term temperature fluctuations on earth bear an apparent relationship, not yet fully explained, to the incidence of visible 'sunspots' on the solar surface, and to related small fluctuations in solar output. Fairly detailed records of sunspots are available from about AD 1600, with more sketchy records from earlier. The incidence of sunspots varies in a regular cycle of 10–12 years, and there are longer-term fluctuations in the peak intensities of successive cycles (see Gribbin 1978; Foukal 1990). It is these longer-term fluctuations in peak intensities which appear to be important. Periods of low peaks – or of a more or less complete absence of sunspots, as in the 'Maunder Minimum' of 1645–1715 – correspond well with the main cold periods of the historical record.

The warm 11th–13th centuries were a time of prolonged intense sunspot activity, as has been the late 20th century (Foukal 1990). Although recent satellite measurements have shown a variation in solar output of only a fraction of one per cent within the short-term sunspot cycle, some argue that the effects could still be appreciable when accumulated over long periods (Daly 1989).

3. Volcanic dust

Dust from volcanic eruptions is the third postulated factor, and is emphasized particularly by Lamb (1982) as contributing to short-term climatic variation. Major eruptions drive large amounts of minute particles high into the stratosphere. Being beyond where they can be washed out by rain, they can stay in suspension, in diminishing quantities, for up to several years. The resulting 'dust veil' reflects away significant amounts of incoming solar radiation, but is ineffective in trapping outgoing long-wave radiation. It therefore causes cooling in proportion to its intensity. Lamb was able to relate a number of historical cool seasons in the Northern Hemisphere to volcanic eruptions, or clusters of eruptions, in that hemisphere. Similarly, discrepancies between Northern and Southern Hemisphere temperature trends in the early to mid 20th century could be explained by the unusual lack of Northern Hemisphere eruptions over the same period.

Schneider and Mass (1975) combined records of inferred solar output (as related by others to sunspot activity) and Lamb's index of volcanic dust veil. Together with a small but gradually increasing allowance for warming correlated with rising atmospheric carbon dioxide concentrations from 1900 onwards (see below), they successfully simulated all the main climatic changes that are generally believed to have occurred over the last 400 years.

The record of the mid to late 20th century suggests that global temperatures, to the extent that they are controlled by medium-term natural factors, must currently be at or close to a peak. The balance of probabilities is therefore that a fall will follow from the high point of the 1980s: possibly in the remaining years of the present century, or else early in the next. The size of the fall, could well be of the order of 0.2–0.5°C. A fall of 1°C or more cannot be ruled out, but is less likely. Recovery towards recent temperature averages might then follow towards the middle of the century. These changes, however, would be complementary to any which might result from rising concentrations of atmospheric 'greenhouse' gases, to which we now turn.

The greenhouse effect

The nature of the greenhouse effect, and possible consequences of it becoming amplified by the increased release of carbon dioxide and other gases into the atmosphere through human activities, has been much debated. Ramanathan (1988), Tucker (1988), Pittock (1988), Schneider (1989a, 1989b), Hume and Cattle (1990) and the World Meteorological Organization (1990) give popular accounts of what has come to be known as the 'consensus' view among viridomologists.[2] This section summarizes, firstly, what I interpret to be the main conclusions from their work, which has been carried out principally with computer models. Secondly, it deals with criticisms by writers such as Idso (1980), Bryant (1987), Lindzen, cited by Kerr (1989), and Daly (1989), all of whom call into question the assumptions on which the models have been based, and thence the size, and even (for practical purposes) the reality, of the global warming they predict.

That the carbon dioxide and water vapour[3] naturally present in the air help to keep the earth's surface and atmosphere warm has been recognized for many years. Researchers estimate that without them the average surface temperature would be about –18°C (255°K, or 255°C above absolute zero), compared with the more clement +15°C it is now. The reason for the difference is that all gases in the atmosphere are fully transparent to the short-wave radiations received from the very hot sun, whereas some of them are opaque to certain wavelengths of the much longer-wave radiation emitted by the relatively cool earth. The energy of these absorbed wavelengths directly warms the air, and thence, indirectly, the earth's surface. If long-wave energy absorption by greenhouse gases were to increase, air and surface temperatures must theoretically rise until the resulting greater output of long-wave radiation just balances the extra absorption by the greenhouse gases.[4]

Of the natural greenhouses gases, the most important is water vapour, followed by carbon dioxide: see Hume and Cattle (1990). Naturally-occurring methane, nitrous oxide and ozone play a much smaller role, although viridomologists think it will be a growing one if their concentrations in the atmosphere continue to increase at accelerating rates as they have up to now. In addition to these the recently synthesized chlorofluorocarbons, used in refrigeration and pressure packs, are especially efficient heat absorbers and greenhouse gases, and are very persistent in the atmosphere. The computer models of the late 1980s predicted that accumulating chlorofluorocarbons, together with increased production of methane and nitrous oxide, could in future play a combined role equal to that of the increased carbon dioxide derived from accelerated burning of fossil fuels; and that whereas atmospheric carbon dioxide may not itself double until late in the 21st century, depending on trends in fuel use, a total greenhouse effect equivalent to a carbon dioxide doubling might be expected some time before AD 2040 (Pearman 1988; Tucker 1988).

The 'consensus' models of the late 1980s predict a resulting global temperature rise of between 1.5 and 4.5°C by that time, with favoured values between 2.0 and 3.0°C (Tucker 1988; Hume and Cattle 1990; World Meteorological Organization 1990).

Broadly explicit in these models, as summarized by Hume and Cattle (1990), is that they incorporate a direct temperature increase of 1.1°C for every doubling of carbon dioxide concentration or its equivalent in other man-produced greenhouse gases, plus a further 0.6°C from the extra water vapour that would evaporate into the air as a result. The total of 1.7°C is then assumed to be amplified up to three-fold by 'positive feedbacks', acting through the earth's snow, ice and cloud covers.

A first reservation must concern the size of the initial response to increased carbon dioxide. All researchers recognize that the existing carbon dioxide concentration is high enough to absorb nearly all the outgoing radiation in those wavelengths it is capable of absorbing at its present and foreseeable concentrations: to the extent that temperature responses to further increases in concentrations are proportional only to the logarithm of the concentration. That is, a similar absolute increase in temperature results from each successive doubling of carbon dioxide. But most published studies that I have been able to locate have estimated temperature increases per effective doubling of carbon dioxide that are much smaller than has been used in the consensus models.

Rasool and Schneider (1971) estimated it to be only 0.6°C, with a further 0.2°C from the resulting increased water vapour. Idso (1980) arrived at a lower figure still, and pointed out further that the contribution of induced extra water vapour would be less than many had assumed, because the wavelengths it absorbs coincide substantially with those already absorbed by carbon dioxide. He calculated a combined effect of only 0.26°C warming from an initial doubling of atmospheric carbon dioxide or its equivalent. Sellers (1973), Newell and Dopplick (1979) and Webster and Stephens (1980) used a variety of approaches and obtained similar, or even lower, estimates of warming. The subsequent literature has been notably silent on the reasons for the much higher values now adopted in the models.

The roles of the other greenhouse gases (besides the chlorofluorocarbons) are equally in dispute. The consensus models appear to assume that increases in these add directly to the effects of carbon dioxide and water vapour. But Daly (1989) points out that the wavelengths they absorb are already very largely absorbed by carbon dioxide and water vapour. To the extent that they have unique ranges of absorbed wavelengths, their own absorption capacities are nearly saturated in the

---

2 From 'viridomology', a term I have coined elsewhere (Gladstones 1991) to describe scientific studies of the greenhouse effect, particularly those using computer modelling. The root comes from the Latin *viridis* (green), and *domus* (a house).

3 That is, water in the strictly gaseous state. Condensed water, as in clouds, has different effects which must be considered separately.

4 The higher its temperature, the faster a solid or liquid body or gas radiates heat. The rate of energy emission from a perfect 'black body' is mathematically proportional to the fourth power of its surface temperature in °K. Therefore any increase in energy absorption by greenhouse gases can be balanced by a relatively small increase in temperature.

same way as that of carbon dioxide. Further increases in them are therefore likely to have at most only very minor effects on temperature. Methane-belching cows would seem to be exculpated.

Only the chlorofluorocarbons absorb earth-emitted wavelengths that are not already largely absorbed by the natural greenhouse gases. Continued increase of these would undoubtedly cause warming, probably in direct proportion to their content in the atmosphere.

But the chlorofluorocarbons are also suspected of contributing to destruction of the upper atmosphere's ozone layer, which shields the earth from harmful ultraviolet rays. Pending better information, their removal from commercial use is rightly considered to be urgent, and is being successfully pursued worldwide. The projections from early trends of their use, as shown by Pearman (1988), are thus most unlikely to be realized.

What, then, of the 'positive feedbacks', which most models say will amplify any primary warming? This remains an area of extreme uncertainty. The range from 1.5 to 4.5°C in the model predictions for global warming by AD 2040 stems more or less entirely from differences in the degree – and indeed, the sign – of the amplification (Cess et al. 1989).

The main source of uncertainty is the effect of clouds, and that of any primary warming on their incidence. Early predictions of Webster and Stephens (1980), now confirmed by direct satellite measurements (Ramanathan et al. 1989), showed clouds to have a strong net cooling effect. That is, they reflect away more incoming short-wave radiation energy during the day than they capture from outgoing long-wave radiation during the day and night.

For changes in cloudiness to act as a positive feedback, initial warming by greenhouse gases must therefore reduce effective cloud cover. That is likely enough over the dry continents; but it seems much less probable over the oceans, which cover over 70 per cent of the earth's surface. It has also to be reconciled with the simultaneous model predictions of global increases in rainfall. Estimated rainfall increases from a carbon dioxide doubling, or its equivalent, are 3 to 11 per cent (Tucker 1988) and 7 to 15 per cent (Hume and Cattle 1990).

A positive feedback through the melting of snow, resulting in less surface reflection of solar radiation back to space, is often mentioned; but it appears to be small and localized compared with that thought to be acting through cloud cover (Ramanathan 1988). In any case that, too, could be negated to the extent that more evaporation from the oceans leads to more total snowfall on the land and polar icecaps.

Daly (1989) and Lindzen, cited by Kerr (1989) argue that the mainstream models neglect a further factor which could be very important. Direct heating of the oceans is balanced (at their equilibrium temperatures) largely by evaporative heat loss, in the form of latent heat of vaporization. This latent heat reappears later as sensible heat in the middle and upper atmosphere, as water vapour condenses to form cloud. The heat released fuels further convection into the upper atmosphere, from which it can radiate away much less hindered by atmospheric greenhouse gases than from the earth's surface and lower atmosphere. Any surface warming intensifies this process, which thus constitutes a *negative* feedback. A reduction in the final net warming results.

Extra heat release could also maintain more water droplets against freezing and precipitation, resulting in slower cloud dissipation and creating a further strong negative feedback (Mitchell et al. 1989; Slingo 1989).[5]

In summary, the assumptions in the current viridomological models – both of primary warming by greenhouse gases and of its subsequent amplification by positive feedbacks acting through the moisture cycle – are questionable, and represent, at best, a 'worst case' scenario. It seems just as likely (indeed, intuitively more likely) that the moisture cycle acts overall to moderate primary temperature changes, not to amplify them. Figures of a few tenths of a degree over the next half century, as estimated or implied by earlier researchers such as Sellers (1973), Newell and Dopplick (1979), Webster and Stephens (1980), Idso (1980) and Lindzen, cited by Kerr (1989), may well be more realistic.

Have temperatures risen already?
Leaving aside the sulphur dioxide hypothesis of Wigley (1989, 1991), just referred to, a direct test of greenhouse warming might be found in the temperature record since extensive thermometer measurements began in the mid 19th century. According to the consensus models, the known increases in greenhouse gases since then should

---

5 Wigley (1989, 1991), on the other hand, suggests that cloudiness may have increased due to extra condensation nuclei, resulting from the emission of sulphur dioxide along with the carbon dioxide and other greenhouse gases produced by human activities. An implication is that any failure of greenhouse warming to appear (see following section) can be thus explained without violating the existing mainstream model assumptions. It also follows that if, for good reason (such as preventing acid rain) it should in time become practicable to remove most sulphur dioxide from the emissions, with carbon dioxide emissions continuing unabated, hitherto masked greenhouse warming will be unleashed in full. Hypotheses such as this cannot, at this stage, be proved one way or the other. However two comments seem relevant. First, it is hard to envisage world-wide application of the necessary technology, if perfected, for at least some decades to come, if ever. Second, if the hypothesis is correct, might technology be able in time to devise alternative ways to maintain the increased cloudiness, with less noxious side effects?

have warmed the earth by between 0.5 and 1.2°C, with a most likely global average of about 1.0°C (Ramanathan 1988; Schneider 1989a, 1989b). This figure allows for a substantial heating lag due to the great heat absorption and storage capacity of the oceans.

The most detailed estimates of recent global temperatures are those of Jones et al. (1986a, 1986b, 1986c), of the University of East Anglia. They concluded that global temperatures have in fact risen by about 0.5°C since 1861. This estimate is now widely accepted as authoritative, and if not actually confirming the computer predictions of global warming, at least to be consistent with the more conservative of them (Jones 1990).

The reality of this apparent warming deserves to be examined more closely.

It has long been known that urban centres create artificial 'heat islands', as measured by the means of their maximum and minimum temperatures, largely through a rise in their minima. This is caused by buildings and pavements absorbing and storing solar heat during the day, and re-radiating it at night, as described in Chapter 4 for rocky soils. Any reduction in maximum temperature due to this absorption is normally insufficient to compensate, so that the mean temperature rises.

The bigger the urban centre, the greater and more intense is its heat island (Karl et al. 1988). Even small towns record significantly higher mean temperatures than the surrounding countryside. Therefore any record which does not allow fully for urban growth around the recording stations will show a spurious rising trend in recorded mean temperature with time. Almost by definition, most of the long-term records come from centres of which the sizes and densities of housing and roads have grown over time: often many-fold. Increasingly, in modern times, there is probably also an effect on real temperatures of urban energy inputs for domestic and industrial uses.

Kukla et al. (1986) examined the differences among North American towns and cities in their mid 20th century trends of recorded temperature with time, and reviewed the results of similar studies from other countries. They concluded that the world's cities had warmed, relative to small rural towns, at an average rate of 0.12°C per decade. A more detailed study of eight city/rural pairs in North America over 1951–1980, with particular care to eliminate factors such as change in thermometer location, showed an average divergence of 0.34°C per decade. A similar study in Australia (Coughlan 1978) showed a divergence since World War II of no less than 0.45°C per decade between the big cities and adjacent country centres. The two latter figures doubtless reflect the rapid post-war growth of North American and Australian cities, and would not be typical of the whole record. Nevertheless, even if one accepts only the most conservative of urban warming estimates, it is clear that the potential biases in the

temperature record are large and growing. And if the discrepancy between large and small centres is great, it must arguably be still greater relative to the true countryside, which constitutes the overwhelming majority of the land surface, and is, of course, the only part that is relevant to present and future commercial viticulture and to the global climate as a whole.

When preparing their estimates of Northern Hemisphere land temperature trends for 1851 to 1984, Jones et al. (1986b) rejected urban warming as a major contributor, and eliminated only 38 out of around 2000 recording sites on that ground. Clearly they could only have removed the most extreme cases. Considerable urban bias must therefore remain in their 'acceptable' records.

A way of testing this would be to look only at recording sites which have remained more or less completely rural. Dronia (1967, cited by Jones et al. 1986b) did this for 163 stations around the world, and estimated a net global *cooling* of 0.11°C between the 1870s and the 1950s. Jones et al. rejected Dronia's work as having 'a number of inconsistencies and methodological deficiencies', but did not elaborate on them.

Another approach is that of Folland et al. (1984), who traced records of sea surface and marine air temperatures between 1856 and 1981. This approach by-passes spurious urban effects, and because of the ocean's temperature inertia, has the further advantage of smoothing out short-term fluctuations.

Problems of changes over time in the methods of marine temperature measurement had to be overcome. For instance, the buckets used before World War II for getting water samples were mostly made of uninsulated canvas. This can depress the recorded temperature by an estimated 0.3–0.7°C, due to evaporative cooling before the reading is taken. Parallel comparisons of the recorded sea surface versus marine air temperatures before and after the war, when insulated buckets or engine-intake measurements were introduced, suggested that the pre-war water temperatures had on average been depressed by 0.3°C. Folland et al. therefore added that amount to all sea surface records up to 1940. They also corrected the marine air temperatures for changes in recording height above sea level, due to the steady growth of ship size. Finally, these authors confined their air temperature records to night readings, because of likely interference by daytime heat reflection from the canvas rigging of the early ships. After allowing for all these factors, they concluded that average marine temperatures in the mid to late 20th century were the same as in the mid 19th century. In between there had been a progressive fall after about 1880, to a minimum about 0.5°C lower in 1904–12. This was followed by recovery to 1940, after which temperatures underwent only minor fluctuations until 1981, apart from a brief peak in 1942–45.

Jones et al. (1986a) argued that the corrections used by Folland et al. were too 'fraught with uncertainty' to be accepted. They chose instead to use the raw marine temperature data, including daytime air readings, and to adjust them by calibrating against nearby land stations. This resulted in all pre–1900 marine air temperatures being adjusted downwards by approximately 0.4°C, to a level close to those of the uncorrected sea surface temperatures and the recorded land temperatures; and those of 1903–41, upwards by 0.17°C. Such, together with the recorded land temperatures themselves, with minimal adjustment for urban warming, was the basis for their conclusion that global temperatures rose by 0.5°C between 1861 and 1984.

But what if the marine air and sea surface temperatures, as logically adjusted by Folland et al. (1984), were really close to correct? A strong case can be mounted to say that this is so, and that more probably the calibration of the marine to the recorded land temperatures, as carried out by Jones et al. (1986a), served only to incorporate the remaining urban bias of their accepted land records into the marine records.

Nor is it only the later land records that may have been at fault. Being most typically associated with towns and cities, a disproportionate number of the recording sites would be in valleys or on flat land. To that extent, and discounting any urban warming effects, their average minimum and (to a lesser extent) mean temperatures would initially have been lower than the overall average for the land, for mesoclimatic reasons (see Chapter 4). Urban warming would then, at first, only have tended to correct a natural site bias. Jones et al. might properly have raised the early recorded land temperatures to match the adjusted marine temperatures, rather than vice-versa.

The combined evidence thus strongly suggests that the supposed rise in temperature since the mid 19th century is no more than an artifact of the average siting of the temperature recording stations on land: the temperature-depressing effects of mainly flat and valley-floor sites being increasingly offset, and finally outweighed, by urban growth and specifically urban warming around them.

A last avenue of evidence may now be examined, to see whether the supposed rise in global temperatures over the last 130 years has been real.

Grapevines as a thermometer

Grapevines provide a unique record of past environments, for the reason that the same varieties are grown on the same sites for many decades. In some famous areas they have been so for centuries. With some reservations as to changing wine styles and wine-making technology, as discussed earlier in this chapter, their times of ripeness are historically well recorded through the dates of beginning of vintage. We now also have ways by which to translate the ripening dates of given grape varieties backwards, with fair accuracy, into average temperatures for the growing season (Chapters 6–8).

The grapevine is most sensitive as a biological thermometer at the cool fringe of viticulture, and most of all in cool maritime climates where mean temperatures remain below 19°C for all months of the growing season, as discussed for Tasmania (Chapter 14) and Oregon (Chapter 22). A rise of 0.5°C should advance ripening in such climates by the equivalent of two vine maturity groups. The difference would be perhaps 1.5 maturity groups in continental cool climates, and less again in climates with progressively hotter mid-summers.

What, then, does the viticultural record show? The vineyards of Germany are an especially sensitive indicator, being at the extreme cool limit for ripening high-quality grape varieties. The Riesling (Maturity Group 4) is the latest-maturing of them grown there, and at least in the late 20th century, can be relied on to ripen only on the warmest and sunniest slopes. Its distribution and performance are therefore a direct measure of climate.

According to Ray (1979), the Riesling was recorded as a German wine-producing grape at least as early as the 15th century. 'By the end of the eighteenth century it was certainly the most important variety in the country, and by the middle of the nineteenth probably the most widely grown'. It was the quality spearhead in the reconstruction and expansion of the German vineyards that started in the 1830s. The climate then must therefore have been at least as warm as in the mid 20th century, if not warmer.

German plantings towards the end of the 19th century were in apparent sympathy with the recorded global cooling which started in the late 1870s (Figure 17). The very early-maturing Müller-Thurgau was launched opportunely in 1883 (Robinson 1986), at a time when the Riesling must have become harder to ripen. Nevertheless Müller-Thurgau was not extensively planted until the present century.

German breeding has continued largely to aim for varieties earlier maturing than Riesling, with further new, very early-maturing varieties such as Bacchus gaining great popularity in recent years. This, admittedly, is partly because they enable viticulture to expand to climatically inferior sites where Riesling could never ripen reliably. Also, such varieties probably allow ripening at higher yield levels for the ordinary commercial grades of wine. At the same time, however, the practice of chaptalization (adding sugar to the fermenting musts, to compensate for lack of natural grape sugars) has become increasingly standard in Germany through the 20th century, for all but the best wines and years.

A rise in the growing season average mean temperature of 0.5°C in Germany should have allowed the growing of grape varieties on any site which are later-maturing by about 1.5 maturity groups compared with the past. Alternatively, the same varieties on their old sites should now ripen, on average, 12–15

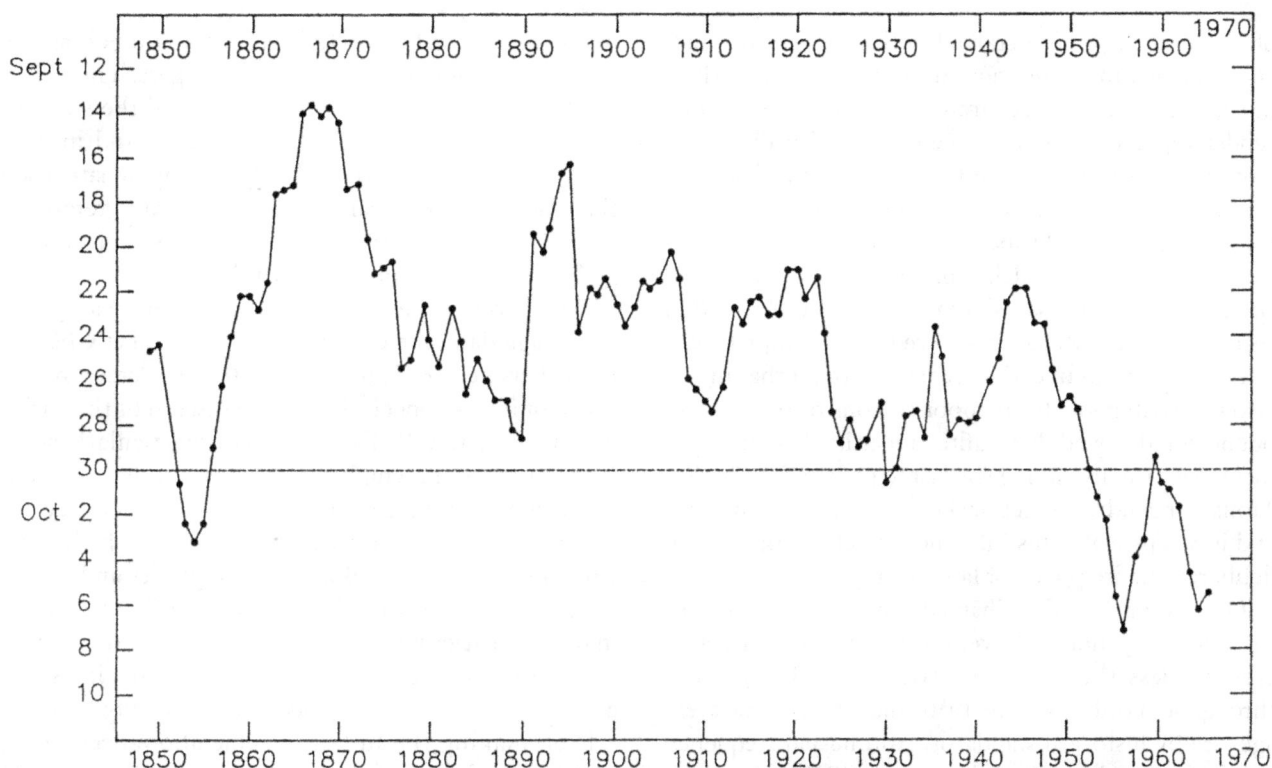

Figure 18. Dates of start of vintage at Château Lafite, Bordeaux, France. Five-year running averages. Based on data of Ray (1968).

days earlier, or to much higher sugar levels. Nothing in the published reports suggests that either of these things has happened: even in the best vineyards, where a jealous regard to quality ought to guarantee that yield levels have not been allowed to increase to the point of delaying maturity. In any case, German grape yields generally did not start to rise markedly until after 1950 (Halliday 1991b).

The German viticultural record of the last 150 years thus gives no evidence of any overall warming of the growing season. If anything, it points to a cooling.

We can now consider Bordeaux over the same period. Two published accounts are of special value here: that of Ray (1968), which includes records of the commencement dates and qualities of the vintages at Château Lafite from 1847 to 1967, and the more general comments of Penning-Rowsell (1976) on the Bordeaux vintages from the late 18th century onwards.

Figure 18 shows the trends with time in the dates of first picking at Château Lafite, expressed as five-year running averages calculated from Ray's data. That is, the point for 1849 is the average for 1847–1851 inclusive, that for 1850 the average for 1848–1852 inclusive, and so on. This method irons out much of the individual year variation, and so highlights medium and long-term trends. That done, we have a continuous record from a single site, growing at least largely the same grape varieties with largely unchanged vineyard management, and with a style of wine which from all accounts has remained essentially consistent for

200 years.[6] Some single-year events nevertheless had enough influence to alter the shape of the graph, and therefore need to be explained.

The sharp dip in the graph in the early to mid 1850s, following a series of fine vintages through the 1840s (Penning-Rowsell 1976), was due in part to the arrival of the disease powdery mildew, or oidium. Control by sulphur was unknown at first, and the 1853 to 1857 vintages were all affected. Temperature records (Figure 17) show that this was also a temporarily cool period in the Northern Hemisphere, so the late vintages were probably caused by a combination of both disease and climate.

The peak in the graph between 1891 and 1895 was due almost entirely to the extraordinarily early vintage

---

6 The modern vineyard composition at Château Lafite is approximately two-thirds Cabernet Sauvignon and one-sixth each of Cabernet Franc and Merlot, which is fairly typical for the Médoc. Merlot is the earliest-maturing of the three, and, as would be standard in the area, is normally the first picked. Malbec (syn. Côt in Bordeaux) was also once included as a fairly major variety, as in many other Médoc vineyards. Johnson (1989) notes that the Malbec was prominent at both Lafite and Latour at the end of the 18th century, and that the oidium (powdery mildew) plague of the 1850s favoured replanting with Malbec and Cabernet rather than Merlot. However Malbec has since greatly diminished, probably due to exacerbation of its coulure problems by grafting on to rootstocks following phylloxera. Malbec is perhaps marginally earlier maturing than Merlot (see Footnote 3 to Table 5, Chapter 7), and so might have advanced the start of picking in the earlier years by up to a few days where present in quantity.

of 1893, when picking started both at Château Lafite and generally in the district on August 14 – the earliest in some 200 years of accurate Bordeaux records. The sudden dip in the graph between 1908 and 1912 is similarly associated with the single, very late vintage of 1910.

Finally, the introduction of better fungicides and generally improved husbandry from about 1950 meant that harvest could be delayed, and fuller grape maturity attained, with less risk than previously. Ray (1968) records that many chateaux were tending to pick later in the 1960s than in earlier years, although that might also be attributable to the poor seasons of the decade; but he noted as well that Lafite, under its then management, was usually among the last to pick in the district. Monsieur André Portet arrived as manager in 1955, and it was probably his influence which prompted the deliberate Lafite policy of late picking.

The last part of the Château Lafite graph is therefore probably not fully representative of Bordeaux. Nevertheless there can be little doubt that picking throughout Bordeaux in the 1950s and 1960s was on average late by historical standards. An unusual frequency of very late, and mostly bad, vintages (1951, 1954, 1956, 1958, 1963, 1965 and 1968) guaranteed this. Average dates of equivalent grape maturity through this period must have been at least as late as in the equally disastrous run of vintages through the late 1920s and 1930s.

How, then, do the vintage dates match the medium-term fluctuations in recorded temperature? Mostly, they do so very well. The earliness of the vintages from the late 1850s to the mid 1870s corresponds exactly with the recorded peak in both land (Jones et al. 1986b) and marine (Folland et al. 1984) temperatures in the Northern Hemisphere generally. The progressively later vintages through the 1870s to 1890, then relatively early vintages to 1906, follow the globally-recorded fluctuations in land temperatures closely. The marine temperatures continued to fall and were less closely related to vintage dates after 1890; but they, too, showed a minor rise about the turn of the century.

The Bordeaux vintages of the 1880s had a poor reputation, generally attributed to the effects of young, grafted vines following the devastation by phylloxera in the previous decade. This undoubtedly contributed; but it seems equally likely that the coincident fall in average temperatures (Figure 17) played a part. With the return to warmth in the 1890s and first few years of the 20th century, there came once again a number of very good vintages.

The late vintages of the late 1920s and 1930s were anomalous, in that both land and sea temperature records at the time showed little more than a pause in their generally upward trend between 1910 and 1945. Vineyard neglect because of depressed economic conditions might have contributed. However the great frequency of reported vintage failures, right through to 1941 (Penning-Rowsell 1976), suggests that the reason was at least partly climatic; true average temperatures in Bordeaux could hardly have risen as much as the global thermometer records might suggest. Thereafter to 1967, the vintage dates again followed the short to medium-term fluctuations in both land and marine temperature records very closely. It may be surmised that the marked upward trend in recorded temperatures since the mid 1970s (Figure 17) has been accompanied by correspondingly early vintages.

Disregarding the short and medium-term variations in vintage date, the clear underlying trend at Château Lafite between 1847 and 1967 was towards later vintage. Some of it was conceivably due to discard of the early-maturing variety Malbec, and some, particularly since 1955, to changing emphases in vineyard management and winemaking. Against this, the present century has seen the substantial discard of Petit Verdot, which is the latest-maturing of the Bordeaux red grapes and is now regarded as a marginal ripener (Dubourdieu 1990); yet in past centuries it was one of the region's most important varieties, and in the mid 17th century its wines commanded the highest prices (Johnson 1989).

Even making the fullest possible allowances, then, nothing can be seen in the Bordeaux record to support any notion of climatic warming over the last century and a half. A 0.5°C rise in temperature – assuming it to be fully reflected in the April-October growing season for vines – should in this climate have advanced ripening by 5–6 days, or by the equivalent of one maturity group. Ripening, especially in the later years with improved fungicides, should have become much safer. The evidence to 1967 suggests neither of these things. If anything, it suggests the opposite.

We may finally look at the data for 1959 to 1990 at Fort-Chabrol, in the Champagne region of France (Panigai and Langellier 1991). May to September temperatures as recorded there showed no trend over the period, and there was a more or less even distribution of warm and cool seasons. Even for making champagne, the seasons giving the best vintages remained invariably the warmest, driest and sunniest. This suggests that the region must now be in one of its cooler climatic phases, and certainly cooler than in its red wine heyday of the mid 17th and 18th centuries and earlier, given that the grape varieties (Pinot Noir and Meunier) are the same and the potential for disease control late in the season, greater. Nor do the seasons show any sign of unusual warmth even in the 1980s.

In summary, the viticultural experience of Germany, and of Bordeaux and Champagne in France, gives no support to the thesis that global warming has taken place since the mid 19th century. There may even have been a slight cooling.

Temperature forecast for the 21st century

We can now assess the probabilities for net temperature change through the rest of the 20th century, and perhaps the first half of the 21st century.

Inspection of Figure 17 suggests, as noted earlier,

that temperatures by 1990 were at or close to a peak in the natural, medium-term fluctuations that have characterized the record for the last 200 years. The greatest immediate likelihood, allowing for a small incremental greenhouse warming, is therefore for global temperatures to fall again within the next decade or so, by perhaps 0.1–0.4°C. A less likely or immediate prospect, but one which remains possible on the basis of past climates, is for a reversion at any time to 'Little Ice Age' conditions, with decadal averages anything up to 1.0°C below the average of early to mid 20th century temperatures, or 1.5°C below that of the 1980s. If this happens, the cold conditions might persist for up to several decades.

Underlying all naturally-caused fluctuations will be any increase in greenhouse warming. The mainstream viridomological models favour a rise of 2 to 4°C by AD 2040, but as discussed earlier, a more realistic estimate is 0.5°C or less, and certainly not much more than 1.0°C. The effects of such a rise would not be detectable above the background of natural variation in the early decades, but could become so towards mid-century.

Taking both sources of variation into account, the prospect is that global average temperatures for the next 60 years will remain close to those of the late 20th century, perhaps within a range of ±0.5°C, and very probably within the range of ±1.0°C. The greatest likelihood is for a slight or moderate fall in the early part of the period, followed by recovery and an increasing, but still only moderate, greenhouse warming by mid century.

But estimates of future temperature change, even if accurate, still do not enable us to plan vine plantings for the 21st century with complete confidence. Some possibility remains of wider temperature variation, in either direction. We must also first consider the direct effect, on vines, of changed atmospheric carbon dioxide concentration in its own right.

Direct effects of carbon dioxide concentration on vine growth, grape quality and optimum temperatures

One forecast is beyond doubt. Atmospheric carbon dioxide concentration will continue to rise for at least some decades to come.

The rate of increase will depend on many things, including especially the economics of fossil fuels and their alternatives. Questions remain over the dynamics of carbon dioxide disappearance into the oceans and other sinks, since it is reckoned that little more than half of the man-caused extra release into the atmosphere since the industrial revolution remains there (Pearman 1988). Nevertheless, even the more conservative of Pearman's projections show a near-doubling of carbon dioxide from its present levels by the end of the next century, compared with only a 25 per cent rise from pre-industrial levels to now. And

since carbon dioxide is the main basic input (together with water, nitrogen and nutrient minerals) into plant growth, such a drastic change cannot fail to have big implications for all forms of plant production.

Photosynthesis and growth, by well nourished and watered plants, can be limited by any or all of three main environmental elements:

- Temperature.
- Light (intensity, spectral quality, and duration of leaf illumination).
- Atmospheric carbon dioxide concentration.

Any one of them can be the strongest, or even sole, limiting factor. For instance, lack of light obviously limits the growth of plants kept in the dark, but under otherwise favourable growing conditions. Light is then the 'primary limiting factor'.

Carbon dioxide, together with light, is essential for photosynthesis. It is the source of the carbon which, with the aid of sunlight energy, combines with water in plant leaves to form sugar, which is the building block and energy source for all plant growth and yield. With ample sunlight and good leaf exposure to it, atmospheric carbon dioxide concentration is normally the primary limitation to photosynthesis. Reduced light, including that within the plant canopy due to leaf crowding, can cause light to become co-limiting. But as stressed by Mortensen (1987), carbon dioxide concentration is now known to play a limiting role even under poor light conditions. More carbon dioxide improves the efficiency with which any available light is used for photosynthesis. In atmosphere-controlled commercial glasshouse horticulture, the optimum carbon dioxide concentration appears to be between two and three times that currently in the external atmosphere (Mortensen, op. cit.).

Temperature does not greatly influence photosynthesis itself, but is important because it controls the speeds with which the immediate products of photosynthesis can diffuse away from their sites of formation and be used for growth and ripening (Chapter 3). Temperatures below the optimum for growth, in the presence of ample light and leaf exposure, lead to a 'banking up' of starch in the leaves and sugar throughout the plant as Acock et al. (1990) have demonstrated in rockmelons. That within the leaves results, in turn, in a biochemical 'back-pressure' which tends to inhibit further photosynthesis.

Inhibition of photosynthesis by surplus starch or sugar at low temperature is reduced by the presence of alternative and readily accessible sinks for the sugar. Direct movement of sugar to the fruit (especially, perhaps, from the nearest leaves) takes place easily, and is less inhibited by sub-optimal temperatures than vegetative growth. Thus sugaring and concurrent photosynthesis can continue relatively unaffected by low temperatures during ripening if there is an adequate crop load.

But, as we also saw in Chapter 3, the enzymic chemical processes of ripening, by which part of that sugar is transformed into pigments, flavours and aromas, are themselves almost certainly inhibited by sub-optimal temperature, in much the same way as those involved in vegetative growth and phenological development. It follows that too low a ripening temperature tips the balance towards simple sugar accumulation in the berries, rather than physiological or flavour ripening, and as a result reduces flavour and colour in grapes of equivalent sugar content, and hence in wines of equivalent alcoholic strength.

These balances are important in the context of what may be expected under rising atmospheric carbon dioxide concentrations.

More carbon dioxide leads to potentially greater production and accumulation of sugar throughout the vine and in the berries. To repeat, other things being equal, more carbon dioxide leads to more photosynthesis and daytime accumulation of sugar or starch in the leaves. But that is advantageous only to the extent that the starch can later be mobilized and the sugar transported to be used productively for growth or transport to the ripening fruit where that is desirable. But it cannot contribute to either if low temperature inhibits starch mobilization and sugar transport out of the leaves, or use of the sugar for growth or enzymic conversion into flavour and aroma compounds in the ripening berries. As Acock et al. (1990) demonstrated, this can only be remedied by an appropriate increase in temperature to accelerate the non-photsynthetic growth and ripening processes. Thus if atmospheric carbon dioxide continues to rise, which it assuredly will for some time to come, there will need to be accompanying rises in temperature if its benefits are to be realized. This has in fact long been known in practice for a wide variety of horticultural crops and flower species in experimental and commercial greenhouse culture (Mortensen 1987; Wittwer 1990).

One can argue still further, on theoretical grounds, that such responses will be greatest where the higher mean temperatures are due mainly to raised night temperatures. Partly that is because excessive day temperatures can cause heat injury to the vine and enhanced loss of volatiles from the ripening berries (Chapter 3). Largely, however, it is because the benefits of enhanced sugar production in the vine – whether for growth or yield, or for the formation and preservation of pigment, flavour and aroma compounds in the berries – can only be fully expressed when night temperatures are high enough, since that is the time low temperature is most most likely to limit metabolism of any kind.

Evidence also exists that a higher atmospheric carbon dioxide concentration can directly enhance the resistance of grape vines to injury by high temperatures, at least during vegetative growth (Kriedemann et al. 1976). This should itself benefit fruit quality, partly because of less defoliation under heat and moisture stress, and partly through other mechanisms.

Enhanced carbon dioxide concentration has a particular benefit for grape quality in arid climates and at high altitudes. More will pass into plant leaves through the open stomata for each unit of water that evaporates out (see discussions of relative humidity, Chapter 3, and altitude, Chapter 4). This means in turn that more dry matter is produced per unit of water transpired, as Downton et al. (1980), Wong (1980), Morison and Gifford (1984a, 1984b) and many others have shown for a variety of plants grown in high carbon dioxide atmospheres. Because leaves then produce more sugar per unit of potassium carried into them from the soil, the ratio of potassium to sugar later passing to the ripening fruit can be expected to fall. Lower must and wine pH result. High atmospheric carbon dioxide concentration in theory exactly negates low atmospheric relative humidity and high altitude in this respect.

On the other hand other vine nutrients will also be diluted, including nitrogen. As atmospheric carbon dioxide rises, more attention will be needed to ensuring that these are in adequate supply. This may be by application to the soil; through irrigation water; or, for certain trace elements, by foliar application at appropriate growth or fruit development stages, as described in Chapter 5. As at present, the need for such enhanced nutrient supply will be least in atmospherically arid climates, and greatest in climates where high relative humidities already maximize dry matter production per unit of water use and nutrient uptake (see, for instance, Mortensen 1987; Gislerod and Mortensen 1990; Conroy et al. 1990). Importantly, the same considerations apply to the contents of nitrogen, and perhaps other essential yeast nutrients, in fermenting musts. The other side of the coin is that the disadvantages of fertile soils for vine balance, and hence for grape and wine quality, will be correspondingly reduced, provided that vine trellising and canopy light relations are adequate.

Two final consequences of rising atmospheric carbon dioxide concentrations may be noted.

First, greater vine fruitfulness may in time be needed if the potential for increased yield and/or fruit quality is to be fully exploited. Lack of adequate fruit 'sink capacity', or crop load, can already inhibit photosynthesis in unfruitful grape varieties. A fortiori, it could prevent increases in photosynthesis and yield at still higher carbon dioxide concentrations. This has been directly demonstrated in other crops, including soybeans (Clough et al. 1981) and Valencia oranges (Downton et al. 1987). More carbon dioxide could even be counter-productive, if the main effect was to stimulate unwanted extra vegetative growth.

On the other hand an increase in photosynthesis may partly or wholly achieve the desired result automatically in grapevines, through improved bud fruitfulness. This will especially be so if temperatures also rise. But genetic selection for greater fruitfulness in high-quality varieties could still be needed for cool and mild maritime climates.

The final consequence is that rising atmospheric carbon dioxide concentrations may well improve the natural resistance of vines (and other plants) to fungal diseases, and possibly some pests, as well as to heat. Resistance mechanisms in plants typically involve the production of substances toxic to the invading organism. Like other secondary metabolites, their production requires the presence of 'surplus' sugar as a substrate. Ample sugar export from plant tops is also needed for the strong production by roots of cytokinins, which possibly play both direct and indirect roles in disease resistance (see Chapter 3, and Appendix 1). Given enough mineral nutrients to avoid deficiencies, the extra carbon dioxide should enhance both these processes.

Viticulture in foreseeable future climates

All the above evidence indicates that increased atmospheric carbon dioxide, given appropriate vine husbandry, will be more or less wholly beneficial to grape yield and to grape and wine quality. Greater yields will be possible without over-cropping and loss of fruit quality. Optimum temperatures for both yield and quality will be higher.

More carbon dioxide will help to offset low relative humidity and high altitude. Nevertheless low areas having moderately high relative humidities will probably retain their advantage.

The optimum temperatures for individual wine styles may rise with increases in atmospheric carbon dioxide concentration, though not as much as for vine growth, yield and absolute fruit and wine quality. The combined result could in time be a tendency for consumer esteem to shift (though not dramatically) from cool towards warmer climate wines.

Summing up the arguments, any changes in the future best locations for commercial viticulture and wine production that are likely to result from future climatic changes will hinge on two factors:

1. The relative sizes and directions of climatic changes resulting from natural temperature fluctuations and any greenhouse warming respectively; and
2. How much the greenhouse warming results directly from increased atmospheric carbon dioxide, and how much from other greenhouse gases.

If global temperatures rise by 2–4°C by mid 21st century, as viridomologists have forecast, and only half of the rise is due to carbon dioxide, any upward shift in the optimum temperatures for yield and quality may not be enough to match the actual temperature rise. Some migration of the commercially optimum areas will then occur towards cooler regions, although it will be less dramatic than writers such as Dry (1988), Boag et al. (1988, 1989) and Smart (1989) have suggested on the basis of previously forecast temperatures alone.

An intermediate scenario is that carbon dioxide will increase as forecast, and temperature with it in accord with the consensus model assumptions; but that the chlorofluorocarbons will be successfully phased out, and other greenhouse gases perhaps prove to be less influential than at first thought. A smaller temperature rise will then be due predominantly to carbon dioxide. The rise in optimum temperatures could in that case easily match the actual rise in temperature, resulting in no change in optimum geographical locations for viticulture. The net result would be a potential improvement in yield and quality in existing grape-growing areas, with perhaps very minor shifts in the optimum locations for particular wine styles.

The third and, I believe, most likely scenario is as the second, but with a temperature rise due to increased carbon dioxide much smaller than in the viridomological models. Any shift in optimum locations will then probably be towards *warmer* climates. This will be especially so for the time being if the natural pattern of medium-term temperature fluctuations brings, as expected, a down-turn in the next decade or so which could overshadow greenhouse warming until several decades into the next century. Under this scenario, the historical southward migration of European vine growing for premium wine production will continue into the 21st century, with regions such as the south of France becoming major new quality centres for table wines. Existing well-established viticulture in northern Europe may not disappear, but will be less competitive because of its relative inability to respond to higher carbon dioxide concentrations. Plantings made during the recent warm phase at the cold limit of viticulture, such as in England, Washington State in the USA, and southern Tasmania and New Zealand, will be vulnerable.

The broad conclusion is that 20th century viticultural and winemaking experience will remain a sound guide in planning for the 21st century, with perhaps some added flexibility into warm and hot areas. Climatic data for the early to mid 20th century, as tabulated in this book, will remain a relevant basis, around which safety margins can now be built to allow for conceivable future climatic deviations.

Regular ripening of Maturity Group 3 (and earlier) grape varieties seems a reasonable minimum criterion for true commercial viticulture at the cool margin, since Group 3 contains the earliest-maturing varieties capable of making still table wines of the highest quality (Chapter 7). Across most vineyard regions this corresponds tolerably with a site-adjusted average mean temperature for the seven-month growing season of 15°C.

Earlier-maturing grape varieties will, of course, ripen regularly in cooler climates than that, together with Group 3, and even later, varieties in some seasons. But such viticulture will tend to be at a commercial

Table 183. Growing season (April-October in the Northern Hemisphere, October-April in the Southern Hemisphere) average mean temperatures for the early to mid 20th century. Raw data, not adjusted for vine sites

°C	Site	Country/State	°C	Site	Country/State	°C	Site	Country/State
13.0	Hamburg	Germany	15.6	*Colmar[128]	France	18.4	*Clare[39]	South Australia
13.0	York	England	15.7	*Angers[120]	France	18.4	*Busselton[10]	Western Australia
13.1	Ross-on-Wye	England	15.7	*Tours[141]	France	18.5	*Verona[160]	Italy
13.2	Bremen	Germany	15.9	*Healesville[62]	Victoria	18.5	*Siena[158]	Italy
13.5	*Oxford[172]	England	15.9	*Eger[151]	Hungary	18.6	*Bega[79]	New South Wales
13.6	Berlin	Germany				18.8	*Plovdiv[155]	Bulgaria
13.8	*Southampton[173]	England	16.2	*Eden Valley[41]	South Australia	18.8	Perpignan	France
13.9	*Greenwich[171]	England	16.3	*Prosser[180]	Washington, USA	18.9	*Stanthorpe[109]	Queensland
13.9	Vancouver	B.C., Canada	16.4	*Keszthely[152]	Hungary	18.9	*Bunbury[9]	Western Australia
			16.4	*Napier[119]	New Zealand			
14.0	Leipzig	Germany	16.4	*Ararat[51]	Victoria	19.2	*Waite Institute[35]	South Australia
14.1	Nancy	France	16.5	*Coonawarra[40]	South Australia	19.2	*Wangaratta[75]	Victoria
14.2	Christchurch	New Zealand	16.7	*Zagreb[150]	Croatia	19.3	*Ciudad Real[169]	Spain
14.3	Dresden	Germany	16.8	*Mount Barker[23]	Western Australia	19.5	*Régua[166]	Portugal
14.4	*Bushy Park[112]	Tasmania	16.8	*Manjimup[21]	Western Australia	19.6	*Lisbon[164]	Portugal
14.5	Prague	Czechoslovakia	16.8	*Mornington[68]	Victoria	19.6	*Inverell[92]	New South Wales
14.5	*Seattle[181]	Washington, USA	16.9	*Bordeaux[125]	France	19.7	*Wokalup[31]	Western Australia
14.6	*Geisenheim[145]	Germany	16.9	*Valence[142]	France	19.7	Sacramento	California, USA
14.6	*Stuttgart[146]	Germany	16.9	*San Jose[177]	California, USA	19.9	*Mudgee[98]	New South Wales
14.7	*Reims[137]	France						
14.7	Cologne	Germany	17.0	*Napa[175]	California, USA	20.0	*Corowa[86]	New South Wales
14.7	Rennes	France	17.0	*Geelong[60]	Victoria	20.1	*Berri[38]	South Australia
14.8	Lille	France	17.1	*Albany[6]	Western Australia	20.3	*Merbein[67]	Victoria
14.8	*Hobart[113]	Tasmania	17.2	*Lugarno[148]	Switzerland	20.3	*Adelaide[34]	South Australia
14.8	*Freiburg[144]	Germany	17.2	*Logroño[170]	Spain	20.5	*Cowra[87]	New South Wales
14.8	Innsbruck	Austria	17.2	*Auckland[117]	New Zealand	20.6	Naples	Italy
14.8	Paris	France	17.2	*Karridale[17]	Western Australia	20.6	Brindisi	Italy
14.9	*Orléans[135]	France	17.3	*Avoca[52]	Victoria	20.7	*Pinhão[165]	Portugal
14.9	*Vienna[149]	Austria	17.4	*Padthaway[44]	South Australia	20.8	*Griffith[91]	New South Wales
			17.5	*Armidale[77]	New South Wales	20.8	*Perth[26]	Western Australia
15.0	*Bad Durkheim[143]	Germany	17.5	*Braga[161]	Portugal			
15.0	*Cosne[129]	France	17.6	*Vila Real[167]	Portugal	21.0	Badajoz	Spain
15.1	*Ballarat[54]	Victoria	17.6	*Margaret River[22]	Western Australia	21.1	*Cessnock[83]	New South Wales
15.2	*Blenheim[118]	New Zealand	17.8	*Strathalbyn[48]	South Australia	21.1	*Jerrys Plains[93]	New South Wales
15.2	*Geneva[147]	Switzerland				21.3	*Guildford[15]	Western Australia
15.3	*Auxerre[122]	France	18.0	*Seymour[72]	Victoria	21.8	Almeria	Spain
15.4	*Dijon[130]	France	18.0	*Bolzano[157]	Italy	21.9	Malaga	Spain
15.4	*Nantes[133]	France	18.2	*Belair[37]	South Australia	22.1	Palermo	Sicily
15.4	*Salinas[176]	California, USA	18.2	*Bendigo[56]	Victoria	22.2	Tunis	Tunisia
15.5	*Launceston[114]	Tasmania	18.3	*Turin[159]	Italy	22.2	Fès	Morocco
15.5	*Roseburg[179]	Oregon, USA	18.3	*Sliven[156]	Bulgaria	22.2	*Fresno[174]	California, USA
15.6	*Portland[178]	Oregon, USA	18.3	*Pleven[154]	Bulgaria	22.3	Zante	Greece
15.6	*Stirling[47]	South Australia	18.4	*Dão (region)[163]	Portugal	22.3	Samos	Greece
15.6	Heywood[63]	Victoria	18.4	*Montpellier[132]	France	23.6	*Narrabri[100]	New South Wales

* Detailed climatic tables are included for these stations in relation to Chapters 9–22. The superscript indicates table number. Growing season average mean temperatures at representative actual vineyard sites in these locations can be estimated by applying the net adjustments to the means as indicated in the commentaries to the individual tables; and for the remaining localities listed, by adjusting for actual or potential vineyard altitudes, topography, aspects and soil types as explained in Chapter 4.

disadvantage through an unacceptable risk of vintage failure, or through growing inferior-quality, very early-maturing varieties and/or chaptalizing excessively. The relative disadvantage will grow as atmospheric carbon dioxide concentration rises. (A case might be made for making principally sparkling wines in climates a little colder than this, but entails reduced commercial flexibility and greater risk.)

The upper limit of desirable temperatures is harder to define. I suggest that any substantial plantings of winegrapes in warm areas need to be able to make good table as well as fortified wines, in order to retain commercial flexibility and ensure an adequate market size. Quality in table wines under high ripening temperatures depends on there being high enough relative humidities or, especially, regular afternoon sea breezes (Chapter 3). Given these, the upper desirable limit of average mean temperatures over the seven-month growing season appears to be about 21°C.

Combining the two limits with an allowance of 1°C as a reasonable buffer against medium-term climatic changes in either direction, we arrive at a very broadly defined optimum of 16–20°C for the growing season average mean temperature, as measured in the early to mid 20th century.

Table 183 lists seasonal average means, not adjusted for vineyard sites, for the main viticultural regions of Europe, Australasia and west coastal North America. Actual vineyard mean temperatures can be up to 0.5°C higher, particularly (through necessary site selection) at the cold margin of viticulture. The figures allow crude temperature comparisons among viticultural regions. They also give a basis for estimating the geographic shifts that would be needed to maintain present temperature regimes following any change in climate.

Geographic shifts in temperature within southern Australia in response to warming or cooling can be judged in greater detail by reference to Figures 11–16 in Chapters 9–14. These show the early to mid 20th century isotherms for the October-April average mean temperatures, reduced to sea level by adding 0.6°C per 100 metres of actual recording station altitude. A change of 1.0°C will shift a given temperature regime, assuming a constant altitude and site topography, by the local distance between two whole-degree isotherms. As a example, in Victoria (Chapter 11, Figure 13), a rise of 1°C would transfer a temperature regime by the distance of one isotherm interval southwards. A fall of 1°C would transfer it by one isotherm interval northwards.

Temperature variability plays a complementary part in determining geographical limits for viticulture, whether they be absolute limits as described in Chapter 3, or limits of the optimum for winegrape production as discussed here. The less the temperature variability, the wider will be the range of mean temperatures tolerated.

The principle applies not only to temperature variability from night to day and day to day, as measured by the Temperature Variability Index (Chapter 3); it applies equally to variability from year to year and decade to decade. In fact all four scales of variability are probably correlated, because all are governed largely by the degree of marine influence over climate. Regions with low temperature variability indices will tend also to be those with least year to year and decade to decade temperature variation. They will be less subject to any greenhouse warming than the centres of the dry continents. This is a further reason for favouring coastal and near-coastal locations in Australia.

The message for Australia can be summarized as follows.

1. Any 'greenhouse' warming is unlikely to cause viticultural migration into present cooler areas. If anything, rising atmospheric carbon dioxide concentrations will compensate, and perhaps more than compensate, for any warming, to the relative advantage of present intermediate and warm areas.
2. Future vine plantings for wine production should preferably be in sunny, but reasonably humid, coastal and near-coastal areas at altitudes not greater than 500 metres.
3. Within all regions, site and soil selection for warm nights and least temperature variability is critical, and will become more so.

It remains now to make specific recommendations for individual regions of Australia, and for future research. These are the subjects of the final chapter.

24

Summary and Practical Recommendations for Australia

The following recommendations assume that:

- Production of winegrapes will be for a combination of fortified and table wines, or table wines only (Dried and fresh fruit production lie outside the main focus of the study, while specialist production of fortified wines seems doubtfully economic in foreseeable markets.)
- Distance from large population centres, within reason, will not matter. Any disadvantages of isolation will usually be offset by a greater and cheaper availability of suitable land.

Three categories of regional suitability are defined. They are based mainly on climate, but also take into account the presence of suitable soils and topography, and whether enough good-quality water is available from rainfall or irrigation. Locations indicated on Figures 19–23 are the same as those shown, with names, on the corresponding maps in Chapters 9–14.

Category 1: Regions with the best and safest climates, with good water supplies and/or rainfall, and enough suitable soils and topography to support substantial vine industries producing consistently very high quality wines at minimum cost.

Category 2: Regions which are capable of producing high to very high quality wines, but over smaller areas and perhaps less regularly or with greater difficulty and risk. Extra care is needed in site selection and vine management.

Category 3: Viticulture is possible, but the environment is clearly marginal, or risky, or both. A few carefully selected sites may give good wines on a small to medium scale. Major commercial development is not recommended.

New South Wales and South-East Queensland
(Figure 19)

No region is in Category 1, the main reason being that spring frosts and/or excessive summer rain and humidity are to varying degrees serious hazards throughout.

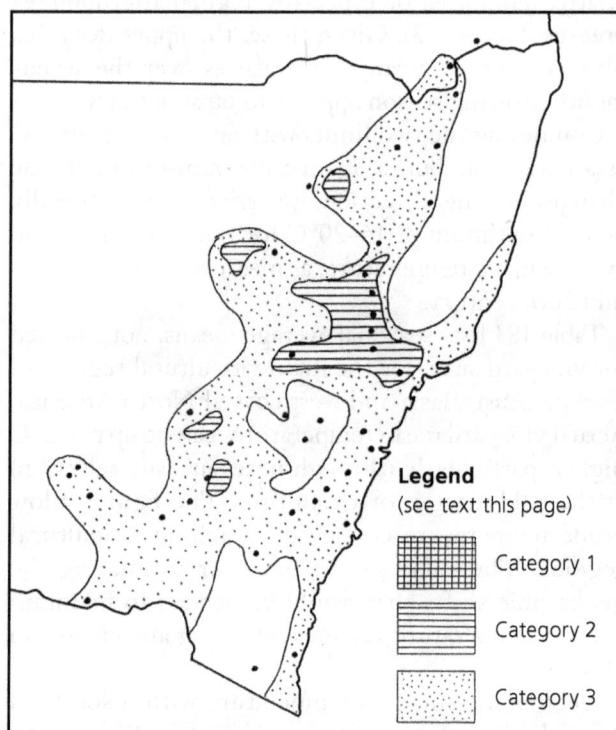

Figure 19. Regional averages of environmental suitability for commercial winegrape growing in New South Wales and south-east Queensland

Present planted areas are mostly in Category 2. Fruit and wine quality can be very good in good seasons, but yields and quality are variable.

Accepting the risks, otherwise favourable climatic conditions extend throughout the whole of the Hunter-Goulburn Valley and its slopes, as far inland as the Liverpool Range; thence perhaps a little way northwards along the western slopes of the Liverpool and Hastings ranges, and southwards to take in Mudgee. Of these areas the middle and lower Goulburn and Hunter valleys, and probably the south-eastern slope of the Liverpool Range, are the best because of their substantial freedom from spring frosts and the strong influence of sea breezes in summer. New irrigated plantings just over the range at Quirindi lack

272

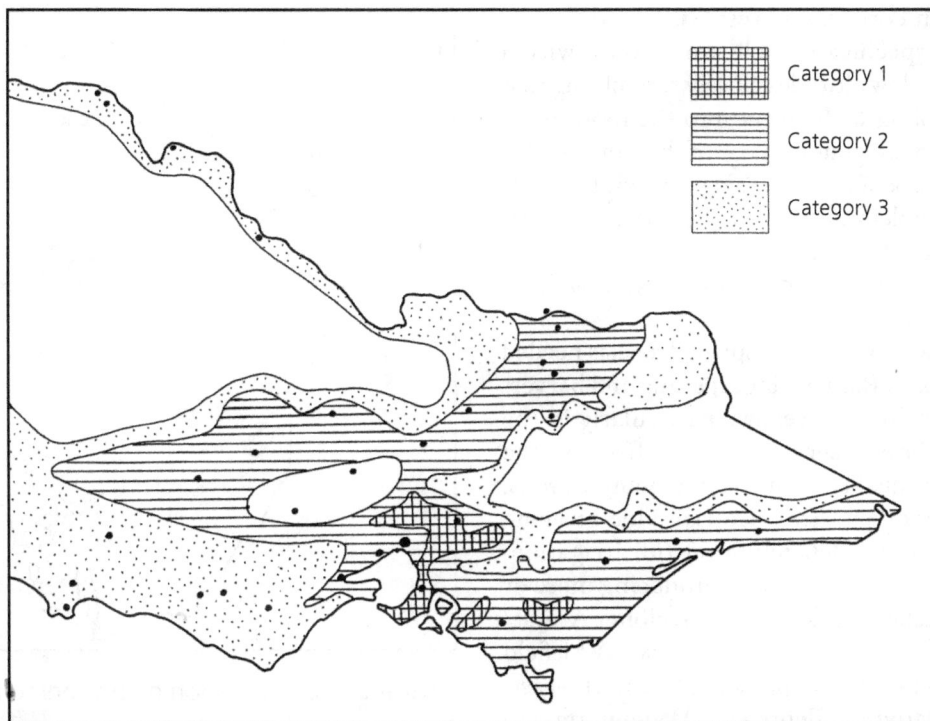

Figure 20. Regional averages of environmental suitability for commercial winegrape growing in Victoria.

this influence, and have a fairly typical inland hot climate; nevertheless their average ripening conditions remain clearly more favourable for table wine quality than those of the southern irrigation areas on the Murrumbidgee and Murray rivers.

Small pockets of reasonably suitable land exist further north along the western edge and upper western slopes of the New England Plateau, as far as the Granite Belt of south-eastern Queensland. To these may be added the slopes of isolated hills further west, such as those of the Warrumbungle Range near Coonabarabran, and of Mount Kaputar near Narrabri, which in addition to fair climates have excellent basaltic soils. I have tentatively placed them in Category 2 as well. However the risks from spring frosts and/or heavy summer rains throughout, together with greater than desirable altitudes where temperatures are most suitable for table wines, mean that most of the region can only be in Category 3.

The immediate coastal areas of New South Wales all suffer from excessive rainfall and humidity in summer. Some viticulture still exists close to Sydney and as far north as the Hastings Valley, and good quality is attainable so long as vine diseases can be controlled. This area must also be placed in Category 3, although vine breeding may one day improve its rating, especially with better species hybrid varieties such as Chambourcin.

The coast southwards from Nowra possibly belongs high in Category 3, because of its lower temperatures and more growing season sunshine compared with the coast further north. However the extreme variability of the climate, with more or less equal dangers of

drought and too much rain and humidity during ripening, precludes a higher rating.

Remaining parts of New South Wales with enough rainfall or irrigation water likewise fall mostly into Category 3, or also are too cold for any viticulture. Some isolated or projecting hills with outstanding air drainage towards the western edge of the central tablelands, for instance near Orange, probably belong in Category 2 (see more detailed discussion on page 159). The Hilltops area around Young, further south, also fairly clearly falls into Category 2, as do the slopes of Sugarloaf Mountain, which rises from the western slopes to the west of Cowra. However, the main southwestern slopes are hot and become increasingly arid in summer from north to south, while still suffering from periodic killing frosts in spring. Lacking the humidity or sea breezes of the coast and its valleys, these inland areas have marginal ripening conditions for table wine quality in most seasons. On the other hand they should, like neighbouring north-east Victoria, be able to make outstanding sweet fortified wines.

Victoria (Figure 20)
The Mornington Peninsula, south of Melbourne, probably has Victoria's best climatic conditions for European-style table wines. The whole of it belongs in Category 1, together with frost-avoiding slopes of the Dandenong Range northwards to the Yarra Valley, and thence, on suitable sun-facing exposures, along the southern slopes of the Great Dividing Range as far west as Sunbury. I more speculatively add parts of the north-facing foothill slopes which define the southern edge of the La Trobe Valley in Gippsland, together

with the Western Port area in west Gippsland.

These are all specifically table wine areas, with a growing season at low altitudes closely resembling that of Bordeaux, cooling with ascent into the foothills to growing seasons analogous to those of the Loire Valley and Burgundy. Lack of continentality and winter chilling may be a problem nearest to the coast and Port Phillip Bay, but I assume that this can largely be overcome by selecting suitable vine varieties and by vineyard management.

Careful site selection to avoid spring frosts is needed away from the coast. Rain during ripening and harvest can also occasionally be excessive, particularly in the Dandenongs. These reservations are offset by the potentially superb quality of the fruit in many seasons.

The remaining established Victorian vine-growing areas, excluding those depending on full irrigation, fall mostly into Category 2. Those producing more or less exclusively table wines include Geelong, Ararat/Great Western, Avoca, Bendigo, Euroa, Whitfield and (potentially) much of Gippsland. The north-eastern warm areas around Glenrowan, Wangaratta and Rutherglen are optimal for fortified wines, but with careful site selection can still produce some good table wines.

A small area around Geelong approaches Category 1, and has very good quality potential; but rainfall is barely enough, spring frosts become common with distance away from Port Phillip Bay, and the area west of Geelong lacks the influence of Port Phillip Bay over the hot north and north-west winds of summer, such as is enjoyed by the Mornington Peninsula.

The La Trobe Valley and East Gippsland have good average conditions for making medium-bodied table wines, but suffer from excessive annual and seasonal variability of climate. Vintages as a result will be variable, although potentially outstanding in good years.

The inland slopes of the Great Dividing Range suffer, to varying degrees, from marginal rainfall and from fairly low afternoon relative humidities, mostly not relieved by sea breezes. The greater relative humidities at the highest altitudes are an advantage, possibly offset to some extent by reduced atmospheric carbon dioxide. Table wines from the intermediate and lower altitudes are of the robust, traditional Australian style. Although often very good they tend to lack the ultimate freshness and finesse of those grown under more coastal influences. Suitable sunny but sheltered exposures among the coastward slopes of the range's western arm could have hitherto unrealized potential.

Tasmania (Figure 21)

Low-altitude sites in the Tamar Estuary and Pipers Brook areas of north-east Tasmania belong in Category 1, having arguably the best climate in Australia for making true cool-climate styles of table wine. Ample sunshine hours, extending well into the autumn, are combined with minimal temperature

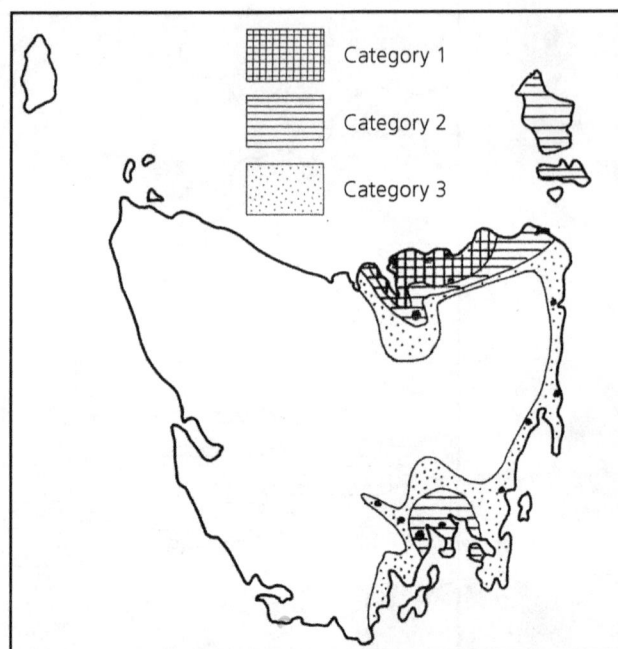

Figure 21. Classification of environmental suitability for commercial winegrape growing in Tasmania.

variability and low extreme maxima, associated with the north-coastal location. Relative humidities are optimal. Limitations are spring frosts away from the coast or estuary, lack of continentality, and strong winds on exposed sites. Temperatures at the intermediate and lower altitudes are high enough to ripen a reasonable range of premium grape varieties, including Bordeaux red varieties on the warmest sites at low altitudes close to the Tamar.

I very tentatively include Flinders and Cape Barren Islands under Category 2, because of their moderately hilly topographies and undoubtedly equable temperatures. Spots that are sheltered enough from the west and south could provide excellent climatic conditions for light table wines.

Projections for the Hobart-Richmond area of south-east Tasmania, and, still more, for the east coast, depend in part on the prospects for future temperature change. A significant and permanent temperature rise could transform the Hobart-Richmond area from one strictly confined to cool-climate wine styles to one more widely based both in area and in styles and grape varieties. If, on the other hand, temperatures were to fall from their present peak before starting to rise again towards the mid 21st century, as I tentatively project in Chapter 23, there could for some time be a predominance of cold seasons and insufficient ripening of any but the earliest varieties or for sparkling wines. Thus for the time being much of the south-east must remain as Category 3, with only low-altitude areas of the lower Derwent Valley and the Coal River basin, together with the small low-altitude pocket inland from the east coast between Bicheno and Swansea (not shown in Figure 21, but discussed in Chapter 14) reliably in Category 2. Warming may in time raise these to

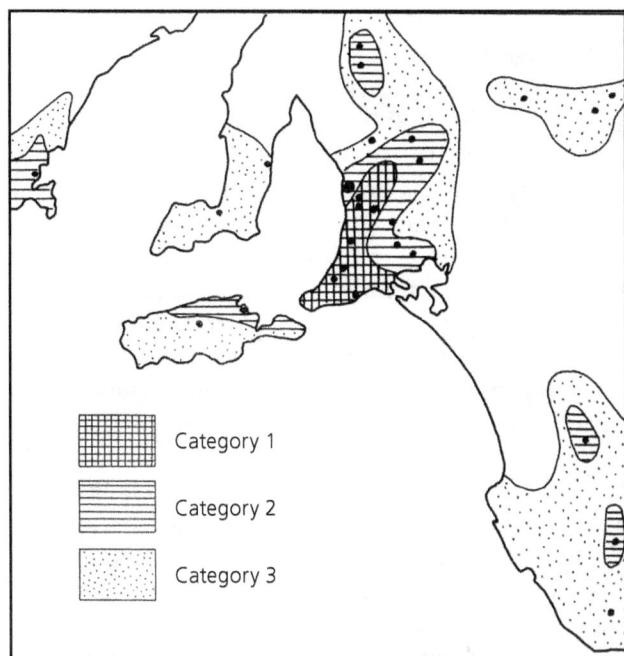

Figure 22. Classification of environmental suitability for commercial winegrape growing in South Australia.

Category 1, although the south coast will still suffer in comparison with the north coast because of its greater temperature variability.

South Australia (Figure 22)

The lower Fleurieu Peninsula has what appears to be the best climate in mainland South Australia for making table wines. Temperatures, depending on altitude, are very suitable for light to full-bodied styles. Rainfall, sunshine and relative humidities are close to optimal over much of the area. Clearly it belongs in Category 1, with actual sites depending mainly on local topography and the availability of suitable soils. A further narrow strip of Category 1 land extends northwards along the foothills and western slopes of the Mount Lofty Range, from McLaren Vale nearly to the Barossa Valley. With further distance north, and eastwards from the crest of the range, there are increasing limitations of excessive temperature variability (especially high temperature extremes) and low relative humidities during ripening.

The eastern and northern parts of Kangaroo Island have an excellent growing season climate for table wine production, given shelter from wind and supplementary watering as needed. Climatically they belong in Category 1, but other factors conspire very largely to rule viticulture out. Lack of suitable soils and sheltered slopes is a main limitation, perhaps with lack of continentality for some grape varieties. Given good enough soils and topography, this could be an interesting area for table wines.

Many of the established, previously non-irrigated viticultural areas of South Australia belong in Category 2. The Adelaide Plains and some of the more sheltered parts of McLaren Vale do not get early enough sea breezes to avoid day-time stresses associated with low relative humidities. This is nevertheless an excellent environment for fortified wines. The Barossa Valley, Barossa Range, Clare and Padthaway are all cooler and mostly do get some benefit from sea breezes, but again, these in general come too late to avoid substantial mid-day stresses. Temperature variabilities are well above optimum. All of these areas produce robust, Australian-style table wines which can be very good, but tend to lack the finesse of those from more humid and equable climates. The best, especially the whites, are mostly from the cooler high altitudes, where ripening is delayed until the cool but generally equable conditions of late autumn. Some grape varieties, notably Riesling, benefit from the late spring-summer stress, if not too extreme, because it helps to subdue vegetative growth and keep them in fruiting balance with simple trellising.

Langhorne Creek and Coonawarra have more favourable afternoon relative humidities than further north, due to earlier sea breezes, but are subject to excessive temperature variability and periodic exposure in summer to hot, desiccating winds from the north. Temperature variability diminishes greatly, and relative humidities rise, during the relatively late ripening at Coonawarra; but that advantage is offset by the danger of late autumn rain and partial failure of ripening.

The adjacent coastal strip, around Robe and southwards, is clearly more equable than Coonawarra, with earlier summer sea breezes and less frost, but with similar ripening potential. Individual sheltered spots with suitable terra rossa soils (not shown in Figure 22) are probably Category 2: see page 112.

The southern tip of Eyre Peninsula constitutes another Category 2 region. Pockets of terra rossa soils with good underground water, similar to those at Coonawarra and Padthaway, might rank better were it not for their flatness and exposure to strong winds.

I have ranked the southern parts of Yorke Peninsula as Category 3, mainly because of limited rainfall and flat topography. Sheltered sites with good air drainage and available supplementary water, if they exist, might be quite suitable for viticulture.

Western Australia (Figure 23)

Western Australia has few locations ideal for making true cool-climate wines, but extensive areas are climatically close to ideal for table wines of medium to full body. These lie in an arc running westwards from Albany and the Porongurup Range, through Mount Barker and Manjimup to Karridale, thence northwards along the west coast and its near hinterland through Margaret River, Busselton and Bunbury to about as far north as Harvey. River drainage systems with

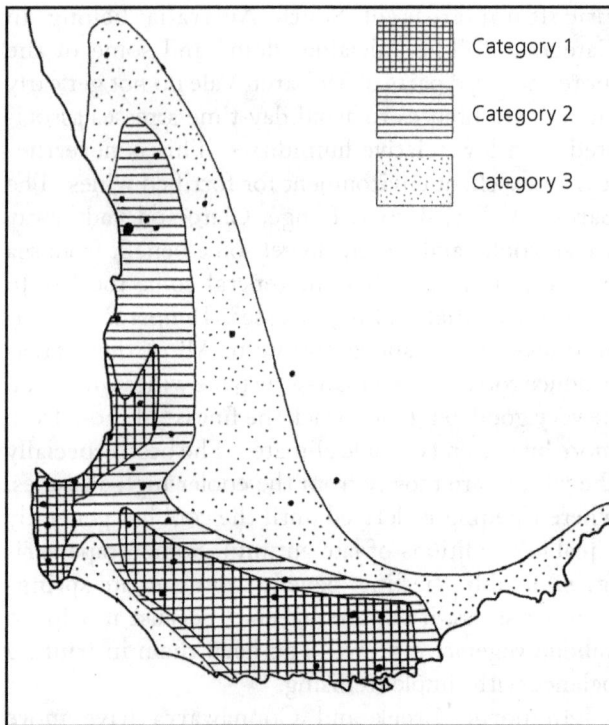

Figure 23. Classification of environmental suitability for commercial winegrape growing in Western Australia.

suitable undulating topography and sandy or gravelly loam soils, or else coastal limestone or alluvial soils on the west-coastal plain, are interspersed through the region. Frost risk is mostly low, and sunshine ample except right on the south coast. Average afternoon relative humidities are optimal near the coast, due to mostly regular afternoon sea breezes in summer.

The south coastal part of the arc has close climatic analogies with the cooler parts of Bordeaux, such as the Médoc, apart from having a more maritime seasonal temperature pattern, and more rain in winter but less in summer.

Growing season temperatures at Margaret River equate with the warmer Bordeaux areas such as Saint Emilion and Pomerol. Compared with Bordeaux and the south coast of Western Australia, there is still more rain in winter, and less rain, together with more sunshine, in summer.

From Busselton and Capel northwards to Harvey the growing season is warmer and still sunnier, but with fairly high afternoon relative humidities because of early sea breezes throughout the summer. Growing conditions closest to the coast resemble those of the Mediterranean coast of France, while the higher slopes of the Darling Range have conditions nearly approaching those of Margaret River, and well suited to a wide variety of table wines.

All these areas belong in Category 1, and the aggregate suitable for viticulture is large. Of all the viticultural regions of Australia, south-western Australia probably has the greatest unused potential to produce medium to full-bodied table wines and light to medium

fortified wines of the highest quality.

The coastal strip and western slopes of the Darling Range northwards from Harvey to Gingin and Bindoon belong in Category 2. Although quite hot as measured by mean temperatures, they benefit greatly from afternoon sea breezes in summer. Intermittent spells of summer heat and low relative humidity, when the sea breezes fail and land winds predominate, are the main climatic limitation. Conditions are excellent for making medium to full-bodied fortified wines, yet sites with good sea breeze exposure can still make very good, full-bodied table wines from heat-adapted grape varieties in most seasons.

The viticultural climate deteriorates markedly beyond about 60 kilometres inland from both south and west coasts of south-western Australia, mainly because of diminishing rainfall and afternoon relative humidities, combined with more spring frosts and summer heat. Viticulture remains possible on carefully selected sites down to about 550 or 600 millimetres annual rainfall; but even with the benefit of rising atmospheric carbon dioxide concentrations, it is unlikely ever to become commercially competitive with an industry based in the more favoured coastal and near-coastal climates.

Research needs for Australia

The traditional European idea that 'great wines are made in the vineyard' is now widely accepted in Australia. Wine industry research, while not neglecting winemaking processes and marketing, therefore needs to focus strongly on the vine and the grape.

A substantial part of that focus needs to be regional. While important roles certainly exist for a strong central institution for oenological and marketing research and teaching, and for some of the more basic research pertaining to viticulture, nevertheless the final translation into practice will only come when the intimate interactions of viticulture with the whole range of environments are better understood. Progress will depend on a continuance of vigorous regional applied research, both by professional researchers and by innovative commercial viticulturists.

One already clear conclusion is that Australia's future viticultural research, at least for winemaking, should be mainly in its coastal and near-coastal regions. These are where the greatest potential exists for both quality and productivity. It is where the industry will increasingly be located, regardless of any future climatic changes, provided that suitable land remains available.

What are the most critical individual fields for viticultural research in Australia? The following come to mind, but the list is not exhaustive.

- The direct effects of rising atmospheric carbon dioxide concentrations on the vines, on fruit qualities, on optimum growing and ripening temperatures, and on management strategies.

- Possible effects of altitude, through differences in carbon dioxide partial pressure and perhaps other factors.
- Soil management and mulching, so as to optimize soil health and vine nutrition on the one hand, and the temperature microclimates of both vine tops and roots on the other.
- Amounts, methods, and especially timing, of nitrogen supply to give optimum contents of natural nitrogen compounds in the berries for rapid fermentation and wine quality.
- How far such compounds can be optimally supplemented or replaced by direct additions to the must, and best formulations of such additions for wine quality.
- Complementary roles of trace and other elements besides nitrogen in fermentation, both via vine nutrition and as fermentation additives.
- The contribution, if any, of soil organic matter and microbiological activity to carbon dioxide supply within the vine canopy.
- Ways of manipulating light quality within the vine canopy to improve yield and/or fruit quality, including soil surface colouring and the use of reflectors, or artificial sources of specific wavelengths.
- Optimum timing and control of water supply in different environments, especially during ripening.
- Strategies of bunch thinning, to optimize quality x yield in fruitful grape varieties and environments.
- Relationship of flavour and aroma accumulation to rate of grape maturation, as governed by grape variety, ripening environment, and sugar flux to the berries.
- Problems of premature or uneven budburst in maritime environments, and management strategies to avoid them.
- Wind control in the vineyard.
- Rootstocks for different environments and purposes: particularly, dwarfing rootstocks to control vegetative growth in maritime environments.
- Interactions of individual grape varieties with Australian climates and soil types.
- Reducing and, if possible, eliminating the use of toxic or suspect chemicals in the vineyard.
- Natural methods of bird control, for instance by providing alternative food or water supply.

Vine breeding

As with general viticultural research, Australian winegrape breeding needs to be transferred more to coastal regions.

Good prospects exist for genetic improvement in these regions, for two reasons. First, our main premium winegrape varieties come from much more continental climates (Chapter 3), and are not necessarily well adapted to the highly maritime temperature regimes of Australia's prime viticultural regions. Second, many of the needed changes are in readily visible vine characteristics which can easily be selected for, such as reducing vegetative rankness and premature or irregular budburst. This means that vine numbers can be greatly reduced by simple visual means in the early breeding stages, before expensive quality assessments have to be made.

The necessary changes in vine architecture may in any case be fairly universally desirable, like those already achieved in many other horticultural and field crops. Olmo (1979) set them out clearly in his vision of vineyards in the year 2000. 'The tendency in breeding and selection of new varieties has been to strive for great vigour and increased production. In another 20 years we will probably have conclusively demonstrated that fruit quality and high vegetative development are negatively correlated. Thus the development of dwarf or semi dwarf plants (or the selection of dwarfing rootstocks to accomplish the same purpose) is a good possibility. Smaller plants would allow closer spacing, better utilization of sunlight, and higher yields per unit area.' Nowhere would these changes be more beneficial, or more effectively selected for, than in Australia's mild coastal environments.

By analogy with other crops, the needed changes could well be associated with reduced responsiveness to the vine's own gibberellins, which could itself be fairly readily screened for. Simultaneous selection may be needed towards looser bunches, since gibberellin insensitivity could result in over-tight bunches and berry splitting.

To the general dwarfing mentioned by Olmo may be added smaller, perhaps thicker, more divided and erectly-held leaves, resulting in more acutely angled presentation of the external leaves to sunlight and greater penetration of sunlight to the lower and internal leaves. Not only could this greatly increase photosynthesis and yield potential; it should also improve fruit quality for winemaking, by increasing photosynthesis relative to the uptake of water and potassium. By the same token it could increase still further the need for late availability of nitrogen, and perhaps of some other elements needed for efficient fermentation and wine quality.

A characteristic which could be worth seeking is genetically controlled, natural self-termination of shoot growth. Such determinate growth patterns are common among crop plants, and are usually under simple genetic control. A natural termination of growth at, say, the 10th to 12th node of each shoot would answer many of the problems of vine training.

Breeding for warm and hot areas of Australia is still justified, but might profitably turn more attention to early or midseason varieties which retain their winemaking qualities under hot ripening conditions, rather than seeking lateness of maturity to escape heat during ripening as emphasised in Australian breeding hitherto (Clingeleffer 1985). The relationship between lateness and poor winemaking quality (Chapter 7) suggests that the scope for quality improvement in that direction is limited. Conversely, the proven performance of quite early varieties like Chardonnay and Verdelho in hot climates, and the fact that modern breeding of early-maturing varieties has mostly been in very cool viticultural environments, such as Germany, suggests that unexploited opportunity remains for their improvement in hot climates.

Existing vine improvement programmes in the Victorian-South Australian inland irrigation areas are well located to meet the needs of most hot vine-growing regions of Australia. A separate programme is indicated for the humid sub-tropics. This would need to place emphasis on resistance to vine diseases, and might seek to exploit further the potential of species hybrid grape varieties. The Hastings Valley of New South Wales typifies the environments suitable for such a programme.

But as already emphasized, Australia's greatest need for future vine breeding lies in the cool-temperate and Mediterranean-type climates along the south coast of the mainland, and in the warmest part of north-eastern Tasmania. Suitable localities for such a programme, typifying both the commercial potential and the problems, include the Mornington Peninsula in Victoria, the lower Fleurieu Peninsula in South Australia, and Margaret River in Western Australia.

APPENDIX 1

Plant Hormones:
Their Functions and Interactions with Environment

Hormones are 'chemical messengers'. Produced in one or more organs or tissues, they are transported to other organs where they influence growth or other processes. In plants their movement is either upwards in the sap from the roots, through the woody xylem system, or in either direction with sugar and other solutes through the phloem system of the inner bark. The production, transport and activities of the individual hormones interact with elements of the environment, and the hormones interact amongst themselves, in a variety of ways which together forms a subtly effective system for mediating plant responses to environment.

Four groups of hormones, and their interactions, appear to exercise the greater part of this control. But while the literature on them individually is vast, few accounts integrate their respective roles within the plant. One of the best available is that of Davies (1987). Still less is there yet a satisfactory account of how the hormones, individually and collectively, interact with the total environment in shaping plant growth and fruiting. The following brief summary attempts to do this with particular reference to grapevines.

1. Auxins
These are formed in vegetative growing points and young growing leaves, and in growing seeds. In stems they are transported only downwards from the growth points, via the phloem system. Their main functions in stems are: to attract sugar and other nutrients back towards their sites of production; to stimulate cell growth in young stem tissue; and to inhibit lateral budburst and branch development further down the stem. This ensures the 'apical dominance' of terminal growing points and shoots during their most active growth. Probably by drawing away nutrients from fruiting structures, shoot-produced auxins also reduce fruit set and development in vegetatively vigorous plants.

While moving down a stem, auxins migrate to its less strongly illuminated side. This causes:

1. preferential cell growth on that side, resulting in young stems bending to grow towards the source of brightest light; and
2. further down, preferential budburst and growth of

lateral branches from the side facing that light. The two mechanisms ensure that the strongest growth is always towards the brightest light.

In fruits, auxins produced by growing seed embryos attract nutrients and cytokinins (see below) which further stimulate growth in the seeds and adjacent flesh tissue. (The effects of lack of auxin can be seen, for instance, in the failure to grow of individual sectors in apples and similar fruits when the seeds in those sectors have been destroyed early in fruit development. Similarly, it can be seen in the small size of seedless grapes.) Auxins from the seeds also stop fruit from ripening, for as long as the seeds are still growing and producing them.

2. Gibberellins
The gibberellins are formed in young growing shoots and leaves, and probably in root tips. They can travel in any direction within the plant, and via either xylem or phloem system.

Opinion differs as to the importance of formation in the roots, but there is evidence at least that shoot-produced gibberellins may be modified there, and the changed forms re-distributed to the tops through the xylem system. The soil environment of the roots can strongly influence some gibberellin effects on growth (Reid and Crozier 1971; Torrey 1976; Richards 1983).

Gibberellins act most conspicuously by promoting the growth and, especially, elongation of cells in growing stems and fruits. They overcome many genetic types of dwarfness. At the same time they tend to suppress flower bud initiation and fruitfulness in vines and fruit trees (Srinivasan and Mullins 1978, 1980, 1981; Pharis and King 1985). Various natural and synthetic 'anti-gibberellins' can antagonize them and cause dwarfing, and sometimes earlier and better fruitfulness, in much the same way as dwarfing rootstocks.

Like auxins, the gibberellins promote growth of cells partly by attracting nutrients towards them. Imported natural gibberellins, or synthetic gibberellins applied to the clusters at the right time, can replace seed-produced auxins in promoting the setting and growth of seedless or seed-aborting grape varieties such as Currant and Sultana (Coombe 1960, 1965). Applied otherwise, however, gibberellins can have an opposite effect through the promotion of competing stem growth.

Plants are more responsive to their own or applied gibberellins at low than at high temperature. This is one of the reasons for the rank stemminess, and often poor set, of vines under moderately cool conditions, as described in Chapter 3. Supplementary gibberellins can substitute for low temperatures in plants, bulbs or seeds that need cold exposure, or 'stratification', to overcome growth inhibitions.

Gibberellin activity is also enhanced by a low ratio of red to far red (or near-infrared) light wavelengths, such as is found in the light reflected from green leaves and inside plant canopies. This reaction causes crowded or shaded stems to grow longer and more spindly than those exposed to light of normal spectral quality, thereby giving them a better chance to reach normal light. Buds on these stems tend to be poorly fruitful. Once stems reach full and normal light, however, their subsequent internodes become shorter and the associated buds more fruitful.

3. Cytokinins

These are formed in actively growing root tips, and possibly in growing seed embryos. They ascend from roots to the upper plant parts via the xylem system, and have two major functions there: a) promoting cell multiplication in newly differentiating tissues; and b) attracting, like auxins and gibberellins, sugar and other nutrients to where they are in greatest concentration.

Ample cytokinin promotes budburst and lateral branching, the development of leaves and fruiting structures, and fruit set. The effect on budburst is antagonized by auxins descending from the young growing shoots. Lateral budburst depends largely on the local balance between the two hormones (Thimann 1972; Phillips 1975).

Later, in grapevines, cytokinin promotes fruitfulness of the newly forming lateral buds on current season's shoots. These results are the opposite of those induced by gibberellins, with the fruitfulness of lateral buds depending on the local balance between cytokinins and gibberellins during their initiation (Srinivasan and Mullins 1980, 1981).

In all these respects cytokinins mimic the effects of, and can substitute for, a high ratio of red to far red light wavelengths (as in normal sunlight), just as gibberellins can mimic and substitute for a low ratio as normally experienced in the shade within plant canopies. Likewise optimally high temperature promotes cytokinin production, whereas low temperature promotes plant responsiveness to gibberellins.

Cytokinin production and export by roots are favoured by ample sunshine and leaf exposure, and a consequent ample supply of sugar to the roots, as well as by a warm, well aerated root environment. These conditions encourage early and uniform budburst in spring, vigorous branching and leafy growth, and fruitfulness. Continued strong cytokinin export through the season helps the leaves to maintain their functional integrity and resistance to adverse conditions and disease, and delays their senescence. Finally, a good supply to the fruit through ripening helps to maintain the berries as a strong sink for sugar, and to keep them in good health and non-senescent through to full physiological maturity. Cytokinin may also contribute directly to ripening by participating in the biochemistry of pigment and flavourant formation.

4. Abscisic acid

Abscisic acid (ABA), or abscisin, is the plant's stress signal. It forms mainly in mature leaves, and probably roots, in response to stresses due to lack of water or to heat, cold or disease. Shortening days in late summer and autumn probably increase its production.

Abscisic acid counteracts all the other three hormone groups, causing the leaf stomata to close, and therefore both water loss and photosynthesis to cease; vegetative growth to slow dramatically (particularly internode extension, as promoted by gibberellins); sugar, amino acids and other soluble nutrients to be mobilized from the leaves to fruits and other storage depots in the plant; remaining cytokinin supplies from the roots to be diverted or passed on largely to the fruits; and the leaves, finally, to fall.

Flux of abscisin into the berries, combined with waning and cessation of auxin production by the maturing seeds, seems to provide a reasonable explanation for the triggering of ripening in grapes.

Summary

The outstanding thing to emerge is the extent to which the effects of environment on plants, under non-stress conditions, act through a balance between auxins and gibberellins on the one hand, and cytokinins on the other.

The auxins and gibberellins act in a complementary manner, to promote vegetative extension growth under conditions of low temperature, low light intensity, and particularly, low red to far red wavelength ratios as found in crowded plant stands or within leaf canopies. Between then they represent the 'earthly forces' of the biodynamic literature (Koepf et al. 1976), and carry the message that the time and conditions are not yet right for fruiting. Cytokinins are elicited most strongly under the opposite conditions of soil warmth, ample sunshine, and good leaf exposure, and their dominance carries the message that the time and conditions for fruiting have arrived, and that the plant can carry a heavy fruit load. They represent the 'cosmic forces' of the biodynamic literature. Abscisin, when its production becomes strong enough, carries the final message that it is time to ripen the fruit or seed crop, in preparation for plant death (in annuals) or seasonal dormancy (in deciduous perennials).

APPENDIX 2

Sources and Treatment
of the Climate Data

Sources of climate data

Australia

Bureau of Meteorology, Australia (1953) for monthly average sunshine hours, by interpolation and smoothing; Bureau of Meteorology, Australia, 'Climatic Averages Australia' (1956) as the main source of monthly average maximum, minimum and mean temperatures, relative humidities and rainfall; Bureau of Meteorology, Australia (1962, 1965, 1966 and undated) for more detailed or longer-term data on south-western Australia; Bureau of Meteorology, Australia, 'Climatic Averages Australia', Metric Edition (1975) for limited data not available in the 1956 edition: see discussion on period adjustment below; and Bureau of Meteorology, Australia, unpublished records of monthly highest maximum and lowest minimum temperatures, to calculate averages of these together with the frequencies of months experiencing 'light' and 'heavy' frosts in spring.

France

Sanson (1945) for monthly average maximum, minimum and mean temperatures, relative humidities and sunshine hours; Garnier (1967) and interpolations from Direction de la Météorologie Nationale, France (1969) for supplementary estimates of sunshine hours; and Meteorological Office, Great Britain (1967) for monthly average extreme maximum and minimum temperatures for some stations: used also to estimate average monthly extremes for the remaining stations (see later discussion in this Appendix), and temperature variability indices.[1]

Portugal

Servico Meteorologico Nacional, Portugal (1965) for monthly average maximum, minimum and mean temperatures, rainfall, relative humidities and sunshine hours; supplemented by figures from the

---

1 All the published data of the British Meteorological Office are in whole degrees Fahrenheit. To minimize error in converting to Celsius, I interpolated values, within limits consistent with the whole Fahrenheit digits, to give as smooth as possible seasonal temperature curves, typical of the respective areas; and at the same time, to give unchanged average temperatures for the whole growing season.

Meteorological Office, Great Britain (1967), used as for France.

Spain

Servicio Meteorologico Nacional, Spain (1946) for some monthly average minimum and mean temperatures and rainfall; Meteorological Office, Great Britain (1967) for all other data, used as for France.

Italy

Menella (1967) for monthly average maximum, minimum and mean temperatures and rainfall; Meteorological Office, Great Britain (1967) for full or supplementary data for some sites, used as for France; sunshine hours as under 'Remaining European countries' (below).

Remaining European countries

Meteorological Office, Great Britain (1967) for all data other than sunshine hours; sunshine hours from Automobile Association, Great Britain (1963), which cites figures supplied by the Meteorological Office, Great Britain.

New Zealand

Garnier (1958).

North America

Weather Bureau, Department of Commerce, USA (1953); Meteorological Office, Great Britain (1958a); and Nelson (1968), for monthly average maximum, minimum and mean temperatures, rainfall and relative humidities. Department of Agriculture, USA (1936) for average monthly sunshine hours (interpolated and smoothed). Meteorological Office, Great Britain (1958a) for average monthly extreme maxima and minima, used as for France.

Periods of climatic records, and period adjustments

The temperature data for Australia were mostly from the 30 years 1911–1940. For Western Australia I used longer periods, often starting earlier and extending up to about 1960, from the regional climatic surveys where available. Data for France were for the 40 years

1891–1930; for Portugal, the 30 years 1931–1960; for the remaining European countries and New Zealand, various periods mostly up to 1950; and for the USA, the 30 years 1921–1950.

The average monthly highest maximum and lowest minimum temperatures for Australia were calculated from raw data supplied by the Australian Bureau of Meteorology, Melbourne, and mostly covered the 30 years 1939–1968. I also counted the numbers of growing season months which had experienced at least one day with a screen temperature of < 2.5°C ('light frost') or < 0.3°C ('heavy frost') over the same 30 years, or over other shorter periods where records for the full standard period were not available; also the lowest temperature ever recorded in each month, over the full period of available records. The number of years of records in each case is specified in the headings of the tables. The latest year used for any of these records was 1976.

The climatic periods used were largely dictated by data availability when the work started (early 1970s). The use of older data also has the theoretical advantage that towns were then less built-up, so that temperature discrepancies between the countryside and town recording sites due to urban warming (see Chapter 23) would have been less than in more recent times. Similarly, standardization of the recording period as far as possible should minimize discrepancies, if any, due to increases in the 'greenhouse effect', as also discussed in Chapter 23. Where it was possible to compare various recording periods for southern Australia against the long-term data, of the order of 100 years or more, I found that the temperature averages for 1911–1940 agreed very closely, and can thus be accepted as a representative benchmark.

I specifically reject the temperature data published in the 1975 and 1988 (metric) editions of 'Climatic Averages Australia' (Bureau of Meteorology, Australia: Melbourne). These used only the temperatures recorded since the beginning of computer data storage. Being more recent, they are not comparable with most of the available overseas data. Moreover, because computer storage started at different times across Australia, they refer to varying and sometimes very short periods. The periods were not specified in the 1975 metric edition, which was the one available when my climatic tables were prepared, but internal irregularity showed clearly enough that some, e.g. Coonawarra and Margaret River, could hardly have been longer than five to ten years. Such data were clearly not reliable enough for the present purpose.

The second metric edition of Climatic Averages Australia (1988) covers rather longer periods, and rectifies the first metric edition's lack of information on length of recording period; but it, too, used only temperature records from after the start of computer information storage. These, although from mostly longer periods, were still further removed in time

from those I have had available, and have used, from overseas. This is important because Australia's rate of specifically urban warming since World War II appears to have been especially high by world standards (Coughlan 1978): a finding which can readily be related to its exceptional rate of urban growth. Inclusion of the most recent temperature figures would therefore have increased the distortions in the comparisons among sites, both within Australia and against overseas sites.

Moreover, the 1980s appear genuinely to have been the warmest decade so far of the 20th century, perhaps largely as part of the normal medium-term pattern of temperature fluctuation (Chapter 23). To include the 1980s temperatures for Australia alone would have distorted the climatic comparisons with overseas viticultural regions still further.

For both these reasons, I chose to disregard the 1988 edition of Climatic Averages Australia as well as the 1975 edition, and instead to continue relying as far as possible on the 1956 edition.

Some resort to the 1975 edition of Climatic Averages Australia was unavoidable for a few locations which lacked earlier temperature records more or less entirely. Where this was done I attempted back-adjustments for the differing recording periods, parallel to the average differences evident for nearby or geographically related recording stations having long enough records. The adjustments were made individually for each of the seven months of the growing season.

South coastal areas of Australia showed almost universal increases in recorded mean temperatures between the 1956 and 1975 editions of Climatic Averages Australia, mostly about 0.2 to 0.4°C when averaged over the growing season. Nearly all the apparent increase was concentrated in the summer months of December to February. Differences for the spring and autumn months were negligible. Recorded summer and growing season temperatures in central and northern New South Wales, on the other hand, remained more or less unchanged between the two editions, or even fell slightly.

The calculated sea-level temperature trends (isotherms) for the growing season based on 1911–1940 records (Figures 11–15) provided an additional check for individual localities. The sea-level isotherms proved to be particularly useful for estimating temperatures in areas for which there were no published records at all, such as the Keppoch Valley in South Australia. Finally, any apparently very short-term data from the 1975 edition, where they had to be used, were smoothed to conform to normally-shaped seasonal temperature curves for their general regions.

A potential source of error in the Australian data for the temperature variability and extremes stems from the unavoidable use of average monthly highest maxima and lowest minima calculated from 1939–1968, against average maxima and minima, and mean temperatures, from 1911–1940. Detailed records of monthly extremes before 1938 were no longer available for most sites, or had never been obtained. The effect of any overall rise in temperature between the two periods would be to reduce the apparent gap (column 5 in the full climatic tables for the Southern Hemisphere) between the average lowest minimum and the average minimum temperatures, and to increase that between the average maximum and the average highest maximum temperatures (column 9) by a corresponding amount. I did not attempt any period adjustment for these figures, so some slight discrepancies may be present in the Australian tables between columns 5 and 9 and the values for mean temperature. However the discrepancies in the two columns cancel each other out in calculating the temperature variability index, which is therefore unaffected.

A relative immunity of the temperature variability index to period changes in average mean temperature is in any case not unexpected, because temperature variability depends on major geographic and topographic features which are themselves unlikely to change much with time. (Urban development and large-scale drainage or dam-building could have some local effects.) The variabilities of the maximum and minimum temperatures (columns 5 and 9) proved indeed to be very consistent across areas of similar geography and topography. They could therefore be extrapolated and applied with some confidence to geographically related locations having records only of average maxima and minima, and thence used to estimate monthly average extreme maxima and minima and temperature variability indices for them. I used this method extensively for overseas locations where full raw data were not available. Regional similarity of the variability indices for topographically similar sites also formed a useful guide, or check, in estimating the topographical adjustments for some individual sites.

In the event that long-term changes in temperature are occurring (Chapter 23), it should be noted that in calculating the day-degree requirements for maturation of the different grape maturity groups, the climatic figures used were calibrated against mainly recent vine phenological experience. Therefore the effects of any climatic changes between the actual temperature recording periods and now are substantially discounted. Extrapolations to new areas, or for viticulture under further-changed future climates, can work safely enough from a baseline of recent (say, 1950–1990) viticultural experience.

APPENDIX 3

Critique of the Papers of Jackson and Cherry (1988) and Moncur et al. (1989)

Jackson and Cherry (1988)

These authors reject simple temperature summations as a basis for predicting grape vine behaviour. They also reject the temperature x daylength, or 'heliothermic' index used by Branas (1946) and Huglin (1978, 1983) in France, and in a modified form by myself in Chapter 6. On the basis of statistical studies they suggest, instead, a latitude-temperature index which is produced by multiplying the average mean temperature of the warmest month by (60 minus location latitude in degrees). Ripening capacity for a given temperature of the warmest month is thus held to increase with lower latitude: a conclusion which at first sight contradicts that of all previous authors.

The reasons for the apparent contradiction are two-fold: first, Jackson and Cherry's choice of warmest-month average mean temperature as the thermal measure, rather than the seasonal average mean temperature or temperature summation; and second – as they themselves point out – the fact that there is a statistical trend to reduced continentality (see Chapter 3) with lower latitude. At low latitudes the periods above a given base temperature are longer on average, *for a given warmest-month temperature*, than those at high latitudes. This gives a theoretically longer available growing season, which in an overall statistical analysis is enough to override the positive multiplier effect found by other authors between summer daylength and temperature in promoting vine development. In fact, Jackson and Cherry's latitude-temperature index gave only a marginally better prediction of grape ripening than latitude alone. Neither is good enough to be of practical use, or to improve on the simple temperature summations or warmest month average mean temperatures of most previous authors.

The temperature summations which Jackson and Cherry cite and dismiss are in any case dubiously based. For instance they show Christchurch, New Zealand, as having 1088 day degrees over a base 10°C. The equivalent figure according to my calculations, from the 52-year data of the Meteorological Office, Great Britain (1958b) is barely more than 900. Some of the difference could have resulted from their using an unrepresentative recording period: see Chapter 23 and Appendix 2. Much of the problem, however, appears to stem from two factors: first, their inappropriate use of full-year temperature summations, rather than those of the real potential growing season for vines, and second, their equally inappropriate adjustment of the temperature data using the formula of Wendland (1983).

Most authors, including myself, have based temperature summations for grapevines on an arbitrary seven-month growing season. In cool, continental climates, this more than covers the actual period above a 10°C base mean temperature. Warm climates hasten maturity enough for the full seven months not to be needed. Vines in cool maritime climates, on the other hand, need the full seven months or even slightly more. In particular, ripening may be able to continue a little beyond the arbitrary October 31 or April 30 end of season: see, for instance, the temperature curve for Hobart in Chapter 3, Figure 1, and the various data for Tasmania in Chapter 14. But as I argue in Chapter 3, ripening at such low temperatures is likely to consist more of simple sugar accumulation than true physiological ripening. In any case the extra accumulation of day degrees is usually minimal, and the loss through their neglect is probably compensated for in most such cases when the normally narrow spring and autumn diurnal temperature range of these environments is taken into account (Chapter 6).

Thus the arbitrary seven months, although not perfect, do fairly adequately describe the limits of the growing season as usefully exploited by the vine. To include summations for the remaining five months, which vary enormously among climates, and are almost entirely irrelevant to vine phenology, is pointless and potentially misleading.

Wendland's method of adjusting monthly average means was designed to allow for the normal variation among the means for individual days, within a month, above or below whatever base mean temperature is being used. The aim is to estimate and include those days with temperature means above the base, but to exclude those with means below it. I make an analogous adjustment by graphic methods in Chapter 6, for the special situation of spring and autumn in cool, continental climates where the trend line of mean temperature crosses the 10°C base line within the months of April and/or October. It allows the part of the month with average mean temperatures above 10°C to be credited with its appropriate day degrees, while disregarding the rest of the month with average means below 10°C.

Such an adjustment makes biological sense. But the ecological relevance of allowing as well for random day to day changes in temperature, as Wendland's formula does, is much more questionable. Why, for instance, should successive days with means of 12 and 8°C accumulate two day degrees, but successive days both of 10°C, none? It is most unlikely that vines, or any other plants, could respond so quickly and selectively. Jackson and Cherry appear also to have applied Wendland's adjustment in full to *all* months of the year with conventionally estimated Celsius summations of less than 110, regardless of whether or not their trend lines passed through the mean temperature base of 10°C within the month, and certainly regardless of whether the months fell within any conceivable vine growing season. The resulting summations overestimate the heat usable by vines quite substantially in cool and mild maritime climates, where the monthly average mean temperatures are often close to 10°C for several months of the year, both inside and outside the true growing season.

Moncur et al. (1989)

From studies on budburst and early growth of vine cuttings under controlled temperature conditions, Moncur et al. concluded that a base average temperature of 4°C was more appropriate for budburst, and 7°C for rate of leaf appearance, than the mean temperature of 10°C that viticultural researchers have traditionally used for calculating summations. However while these figures may have held true for the experimental conditions used, there are good grounds for querying their direct applicability to the field.

Firstly, the experiments used a diurnal temperature range in each treatment of only 5°C, whereas in nature it is more typically around 10°C over the budburst-early growth period. The experimental temperatures may indeed reflect the range between average night and average day temperatures in the field, but they do not reflect average or extreme minimum temperatures, which are clearly much more important in the physiological and ecological context of avoiding frost injury in spring (Chapter 6).[2] Moreover, for unknown reasons the lower temperatures in the experiment were maintained for 16 hours of the day and the upper temperatures for only eight, so that the computed average temperatures were only 1.7°C higher than the respective minima. Therefore, to have the same effect as the 10/5°C (average 6.7°C) experimental treatment, one might need realistic field temperatures of about 15/5°C (mean 10°C). In other words, the disagreement with the traditional 10°C base mean temperature could be more apparent than real.

Secondly, Moncur et al. used temperature regimes which were constant throughout the experimental growth period, whereas in nature, temperatures rise over the same period. Some lag can be expected in the field before visible or measurable responses to given temperatures appear, especially in the lower temperature range where all processes are slow. Therefore, even if responses leading to budburst or growth do occur below 10°C mean temperature, they are unlikely to – and in practice do not – attain expression until the means have reached or exceeded 10°C. A base of 10°C is therefore still potentially appropriate *in practice*, for defining the season of actual, visible growth.

---

2 I might be accused of inconsistency here, because I use estimated *average* night temperatures in Chapter 6 as a basis on which to adjust for diurnal temperature range in calculating effective temperature summations; and true effective temperature summations, in part, would depend in the field on the timing of budburst. It should be pointed out, however, that variations in the estimated average night temperatures in my case directly reflect those in the full diurnal temperature range, so the adjustments remain appropriate for phenomena related to both average night and average minimum temperatures.

References

Abbott, Nerida A., Coombe, B.G. and Williams, P.J. (1990a). The flavour of Shiraz as characterized by chemical and sensory analysis. Australian and New Zealand Wine Industry Journal 5(4), 31–59.

Abbott, Nerida A., Coombe, B.G., Sefton, M.A. and Williams, P.J. (1990b). The secondary metabolites of Shiraz grapes as an index of table wine quality. Proceedings, Seventh Australian Wine Industry Technical Conference, Adelaide 1989, 117–20.

Acock, B., Acock, M.C. and Pasternak, D. (1990). Interactions of CO_2 enrichment and temperature on carbohydrate production and accumulation in muskmelon leaves. Journal of the American Society for Horticultural Science 115, 525–9.

Adelsheim, A. (1988). Oregon experiences with Pinot Noir and Chardonnay. Proceedings, Second International Symposium for Cool Climate Viticulture and Oenology, Auckland 1988, 26–49.

Allen, M.S., Lacey, M.J., Harris, R.L.N. and Brown, W.V. (1988). Sauvignon Blanc and varietal aroma. Australian Grapegrower and Winemaker, No. 292 April 1988, 51–6.

Allen, M.S., Lacey, M.J., Brown, W.V. and Harris, R.L.N. (1989). Varietal aromas in Sauvignon Blanc wines. Practical Winery and Vineyard 10(1), 15–20.

Allen, M.S., Lacey, M.J., Brown, W.V. and Harris, R.L.N. (1990a). Contribution of methoxypyrazines to the flavour of Cabernet Sauvignon and Sauvignon Blanc grapes and wines. Proceedings, Seventh Australian Wine Industry Technical Conference, Adelaide 1989, 113–6.

Allen, M.S., Lacey, M.J., Harris, R.L.N. and Brown, W.V. (1990b). Viticultural influences in methoxypyrazines in Sauvignon Blanc. Australian and New Zealand Wine Industry Journal 5(1), 44–6.

Alleweldt, G., Düring, H. and Jung, K.H. (1984). The effects of climate on berry development, yield and quality of grapevines: results of a seven-year factor analysis. Vitis 23, 127–42.

Amerine, M.A. and Joslyn, M.A. (1970). Table Wines: the Technology of their Production. Ed. 2. (University of California Press: Berkeley)

Amerine, M.A. and Winkler, A.J. (1944). Composition and quality of musts and wines of California grapes. Hilgardia 15, 493–675.

Aney, W.A. (1974). Oregon climates exhibiting adaptation potential for vinifera. American Journal of Enology and Viticulture 25, 212–8.

Anon. (1987). Regional report: Mornington Peninsula. Australian and New Zealand Wine Industry Journal 1(4), 11–23.

Anon. (1988). The southern-most wines in the world. Queenstown comes of age. Australian Beverage Review No. 23, August-September 1988, 24.

Anon. (1990). Regional report: Macedon region. Australian and New Zealand Wine Industry Journal 5(1), 17–25.

Antcliff, A.J. and May, P. (1961). Dormancy and bud burst in Sultana vines. Vitis 3, 1–14.

Asher, Gerald (1986). On Wine. (Vintage Books: New York)

Automobile Association, Great Britain (1963). Continental Motoring. (The Automobile Association: London)

Baldwin, J.G. (1965). The relation between weather and fruitfulness of the Sultana vine. Australian Journal of Agricultural Research 15, 902–8.

Baldwin, J.G. (1966). Dormancy and time of bud burst in the Sultana vine. Australian Journal of Agricultural Research 17, 55–68.

Barber, S.A. (1985). Potassium availability at the soil-root interface and factors influencing potassium uptake. In 'Potassium in Agriculture', ed. R.D. Munson, 309–24. (American Society of Agronomy et al.: Madison)

Bavaresco, L. (1989). Nutrizione minerale e resistenza alle malattie (dovute a fattori biotici) della vite. Vignevini 9, 25–35.

Becker, N.J. (1977). Selection of vineyard sites in cool climates. Proceedings, Third Australian Wine Industry Technical Conference, Albury 1977, 25–30.

Becker, N.J. (1988). Wine in the history of civilization and its position in modern society. Proceedings, Second International Symposium for Cool Climate Viticulture and Oenology, Auckland 1988, 289–93.

Bell, A.A., Ough, C.S. and Kliewer, W.M. (1979). Effects on must and wine composition, rates of fermentation, and wine quality of nitrogen fertilization of Vitis vinifera var. Thompson Seedless grapevines. American Journal of Enology and Viticulture 30, 124–9.

Bell, SallyAnn (1991). The effect of nitrogen fertilization on growth, yield and juice composition of Vitis vinifera cv. Cabernet Sauvignon grapevines. Proceedings, International Symposium on Nitrogen in Grapes and Wine, Seattle 1991, 206–10.

Benwell, W.S. (1976). Journey to Wine in Victoria. Ed. 2. (Pitman: Melbourne)

Bernstein, Z. (1984). L'amélioration de la regularité de débourrement dans les regions à hiver doux. Bulletin de l'O.I.V. 57, 480–8.

Bertrand, A. Ingargiola, M.C. and Delas, J. (1991). Effects of nitrogen fertilization and grafting on the composition of must and wine from Merlot grapes, particularly on the presence of ethyl carbamate. Proceedings, International Symposium on Nitrogen in Grapes and Wine, Seattle 1991, 215–20.

Bettenay, E., McArthur, W.M. and Hingston, F.J. (1960). The soil associations of part of the Swan coastal plain, Western Australia. Division of Soils, CSIRO, Australia, Soil and Land Use Series No. 35. (Melbourne)

Bissell, P., Steans, G. and Ewart, A. (1989). A study of colour development in Pinot Noir wines. Australian and New Zealand Wine Industry Journal 4(1), 58–61.

Bisson, Linda F. (1991). Influence of nitrogen on yeast and fermentation of grapes. Proceedings, International Symposium on Nitrogen in Grapes and Wine, Seattle 1991, 78–89.

Bledsoe, A.M., Kliewer, W.M. and Marois, J.J. (1988). Effects of timing and severity of leaf removal on yield and fruit composition of Sauvignon Blanc grapevines. American Journal of Enology and Viticulture 39, 49–54.

Boag, T.S., Tassie, E. and Hubick, K. (1988). The greenhouse effect – implications for the Australian grape and wine industry. Australian and New Zealand Wine Industry Journal 3(2), 30–5.

Boag, S., Hubick, K. and Freeman, B. (1989). Grapegrowing and the greenhouse effect. Practical Winery and Vineyard 10(1), 76–9.

Boehm, E.W. (1970). Vine development and temperature in the Barossa district, South Australia. Department of Agriculture, South Australia, Experimental Record No. 5, 16–24.

Boulton, R. (1980). The general relationship between potassium, sodium and pH in grape juice and wine. American Journal of Enology and Viticulture 31, 182–6.

Brady, Barry (1982). Wines, Winemakers and Vignerons of the South-West and Great Southern Areas of Western Australia. (Apollo Press: Perth)

Brakjovich, M.J. (1988). Cabernet Sauvignon, Merlot and Sauvignon Blanc in Kumeu. Proceedings, Second International Symposium for Cool Climate Viticulture and Oenology, Auckland 1988, 187.

Branas, J., Bernon, G. and Levadoux, L. (1946). In 'Eléments de Viticulture Générale', 41–134. (Cited by Huglin 1983.)

Bryant, E. (1987). CO_2-warming, rising sea level and retreating coasts: review and critique. Australian Geographer 18(2), 101–13.

Bulleid, Nick (1987). Tasmania. Wine and Spirit, August 1987, supplement, pp. 20.

Burden, Rosemary (1976). Wines and Wineries of the Southern Vales. (Rigby: Adelaide)

Bureau of Meteorology, Australia (1953). Sunshine maps of Australia. (Melbourne)

Bureau of Meteorology, Australia (1956). Climatic Averages Australia. (Melbourne)

Bureau of Meteorology, Australia (1962). Climatological Survey Region 12 – Albany, Western Australia. (Melbourne)

Bureau of Meteorology, Australia (1965). Climatic Survey Region 16 – Southwest Western Australia. (Melbourne)

Bureau of Meteorology, Australia (1966). Climatic Survey Region 15 – Metropolitan Western Australia. (Melbourne)

Bureau of Meteorology, Australia (1975). Climatic Averages Australia: New Metric Edition. (Australian Government Publishing Service: Canberra)

Bureau of Meteorology, Australia (1988). Climatic Averages Australia: Second Metric Edition. (Australian Government Publishing Service: Canberra)

Bureau of Meteorology, Australia (undated). Climatic Survey Region 11 – Great Southern, Western Australia. (Melbourne)

Burt, J. and Parr, D. (1988). Marketing quality tomatoes. Department of Agriculture, Western Australia, Farmnote 127/88.

Busby, James (1825). A Treatise on the Culture of the Vine, and the Art of Making Wine. (Government Printer: Australia). Facsimile edition 1979. (David Ell Press: Sydney)

Butcher, E., Cullen, Diana and Cullen, K.J. (1982). Crop yields and viticultural experiments from the Margaret River region of W.A. Australian Grapegrower and Winemaker, No. 224 August 1982, 9–10.

Buttrose, M.S. (1968). Some effects of light intensity and temperature on dry weight and shoot growth of grape vine. Annals of Botany 32, 753–65.

Buttrose, M.S. (1969). Vegetative growth of grapevine varieties under controlled temperature and light intensity. Vitis 8, 280–5.

Buttrose, M.S. (1970). Fruitfulness in grape vines: the response of different cultivars to light, temperature and daylength. Vitis 9, 121–5.

Buttrose, M.S. (1974). Climatic factors and fruitfulness in grapevines. C.A.B. Horticultural Abstracts 44, 319–26.

Buttrose, M.S. and Hale, C.R. (1973). Effect of temperature on development of the grapevine inflorescence after bud burst. American Journal of Enology and Viticulture 24, 14–6.

Buttrose, M.S., Hale, C.R. and Kliewer, W.M. (1971). Effect of temperature on the composition of 'Cabernet Sauvignon' berries. American Journal of Enology and Viticulture 22, 71–5.

Camden, William. See Trevelyan (1944).

Carbonneau, A.P. and Casteran, P. (1987a). Optimization of vine performance by the lyre training systems. Proceedings, Sixth Australian Wine Industry Technical Conference, Adelaide 1986, 194–204.

Carbonneau, A.P. and Casteran, P. (1987b). Interactions 'training system x soil x rootstock' with regard to vine ecophysiology, vigour, yield and red wine quality in the Bordeaux area. Acta Horticulturae 206, 119–40.

Carbonneau, A.P. and Huglin, P. (1982). Adaptation of training systems to French regions. Proceedings, Grape and Wine Centennial Symposium, University of California, Davis 1980, 376–85.

Carbonneau, A.P., Casteran, P. and Leclaire, P. (1978). Essai de détermination en biologie de la plante entière de relations essentielles entre le bioclimat naturel, la physiologie de la vigne et la composition du raisin. Ann. Anal. Plantes 28, 195–221. (Cited by Smart 1982.)

Cawthon, D.L. and Morris, J.R. (1982). Relationship of seed number and maturity to berry development, fruit maturation, hormonal changes and uneven ripening of 'Concord' (Vitis labrusca L.) grapes. Journal of the American Society for Horticultural Science 107, 1097–1104.

Cess, R.D. and 19 others (1989). Interpretation of cloud-climate feedback as produced by 14 atmospheric general circulation models. Science 245, 513–6.

Champagnol, F. (1984). Eléments de Physiologie de la Vigne et de Viticulture Générale. (Déhan: Montpellier)

Chaptal, L. (1943). Contribution à l'étude de la température de l'air dans les couches inférieures de la biosphère. Annales Agronomiques, 427–37. (Cited by Champagnol 1984.)

Clarke, Oz (1991). New Classic Wines. (Websters/Mitchell Beazley: London)

Clingeleffer, P.R. (1984). Minimal pruning – its role in canopy management and implications of its use for the wine industry. Proceedings, Fifth Australian Wine Industry Technical Conference, Perth 1983, 133–45.

Clingeleffer, P.R. (1985). Breeding of grapevines for hot climates. Australian Grapegrower and Winemaker, No. 244 April 1985, 99–104.

Clingeleffer, P.R. (1989). Update: minimal pruning of cordon trained vines (MPCT). Australian Grapegrower and Winemaker, No. 304 April 1989, 78–83.

Clingeleffer, P.R. and Possingham, J.V. (1987). The role of minimum pruning of cordon trained vines (MPCT) in canopy management and its adoption in Australian viticulture. Australian Grapegrower and Winemaker, No. 280 April 1987, 7–11.

Clough, J.M., Peet, M.M. and Kramer, P.J. (1981). Effects of high atmospheric CO_2 and sink size on rates of photosynthesis of a soybean cultivar. Plant Physiology 67, 1007–10.

Cocks, P.S. (1973). The influence of temperature and density on the growth of communities of subterranean clover (Trifolium subterraneum L. cv. Mount Barker). Australian Journal of Agricultural Research 24, 479–95.

Collander, R. (1959). Cell membranes: their resistance to penetration and their capacity for transport. In 'Plant Physiology: a Treatise', ed. F.C. Steward, Vol. II, 3–102. (Academic Press: New York)

Conradie, W.J. (1986). Utilization of nitrogen by the grape vine as affected by time of application and soil type. South African Journal of Enology and Viticulture 7, 76–83.

Conradie, W.J. (1991). Translocation and storage of nitrogen by grapevines as affected by time of application. Proceedings, International Symposium on Nitrogen in Grapes and Wine, Seattle 1991, 32–42.

Conroy, J.P., Milham, P.J., Reed, M.L. and Barlow, E.W. (1990). Increases in phosphorus requirements for CO_2-enriched pine species. Plant Physiology 92, 977–82.

Cook, E., Bird, T., Peterson, M., Barbetti, M., Buckley, B., D'Arrigo, R., Francey, R. and Tans, P. (1991). Climatic change in Tasmania inferred from a 1089-year tree-ring chronology of Huon pine. Science 253, 1266–8.

Coombe, B.G. (1950). Artificial parthenocarpy in grape vines. Journal of the Australian Institute of Agricultural Science 16, 69–70.

Coombe, B.G. (1960). The relationship of growth and development to changes in sugars, auxins, and gibberellins in fruit of seeded and seedless varieties of Vitis vinifera. Plant Physiology 35, 241–50.

Coombe, B.G. (1962). The effect of removing leaves, flowers and shoot tips on fruit set in Vitis vinifera L. Journal of Horticultural Science 37, 1–15.

Coombe, B.G. (1965). The effect of growth substances and leaf number on fruit set and size in Corinth and Sultanina grapes. Journal of Horticultural Science 40, 307–16.

Coombe, B.G. (1973). The regulation of set and development of the grape berry. Acta Horticulturae 34, 261–73.

Coombe, B.G. (1976). The development of fleshy fruits. Annual Review of Plant Physiology 27, 507–28.

Coombe, B.G. (1987). Influence of temperature on composition and quality of grapes. Acta Horticulturae 206, 23–35.

Coombe, B.G. (1988). Grape phenology. In 'Viticulture', Vol. 1: 'Resources in Australia', ed. B.G. Coombe and P.R. Dry, 139–53. (Australian Industrial Publishers: Adelaide)

Coombe, B.G. (1990). Research on berry composition and ripening. Proceedings, Seventh Australian Wine Industry Technical Conference, Adelaide 1989, 150–2.

Coombe, B.G. and Hale, C.R. (1973). The hormone content of ripening grape berries and the effects of growth substance treatments. Plant Physiology 51, 629–34.

Coombe, B.G. and Iland, P.G. (1987). Grape berry development. Proceedings, Sixth Australian Wine Industry Technical Conference, Adelaide 1986, 50–4.

Coughlan, M.J. (1978). Changes in Australian rainfall and temperature. In 'Climatic Change and Variability: a Southern Perspective', ed. A.B. Pittock, L.A. Frakes, D. Jenssen, J.A. Peterson and J.W. Zillman, 194–9. (Cambridge University Press: Cambridge)

Cox, Harry (1967). The Wines of Australia. (Hodder and Stoughton: London)

Creveling, R.K. and Jennings, W.G. (1970). Volatile components of Bartlett pear: higher boiling esters. Journal of Agricultural and Food Chemistry 18, 19–24.

Cripps, J.E.L. and Goldspink, B.H. (1983). Fertilizers for winegrapes. Western Australian Department of Agriculture, Farmnote 116/83.

Croser, B.J. (1987). Cool area winemaking. Proceedings, Sixth Australian Wine Industry Technical Conference, Adelaide 1986, 34–9.

Cullen, Vanya, A. (1988). Canopy management for Sauvignon Blanc vines at Margaret River. Proceedings, Second International Symposium for Cool Climate Viticulture and Oenology, Auckland 1988, 139–43.

Daly, J.L. (1989). The Greenhouse Trap. (Bantam: Moorebank, N.S.W.)

Darst, B.C. and Wallingford, G.W. (1985). Interrelationships of potassium with cultural and management practices. In 'Potassium in Agriculture', ed. R.D. Munson, 559–73. (American Society of Agronomy et al.: Madison)

Daudt, C.E. and Ough, C.S. (1973). Variations in some volatile acetate esters formed during grape juice fermentation. Effects of fermentation temperature, SO_2, yeast strain, and grape variety. American Journal of Enology and Viticulture 24, 130–5.

Davies, P.J. (1987). The plant hormones: their nature, occurrence, and functions. In 'Plant Hormones and Their Role in Plant Growth and Development', ed. P.J. Davies, 1–11. (Martinus Nijhoff: Dordrecht)

Debuigne, Gérard (1976). Larousse Dictionary of Wines of the World. (Hamlyn: London)

DeCandolle, A.P. (1855). Géographie Botanique Raisonée. (Paris, 2 vols). Cited by Prescott (1969b).

De Castella, H. (1886). John Bull's Vineyard, Australian Sketches. (Melbourne). Cited by Peel (1965).

Delas, J., Molot, C. and Soyer, J.P. (1991). Effects of nitrogen fertilization and grafting on the yield and quality of the crop of Vitis vinifera cv. Merlot. Proceedings, International Symposium on Nitrogen in Grapes and Wine, Seattle 1991, 242–8.

Department of Agriculture, USA (1936). Atlas of American Agriculture. (U.S. Government Printing Office: Washington)

Despeissis, A. (1895, 1902, 1921). The Handbook of Horticulture and Viticulture of Western Australia. Ed. 1 1895 (City Press: Perth); Ed. 2 1902 and Ed. 3 1921 (Government Printer: Perth)

Despeissis, A. (1897). See LindleyCowen (1897).

Direction de la Météorologie Nationale, France (1969). Atlas Climatique de la France. (Paris)

Division of National Mapping, Australia (1986). Atlas of Australian Resources, Third Series, Vol. 4: Climate. (Commonwealth Government Printer: Canberra)

Downes, R.W. and Gladstones, J.S. (1984). Physiology of growth and seed production in Lupinus angustifolius L. II. Effect of temperature before and after flowering. Australian Journal of Agricultural Research 35, 501–9.

Downton, W.J.S. (1977). The influence of rootstocks on the accumulation of chloride, sodium and potassium in grapevines. Australian Journal of Agricultural Research 28, 879–89.

Downton, W.J.S., Björkman, O. and Pike, C.S. (1980). Consequences of increased atmospheric concentrations of carbon dioxide for growth and photosynthesis of higher plants. In 'Carbon Dioxide and Climate: Australian Research', ed. G.I. Pearman, 143–51. (Australian Academy of Science: Canberra)

Downton, W.J.S., Grant, W.J.R. and Loveys, B.R. (1987). Carbon dioxide enrichment increases yield of Valencia orange. Australian Journal of Plant Physiology 14, 493–501.

Doyle, S. and Hedberg, P. (1990). The Central Highlands of New South Wales – a premium cool climate growing region. Australian Grapegrower and Winemaker, No. 316 April 1990, 125–9.

Dronia, H. (1967). See Jones et al. (1986b).

Drouhin, R.J. (1988). Burgundy experiences with Pinot Noir and Chardonnay. Proceedings, Second International Symposium for Cool Climate Viticulture and Oenology, Auckland 1988, 270–2.

Dry, P.R. (1984). Recent advances in Australian viticulture. Proceedings, Fifth Australian Wine Industry Technical Conference, Perth 1983, 9–23.

Dry, P.R. (1988). Climate change and the Australian grape and wine industry. Australian Grapegrower and Winemaker, No. 300 December 1988, 14–5.

Dry, P.R. (1990). Observations on Californian viticulture. Australian Grapegrower and Winemaker, No. 321 September 1990, 16–7.

Dry, P.R. and Gregory, G.R. (1988). Grapevine varieties. In 'Viticulture', Vol. 1: 'Resources in Australia', ed. B.G. Coombe and P.R. Dry, 119–38. (Australian Industrial Publishers: Adelaide)

Dry, P.R. and Smart, R.E. (1988a). The grapegrowing regions of Australia. In 'Viticulture', Vol. 1: 'Resources in Australia', ed. B.G. Coombe and P.R. Dry, 37–60. (Australian Industrial Publishers: Adelaide)

Dry, P.R. and Smart, R.E. (1988b). Vineyard site selection. In 'Viticulture', Vol. 1: 'Resources in Australia', ed. B.G. Coombe and P.R. Dry, 190–204. (Australian Industrial Publishers: Adelaide)

Dubourdieu, D. (1990). Complementarity of grape varieties and their sensory influence in the style of red and white great Bordeaux wines. Proceedings, Seventh Australian Wine Industry Technical Conference, Adelaide 1989, 5–7.

Due, G. (1988). Excessive vine vigour in mild climates. Australian and New Zealand Wine Industry Journal 3(1), 37–41.

Dukes, B.C., Goldspink, B.H., Elliott, J.F. and Frayne, R.F. (1991). Timing of nitrogenous fertilization can reduce fermentation time and improve wine quality. Proceedings, International Symposium on Nitrogen in Grapes and Wine, Seattle 1991, 249–54.

Düring, H., Alleweldt, G. and Koch, R. (1978). Studies on hormonal control of ripening in berries of grape vines. Acta Horticulturae 80, 397–405.

Duval, J.R. (1988). Experience with Cabernet Sauvignon at Coonawarra. Proceedings, Second International Symposium for Cool Climate Viticulture and Oenology, Auckland 1988, 186.

Ewart, A.J.W., Iland, P.G. and Sitters, J.H. (1987). The use of shelter in vineyards. Australian Grapegrower and Winemaker, No. 280 April 1987, 19–21.

Fear, C.D. and Nonnecke, Gail R. (1989). Soil mulches influence reproductive and vegetative growth of 'Fern' and 'Tristar' day-neutral strawberries. HortScience 24, 912–3.

Feuillat, M. and Charpentier, C. (1982). Autolysis of yeasts in champagne. American Journal of Enology and Viticulture 33, 6–13.

Folland, C.K., Parker, D.E. and Kates, F.E. (1984). Worldwide marine temperature fluctuations 1856–1981. Nature 310, 670–3.

Foukal, P.V. (1990). The variable sun. Scientific American 262(2), 26–33.

Francis, A.D. (1972). The Wine Trade. (Constable: Edinburgh)

Freeman, B.M. and Kliewer, W.M. (1984). Regulation of grapevine potassium by shoot development. Proceedings, Fifth Australian Wine Industry Technical Conference, Perth 1983, 147–61.

Freeman, B.M., Lee, T.H. and Turkington, C.R. (1980). Interaction of irrigation and pruning level on grape and wine quality of Shiraz vines. American Journal of Enology and Viticulture 31, 124–35.

Freeman, B.M., Kliewer, W.M. and Stern, P. (1982). Influence of wind breaks and climatic region on diurnal fluctuation of leaf water potential, stomatal conductance, and leaf temperature of grapevines. American Journal of Enology and Viticulture 33, 233–6.

Freese, P.K. (1988). Canopy modification and fruit composition. Proceedings, Second International Symposium for Cool Climate Viticulture and Oenology, Auckland 1988, 134–6.

Gadille, Rolande (1967). Le Vignoble de la Côte Bourguignonne. (Paris). Cited by Johnson (1971).

Galet, P. (1958, 1962). Cépages et Vignobles de France. Tomes II and III. (Paul Déhan: Montpellier)

Galet, P. (1979). A Practical Ampelography. Translated from the French by Lucie Morton. (Cornell University Press: London)

Garnier, B.J. (1958). The Climate of New Zealand. (Edward Arnold: London)

Garnier, M. (1967). Climatologie de la France: Selection de Données Statistiques. Mémorial de la Météorologie. (Paris)

Geiger, R. (1965). The Climate near the Ground. English translation of the 4th German edition. (Harvard University Press: Cambridge, Mass.)

Gislerod, H.R. and Mortensen, L.M. (1990). Relative humidity and nutrient concentration affect nutrient uptake and growth of Begonia x hiemalis. HortScience 25, 524–6.

Gladstones, J.S. (1965). The climate and soils of south-western Australia in relation to vine growing. Journal of the Australian Institute of Agricultural Science 31, 275–88.

Gladstones, J.S. (1966). Soils and climate of the Margaret River-Busselton area: their suitability for wine grape production. Department of Agronomy, University of Western Australia, mimeo, pp.11.

Gladstones, J.S. (1976). Criteria for assessing environmental suitability for wine grape growing. Proceedings, Continuing Education Conference, Busselton, Western Australia. (University of Western Australia and Australian Institute of Agricultural Science). Mimeo, pp. 27, tab. 22.

Gladstones, J.S. (1977). Temperature and wine grape quality in European vineyards. Proceedings, Third Australian Wine Industry Technical Conference, Albury 1977, 7–11.

Gladstones, J.S. (1991). Western Australian viticulture – then, now and in the future. Inaugural Wine Baron Lecture, University of Western Australia, 4 October 1990. (Wine Industry Association of Western Australia: Perth)

Gladstones, J.S. (1994). Climatology of cool viticultural environments: reply to criticism and a further examination of Tasmania and New Zealand. Australian and New Zealand Wine Industry Journal 9(4), 349–61.

Goldspink, B.H. (1990). What are your fertilizer costs? Australian and New Zealand Wine Industry Journal 5(1), 31–3.

Goldspink, B.H. and Gordon, C.S. (1991). Response of Vitis vinifera cv. Sauvignon Blanc grapevines to timed applications of nitrogen fertilizers. Proceedings, International Symposium on Nitrogen in Grapes and Wine, Seattle 1991, 255–8.

Goldspink, B.H., Cripps, J.E.L. and Frayne, R.F. (1990). The influence of nitrogen and potassium nutrition on winegrape juice composition. Proceedings, Seventh Australian Wine Industry Technical Conference, Adelaide 1989, 29–34.

Goldspink, B.H., Frayne, R.F. and Dukes, B. (1991). Unpublished data. Department of Agriculture, Western Australia, Baron-Hay Court, South Perth, W.A. 6151, Australia.

Grape Industry Committee, Western Australia (1964). Report to the Minister for Industrial Development. (Department of Agriculture, Western Australia: Perth)

Gregory, G.R. (1963). Soil requirements for grapegrowing. Agricultural Gazette of New South Wales 74, 434–8, 465–7.

Gregory, G.R. (1988). The development and status of Australian Viticulture. In 'Viticulture', Vol. 1: 'Resources in Australia', ed. B.G. Coombe and P.R. Dry, 1–36. (Australian Industrial Publishers: Adelaide)

Gribbin, J. (1978). The Climatic Threat. (Fontana: Glasgow)

Guyot, J. (1865). Culture of the Vine and Wine Making. Translated from the French by L. Marie. (Walker, May & Co.: Melbourne)

Hale, C.R. (1968). Growth and senescence of the grape berry. Australian Journal of Agricultural Research 19, 939–45.

Hale, C.R. (1977). Relation between potassium and the malate and tartrate contents of grape berries. Vitis 16, 9–19.

Hale, C.R. and Buttrose, M.S. (1973). Effect of temperature on anthocyanin content of Cabernet Sauvignon berries. Division of Horticultural Research, CSIRO, Adelaide, Report for 1971–73, 98–9.

Hale, C.R. and Buttrose, M.S. (1974). Effect of temperature on ontogeny of berries of Vitis vinifera L. cv. Cabernet Sauvignon. Journal of the American Society for Horticultural Science 99, 390–4.

Hallgarten, S.F. (1965). Rhineland Wineland. Ed. 4. (Arlington: London)

Halliday, James (1980). Wines and Wineries of New South Wales. (University of Queensland Press: St Lucia)

Halliday, James (1981). Wines and Wineries of South Australia. (University of Queensland Press: St Lucia)

Halliday, James (1982a). Wines and Wineries of Victoria. (University of Queensland Press: St Lucia)

Halliday, James (1982b). Wines and Wineries of Western Australia. (University of Queensland Press: St Lucia)

Halliday, James (1983). Coonawarra: the History, the Vignerons and the Wines. (Yenisey: Sydney)

Halliday, James (1985a). The Australian Wine Compendium. (Angus and Robertson: Sydney)

Halliday, James (1985b). Clare Valley: The History, the Vignerons and the Wines. (Vin Publications: South Yarra)

Halliday, James (1990). Cassegrains seed will bear Clos-al fruits. Weekend Australian, Review p. 15, 6 October 1990.

Halliday, James (1991a). Wine Atlas of Australia and New Zealand. (Angus and Robertson: North Ryde, N.S.W.)

Halliday, James (1991b). Yielding to a demand for quality. Weekend Australian, Review p. 13, 6 April 1991.

Halliday, James and Jarratt, Ray (1979). The Wines and History of the Hunter Valley. (McGraw-Hill: Sydney)

Hamilton, R.P. (1986). Severe wine grape losses in the Margaret River region of Western Australia. Report of a study undertaken from June 1982 to April 1984. Department of Agriculture, Western Australia, mimeo, pp.32.

Hamilton, R.P. (1988). Wind effects on grape vines. Proceedings, Second International Symposium for Cool Climate Viticulture and Oenology, Auckland 1988, 65–8.

Happ, E. (1987). Trellis and vine spacing. Unpublished report. Mimeo, pp. 4.

Happ, E. (1989). Thoughts following a seminar: vine nutrition. Wine Industry Newsletter (Western Australia), No. 21, July 1989, 7–8.

Hardie, W.J. and Cirami, R.M. (1988). Grapevine rootstocks. In 'Viticulture', Vol. 1: 'Resources in Australia', ed. B.G. Coombe and P.R. Dry, 154–76. (Australian Industrial Publishers: Adelaide)

Hardie, W.J. and Considine, J.A. (1976). Response of grapes to water-deficit stress in particular stages of development. American Journal of Enology and Viticulture 27, 55–61.

Hardie, W.J. and Martin, S.R. (1990). A strategy for vine growth regulation by soil water management. Proceedings, Seventh Australian Wine Industry Technical Conference, Adelaide 1989, 51–7.

Hedberg, P.R. and Raison, Joy (1982). The effect of vine spacing and trellising on yield and fruit quality of Shiraz grapes. American Journal of Enology and Viticulture 33, 20–30.

Heinze, R.A. (1977). Regional effects on vineyard development and must composition. Proceedings, Third

Australian Wine Industry Technical Conference, Albury 1977, 18–25.

Helm, K.F. and Cambourne, B. (1988). The influence of climatic variability on the production of quality wines in the Canberra district of South Eastern Australia. Proceedings, Second International Symposium for Cool Climate Viticulture and Oenology, Auckland 1988, 17–20.

Henschke, P.A. and Jiranek, V. (1991). Hydrogen sulphide formation during fermentation: effect of nitrogen composition in model grape must. Proceedings, International Symposium on Nitrogen in Grapes and Wine, Seattle 1991, 172–83.

Henschke, P.M., Young, D., Maggs, J., Iland, P. and Gawel, R. (1990). The effect of canopy management on grape and wine composition and style: experiences at Henschke's Mount Edelstone vineyard. Australian and New Zealand Wine Industry Journal 5(4), 309–14.

Hubble, G.D. (1970). Soils. In 'Australian Grasslands', ed. R.M. Moore, 44–58. (Australian National University Press: Canberra)

Huber, D.M. and Arny, D.C. (1985). Interactions of potassium with plant disease. In 'Potassium in Agriculture', ed. R.D. Munson, 467–88. (American Society of Agronomy et al.: Madison)

Huglin, P. (1978). Nouveau mode d'évaluation des possibilités héliothermiques d'un milieu viticole. C.R. Acad. Agr. France, 1117–26. (Cited by Huglin 1983.)

Huglin, P. (1983). Possibilités d'appréciation objective du milieu viticole. Bulletin de l'O.I.V. 56, 823–33.

Hume, C.J. and Cattle, H. (1990). The greenhouse effect – meteorological mechanisms and models. Outlook on Agriculture 19, 17–23.

Hunt, P.G., Kasperbauer, M.J. and Matheny, T.A. (1990). Influence of Bradyrhizobium japonicum strain and far-red/red canopy light ratios on nodulation of soybean. Crop Science 30, 1306–8.

Hyams, Edward (1987). Dionysus: a Social History of the Wine Vine. (Sidgwick and Jackson: London)

Idso, S.B. (1980). The climatological significance of a doubling of Earth's atmospheric carbon dioxide concentration. Science 207, 1462–3.

Iland, P.G. (1987). Interpretation of acidity parameters in grapes and wine. Australian Grapegrower and Winemaker, No. 280 April 1987, 81–5.

Iland, P.G. (1988). Grape berry ripening: the potassium story. Australian Grapegrower and Winemaker, No. 289 January 1988, 22–4

Iland, P.G. and Marquis, N. (1990). Pinot Noir – viticultural directions for improving wine quality. Proceedings, Seventh Australian Wine Industry Technical Conference, Adelaide 1989, 233.

Jackson, D.I. and Cherry, N.J. (1988). Prediction of a district's grape-ripening capacity using a latitude-temperature index (LTI). American Journal of Enology and Viticulture 39, 19–26.

Jackson, D.I. and Spurling, M.B. (1988). Climate and viticulture in Australia. In 'Viticulture', Vol. 1: 'Resources in Australia', ed. B.G. Coombe and P.R. Dry, 91–106. (Australian Industrial Publishers: Adelaide)

Jackson, M.G., Timberlake, C.F., Bridle, P. and Vallis, L. (1978). Red wine quality: correlations between colour, aroma and flavour and pigment and other parameters of young beaujolais. Journal of the Science of Food and Agriculture 29, 715–27.

Jeffs, Julian (1961). Sherry. (Faber and Faber: London)

Jiranek, V., Langridge, P. and Henschke, P.A. (1990). Nitrogen requirement of yeast during wine fermentation. Proceedings, Seventh Australian Wine Industry Technical Conference, Adelaide 1989, 166–71.

Johnson, Hugh (1971). The World Atlas of Wine. (Mitchell Beazley: London)

Johnson, Hugh (1989). The Story of Wine. (Mitchell Beazley: London)

Jones, P.D. (1990). The climate of the last 1000 years. Endeavour 14(3), 129–36.

Jones, P.D. and Wigley, T.M.L. (1990). Global warming trends. Scientific American, August 1990, 66–73.

Jones, P.D., Wigley, T.M.L. and Wright, P.B. (1986a). Global temperature variations between 1861 and 1984. Nature 322, 430–4.

Jones, P.D., Raper, S.C.B., Bradley, R.S., Diaz, H.F., Kelly, P.M. and Wigley, T.M.L. (1986b). Northern Hemisphere surface air temperature variations: 1851–1984. Journal of Climate and Applied Meteorology 25 : 161–79.

Jones, P.D., Raper, S.C.B. and Wigley, T.M.L. (1986c). Southern Hemisphere surface air temperature variations: 1851–1984. Journal of Climate and Applied Meteorology 25, 1213–30.

Jordan, A.D. and Croser, B.J. (1984). Determination of grape maturity by aroma/flavour assessment. Proceedings, Fifth Australian Wine Industry Technical Conference, Perth 1983, 261–74.

Joslin, W.S. and Ough, C.S. (1978). Cause and fate of certain C6 compounds formed enzymatically in macerated grape leaves during harvest and wine fermentation. American Journal of Enology and Viticulture 29, 11–7.

Judd, K. (1990). An overview of New Zealand Sauvignon Blanc – 'herbaceous or not herbaceous'. Australian and New Zealand Wine Industry Journal 5(1), 47–50.

Kadar, A.A. (1985). Postharvest biology and technology: an overview. In 'Postharvest Technology of Horticultural Crops', University of California, Division of Agriculture and Natural Resources, Special Publication 3311, 3–7.

Karl, T.R., Diaz, H.F. and Kukla, G. (1988). Urbanization: its detection and effect in the United States climate record. Journal of Climate 1, 1099–123.

Kasperbauer, M.J. (1987). Far-red light reflection from green leaves and effects on phytochrome-mediated assimilate partitioning under field conditions. Plant Physiology 85, 350–4.

Kasperbauer, M.J., Hunt, P.G. and Sojka, R.E. (1984). Photosynthate partitioning and nodule formation in soybean plants that received red or far-red light at the end of the photosynthetic period. Physiologia Plantarum 61, 549–54.

Kataoka, I., Sugiura, A., Utsunomiya, N. and Tomana, T. (1982). Effect of abscisic acid and defoliation on anthocyanin accumulation in Kyoho grapes (Vitis vinifera L. x V. labruscana Bailey). Vitis 21, 325–32.

Kelly, A.C. (1867). Wine Growing in Australia. (E.S. Wigg: Adelaide)

Kelly-Treadwell, P.H. (1988). Protease activity in yeast: its relationship to autolysis and champagne character. Australian Grapegrower and Winemaker, No. 292 April 1988, 58–66.

Kerr, R.A. (1989). Greenhouse sceptic out in the cold. Science 246, 1118–9.

Kerr, R.A. (1990). New greenhouse report puts down dissenters. Science 249, 481–2.

Killian, E. and Ough, C.S. (1979). Fermentation esters – formation and retention as affected by fermentation temperature. American Journal of Enology and Viticulture 30, 301–5.

Kirk, J.T.O. (1986). Application of a revised temperature summation system to Australian viticultural regions. Australian Grapegrower and Winemaker, No. 268 April 1986, 48–52.

Kliewer, W.M. (1968). Changes in concentration of free amino acids in grape berries during maturation. American Journal of Enology and Viticulture 19, 166–74.

Kliewer, W.M. (1975). Effect of root temperature on budbreak, shoot growth, and fruit set of 'Cabernet Sauvignon' grape vines. American Journal of Enology and Viticulture 26, 82–9.

Kliewer, W.M. (1977). Influence of temperature, solar radiation and nitrogen on coloration and composition of Emperor grapes. American Journal of Enology and Viticulture 28, 96–103.

Kliewer, W.M. and Antcliff, A.J. (1970). Influence of defoliation, leaf darkening, and cluster shading on the growth and composition of sultana grapes. American Journal of Enology and Viticulture 21, 26–36.

Kliewer, W.M. and Bledsoe, A.M. (1987). Influence of hedging and leaf removal on canopy microclimate, grape composition, and wine quality under California conditions. Acta Horticulturae 206, 157–68.

Kliewer, W.M. and Gates, D. (1987). Wind effects on grapevine growth, yield and fruit composition. Australian and New Zealand Wine Industry Journal 2(1), 30–7.

Kliewer, W.M. and Ough, C.S. (1970). The effect of leaf area and crop level on the concentration of amino acids and total nitrogen in 'Thompson Seedless' grapes. Vitis 9, 196–206.

Kliewer, W.M. and Torres, R.E. (1972). Effect of controlled day and night temperatures on grape coloration. American Journal of Enology and Viticulture 23, 71–7.

Kliewer, M.W., Marois, J.J., Bledsoe, A.M., Smith, S.P., Benz, M.J. and Silvestroni, Oriana (1988). Relative effectiveness of leaf removal, shoot positioning, and trellising for improving winegrape composition. Proceedings, Second International Symposium for Cool Climate Viticulture and Oenology, Auckland 1988, 123–6.

Kobayoshi, A., Fukushima, T., Nii, N. and Harada, K. (1967). Studies on the thermal conditions of grapes. VI. Effects of day and night temperatures on yield and quality of Delaware grapes. Journal of the Japanese Society for Horticultural Science 36, 373–9.

Koblet, W. (1987). Effectiveness of shoot topping and leaf removal as a means of improving quality. Acta Horticulturae 206, 141–56.

Kobriger, J.M., Kliewer, W.M. and Latier, S.T. (1984). Effects of wind on water relations of several grapevine cultivars. American Journal of Enology and Viticulture 35, 164–9.

Kodama, S., Suzuki, T., Fujinawa, S., Del La Teja, P. and Yotsuzuka, F. (1991). Prevention of ethyl carbamate formation in wine by urea degradation using acid urease. Proceedings, International Symposium on Nitrogen in Grapes and Wine, Seattle 1991, 270–3.

Koepf, H.H., Pettersson, B.D. and Schaumann, W. (1976). Bio-dynamic Agriculture. (Anthroposophic Press: Spring Valley, N.Y.)

Köppen, W. (1931). Grundriss der Klimakunde. (Walter de Gruyter & Co: Berlin)

Kriedemann, P.E. (1968). Photosynthesis in vine leaves as a function of light intensity, temperature and leaf age. Vitis 7, 213–20.

Kriedemann, P.E., Torokfalvy, E. and Smart, R.E. (1973). Natural occurrence and photosynthetic utilization of sun-flecks by grapevine leaves. Photosynthetica 7, 18–27. (Cited by Champagnol 1984.)

Kriedemann, P.E., Sward, R.J. and Downton, W.J.S. (1976). Vine response to carbon dioxide enrichment during heat therapy. Australian Journal of Plant Physiology 3, 605–18.

Kukla, G., Gavin, J. and Karl, T.R. (1986). Urban warming. Journal of Climate and Applied Meteorology 25, 1265–70.

Kunkee, R.E. (1991). Relationship between nitrogen content of must and sluggish fermentation. Proceedings, International Symposium on Nitrogen in Grapes and Wine, Seattle 1991, 148–55.

Lacey, M.J., Brown, W.V., Allen, M.S. and Harris, R.L.N. (1988). Alkyl methoxypyrazines and Sauvignon Blanc character. Proceedings, Second International Symposium for Cool Climate Viticulture and Oenology, Auckland 1988, 344–5.

Ladurie, E. LeR. (1972). Times of Feast, Times of Famine. A History of Climate since the Year 1000. Translated from the French by Barbara Bray. (Allen and Unwin: London)

Lake, Max (1964). Hunter Wine. (Jacaranda Press: Brisbane)

Lake, Max (1970). Hunter Winemakers. (Jacaranda Press: Brisbane)

Lamb, H.H. (1982). Climate, History and the Modern World. (Methuen: London)

Lamb, H.H. (1984). Climate of the last thousand years: natural climatic fluctuations and change. In 'The Climate of Europe: Past, Present and Future', ed. H. Flohn and R. Fantechi, 25–64. (Reidel: Dordrecht)

Lang, A. and Thorpe, M. (1988). Why do grape berries split? Proceedings, Second International Symposium for Cool Climate Viticulture and Oenology, Auckland 1988, 69–71.

Langridge, J. and McWilliam, J.R. (1967). Heat responses of higher plants. In 'Thermobiology', ed. A.H. Rose, 231–92. (Academic Press: London)

Lavee, S. (1990). Chemical growth regulation as a tool for controlling vineyard development and production. Proceedings, Seventh Australian Wine Industry Technical Conference, Adelaide 1989, 142–9.

Leforestier, C. (1987). La charte du Bourgogne Pinot: recherche de correlations entre variables sensorielles et analytiques. Mémoire, Diplome d'Ingénieur Esbana, Université de Bourgogne. (Cited by Bissell et al. 1989.)

Lilov, D. and Temenuschka, Andonova (1976). Cytokinins, growth, flower and fruit formation in Vitis vinifera. Vitis 15, 160–70.

Lindley-Cowen, L. (Ed.) (1897). The West Australian Settler's Guide and Farmer's Handbook. (Western Australian Department of Agriculture: Perth)

Lindzen, R. See Kerr (1989).

Löhnertz, O. (1991). Soil nitrogen and the uptake of nitrogen in grapevines. Proceedings, International Symposium on Nitrogen in Grapes and Wine, Seattle 1991, 1–11.

Long, Z.R. (1987). Manipulation of grape flavour in the vineyard: California, North Coast region. Proceedings, Sixth Australian Wine Industry Technical Conference, Adelaide 1986, 82–8.

Ludvigsen, R.K. (1987a). Wise use of irrigation could improve yield without harming wine quality. Australian Grapegrower and Winemaker, No. 280 April 1987, 71–2.

Ludvigsen, R.K. (1987b). Vineyard soil management: use of cover crops. Australian Grapegrower and Winemaker, No. 280 April 1987, 102–8.

McArthur, W.M. and Bettenay, E. (1958). The soils of the Busselton area, Western Australia. Division of Soils, CSIRO, Australia, Divisional Report No. 3/58.

McCarthy, M.G. (1986). Influence of Irrigation, Crop Thinning and Canopy Manipulation on Composition and Aroma of Riesling Grapes. M.Ag.Sc. Thesis, University of Adelaide. (Cited by Coombe and Iland 1987.)

McCarthy, M.G., Cirami, R.M. and Furkaliev, D.G. (1987). Effect of crop load and vegetation growth control on wine quality. Proceedings, Sixth Australian Wine Industry Technical Conference, Adelaide 1986, 75–7.

McIntyre, G.N. (1982). Grapevine Phenology. Ph.D. Thesis, University of Newcastle, New South Wales. (Cited by Coombe 1987.)

McIntyre, G.N., Kliewer, W.M. and Lider, L.A. (1987). Some limitations of the degree day system as used in viticulture in California. American Journal of Enology and Viticulture 38, 128–32.

Marais, J. (1983). Terpenes in the aroma of grapes and wines: a review. South African Journal of Enology and Viticulture 4, 49–58.

Marès, H. (1890). Description des Cépages Principaux de la Région Mediterranéenne de la France. (Camille Coulet: Montpellier)

Matthews, M.A. and Anderson, M.M. (1988). Fruit ripening in *Vitis vinifera* L.: responses to seasonal water deficits. American Journal of Enology and Viticulture 39, 313–20.

May, P. (1965). Reducing inflorescence formation by shading individual sultana buds. Australian Journal of Biological Sciences 18, 463–73.

May, P. (1987). The grapevine as a perennial, plastic and productive plant. Proceedings, Sixth Australian Wine Industry Technical Conference, Adelaide 1986, 40–9.

May, P. and Antcliff, A.J. (1963). The effect of shading on fruitfulness and yield in the Sultana. Journal of Horticultural Science 38, 85–94.

Mazliak, P. (1970). Lipids. In 'The Biochemistry of Fruits and their Products', ed. A.C. Hulme, Vol. 1, 209–38. (Academic Press: London)

Menella, C. (1967). Il Clima D'Italia. Vol. 1: Il Clima D'Italia in Generale. (Fratelli Conte: Naples)

Meteorological Office, Air Ministry, Great Britain (1952). Climatological Atlas of the British Isles. (HMSO: London)

Meteorological Office, Great Britain (1958a). Tables of Temperature, Relative Humidity and Precipitation for the World. Part I: North America, Greenland and the North Pacific Ocean. (HMSO: London)

Meteorological Office, Great Britain (1958b). Tables of Temperature, Relative Humidity and Precipitation for the World. Part VI: Australasia and the South Pacific Ocean. (HMSO: London)

Meteorological Office, Great Britain (1967). Tables of Temperature, Relative Humidity and Precipitation for the World. Part III: Europe and the Atlantic Ocean north of 35°N. Ed. 2. (HMSO: London)

Milthorpe, F.L. and Moorby, J. (1979). An Introduction to Plant Physiology. Ed 2. (Cambridge University Press: Cambridge)

Mitchell, J.F.B., Senior, C.A. and Ingram, W.J. (1989). CO_2 and climate: a missing feedback? Nature 341, 132–4.

Moncur, M.W., Rattigan, K., Mackenzie, D.H. and McIntyre, G.N. (1989). Base temperatures for budbreak and leaf appearance of grapevines. American Journal of Enology and Viticulture 40, 216.

Morgan, D.C. and Smith, H. (1981). Non-photosynthetic responses to light energy. In 'Physiological Plant Ecology. I. Responses to the Physical Environment', ed. O.L. Lange, P.S. Nobel, C.B. Osmond and H. Ziegler, 109–34 (Springer-Verlag: Berlin)

Morgan, D.C., Stanley, C.J. and Warrington, I.J. (1985). The effects of simulated daylight and shade-light on vegetative and reproductive growth in kiwifruit and grapevine. Journal of Horticultural Science 60, 473–84.

Morison, J.I.L. and Gifford, R.M. (1984a). Plant growth and water use with limited water supply in high CO_2 concentrations. I. Leaf area, water use and transpiration. Australian Journal of Plant Physiology 11, 361–74.

Morison, J.I.L. and Gifford, R.M. (1984b). Plant growth and water use with limited water supply in high CO_2 concentrations. II. Plant dry weight, partitioning and water use efficiency. Australian Journal of Plant Physiology 11, 375–84.

Morrison, Janice C. (1988). The effects of shading on the composition of Cabernet Sauvignon grape berries. Proceedings, Second International Symposium for Cool Climate Viticulture and Oenology, Auckland 1988, 144–6.

Mortensen, L.M. (1987). Review: CO_2 enrichment in glasshouses. Crop responses. Scientia Horticulturae 33, 1–25.

Mullins, M.G. (1967). Morphogenetic effects of roots and of some synthetic cytokinins in *Vitis vinifera* L. Journal of Experimental Botany 18, 206–14.

Mullins, M.G. (1968). Regulation of inflorescence growth in cuttings of the grape vine (*Vitis vinifera* L.). Journal of Experimental Botany 19, 532–43.

Mullins, M.G. and Osborne, Daphne J. (1970). Effect of abscisic acid on growth correlation in *Vitis vinifera* L. Australian Journal of Biological Sciences 23, 479–83.

Mullins, M.G. and Rajasekaran, K. (1981). Fruit cuttings: revised method for producing test plants of grapevine cultivars. American Journal of Enology and Viticulture 32, 35–40.

Nelson, C.E. (1968). Data on weather from 1924–1967, irrigated agriculture research and extension center near Prosser, Washington. Washington Agricultural Experiment Station, Bulletin 703.

Newell, R.E. and Dopplick, T.G. (1979). Questions concerning the possible influence of anthropogenic CO_2 on atmospheric temperature. Journal of Applied Meteorology 18, 822–5.

Niimi, Y. and Torikata, H. (1978). Changes in endogenous plant hormones in the xylem sap of grapevines during development. Journal of the Japanese Society for Horticultural Science 47, 181–7.

Nordstrom, K. (1965). Possible control of volatile ester formation in brewing. European Brewery Convention

Proceedings, 10th Congress 1965, 195–208. (Cited by Daudt and Ough 1973.)

Norton, R.A. (1979). Growing grapes for wine and table in the Puget Sound region. Extension Bulletin 0775, Washington State University, Pullman, pp. 7.

Olmo, H.P. (1956). A Survey of the Grape Industry of Western Australia. (Vine Fruits Research Trust: Perth)

Olmo, H.P. (1979). Vineyards in the year 2000: technical pressures. Acta Horticulturae 104, 11–9.

Ough, C.S. (1968). Proline content of grapes and wines. Vitis 7, 321–31.

Ough, C.S. (1991). Influence of nitrogen compounds in grapes on ethyl carbamate formation in wines. Proceedings, International Symposium on Nitrogen in Grapes and Wine, Seattle 1991, 165–71.

Ough, C.S. and Amerine, M.A. (1966). Effects of temperature on winemaking. California Agricultural Experiment Station, University of California, Bulletin 827.

Ough, C.S. and Bell, A.A. (1980). Effects of nitrogen fertilization of grapevines on amino acid metabolism and higher-alcohol formation during grape juice fermentation. American Journal of Enology and Viticulture 31, 122–3.

Ough, C.S. and Lee, T.H. (1981). The effect of vineyard fertilization level on the formation of some fermentation esters. American Journal of Enology and Viticulture 32, 125–7.

Ough, C.S., Stevens, D. and Almy, J. (1989). Preliminary comments on the effects of grape vineyard nitrogen fertilization on the subsequent ethyl carbamate formation in wines. American Journal of Enology and Viticulture 40, 219–20.

Palejwala, V.A., Parikh, H.R. and Modi, V.V. (1985). The role of abscisic acid in ripening grapes. Physiologia Plantarum 65, 498–502.

Panigai, L. and Langellier, F. (1991). Deux années chaudes 1989 et 1990. Le Vigneron Champenois 112(5), 44–61.

Peacock, W.L., Christensen, L.P., Hirschfelt, Donna J., Broadbent, F.E. and Stevens, R.G. (1991). Efficient uptake and utilization of nitrogen in drip- and furrow-irrigated vineyards. Proceedings, International Symposium on Nitrogen in Grapes and Wine, Seattle 1991, 116–9.

Pearman, G.I. (1988). Greenhouse gases: evidence for atmospheric changes and anthropogenic causes. In 'Greenhouse: Planning for Climate Change', ed. G.I. Pearman, 3–21. (CSIRO Australia: Melbourne)

Peel, Lynette J. (1965). Viticulture at Geelong and Lillydale. The Victorian Historical Magazine 36(4), 154–73.

Peel, Lynette J. (1972). Novel primary industries in Victoria: a vineyard at Sunbury in the 1880s. Journal of the Australian Institute of Agricultural Science 38, 76–80.

Penning-Rowsell, E. (1976). The Wines of Bordeaux. (Penguin: Harmondsworth)

Perold, A.I. (1926). A Treatise on Viticulture. (Macmillan: London)

Peynaud, E. and Ribéreau-Gayon, P. (1971). The grape. In 'The Biochemistry of Fruits and their Products', ed. A.C. Hulme, Vol. 2, 171–205. (Academic Press: London)

Pfister, C. (1984). Klimageschichte der Schweiz 1525–1860. Das Klima der Schweiz von 1525–1860 und seine Bedeutung in der Geschichte von Bevölkerung und Landwirtschaft. Vol. 1. (Haupt: Bern). (Cited by Pfister 1988.)

Pfister, C. (1988). Variations in the spring-summer climate of central Europe from the High Middle Ages to 1850.

In 'Long and Short Term Variability of Climate', ed. H. Wanner and U. Siegenthaler, 57–82. (Springer-Verlag: Berlin)

Pharis, R.P. and King, R.W. (1985). Gibberellins and reproductive development in seed plants. Annual Review of Plant Physiology 36, 517–68.

Phillips, I.D.J. (1975). Apical dominance. Annual Review of Plant Physiology 26, 341–67.

Pirie, A.J.G. (1978a). Comparison of the climates of selected Australian, French and Californian wine producing areas. Australian Grapegrower and Winemaker, No. 172 April 1978, 74–8.

Pirie, A.J.G. (1978b). Phenolics Accumulation in Red Wine Grapes (Vitis vinifera). Ph.D. Thesis, University of Sydney.

Pirie, A.J.G. (1979). Red pigment content of winegrapes. Australian Grapegrower and Winemaker, No. 189 September 1979, 10–2.

Pirie, A.J.G. (1982). Interview, reported in The Australian Wine Newsletter, No. 6 August 1982, 2–6, and No. 7 September 1982, 4–7.

Pirie, A.J.G. (1988). Tasmania – recent developments and prospects for expansion. Australian and New Zealand Wine Industry Journal 3(1), 54–7.

Pirie, A.J.G. and Mullins, M.G. (1976). Changes in anthocyanin and phenolics content of grapevine leaf and fruit tissues treated with sucrose, nitrate and abscisic acid. Plant Physiology 58, 468–72.

Pirie, A.J.G. and Mullins, M.G. (1977). Interrelationships of sugars, anthocyanins, total phenols and dry weight in the skin of grape berries during ripening. American Journal of Enology and Viticulture 28, 204–9.

Pirie, A.J.G. and Mullins, M.G. (1980). Concentration of phenolics in the skin of grape berries during fruit development and ripening. American Journal of Enology and Viticulture 31, 34–6.

Pittock, A.B. (1988). Actual and anticipated changes in Australia's climate. In 'Greenhouse: Planning for Climate Change', ed. G.I. Pearman, 35–51. (CSIRO Australia: Melbourne)

Portes, L. and Ruyssen, F. (1886). Traité de la Vigne et de ses Produits. Tome 1. (Octave Doin: Paris)

Possingham, J.V. (1970). Aspects of the physiology of grape vines. In 'Physiology of Tree Crops', ed. L.C. Luckwill and C.V. Cutting, 335–49. (Academic Press: London)

Possingham, J.V., Clingeleffer, P.R. and Cooper, A.M. (1990). Vine management techniques for the wine industry. Australian Grapegrower and Winemaker, No. 316 April 1990, 85–9.

Prescott, J.A. (1965). The climatology of the vine – Vitis vinifera L. The cool limit of cultivation. Transactions of the Royal Society of South Australia 89, 5–23.

Prescott, J.A. (1969a). The climatology of the vine (Vitis vinifera L.). 2. A comparison of temperature regimes in the Australian and Mediterranean regions. Transactions of the Royal Society of South Australia 93, 1–6.

Prescott, J.A. (1969b). The climatology of the vine (Vitis vinifera L.). 3. A comparison of France and Australia on the basis of the warmest month. Transactions of the Royal Society of South Australia 93, 7–15.

Puillat, V. (1888). Mille Variétés de Vignes. Ed. 3. (Camille Coulet: Montpellier)

Pym, L.W. (1955). Soils of the Swan Valley vineyard area, Western Australia. Division of Soils, CSIRO, Australia, Soils and Land Use Series No. 15. (Melbourne)

Quintana Gana, M. and Gomez Pinol, J.M. (1989). Influence des sols, de la climatologie et d'autres facteurs sur le contenu phénolique du raisin de la variété Xarello. Bulletin de l'O.I.V. 62, 485–97.

Radler, F. and Fäth, K.-P. (1991). Histamines and other biogenic amines in wines. Proceedings, International Symposium on Nitrogen in Grapes and Wine, Seattle 1991, 185–95.

Ramanathan, V. (1988). The greenhouse theory of climate change: a test by an inadvertent global experiment. Science 240, 293–9.

Ramanathan, V., Cess, R.D., Harrison, E.F., Minnis, P., Barkstrom, B.R., Ahmad, E. and Hartmann, D. (1989). Cloud-radiative forcing and climate: results from the Earth Radiation Budget Experiment. Science 243, 57–63.

Rankine, Bryce (1971). Wines and Wineries of the Barossa Valley. (Jacaranda Press: Brisbane)

Rankine, Bryce (1989). Making Good Wine: a Manual of Winemaking Practice for Australia and New Zealand. (Macmillan: Melbourne)

Rapp, A. and Versini, G. (1991). Influence of nitrogen compounds in grapes on aroma compounds in wines. Proceedings, International Symposium on Nitrogen in Grapes and Wine, Seattle 1991, 156–64.

Rasool, S.I. and Schneider, S.H. (1971). Atmospheric carbon dioxide and aerosols: effects of large increases on global climate. Science 173, 138–41.

Rawson, H.M., Begg, J.E. and Woodward, R.G. (1977). The effect of atmospheric humidity on photosynthesis, transpiration and water use efficiency of leaves of several plant species. Planta 134, 5–10.

Ray, Cyril (1968). Lafite. (Peter Davies: London)

Ray, Cyril (1979). The Wines of Germany. (Penguin: Harmondsworth)

Reid, D.M. and Crozier, A. (1971). Effects of waterlogging on the gibberellin content and growth of tomato plants. Journal of Experimental Botany 22, 39–48.

Rendu, V. (1857). Ampélographie Francaise. (Victor Masson: Paris)

Reynolds, A.G. and Wardle, D.A. (1988). Canopy microclimate of Gewürztraminer and monoterpene levels. Proceedings, Second International Symposium for Cool Climate Viticulture and Oenology, Auckland 1988, 116–22.

Richards, D. (1983). The grape root system. Horticultural Reviews 5, 127–68.

Robertson, George (1987). Port. (Faber and Faber: London)

Robinson, Jancis (1986). Vines, Grapes and Wines. (Mitchell Beazley: London)

Royal Commission on Vegetable Products, Victoria (1891). Handbook on Viticulture. (Government Printer: Melbourne)

Ruhl, E.H. (1988). Experience with irrigation of winegrapes in West Germany. Australian Grapegrower and Winemaker, No. 292 April 1988, 99–101, 104.

Ruhl, E.H. and Walker, R.R. (1990). Rootstock effects on grape juice potassium concentration and pH. Proceedings, Seventh Australian Wine Industry Technical Conference, Adelaide 1989, 75–8.

Russell, R.S. and Barber, D.A. (1960). The relationship between salt uptake and the absorption of water by intact plants. Annual Review of Plant Physiology 11, 127–40.

Saayman, D. (1982). Soil preparation studies. II. The effect of depth and method of soil preparation and of organic material on the performance of Vitis vinifera (var. Colombar) on Clovelly/Hutton soil. South African Journal of Enology and Viticulture 3, 61–74.

Saayman, D. and Kleynhans, P.H. (1978). The effect of soil type on wine quality. Proceedings, South African Society of Enology and Viticulture, 21st Annual Meeting, Cape Town 1978, 105–19.

Sanson, J. (1945). Recueil de Données Statistiques Relatives à la Climatologie de la France. Mémorial de la Météorologie Nationale, No. 30. (Paris)

Saugier, B. (1977). Micrometeorology on crops and grasslands. In 'Environmental Effects on Crop Physiology', ed. J.J. Landsberg and C.V. Cutting, 39–55. (Academic Press: London)

Schneider, S.H. (1989a). The greenhouse effect: science and policy. Science 243, 771–81.

Schneider, S.H. (1989b). The changing climate. Scientific American, September 1989, 70–9.

Schneider, S.H. and Mass, C. (1975). Volcanic dust, sunspots, and temperature trends. Science 190, 741–6.

Scudamore-Smith, P.D. (1988). Oenological evaluation of experimental wine grape cultivars grown in south-east Queensland. Queensland Journal of Agricultural and Animal Sciences 45, 195–204.

Seguin, G. (1983). Influence des terroirs viticoles. Bulletin de l'O.I.V. 56, 3–18.

Seguin, G. (1986). 'Terroirs' and pedology of wine growing. Experientia 42, 861–72.

Sellers, W.D. (1973). A new global climatic model. Journal of Applied Meteorology 12, 241–54.

Servicio Meteorologico Nacional, Spain (1946). Calendario Meteorofenologico. (Madrid)

Servico Meteorologico Nacional, Portugal (1965). Valores Medios dos Elementos Climaticos no Territorio Nacional em 1931–1960. (Lisbon)

Seward, Desmond (1979). Monks and Wine (Mitchell Beazley: London).

Shand, P. Morton (1964). A Book of French Wines. Ed. 2, revised and edited by Cyril Ray. (Penguin: Harmondsworth)

Shaulis, N., Amberg, H. and Crowe, D. (1966). Response of Concord grapes to light, exposure and Geneva double curtain training. Proceedings of the American Society for Horticultural Science 89, 268–80.

Siegelman, H.W. (1969). Phytochrome. In 'Physiology of Plant Growth and Development', ed. M.B. Wilkins, 489–506. (McGraw-Hill: London)

Simon, André L. (1971). The History of Champagne. (Octopus: London)

Sinton, T.H., Ough, C.S., Kissler, J.J. and Kasimatis, A.N. (1978). Grape juice indicators for prediction of potential wine quality. I. Relationship between crop level, juice and wine composition, and wine sensory ratings and scores. American Journal of Enology and Viticulture 29, 267–71.

Skene, K.G.M. and Kerridge, G.H. (1967). Effect of root temperature on cytokinin activity in root exudates of Vitis vinifera L. Plant Physiology 42, 1131–9.

Skoog, F. and Schmitz, Ruth Y. (1972). Cytokinins. In 'Plant Physiology: a Treatise', ed. F.C. Steward, Vol. VIB, 181–213. (Academic Press: New York)

Slingo, Tony (1989). Wetter clouds dampen greenhouse warming. Nature 341, 104.

Smart, R.E. (1973). Sunlight interception by vineyards. American Journal of Enology and Viticulture 24, 141–7.

Smart, R.E. (1977). Climate and grapegrowing in Australia. Proceedings, Third Australian Wine Industry Technical Conference, Albury 1977, 12–18.

Smart, R.E. (1982). Vine manipulation to improve wine grape quality. Proceedings, Grape and Wine Centennial Symposium, University of California, Davis, 1980, 362–75.

Smart, R.E. (1984). Canopy microclimates and effects on wine quality. Proceedings, Fifth Australian Wine Industry Technical Conference, Perth 1983, 113–32.

Smart, R.E. (1985a). Principles of grapevine canopy microclimate manipulation with implications for yield and quality. American Journal of Enology and Viticulture 36, 230–9.

Smart, R.E. (1985b). Trellis trial improves yield and quality with Gewürztraminer. Australian Grapegrower and Winemaker, No. 263 November 1985, 10.

Smart, R.E. (1987a). Canopy management to improve yield, fruit composition and vineyard mechanization – a review. Proceedings, Sixth Australian Wine Industry Technical Conference, Adelaide 1986, 205–11.

Smart, R.E. (1987b). Influence of light on composition and quality of grapes. Acta Horticulturae 206, 37–47.

Smart, R.E. (1988). Will the real winemakers please stand up? Australian and New Zealand Wine Industry Journal 3(3), 3–7.

Smart, R.E. (1989). Climate change and the New Zealand wine industry – prospects for the third millenium. Australian and New Zealand Wine Industry Journal 4(1), 8–11.

Smart, R.E. (1991). Canopy microclimate implications for nitrogen effects on yield and quality. Proceedings, International Symposium on Nitrogen in Grapes and Wine, Seattle 1991, 90–101.

Smart, R.E. and Dry, P.R. (1980). A climatic classification for Australian viticultural regions. Australian Grapegrower and Winemaker, No. 196 April 1980, 8, 10, 16.

Smart, R.E. and Robinson, M. (1991). Sunlight into Wine: a Handbook for Winegrape Canopy Management. (Winetitles: Adelaide)

Smart, R.E. and Smith, S.M. (1988). Canopy management: identifying the problems and practical solutions. Proceedings, Second International Symposium for Cool Climate Viticulture and Oenology, Auckland 1988, 109–15.

Smart, R.E., Alcorso, C. and Hornsby, D.A. (1980). A comparison of winegrape performance at the present limits of Australian viticultural climates – Alice Springs and Hobart. Australian Grapegrower and Winemaker, No. 184 April 1980, 28–30.

Smart, R.E., Shaulis, N.J. and Lemon, E.R. (1982a). The effect of Concord vineyard microclimate on yield. I. The effects of pruning, training and shoot positioning on radiation microclimate. American Journal of Enology and Viticulture 33, 99–108.

Smart, R.E., Shaulis, N.J. and Lemon, E.R. (1982b). The effect of Concord vineyard microclimate on yield. II. The interrelations between microclimate and yield expression. American Journal of Enology and Viticulture 33, 109–16.

Smart, R.E., Robinson, J.B., Due, G.R. and Brien, C.J. (1985a). Canopy microclimate modification for the cultivar Shiraz. I. Definition of canopy microclimate. Vitis 24, 17–31.

Smart, R.E., Robinson, J.B. Due, G.R. and Brien, C.J. (1985b). Canopy microclimate modification for the cultivar Shiraz. II. Effects on must and wine composition. Vitis 24, 119–28.

Smart, R.E., Dick, Joy K. and Gravett, Isabella M. (1990a). Shoot devigoration by natural means. Proceedings, Seventh Australian Wine Industry Technical Conference, Adelaide 1989, 58–65.

Smart, R.E., Dick, Joy K., Gravett, Isabella M. and Fisher, B.M. (1990b). Canopy management to improve grape yield and wine quality – principles and practice. South African Journal of Enology and Viticulture 11(1), 3–17.

Smart, R.E., Kirchhof, G. and Blackwell, J. (1991). The diagnosis and treatment of acidic vineyard soil. Australian and New Zealand Wine Industry Journal 6 (1), 35–8, 40.

Smith, H. (1972). The photocontrol of flavonoid biosynthesis. In 'Phytochrome', ed. K. Mitrakos and W. Shropshire, 433–81. (Academic Press: London)

Smith, H. (1982). Light quality, photoperception, and plant strategy. Annual Review of Plant Physiology 33, 481–518.

Smith, R. (1948). Soil types of the Margaret River district. Journal of Agriculture, Western Australia (2)25, 426–37.

Smith, R. (1951). Soils of the Margaret River – Lower Blackwood districts, Western Australia. CSIRO, Australia, Bulletin 262.

Smith, S.M., Codrington, I.C., Robertson, M. and Smart, R.E. (1988). Viticultural and oenological implications of leaf removal for New Zealand vineyards. Proceedings, Second International Symposium for Cool Climate Viticulture and Oenology, Auckland 1988, 127–33.

Smithsonian Institute (1951). The Smithsonian Meteorological Tables, 6th Revised Edition. (Washington)

Somers, T.C. (1975). In search of quality for red wines. Food Technology in Australia 27, 49–56.

Somers, T.C. (1977). A connection between potassium levels in the harvest and relative quality in Australian red wines. Australian Wine, Brewing and Spirit Review, 24 May 1977, 32–4.

Somers, T.C. and Evans, M.E. (1974). Wine quality: correlations with colour density and anthocyanin equilibria in a group of young red wines. Journal of the Science of Food and Agriculture 25, 1369–79.

Spencer, Mary (1965). Fruit ripening. In 'Plant Biochemistry', ed. J. Bonner and J.E. Varner, 793–825. (Academic Press: New York)

Sponholz, W.R. (1991). Nitrogen compounds in grapes, must and wine. Proceedings, International Symposium on Nitrogen in Grapes and Wine, Seattle 1991, 67–77.

Srinivasan, C. and Mullins, M.G. (1976). Reproductive anatomy of the grapevine (Vitis vinifera L.): origin and development of the anlage and its derivatives. Annals of Botany 38, 1079–84.

Srinivasan, C. and Mullins, M.G. (1978). Control of flowering in the grapevine (Vitis vinifera L.). Formation of inflorescences in vitro by isolated tendrils. Plant Physiology 61, 127–30.

Srinivasan, C. and Mullins, M.G. (1980). Effects of temperature and growth regulators on formation of anlagen, tendrils and inflorescences in *Vitis vinifera* L. Annals of Botany 45, 439–46.

Srinivasan, C. and Mullins, M.G. (1981). Physiology of flowering in the grapevine – a review. American Journal of Enology and Viticulture 32, 47–63.

Stevens, K.L., Bomben, J.L. and McFadden, W.H. (1967). Volatiles from grapes, *Vitis vinifera* (Linn.) cultivar Grenache. Journal of Agricultural and Food Chemistry 15, 378–80.

Strauss, C.R., Wilson, B. and Williams, P.J. (1987). Flavour of non-Muscat varieties. Proceedings, Sixth Australian Wine Industry Technical Conference, Adelaide 1986, 117–20.

Stuckey, Wendy, Iland, P., Henschke, P.A. and Gawel, R. (1991a). Influence of lees contact on quality and composition of Chardonnay wine. Australian and New Zealand Wine Industry Journal 6(4), 281–3, 285.

Stuckey, Wendy, Iland, P., Henschke, P.A. and Gawel, R. (1991b). The effect of lees contact time on Chardonnay wine composition. Proceedings, International Symposium on Nitrogen in Grapes and Wine, Seattle 1991, 315–9.

Thimann, K.V. (1972). The natural plant hormones. In 'Plant Physiology: a Treatise', ed. F.C. Steward, Vol. VIB, 3–332. (Academic Press: London)

Thorpy, Frank (1973). Wine of New Zealand. In 'Australia and New Zealand Complete Book of Wine', ed. Len Evans, 475–513. (Paul Hamlyn: Sydney)

Torrey, J.G. (1976). Root hormones and plant growth. Annual Review of Plant Physiology 27, 435–59.

Trevelyan, G.M. (1944). English Social History. (Longman, Green and Co.: London). Reprint Society edition 1948.

Tromp, A. (1984). The effect of yeast strain, grape solids, nitrogen and temperature on fermentation rate and wine quality. South African Journal of Enology and Viticulture 5, 1–6.

Truscott, J.H.L. and Warner, J. (1967). Report of the Horticultural Experiment Station, Ontario, for 1966, p. 96. (Cited by Hobson and Davies 1970.)

Tucker, G.B. (1988). Climatic modelling: how does it work? In 'Greenhouse: Planning for Climate Change', ed. G.I. Pearman, 22–34. (CSIRO, Australia: Melbourne)

Usherwood, N.R. (1985). The role of potassium in crop quality. In 'Potassium in Agriculture', ed. R.D. Munson, 489–513. (American Society of Agronomy et al.: Madison)

Van Eimern, J. (1966). Soil climate relations. The dependence of soil temperature on radiation, albedo, moisture and agricultural practices. Proceedings, World Meteorological Seminar, Melbourne 1966, 491–520.

Van Huyssteen, L. (1990). The effect of soil management and fertilization on grape composition and wine quality with special reference to South African conditions. Proceedings, Seventh Australian Wine Industry Technical Conference, Adelaide 1989, 16–25.

Van Overbeek, J. (1962). Proceedings, Plant Sciences Symposium, Campbell Soup Co., 37–56. (Cited by Skoog and Schmitz 1972.)

Van Rooyen, P.C. and Tromp, A. (1982). The effect of fermentation time (as induced by fermentation and must conditions) on the chemical profile and quality of Chenin Blanc wine. South African Journal of Enology and Viticulture 3, 75–80.

Van Zyl, J.L. and van Huyssteen, L. (1984). Soil and water management for optimum grape yield and quality under conditions of limited or no irrigation. Proceedings, Fifth Australian Wine Industry Technical Conference, Perth 1983, 25–66.

Viala, P. and Vermorel, V. (1901–1904). Traité Général de Viticulture: Ampélographie. (Masson et Cie: Paris)

Vos, P.J.A. and Gray, R.S. (1979). The origin and control of hydrogen sulfide during fermentation of grape must. American Journal of Enology and Viticulture 30, 187–97.

Weast, R.C. and Astle, M.J. (Eds) (1982). CRC Handbook of Chemistry and Physics. Ed. 63. (CRC Press: Boca Raton, Florida)

Weather Bureau, Department of Commerce, USA (1953). Normals for 1921–1950.

Weaver, R.J. (1962). The effect of benzo-thiazole–2-oxyacetic acid on maturation of seeded varieties of grapes. American Journal of Enology and Viticulture 13, 141–9.

Weaver, R.J. and Williams, W.O. (1951). Response of certain varieties of grapes to plant growth-regulators. Botanical Gazette 113, 75–85.

Weaver, R.J., van Overbeek, J. and Pool, R.M. (1965). Induction of fruit set in *Vitis vinifera* L. by a kinin. Nature 206, 952–3.

Weaver, R.J., van Overbeek, J. and Pool, R.M. (1966). Effect of kinins on fruit set and development in *Vitis vinifera*. Hilgardia 37, 181–201.

Webster, P.J. and Stephens, L. (1980). Gleaning CO_2-climate relations from model calculations. In 'Carbon Dioxide and Climate: Australian Research', ed. G.I. Pearman, 185–95. (Australian Academy of Science: Canberra)

Weinhold, Rudolf (1978). Vivat Bacchus: a History of the Vine and its Wine. Translated from the German by Neil Jones. (Argus: Watford)

Wendland, W.M. (1983). A fast method to calculate monthly day degrees. Bulletin of the American Meteorological Society 64, 279–81.

Went, F.W. (1953). The effect of temperature on plant growth. Annual Review of Plant Physiology 4, 347–62.

Went, F.W. (1957). Experimental Control of Plant Growth. (Chronica Botanica: Waltham, Mass.)

Went, F.W. and Sheps, Lillian O. (1969). Environmental factors in regulation of growth and development: ecological factors. In 'Plant Physiology, a Treatise', ed. F.C. Steward, vol. VA, 299–406. (Academic Press: New York)

Wermelinger, B. (1991). Nitrogen dynamics in grapevine: physiology and modelling. Proceedings, International Symposium on Nitrogen in Grapes and Wine, Seattle 1991, 23–31.

Westwood, M.N. (1978). Temperate Zone Pomology. (W.H. Freeman: San Francisco)

Whiting, J.R. (1988). Influences of rootstocks on yield, juice composition and growth of Chardonnay. Proceedings, Second International Symposium for Cool Climate Viticulture and Oenology, Auckland 1988, 48–50.

Wigley, T.M.L. (1989). Possible climate change due to SO_2-derived cloud condensation nuclei. Nature 339, 365–7.

Wigley, T.M.L. (1991). Could reducing fossil fuel emissions cause global warming? Nature 349, 503–6.

Williams, L.E. (1991). Vine nitrogen requirements – utilization of N sources from soils, fertilizers and reserves. Proceedings, International Symposium on Nitrogen in Grapes and Wine, Seattle 1991, 62–6.

Williams, P.J., Strauss, C.R. and Wilson, B. (1988). Developments in flavour research on premium varieties. Proceedings, Second International Symposium for Cool Climate Viticulture and Oenology, Auckland 1988, 331–4.

Williams, P.J., Strauss, C.R., Aryan, A.P. and Wilson, B. (1987). Grape flavour – a review of some pre- and post-harvest influences. Proceedings, Sixth Australian Wine Industry Technical Conference, Adelaide 1986, 111–6.

Windholz, Martha; Budavari, Susan; Blumetti, Rosemary F. and Otterbein, Elizabeth S. (Eds) (1983). The Merck Index. Ed. 10. (Merck & Co.: Rahway, N.J.)

Winkler, A.J. (1959). The effect of vine spacing at Oakville on yields, fruit composition, and wine quality. American Journal of Enology and Viticulture 10, 39–43.

Winkler, A.J. (1962). General Viticulture. (University of California Press: Berkeley). Australian Edition 1963. (Jacaranda Press: Brisbane)

Wittwer, S.H. (1990). Implications of the greenhouse effect on crop productivity. HortScience 25, 1560–7.

Wong, S.C. (1980). Effects of elevated partial pressure of CO_2 on rate of CO_2 assimilation and water use efficiency in plants. In 'Carbon Dioxide and Climate: Australian Research', ed. G.I. Pearman, 159–66. (Australian Academy of Science: Canberra)

Woodham, R.C. and Alexander, D.McE. (1966). The effect of root temperature on development of small fruiting Sultana vines. Vitis 5, 345–50.

Woodward, F.I. (1987). Stomatal numbers are sensitive to increases in CO_2 from pre-industrial levels. Nature 327, 617–8.

World Meteorological Organization/United Nations Environment Programme (1990). Impacts Assessment of Climate Change: the Policymakers' Summary of the Report of Working Group II to the Intergovernmental Panel on Climate Change. (Australian Government Publishing Service: Canberra)

Zekulich, Mike (1990). Wines and Wineries of the West. (St George Books: Perth)

Zelleke, A. and Kliewer, W.M. (1979). Influence of root temperature and rootstock on budbreak, shoot growth, and fruit composition of Cabernet Sauvignon grapevines grown under controlled conditions. American Journal of Enology and Viticulture 30, 312–7.

Zelleke, A. and Kliewer, W.M. (1981). Factors affecting the qualitative and quantitative levels of cytokinin in xylem sap of grapevines. Vitis 20, 93–104.

Author Index

Abbott, Nerida 10
Acock, B. 267–8
Adelsheim, A. 253
Alexander, D.McE. 13
Allen, M.S. 10, 37
Alleweldt, G. 9
Amerine, M.A. 4, 9, 56, 62, 128, 246
Anderson, M.M. 27
Aney, W.A. 253
Antcliff, A.J. 18, 20
Arny, D.C. 37
Asher, Gerald 41, 258
Astle, M.J. 10
Automobile Association, Great
 Britain 281
Baldwin, J.G. 12, 13, 18, 20
Barber, S.A. 29
Bavaresco, L. 37–9
Becker, N.J. 3, 21, 216
Bell, A.A. 55
Bell, Sally-Ann 54
Benwell, W.S. 124
Bernstein, Z. 60
Bertrand, A. 56
Bettenay, E. 79, 98
Bissell, P. 55
Bisson, Linda 55–6
Bledsoe, A.M. 27, 46
Boag, T.S. 269
Boehm, E.W. 107, 121–2
Boulton, R. 29
Brady, Barry 72
Brajkovich, M.J. 37
Branas, J. 64, 284
Bryant, E. 261
Bulleid, Nick 183
Burden, Rosemary 104
Bureau of Meteorology, Australia 69,
 101–2, 111–2, 121–3, 147–8, 150, 158,
 176–7, 184, 191, 281–3
Burt, J. 9
Busby, James 4, 33, 46
Butcher, E. 58
Buttrose, M.S. 9, 11–3, 17, 19–20, 62
Cambourne, B. 63
Camden, William 259
Cameron, Ian 80
Carbonneau, A.P. 2, 21, 28, 33, 46, 48
Casteran, P. 33, 46, 48
Cattle, H. 260–2
Cawthon, D.L. 14
Cess, R.D. 262
Champagnol, F. 33, 37–8, 46, 58
Chaptal, L. 47
Charpentier, C. 56
Cherry, N.J. 6, 7, 64, 192, 285–6
Cirami, R.M. 60
Clarke, Oz 130
Clingeleffer, P.R. 49, 277
Clough, J.M. 268
Cocks, P.S. 51

Collander, R. 10
Conradie, W.J. 57
Conroy, J.P. 268
Considine, J.A. 27
Cook, E. 186
Coombe, B.G. 6, 13–4, 61, 63, 66, 279
Coughlan, M.J. 263, 282
Cox, Harry 151, 157
Creveling, R.K. 10
Cripps, J.E.L. 54, 75
Croser, B.J. 1, 37, 61
Crozier, A. 279
Cullen, Vanya 48
Daly, J.L. 260–2
Darst, B.C. 58
Daudt, C.E. 9
Davies, P.J. 279
Debuigne, Gérard 211, 257
De Candolle, A.P. 5
De Castella, H. 124
Delas, J. 54
Department of Agriculture, USA 281
Despeissis, A. 4, 33, 80, 82
Direction de la Météorologie
 Nationale, France 281
Division of National Mapping,
 Australia 111, 125–6, 154
Dopplick, T.G. 261–2
Downes, R.W. 11
Downton, W.J.S. 60, 268
Doyle, S. 158
Dronia, H. 263
Drouhin, R.J. 46
Dry, P.R. 6–7, 15, 30, 32, 53, 66, 269
Dubourdieu, D. 67, 78, 266
Due, G. 11
Dukes, B.C. 54, 57
Duval, J.R. 37
Evans, M.E. 9, 50
Ewart, A.J.W. 31, 59
Falstaff, Sir John 259
Fäth, K.P. 56
Fear, C.D. 15
Feuillat, M. 56
Fitzpatrick, E.A. 62
Folland, C.K. 185, 256, 263–4, 266
Foukal, P.V. 260
Francis, A.D. 257, 259
Freeman, B.M. 28–9, 31, 49, 51
Freese, P.K. 46
Gadille, Rolande 17, 20–1
Galet, P. 66
Garnier, B.J. 281
Garnier, M. 281
Gates, D. 31
Geiger, R. 36, 41, 44
Gifford, R.M. 268
Gislerod, H.R. 268
Gladstones, J.S. 5–6, 8, 11, 41, 62–63,
 69, 72, 186, 191, 261
Goldspink, B.H. 29, 53–5, 57

Gomez Pinol, J.M. 33
Gordon, C.S. 54, 57
Grape Industry Committee, Western
 Australia 5
Gray, R.S. 55
Gregory, G.R. 4, 33, 66
Gribbin, J. 242, 255, 260
Guyot, J. 33, 46, 58
Hale, C.R. 9, 13, 17, 49, 62
Hallgarten, S.F. 219, 259
Halliday, James 4, 72, 104, 106,
 108–10, 124, 126, 129, 132, 151, 154,
 157, 179, 183, 186, 256, 258, 265
Hamilton, R.P. 17, 30–1, 59–60, 78,
 127
Happ, Erland 17, 28, 47, 53, 58
Hardie, W.J. 17, 27, 52, 60
Hedberg, P.R. 48, 158
Heinze, R.A. 127
Helm, K.F. 63
Henschke, P.A. 55
Henschke, P.M. 49
Hingston, F.J. 79, 98
Hubble, G.D. 112, 133
Huber, D.M. 37
Huglin, P. 21, 28, 48, 64, 284
Hume, C.J. 260–2
Hunt, P.G. 47, 59
Hyams, Edward 257
Idso, S.B. 261–2
Iland, P.G. 28–9, 50, 66
Jackson, D.I. 6–7, 64, 192, 284–5
Jackson, M.G. 9, 50
Jarratt, Ray 151
Jeffs, Julian 259
Jennings, W.G. 10
Jiranek, V. 55–6
Johnson, Hugh 2, 6, 12, 17, 19, 21, 34,
 41, 44–5, 209, 211, 214, 227–8, 239–41,
 255–6, 258–9, 265–6
Jones, P.D. 256, 263–4, 266
Jordan, A.D. 37
Joslin, W.S. 10
Joslyn, M.A. 56
Judd, K. 37
Kadar, A.A. 9
Karl, T.R. 185, 263
Kasperbauer, M.J. 27, 47
Kataoka, I. 14
Kelly, Dr. A.C. 4
Kelly-Treadwell, P.H. 56
Kerr, R.A. 256, 261–2
Kerridge, G.H. 13
Killian, E. 9
King, R.W. 279
Kirk, J.T.O. 62
Kleynhans, P.H. 33–4
Kliewer, W.M. 8–9, 13–4, 19–20, 27,
 31, 41, 46, 49, 55.
Kobayoshi, A. 9, 19
Koblet, W. 27, 46

Kobriger, J.M. 31, 59
Kodama, S. 56-7
Koepf, H.H. 280
Köppen, W. 7, 32
Kosovich, John 19
Kriedemann, P.E. 9, 30, 268
Kukla, G. 185, 263
Kunkee, R.E. 38, 55-6
Lacey, M.J. 10
Ladurie, E.LeR. 257
Lake, Max 33, 151
Lamb, H.H. 256-60
Lang, A. 29
Langellier, F. 266
Langridge, J. 65
Lavee, S. 60
Lee, T.H. 55
Leforestier, C. 9, 55
Lilov, D. 14
Lindzen, R. 261-2
Löhnertz, O. 55
Loneragan, J.F. 29
Long, Z.R. 28
Ludvigsen, R.K. 27, 54, 58
McArthur, W.M. 79, 98
McCarthy, M.G. 51
McIntyre, G.N. 5, 64
McWilliam, J.R. 65
Marais, J. 10
Marès, H. 66
Marquis, N. 50
Martin, S.R. 17, 52
Mass, C. 260
Matthews, M.A. 27
May, P. 18, 20, 60
Mazliak, P. 10
Menella, C. 281
Meteorological Office, Air Ministry,
  Great Britain 212, 241, 281, 284
Milthorpe, F.L. 39
Mitchell, J.F.B. 262
Moncur, M.W. 7, 285
Moorby, J. 39
Morgan, D.C. 26
Morison, J.I.L. 268
Morris, J.R. 14
Morrison, Janice 27
Mortensen, L.M. 267-8
Mullins, M.G. 13-4, 20, 50, 279-80
Nelson, C.E. 253, 281
Newell, R.E. 261-2
Nigond, J. 62
Niimi, Y. 14
Nix, H.A. 62
Nonnecke, Gail 15
Nordstrom, K. 9
Norton, R.A. 254
Olmo, H.P. 4-5, 62, 72, 277
Osborne, Daphne 14
Ough, C.S. 9-10, 55-6
Palejwala, V.A. 14
Panigai, L. 266
Pannell, Dr Bill 26
Parr, D. 9
Peacock, W.L. 57
Pearman, G.I. 261-2, 267

Peel, Lynette 124
Penning-Rowsell, E. 257-8, 265-6
Perold, A.I. 60
Peterkin, Dr Michael 17, 78
Peynaud, E. 8, 33, 66
Pfister, C. 63, 256-7
Pharis, R.P. 279
Phillips, I.D.J. 280
Pirie, A.J.G. 7, 9, 14, 20, 32, 50, 183,
  186-7
Pittock, A.B. 260
Portes, L. 23-4, 36
Possingham, J.V. 12, 49, 62
Prescott, J.A. 6, 216
Puillat, V. 66
Pym, L.W. 81
Quintana Gana, M. 33
Radler, F. 56
Raison, Joy 48
Rajasekaran, K. 13
Ramanathan, V. 260, 262
Rankine, Bryce 1, 29, 54, 56, 104
Rapp, A. 55
Rasool, S.I. 261
Rawson, H.M. 29
Ray, Cyril 259, 264-6
Reid, D.M. 279
Rendu, V. 34, 36, 66
Reynolds, A.G. 10
Ribéreau-Gayon, P. 8, 33, 66
Richards, D. 279
Robertson, George 259
Robinson, Jancis 257, 264
Robinson, M. 48-50
Robson, A.D. 29
Royal Commission on Vegetable
  Products, Victoria 4, 124, 128
Ruhl, E.H. 28, 60
Russell, R.S. 29
Ruyssen, F. 33-4, 36
Saayman, D. 33-4, 59
Sanson, J. 212, 281
Saugier, B. 39
Schmitz, Ruth 14
Schneider, S.H. 256, 260-1, 263
Scudamore-Smith, P.D. 179
Seguin, G. 33-4, 37, 41
Sellers, W.D. 261-2
Servicio Meteorologico Nacional,
  Spain 281
Servico Meteorologico Nacional,
  Portugal 281
Seward, Desmond 242
Shakespeare, William 259
Shand, P. Morton 256
Shaulis, N. 2, 48
Sheps, Lilian 51
Siegelman, H.W. 26
Simon, André 257-8
Sinton, T.H. 51, 55
Skene, K.G.M. 13
Skoog, F. 14
Slingo, Tony 262
Smart, R.E. 1-2, 6-7, 15, 26-7, 30,
  32-3, 47-51, 54, 58, 158, 269
Smith, H. 26, 33, 37

Smith, R. 77
Smith, S.M. 46, 48
Smithsonian Institute,
  Washington 63
Somers, T.C. 9, 29, 50
Spencer, Mary 9
Sponholz, W.R. 55
Spurling, M.B. 6
Srinivasan, C. 13, 20, 279-80
Stephens, L. 261-2
Stevens, K.L. 10
Strauss, C.R. 10
Stuckey, Wendy 56
Temenuschka, Andronova 14
Thimann, K.V. 280
Thorpe, M. 29
Thorpy, Frank 192
Torikata, H. 14
Torres, R.E. 9, 19, 41
Torrey, J.G. 279
Trevelyan, G.M. 255, 259
Tromp, A. 55
Truscott, J.H.L. 9
Tucker, G.B. 260-2
Usherwood, N.R. 37
Van Eimern, J. 36
Van Huyssteen, L. 27, 58-9
Van Overbeek, J. 14
Van Rooyen, P.C. 55
Van Zyl, J.L. 27, 58
Vermorel, V, 66, 68, 212, 239
Versini, G. 55
Viala, P. 66, 68, 212, 239
Vos, P.J.A. 55
Walker, R.R. 60
Wallingford, G.W. 58
Wardle, D.A. 10
Warner, J. 9
Weast, R.C. 10
Weather Bureau, Department of
  Commerce, USA 281
Weaver, R.J. 13-4
Webster, P.J. 261-2
Weinhold, Rudolf 259
Wendland, W.M. 64, 284-5
Went, F.W. 10, 19, 51, 65
Wermelinger, B. 57
Westwood, M.N. 9-10, 14
Whiting, J.R. 60
Wigley, T.M.L. 256, 262
Williams, L.E. 57
Williams, P.J. 10
Williams, W.O. 13
Windholz, Martha 10
Winkler, A.J. 4, 13, 37-8, 48, 50, 62,
  66, 128, 246
Wittwer, S.H. 268
Wong, S.C. 268
Woodham, R.C. 13
Woodward, F.I. 45
World Meteorological
  Organization 260-1
Zekulich, Mike 72
Zelleke, A. 13-4

Locality and Appellation Index

*Entries in **italic text** give the page numbers of the climatic tables.
In these cases the following entry is always that for the commentary on the table.*

Adelaide, South Australia   5, 22, 25, 79, 104–6, *113*, 121, 270, 275

Adelaide, Waite Institute, South Australia   16, 22, 105–6, *113*, 121, 270

Ain (region), Rhône Valley, France   212

Albany, Western Australia   16, 21–2, 73–4, *84*, 97, 270, 275

Almeria, Spain   270

Alsace, France   211

Alto Adige (region), N. Italy   231

Anakie, Victoria   150

Ancenis, Loire Valley, France   209

Angers, Loire Valley, France   23, *197*, 209, 270

Angoulême, Cognac region, France   23, *198*, 209

Anjou (region), Loire Valley, France   209, 215

Ararat, Victoria   16, 22, 127, *133*, 146, 270, 274

Armidale, New South Wales   22, 159, *160*, 175, 270

Asti, Piedmont, Italy   129, 232

Auckland, New Zealand   23, 25–6, 192, *193*, 195, 270

Austria   16, 226–7

Auxerre, Chablis region, France   23, *198*, 209, 270

Avoca, Victoria   16, 21–2, 124, 128, *134*, 146, 270, 274

Badajoz, Spain   270

Bad Durkheim, Rheinpfalz, Germany   16, 23, *216*, 218, 270

Baden, Vienna region, Austria   226

Bairnsdale, Victoria   16, 22, 132, *134*, 146

Bakers Hill, Western Australia   16, 22, 72, 82, *84*, 97

Baldivis, Western Australia   80

Ballarat, Victoria   22, 124, 127, *135*, 147, 270

Bandol, Provence, France   214, 258

Bardolino, Verona region, Italy   232

Barolo, Piedmont, Italy   232

Barossa Hills, South Australia   18, 32, 107–8, 275

Barossa Valley, South Australia (*see also* Nuriootpa)   5, 18, 21–2, 61, 104, 107, *114*, 121, 275

Bathurst, New South Wales   151, 158, *160*, 175

Beachport, South Australia   112

Beaujolais (region), France   23, 50,*199*, 209

Beechworth, Victoria   129, *135*, 147

Bega, New South Wales   16, 22, 154, *161*, 175, 270

Belair, South Australia   22, 106, *114*, 121, 270

Belgium   257

Bendigo, Victoria   5, 21–2, 124, 128, *136*, 147, 270, 274

Bennett Range, Western Australia   44, 75, 99

Bergerac (region), France   23, *199*, 210

Berlin, Germany   270

Berri, South Australia   5, 16, 22, 110, *115*, 121, 154, 270

Bicheno, Tasmania   184, 187, 191, 274

Bindoon, Western Australia   25, 72, 81, 100, 276

Blackwood Valley, Western Australia   76, 98

Blenheim, Marlborough, New Zealand   16, 23, 26, 192, *194*, 195, 270

Bolzano, Alto Adige region, Italy   23, 26, *229*, 231, 270

Bombala, New South Wales   6, 158, *161*, 175

Bordeaux, France   16, 18, 21, 23, 34, 45, 51, 72–8, 107, 124–6, 186, 195–6, *200*, 210, 255, 258, 265–6, 270, 274, 276

Bourgueil, Loire Valley, France   209

Bowral, New South Wales   157–8, *162*, 175

Braga, Minho region, Portugal   23, *233*, 238, 270

Braganca, Portugal   16

Bremen, Germany   270

Bridgetown, Western Australia   16, 22, 80, *85*, 98

Bridport, Tasmania   184

Brindisi, Italy   270

Brokenback Range, Hunter Valley, New South Wales   44, 153

Bukkulla, New South Wales   157

Bulgaria   16, 21, 227–8

Bunbury, Western Australia   16, 21–2, 72, 79–80, *85*, 98, 240, 270, 275

Bunbury Hills, Western Australia   79–80

Burcelas, Lisbon region, Portugal   239

Burgenland (region) Austria   227

Burgundy (region), France   16–7, 21, 27, 46, 107, 125–6, 132, 187, 211, 253, 257, 274

Bushy Park, Tasmania   23, 185, 187, *188*, 190, 270

Busselton, Western Australia   5, 16, 21–2, 78–9, *86*, 98, 270, 276

Cahors, Lot, France   23, *201*, 210

Caldas da Rainha, Estremadura region, Portugal   23, *234*, 238

California (state), USA   4, 16–7, 26, 53, 246, 251–2

Campania, Tasmania   187

Camperdown, Victoria   126, *136*, 147

Canada   16

Canberra, Australia   22, 63, 157, *162*, 176

Canterbury (province), New Zealand   192–3

Cape Barren Island, Tasmania   274

Capel, Western Australia   79, 276

Carbunup, Western Australia   78

Carcavelos, Lisbon region, Portugal   239

Carlotta, Western Australia   76

Cassis, Provence, France   214

Central tablelands, New South Wales   158–9

Cessnock, Hunter Valley, New South Wales   16, 22, 65, 152–3, *163*, 176, 270

Chablais, Upper Rhône Valley, Switzerland   225

Chablis (region), France   209

Champagne (region), France   16, 106, 126–7, 187, 213, 252, 257–8, 266

Chapman's Hill, Western Australia   79

Charente (region), France   109

Château Grillet, Rhône Valley, France   44

Château Lafite, Bordeaux region, France   265–6

Châteauneuf du Pape, Lower Rhône Valley, France   12, 36, 213

Chianti (region), Italy   16, 231

Chiaretto del Garda, Verona region, Italy   232

Chinon, Loire Valley, France   209

Christchurch, Canterbury, New Zealand   192, 270, 284

Ciudad Real, Valdepeñas region, Spain   16, *237*, 241, 270

Clairette du Languedoc, S. France   212

Clare, South Australia   5, 16, 18, 21–2, 44, 61, 79, 104, 108–9, *115*, 121, 239, 270, 275

Coal River, Tasmania   187, 274

Cognac (region), France   209

Colac, Victoria   126, *137*, 147

Colares, Estremadura region, Portugal   238

Collie, Western Australia   22, 82, *86*, 99

Colmar, Alsace, France   23, *201*, 211, 216, 270

Cologne, Rhine Valley, Germany   270

Columbia Valley, Washington/ Oregon 252–4

Constantia, Cape Province, S. Africa 129

Coonabarabran, New South Wales 22, 153, 156, *163*, 176, 273

Coonawarra, South Australia 5, 22, 32, 104, 110, 112, *116*, 121, 270, 275

Cootamundra, New South Wales 155, *164*, 176

Corowa, New South Wales 22, 129, 150, 154, *164*, 176, 270

Corton, Burgundy region, France 44, 211

Cosne, Upper Loire Valley, France 23, *202*, 211, 270

Côte d'Or, Burgundy region, France 211, 214

Côte Rôtie, Rhône Valley, France 44

Côtes du Rhône, Rhône Valley, France 215

Cowra, New South Wales 16, 21–2, 155, *165*, 176, 270, 273

Croatia 16, 226

Crooked Brook, Western Australia 79

Dandaragan, Western Australia 81

Dandenong Range, Victoria 126, 274

Dão (region), Portugal 23, *234*, 239–40, 270

Dardanup, Western Australia 79

Darling Range, Western Australia (*see also* Bunbury Hills, Perth Hills) 72, 76, 79, 81–2, 100, 276

Denmark, Western Australia 16, 22, 73, 75, *87*, 99

Derwent Valley, Tasmania 44, 183, 187, 190, 274

Dijon, Burgundy region, France 16, 23, *202*, 211, 270

Donnybrook, Western Australia 16, 21–2, 80, *87*, 99

Douro Littoral (region), Portugal 238

Douro Valley, Portugal 25, 44, 68, 72, 79, 82, 106, 239, 275

Dresden, Germany 259, 270

Drumborg, Victoria (*see also* Heywood) 5, 106, 127, 148

Duckhole Rivulet, Tasmania 187

Eden Valley, South Australia 5, 22, 107, *116*, 122, 270

Eger (for Tokay etc.), Hungary 16, 23, 220, *222*, 226, 270

England 242–5, 255–6, 258, 269

Esperance, Western Australia 83, *88*, 99

Estremadura (region), Portugal 238

Euroa, Victoria 22, 128, *137*, 147, 274

Eyre Peninsula, South Australia 112

Ferguson Valley, Western Australia 79

Fès, Morocco 270

Fleurieu Peninsula, South Australia 110–2, 123, 275, 278

Flinders Island, Tasmania 274

Forbes, New South Wales 22, 154, *165*, 176

Forest Grove, Western Australia 77

France 16, 197–215, 257–9, 269, 282

Frankfurt on Main, Germany 16

Frankland, Western Australia 5, 18, 22, 72–4, *95*, 103

Freiburg, S. Baden, Germany 23, 216, *217*, 219, 270

Fresno, San Joaquin Valley, California 16, 23, 25, 154, *247*, 251, 270

Frontignan, Languedoc, France 79, 212

Funchal, Madeira 16, *237*, 240

Geelong, Victoria 5, 16, 22, 124–6, *138*, 148, 270, 273–4

Geisenheim, Rheingau, Germany 23, *217*, 219, 270

Geneva, Switzerland 23, 192, *220*, 225, 270

George Town, Tasmania 184

Germany 16–7, 51, 216–9, 258–9, 264–5

Gingin, Western Australia 25, 72, 81, 100, 276

Gippsland, Victoria 124, 132–3, 146, 148–9, 273–4

Gironde Estuary, Bordeaux region, France 45, 210

Glen Innes, New South Wales 22, 159, *166*, 176

Glenrowan, Victoria 45, 129, 274

Goulburn, New South Wales 22, 157–8, *166*, 176

Goulburn Valley, New South Wales 153, 156, 272

Goulburn Valley, Victoria 128, 150

Granite Belt, Queensland 26, 159, 179, 273

Graves, Bordeaux region, France 210, 214

Great Dividing Range, S.E Australia 124, 126–32, 153, 155–9, 272–4

Great Western, Victoria 16, 124, 127, 274

Greenwich, England 23, *243*, 245, 270

Griffith, New South Wales 5, 16, 22, 25, 50, 110, 154, *167*, 177, 270

Guildford, Western Australia 16, 22, 68, 80, *88*, 100, 270

Gyöngyös, Hungary 226

Hamburg, Germany 270

Hamilton, Victoria 127, *138*, 148

Harvey, Western Australia 78–9, 275–6

Hastings Valley, New South Wales 154, 273, 278

Haute Savoie (region), Rhône Valley, France 225

Hawke Bay, New Zealand 192, 196

Healesville, Yarra Valley, Victoria 22, 125, *139*, 148, 270

Henty Brook, Western Australia 79

Hermitage, Rhône Valley, France 21, 44, 215

Heywood, Victoria (*see also* Drumborg) 22, 127, *139*, 148, 270

Hobart, Tasmania 5, 16, 23, 65, 183, 185, 187, *188*, 190–1, 195, 270, 274

Hungary 16, 226–7

Hunter Valley, New South Wales 5, 16, 18, 25, 30, 32, 44, 64–5, 151–3, 155–6, 229, 272

Huon Valley, Tasmania 44, 187

Innsbruck, Austria 270

Inverell, New South Wales 21–2, 157, *167*, 177, 270

Isère (region) Rhône Valley, France 212

Italy 16, 26, 229–32, 282

Jerry's Plains, Hunter Valley, New South Wales 22, 152–3, *168*, 177, 270

Jindong, Western Australia 78

Junee, New South Wales 155, *168*, 177

Kaiser Stuhl, Rhine Valley, Germany 44, 219

Kalamunda, Western Australia 22, 25, 72, 82, *89*, 100

Kangaroo Island, South Australia 112, 275

Karridale, Western Australia 16, 22, 74–5, 77, *89*, 100, 270, 275

Katanning, Western Australia 16, 72, 82, *90*, 100

Keppoch Valley, South Australia *See* Padthaway

Kesthely, Lake Balaton region, Hungary 23, 220, *223*, 227, 270

Keyneton, South Australia 107

Killarney, Queensland 16, 179, *180*, 182

King River Valley, Victoria 130–1

Kirup, Western Australia 80

Kojonup, Western Australia 22, 82, *90*, 100

Kudardup, Western Australia 77

Kybybolite, South Australia 16, 121

Kyneton, Victoria 127

Lake Balaton, Hungary 45, 227

Lake Geneva, Switzerland 45, 225

Lake George, New South Wales 157

Lake Mokoan, Victoria 45, 129

Lake Neuchâtel, Switzerland 45

Langenlois, Krems region, Austria 226

Langhorne Creek, South Australia (*see also* Strathalbyn) 21, 104, 109–10, 123, 275

Languedoc, S. France 212

La Trobe Valley, Victoria 132, 148, 273–4

Launceston, Tasmania 5, 16, 23, 26, 185–6, *189*, 191, 270

Launceston Elphin, Tasmania 184, 191

Launceston Mount Pleasant, Tasmania 184, 191

Launceston 7EX, Tasmania 184, 191

Leeton, New South Wales 22, 110, 154, *169*, 177

Leipzig, Germany 270

Leongatha, Victoria 22, 132, *140*, 148
Lille, France 270
Lilydale, Yarra Valley, Victoria 124
Lirac, Lower Rhône Valley, France 213
Lisbon, Portugal 16, 23, 26, 129, *235*, 239, 270
Lithgow, New South Wales 6, 158, *169*, 177
Liverpool Range, New South Wales 153, 156, 272
Ljutomer, Slovenia 226
Logroño, Rioja region, Spain *238*, 241, 270
Loire Atlantique (region), France 212
Loiret (region), Loire Valley, France 213
Loire Valley, France 126, 132, 187, 209, 211-4, 252, 258, 274
London, Ontario, Canada 16
Lower Great Southern (region), Western Australia 73-5
Lower Savoie (region), Rhône Valley, France 212
Lugarno, Ticino region, Switzerland 23, 179, 220, *221*, 226, 270
Lunel, Languedoc, France 212
Luxembourg 257
Lyon, Rhône Valley, France *203*, 209, 212
McLaren Vale, South Australia (*see also* Southern Vales) 104, 106, 275
Mâcon, France 6
Mâconnais (region), France 214
Madeira 16, 240
Maffra, Victoria 22, 132, *140*, 148
Magill, South Australia 106
Malaga, Spain 16, 270
Mandurah, Western Australia 16, 22, 25, 80, *91*, 100-1
Manjimup, Western Australia 16, 21-2, 73, 75-6, 80, *91*, 101, 270, 275
Margaret River, Western Australia 5, 16, 18, 22, 32, 61, 75-80, *92*, 101, 107, 240, 270, 275-6, 278
Marlborough (province), New Zealand 192, 195
Marybrook, Western Australia 78-9
Mateus, (region), Portugal 240
Médoc, Bordeaux region, France 45, 74, 77, 126, 210, 214, 276
Melbourne, Victoria 22, 124-5, *141*, 149
Merbein, Victoria 16, 22, 25, 128-9, *141*, 149, 270
Milawa, Victoria 79, 129
Mildura, Victoria 128-9, 149
Minho (region), Portugal 238
Mireval, Languedoc, France 212
Monferrato, Piedmont, Italy 232
Montagne de Reims, Champagne region, France 44, 213
Montpellier, Languedoc, France 21, 23, 129, *203*, 212, 270
Mornington, and **Mornington Peninsula**, Victoria 16, 22, 61, 126, *142*, 149, 270, 273-4, 278

Mosel Valley, Germany 45, 259
Mosonmagyaróvár, Hungary 23, 220, *223*, 227
Moss Vale, New South Wales 157-8, *170*, 177
Mount Barker, South Australia 107, *117*, 122
Mount Barker, Western Australia 5, 16, 21-2, 72-4, 80, *92*, 101, 270, 275
Mount Canobolas, New South Wales 158
Mount Gambier, South Australia 16, 112, 121
Mount Kaputar, New South Wales 159, 273
Mount Lofty Range, South Australia (*see also* Adelaide Hills, Barossa Hills) 104-10, 275
Mudgee, New South Wales 5, 16, 21-2, 50, 65, 155, *170*, 177, 229, 270-2
Murray Valley, S.E. Australia 45, 104, 110, 124, 129, 153, 273
Murrumbateman, New South Wales 157
Murrumbidgee Valley, New South Wales 110, 151, 154, 273
Murrurundi, New South Wales 22, 156, *171*, 177
Muscadet, Loire Atlantique, France 212
Muswellbrook, New South Wales 152-3
Myponga, Fleurieu Peninsula, South Australia 111
Nancy, France 270
Nannup, Western Australia 75-6
Nantes, Loire Atlantique, France 23, *204*, 212, 270
Napa, and **Napa Valley**, California 23, *247*, 251, 270
Napier, Hawke Bay, New Zealand 16, 23, 192, *194*, 196, 270
Naples, Italy 270
Narrabri, New South Wales 155, *171*, 177, 270, 273
Narrogin, Western Australia 83, *93*, 102
Neckar Valley, Württemberg, Germany 219
Nelson, New Zealand 16, 192
Neusiedler See, Austria 45, 227
New England (region), New South Wales (*see also* Northern table-lands) 26, 159, 273
New Norfolk, Tasmania 184
New South Wales 16, 30, 51-2, 58, 65, 129, 151-78, 229, 272-3
New Zealand 16, 28, 32, 51, 192-6, 269, 282
Northern Tablelands, New South Wales 159, 273
Nowra, New South Wales 154, 273
Nuriootpa, South Australia (*see also* Barossa Valley) 22, 107, *117*, 122
Oberon, New South Wales 6, 158
Orange, Lower Rhône Valley, France 21, 23, *204*, 213
Orange, New South Wales 158, 273

Orbost, Victoria 22, 132, *142*, 149
Oregon (state), USA 16, 246, 252-3
Orford, Tasmania 184
Orléans, Loiret region, France 23, *205*, 213, 257, 270
Otago (province), New Zealand 192
Otway Range, Victoria 127, 147
Oxford, England *244*, 245, 270
Padthaway, Keppoch Valley, South Australia 5, 21-2, 78, 104, 110, *118*, 122, 270, 275
Pages Creek, Tasmania 187
Palermo, Sicily 270
Paris (city and region), France 257, 270
Parramatta, New South Wales 16, 22, 153-4, *172*, 177
Pemberton, Western Australia 16, 21-2, 73-6, 80, *93*, 102
Perpignan, Roussillon, France 270
Perth, Western Australia 16, 22, 25, 72, 80, *94*, 102, 270
Perth Hills, Western Australia 82
Picton, New South Wales 22, 154, *172*, 177
Piedmont (region), Italy 232
Pinhão, Douro Valley, Portugal 23, 68, 79, *235*, 239, 270
Pipers Brook, Tasmania 44, 183, 186-7, 195, 274
Pitt Water, Tasmania 187
Pleven, Bulgaria 23, 220, *224*, 227, 270
Plovdiv, Bulgaria 16, 23, 220, *224*, 228, 270
Pomerol, Bordeaux region, France 74, 77, *205*, 213, 276
Porongurup (locality and range), Western Australia 22, 44, 73-4, *94*, 102, 275
Portland, Oregon 16, 23, *249*, 252, 270
Portland, Victoria 22, 127, *143*, 149
Port Lincoln, South Australia 16, 22, 112, *118*, 122
Port Phillip Bay, Victoria 4, 21, 124-6, 274
Portugal 16, 68, 233, 238-40, 259, 282
Pouilly-sur-Loire, Upper Loire Valley, France 211
Prague, Czechoslovakia 270
Prosser, Washington 23, *250*, 253, 270
Provence (region), France 214
Pyrenees Range, Victoria 44, 128, 146
Queensland 16, 30, 179-82, 229, 272-3
Queenstown, Otago, New Zealand 192
Quirindi, New South Wales 153, 156, 272
Régua, Douro Valley, Portugal 23, 68, 79, *236*, 239, 270
Reims, Champagne region, France 16, 23, *206*, 213, 270
Rennes, Brittany, France 270

Reynella, South Australia 106
Rheingau, Rhine Valley, Germany 219
Rheinpfalz, Rhine Valley, Germany 218
Rhine Valley, Germany/Alsace 44–5, 107, 126, 187, 211, 216–20, 252, 259
Rhône Valley, France/ Switzerland 44, 212–3, 215, 225
Richmond, New South Wales 22, 154, *173*, 178
Richmond, Tasmania 274
Rioja (region), Spain 241
Risdon, Tasmania 23, 185, 187, *189*, 191
Riverland (region), South Australia 104, 110
Robe, South Australia 112, 275
Rocky Gully, Western Australia 5, 22, 74, *95*, 103
Roseburg, Oregon 23, *249*, 253, 270
Roseworthy, South Australia 107, *119*, 123
Ross–on–Wye, England 270
Rutherglen, Victoria 5, 25, 79, 124, 129, *143*, 150, 228, 274
Sacramento, California 16, 270
Saint Emillion, Bordeaux region, France 21, 23, 74, 77, *206*, 213–4, 276
Saint Helens, Tasmania 16, 185, 187, *190*, 191
Salinas (location and valley), California *248*, 251–2, 270
Samos, Greece 270
Sancerre, Upper Loire Valley, France 211
Sandy Hollow, Hunter Valley, New South Wales 153
San Francisco, California 16
San Joaquin Valley, California 154, 251
San Jose, Santa Clara Valley, California 16, 23, *248*, 252, 270
Santa Clara Valley, California 252
Saône Valley, France 44
Sauternes, Bordeaux region, France 210
Scone, Upper Hunter Valley, New South Wales 22, 152–3, *173*, 178
Scottsdale, Tasmania 184
Seattle, Washington State 23, *250*, 252, 254, 270
Serrières, Mâconnais, France 23, *207*, 209, 214
Setubal, Portugal 129, 239
Seymour, Victoria 5, 16, 21–2, 128, *144*, 150, 270
Sicily 229
Siena, Chianti region, Italy 23, 26, 153, *230*, 231, 270
Sliven, Bulgaria 23, 220, *225*, 228, 270
Slovenia 236
Soave, Verona region, Italy 232,
Sopron, Hungary 227
Southampton, England 23, *244*, 245, 270

South Australia 16, 31–2, 50, 104–23, 274–5
South Baden (region), Germany 219
Southern Tablelands, New South Wales 157–8
Southern Vales, South Australia (*see also* McLaren Vale) 5, 25, 50, 79
South West Coast (region), Western Australia 78–9
Spain 16, 68, 241, 282
Spearwood, Western Australia 80
Springton, South Australia 22, 107, 122
Stanthorpe, Queensland 23, 159, 179, *181*, 182, 270
Stirling, South Australia 22, 106, *119*, 123, 270
Strathalbyn, South Australia 16, 22, 104, 109, *120*, 123, 270
Strathbogie Range, Victoria 21, 44, 128, 147
Stuttgart, Württemberg, Germany 216, *218*, 219, 270
Sugarloaf Mountain, New South Wales 155, 273
Sunbury, Victoria 124, 126, 273
Swan Hill, Victoria 22, 129, *144*, 150
Swansea, Tasmania 16, 185, 187, 274
Swan Valley, Western Australia 2, 4–5, 16, 18, 25, 30, 45, 68, 72–3, 80–1, 100
Switzerland 192, 225–6, 258
Sydney, New South Wales 151, 153–4, 273
Tamar Valley, Tasmania 44, 183, 186–7, 274
Tamworth, New South Wales 153, 156, *174*, 178
Tasmania 7, 16, 28, 32, 50, 65, 183–91, 240, 269, 274
Tavel, Lower Rhône Valley, France 213
Tea Tree Gully, South Australia 106
Tenterfield, New South Wales 22, 159, *174*, 178–9
Terang, Victoria 126, *145*, 150
Ticino (region), Switzerland 226
Tokay, Hungary 226
Toodyay, Western Australia 82
Toowoomba, Queensland 179, *181*, 182
Torres Vedras, Estremadura region, Portugal 238
Toulon, Provence, France 16, 23, 26, *207*, 214
Tours, Loire Valley, France 23, *208*, 215, 270
Tras–Os–Montes (region), Portugal 240
Tunis, Tunisia 270
Turin, Piedmont, Italy 16, 129, *230*, 232, 270
United States of America 16, 246–54, 269, 282
Valais, Upper Rhône Valley, Switzerland 225
Valdepeñas (region), Spain 241

Valence, Rhône Valley, France 21, 23, *208*, 215, 270
Valpolicella, Verona region, Italy 232
Vancouver, British Columbia, Canada 270
Veinviertel, Vienna region, Austria 226
Verona, Lombardy, Italy 16, *231*, 232, 270
Victor Harbor, South Australia 110–1, *120*, 123
Victoria 4, 16, 31, 45, 124–50, 273–4
Vienna, Austria 16, 23, 220, *221*, 226, 270
Vila Real, Portugal 23, *236*, 240, 270
Vöslau, Vienna Region, Austria 226
Vouvray, Loire Valley, France 215
Wachau, Krems region, Austria 226
Waite Institute: *see* Adelaide, Waite Institute
Walla Walla, Washington State 16, 23, *251*, 254
Wandering, Western Australia 22, 82, *95*, 103
Wangaratta, Victoria 16, 21–2, 70, 129–31, *145*, 150, 270, 274
Wanneroo, Western Australia 80, 102
Warby Range, Victoria 44–5, 129
Warren Valley, Western Australia 75–6
Warrumbungle Range, New South Wales 156, 273
Washington (state), USA 16, 246, 252–4, 269
Watervale, South Australia 108
Werribee, Victoria 22, 124, *146*, 150
Western Australia 4, 16, 31, 33, 58, 68, 72–103, 240, 275–6
Western Districts, Victoria 126–7
Western plains, New South Wales 154–5
Western Port, Victoria 133, 274
Western slopes, New South Wales 155–7, 273
Whitfield (also Whitlands), Victoria 70, 130–2, 273
Willamette Valley, Oregon 44, 184, 252–3
Wilyabrup, Western Australia 22, 77–8, *96*, 103
Witchcliffe, Western Australia 77
Wokalup, Western Australia 16, 22, 72, 79, *96*, 103, 270
Württemberg, Germany 219
Yallingup, Western Australia 21–2, 26, 77–8, *97*, 103
Yanchep, Western Australia 80
Yarra Valley, Victoria 5, 44, 124–6, 148, 273
York, England 270
Yorke Peninsula, South Australia 112, 275
Young, New South Wales 22, 155, *175*, 178, 273
Zagreb, Croatia 16, 220, *222*, 226, 270
Zante, Greece 270

Grape Variety Index

Included in the list are common valid synonyms for some of the more important varieties. In most cases I follow the standard Australian nomenclature as listed by Dry and Gregory (1988). Exceptions are Currant, for the sake of simplicity, and Grenache Blanc, where I retain the French form in view of the variety's major role there and the evident confusion over 'White Grenache' and 'Grey Grenache' in Australia.

Airén 241
Aligoté 67, 211-2, 214
Alvarelhão 67, 239–40
Alvarinho 238
Aramon 67, 212
Bacchus 67, 218, 264
Barbera 67, 229, 231-2
Bastardo (syn. Trousseau) 67, 68, 75, 187, 239–40
Biancone 67
Blue Portuguese 67, 218-9, 226
Bual 240
Cabernet Franc 38, 67, 78, 130, 132, 195, 209-10, 212-4, 227, 265
Cabernet Sauvignon 10, 17, 19, 33, 38, 50, 67, 75, 78, 108, 110, 126, 128, 130, 153, 179, 186-7, 192, 195, 210, 212, 214, 231-2, 241, 265
Carignan 38, 59, 67, 212, 214
Chambourcin 154
Chardonnay 18, 38, 59, 67, 72, 74-6, 78, 106-7, 132-3, 192, 195, 209, 211-4, 228, 253, 277
Chasselas 67, 124, 132, 192, 211-2, 219, 225, 228, 257
Chenin Blanc 18, 28, 38, 59, 67, 78, 209, 215
Cinsaut 67, 212, 214
Clairette 67, 212, 214
Colombard 67, 209
Corvina 67, 232
Crouchen 67
Currant (syn. Zante Currant, Black Corinth) 13, 104, 279
Dolcetto 67, 238, 240
Dona Branca 67, 238, 240
Doradillo 67, 128
Durif 67, 187, 212
Elbling 67
Folle Blanche 67, 209, 212
Freisa 67, 232
Frontignac (syn. Brown Muscat, Muscat à Petits Grains) 21, 67, 78-80, 129, 212, 228, 232, 239
Furmint 67, 226-7
Gamay 38, 50, 66-7, 75, 125, 132, 187, 209, 211-2, 225, 252, 257
Garganega 232
Graciano (syn. Morrastel) 67, 241
Green Veltliner 67, 226, 231
Grenache 59, 67, 80, 213-4, 241
Grenache Blanc 67, 241
Grignolino 67, 232
Harslevelu 67, 226

Kadarka 67, 226-7
Kerner 67, 218
Leanyka 67
Madeleine (syn. Early Madeleine, Madeleine Angevine) 67
Madeleine-Sylvaner 67
Malbec 18, 67, 74, 78, 109, 210, 265-6
Malvasia Bianca 67, 240
Marsanne 67, 215
Mataro 67, 212, 214
Melon (syn. Muscadet) 67, 212
Merlot 67, 72, 78, 130, 132, 179, 195, 210, 213-4, 226, 231-2, 265
Meunier (syn. Pinot Meunier) 67, 75-6, 132, 186, 195, 213, 257, 266
Monastrell 67, 241
Mondeuse (syn. Refosco) 67, 75, 179, 212
Montils 67
Morio-Muskat 67, 218
Mourisco Tinto 67, 239
Müller-Thurgau 67, 132, 218-9, 226, 231, 264
Muscadelle 21, 67, 79-80, 129, 210
Muscat à Petits Grains: see Frontignac
Muscat Gordo Blanco (syn. Gordo, Muscat of Alexandria) 12, 67, 104, 128, 239
Muscat Ottonel 67, 211, 218
Nebbiolo 67, 229, 232
Negrara 67, 232
Ohanez (syn. Almeria) 68
Palomino (syn. Listan) 67
Pedro Ximenez 67
Petit Verdot 67, 78, 210, 266
Pinotage 67
Pinot Gris (syn. Rulander) 67, 211, 218-9, 225-6, 257
Pinot Noir 2, 19, 33, 38, 50, 55, 66, 67, 68, 72, 74-6, 106, 108, 110, 124-6, 130-3, 179, 183, 186-7, 195, 211, 213, 219, 225-7, 231, 252-3, 257, 266
Rabigato 67, 240
Ramisco 67, 238
Red Veltliner 67, 226
Riesling (syn. Rhine Riesling) 12, 18, 21, 74-5, 78, 107, 109, 183, 211, 218-9, 226-7, 231, 252, 257, 259, 264-5, 275
Roussanne 67, 215
Ruby Cabernet 67
Sangiovese 67, 153, 229, 223
Sauvignon Blanc 10, 37-8, 67, 72, 78, 106, 132-3, 192, 195, 210-1, 226, 231
Scheurebe 67, 218

Schiava (syn. Trollinger) 67, 231
Sémillon 18, 37, 67, 72, 78, 153, 195, 210
Sercial 67, 240
Shiraz (syn. Syrah) 38, 50, 67, 74, 78, 80, 107, 110, 128-9, 153, 179, 215
Siegerrebe 67
Souzão 67, 238-40
Sultana (syn. Sultanina, Thompson Seedless) 13, 20, 67, 104-5, 129, 279
Sylvaner 67, 211, 218-9, 225-6, 228, 231
Taminga 67
Tannat 67, 179
Tarrango 67
Tempranillo 67, 68, 241
Terret Noir 67, 212
Tinta Amarella 67, 239-40
Tinta Carvalha 67, 239-40
Tinta Madeira 67, 240
Touriga 67, 239-40
Traminer (syn. Gewürztraminer) 18, 67, 72, 75, 78, 132, 183, 192, 211, 218-9, 226-7, 257
Trebbiano (syn. Ugni Blanc) 67, 76, 128, 209, 212, 214, 231-2
Trollinger: see also Schiava 67, 219
Valdiguié 67
Verdelho 21, 67, 68, 78-9, 81, 240, 277
Viognier 67, 215
Welschriesling 67, 226-8, 231
Zinfandel 12, 18, 28, 55, 67, 78, 246

General Index to Main Text

Abscisic acid: abscisin   14–5, 18, 280
Acids: *see* Grape acids; Malic acid; Tartaric acid
Air drainage   40–4, 71, 74, 82, 105–6, 109, 130–2, 153, 155–9, 213, 246, 273
Altitude   40–1, 44–5, 70, 74–5, 79–82, 105–12, 126–32, 152–9, 179, 183–4, 186, 216, 268, 273–7
Animal manures   53
Anthocyanins   9–10, 15, 27, 50, 68
Apical dominance   17, 279
Aroma (*see also* Flavourants)   8–10, 17, 20, 36–7, 55, 66, 68, 186, 195
Aspect   13, 41–2, 44, 153, 155–6, 192, 209, 211, 213, 215–6, 219, 252, 254
Assimilate reserves   20–1, 37, 60
Auxins   11, 14–5, 17, 279–80
Berries:
  Set   14, 17, 20, 28, 183, 187, 279–81
  Growth   13–4, 17, 279–81
  Size   17, 52
  Sugar content   8–9, 19–21, 25, 37, 50, 55–6, 64, 156, 265
  Splitting   28–9, 34, 246, 277
  Senescence   14, 37, 280
Biodynamic forces   280
Bird damage   75
Brandy   76, 128, 209
Budburst   17–9, 32, 60, 64–6, 155–6, 159, 183, 187, 277, 279–80, 285
Bud differentiation   12–3, 17, 192, 279–80
Bunch rot   34, 75, 110, 127, 209, 212
Bunch thinning   12, 52, 60, 277
Calcium, as nutrient   38, 58, 211
Canopy:
  Light relations   25–7, 31, 36–7, 46–52, 56, 58, 109, 267–9, 276–7, 280
  Management   46–52, 56, 59
Carbon dioxide   29, 39, 45, 56, 58, 106, 130, 179, 214, 260–2, 267–9, 271, 277
Champagne (and other sparkling wines)   106, 127–8, 186, 195, 213–4, 257–8, 266
Chaptalization   216, 243, 264
Chlorofluorocarbons   261–2, 269
Climate:
  Definitions of scale   6
  Construction and use of tables   69–71
  Change, general   32, 104, 124, 126, 183–7, 213, 239, 242–3, 255–67, 269, 271, 274, 282–3
  Change, natural causes   260–1
  Change, human causes: *see* Greenhouse effect
Cloudiness   30, 51, 75, 153, 211, 238, 262

Coast:
  Proximity to   30–1, 44–5, 52, 72–82, 105–12, 126–7, 149, 152–3, 157–8, 186–7, 193, 209–10, 212, 214, 238, 240, 246, 252–4, 271–7
  Aspect of   31, 75, 77–8, 110–2, 124, 132, 186–7, 212, 214, 274
Combinations of climatic factors   21–6, 31–2, 157
Continentality   7, 15–8, 32, 69–70, 77–8, 112, 124, 126, 183, 187, 192, 210, 229, 240–1, 246, 273–4, 276–7, 284
Costs of production   2, 61, 127, 151, 159, 272
Coulure   13, 18, 265
Crop load   12, 17, 25, 28, 49–50, 55–6, 60, 268–9
Cytokinins   12–5, 17, 26–7, 36–7, 50, 58, 269, 279–80
Daylength   5, 11, 14, 18, 63–4, 153
Disease:
  Risk   25, 30, 46, 151, 154, 157, 159, 179, 192, 195–6, 266–7, 273
  Susceptibility or resistance to   37, 39, 50, 192, 269, 280
Dried vine fruits: *see* Grapes, drying
Earthworms   54, 58
Enzymes, their functions and temperature responses   8–9, 19, 38, 268
Ethyl carbamate   56–7
Fermentation   9–10, 38, 55–7, 66, 68, 258, 277
Fertigation (*see also* Irrigation, trickle)   38, 57
Fertilizers:
  Deep placement   57–8, 60
  Foliar application   57–8
  General   38–9, 53–8
Flavourants:
  Nature, formation and volatile losses   8–11, 19–20, 36–8, 50, 55, 66, 68
Flavour ripening:
  see Ripening, physiological
Fog level   40, 44
'Fremantle Doctor', Western Australia   45
Frost   6, 17, 19, 32, 45, 54, 58–9, 65, 69–70, 110, 124, 126–7, 129–30, 132, 151, 155–9, 179, 186–7, 195, 209–10, 246, 253, 258, 272–4, 275–6, 281–2
Fruitfulness: *see* Bud differentiation; Vines, fruitfulness; Temperature, effect on fruitfulness
Gibberellins   11–5, 20, 26–7, 277, 279–80
Glycosides   10

Grape:
  Acids (*see also* Malic acid; Tartaric acid)   8, 10, 17, 19–20, 26, 29, 50, 209
  Tannins   8, 10, 26, 28, 66–8, 239
  Quality (*see also* Wine quality)   8, 18, 20, 31, 37–8, 46, 48–50, 52, 58, 60, 66, 68, 108, 155, 246, 268–9
  Yield   17, 28, 37, 48, 52, 61, 66, 78, 265, 269, 272, 276–7
  Table   5, 25, 80, 251
  Drying   4–5, 8, 25, 30, 31, 81, 104, 129, 251
  Varieties, general   12, 18, 21, 26, 61, 67, 71, 74–6, 78, 80–1, 83, 104–5, 109, 179, 186–7, 192–6, 229, 240, 265–6, 278
  Varieties, maturity differences and groups   21, 62, 66–8, 71, 127, 157, 159, 179, 184, 186–7, 195, 212, 243, 246, 252–3, 256–8, 264–5, 269, 274, 277
Greenhouse effect   32, 255, 260–7, 269, 271, 282
Green manuring, and cover crops   48, 53–4, 58–9
Growing season, definition of length   5, 63–4, 284–5
Heat injury   19–20, 25, 28–9, 31, 37, 47–8, 52, 58, 68, 70, 74, 81, 107–8, 110, 125–6, 186–7, 225, 239, 268
Hills, isolated and projecting   40–4, 74–5, 109, 125, 127, 130–2, 153–9, 213, 252–3, 273
Histamine   56–7
History of viticulture   72–3, 104–5, 124, 151, 179, 183, 192, 242, 255–60, 264–7
Homoclimes   6, 158
Hormones   11–15, 279–80
Hybrid varieties: *see* Species hybrid varieties
Hydrogen sulphide   55–6
Irrigation (*see also* Supplementary watering):
  General   29, 34, 50, 52–3, 109–10, 128, 154, 246
  Trickle   52–3, 111–2
Lakes: *see* Rivers and lakes, proximity to
Land breezes   25, 30–1, 44, 157
Latitude   5, 44, 63–4, 70, 252, 284
Leaching, of nutrients   53–4, 57–8
Leaf removal   27, 46–8
Leaf area index   51
Legumes, for green manuring   53–4, 58–9

Light:
  Intensity   25, 51, 64, 280–1
  Spectral quality   26–7, 33, 36, 47–8, 58–9, 211, 277, 279–80
  Supplementation   59, 277
  Relations general (*see also* Sunshine hours; Canopy light relations)   11, 20–7, 267–9, 279–80
Limestone, chalk: *see* Soil, calcareous
Lupins   11, 53–4, 58
Machine harvesting   20, 47, 49
Macroclimate   6
Magnesium, as nutrient   38, 53, 58
Malic acid   8, 49
Marketing   2, 61, 104, 269, 271, 276
Mesoclimate   6, 36, 40–5, 53, 108–9, 121, 125, 129–32
Methane   261–2
Methoxypyrazines   10, 37
Methuen, second Treaty of   259
Microclimate   6, 13, 36, 41, 46–7, 58, 211, 277
Mildew: downy and powdery (Oidium)   151, 157, 265
Moisture relations: *see* Vine water relations
Moisture stress (*see also* Vine water relations)   7, 17, 27–8, 32, 52, 110
Native vegetation, as soil indicator   41, 72, 74, 77, 79, 100, 102
Nitrogen:
  In vine nutrition   37, 53–8, 60–1, 268
  Timing and method of application   57–8, 277
  In musts and fermentation   54–7, 268, 276–7
Noble rot (*Botrytis cinerea*; *see also* Bunch rot)   17, 37, 209–10, 219, 226
Organic methods   53–4, 57
Ozone   261–2
Pest control   277
Petiole analysis   53
pH of musts and wines   10, 20, 28–31, 37, 45, 49–51, 60, 256, 268
Phenology: *see* Vine phenology
Phosphate, as nutrient   53, 57–8
Photosynthesis   9, 21, 26, 28–31, 50–1, 64, 267–9, 277
Phylloxera   72–3, 104, 124, 212, 239, 258, 265–6
Phytochrome   26, 33
Potassium:
  As nutrient   29–30, 37–8, 53, 58
  Uptake and flux to berries   28–9, 37–8, 45, 49–50, 60, 268, 277
  Bitartrate   29, 50
Pruning   46, 48–9, 60–1, 78
Rainfall
  Ripening period   28–9, 31, 75, 107, 110, 151, 153–4, 156–7, 159, 179, 195–6, 212, 215, 226, 246, 253, 272–3, 275
  Variability   74, 132, 151, 153–4, 156–7, 273–4
  General   7, 27–30, 32, 34, 57, 71, 74–6, 82–3, 106–7, 109, 111, 151,

153–9, 179, 195–6, 214, 216, 219, 226, 238, 272–6, 280
Rain shadow   44, 187, 192–3, 211, 216, 219
Red to far red ratios: *see* Light, spectral quality
Regional climate   6
Relative humidity   7, 25–6, 29–32, 45, 51, 71, 107, 129–30, 151, 153–9, 179, 192, 196, 239, 241, 246, 251, 253, 268–76
Research needs   56–7, 276–7, 280
Ripening:
  Sugar accumulation: *see* Sugar, flux to berries
  Physiological, or flavour   8–10, 19–21, 30, 36–7, 66, 81, 110, 268, 277, 280
  Speed of   20, 60
  Processes, general   8–10, 13–15, 19–20, 27, 50, 64, 66, 267–9, 276, 280
Rivers and Lakes, proximity to   40–5, 129, 187, 192, 210, 219, 227
Rootstocks: *see* Vines, rootstocks and grafting
Saturation deficit (*see also* Relative humidity)   29–30, 49–50, 246
Sea breezes   25, 30–1, 43–5, 72, 79–83, 105–10, 128–9, 153, 155–7, 251, 271–6
Sherry sack   259
Site climate (*see also* Mesoclimate)   6
Soil:
  Colour   27, 33, 36, 38, 211, 276
  Clay content   33, 38, 58
  Calcareous   33–4, 38, 211
  pH and liming   58
  Cultivation   54, 58–9, 276
  Organic matter   38–9, 53–4, 58–9, 277
  Mulching   15, 36, 58–9, 277
  Microbiology (*see also* Earthworms)   39, 53, 58–9, 277
  Fertility, general   34, 37–41, 53–4, 57–8, 61, 268
  Physical properties   34–7, 39, 58, 211
  Stone or rock content   12–13, 33–6, 41, 59, 195, 210–1, 246, 252, 263
  Erosion   36, 58
  Drainage   4, 33–4, 36, 38, 59, 211, 246
  Deep ripping   59, 60
  Temperature   12–5, 18, 38–9, 211, 279
  Thermal relations   12–3, 33, 35–44, 46–7, 58–9, 153, 195, 211, 280
  Water relations   28, 34, 38, 53, 58–60, 128–9
  Type, general   33–41, 53, 72–83, 106, 109–12, 128–30, 154, 156–7, 195, 211, 273, 275
Solar radiation, insolation   7, 47, 64
Species hybrid varieties   25, 154, 192, 273, 278

Sugar:
  Content of berries: *see* Berries, sugar content
  Flux to berries   9, 14, 20–1, 28, 37, 45, 49–51, 55, 60, 268, 277, 280
Sunshine:
  Hours   6–7, 17, 20–6, 31–2, 51, 64, 70–1, 74–80, 107, 109, 111, 125–7,132, 153–7, 159, 179, 183, 186, 216, 219–20, 225, 229, 239, 252, 271, 273–6, 280–1
  Ratio of hours to effective temperature summation   21–6, 75, 109, 127, 153–4, 156–7, 179, 195, 214, 229, 246
Supplementary watering   28, 52–3, 57, 78, 104, 107, 109, 111–2, 155–7, 246, 275, 277
Tartaric acid   29, 49
Temperature:
  Definitions of mean and average   5, 64, 108, 153, 285
  Night   19, 30, 40–5, 47, 51–2, 64–5, 69–70, 81–2, 105, 110, 153, 155–6, 195, 251, 263–4, 268, 271, 285
  Day   19–20, 30–1, 40, 44, 47, 64, 70, 108–9
  Diurnal range   18–9, 40–5, 47, 51, 64–5, 70, 107, 130, 155, 157–9, 187, 210, 240, 251–2, 284–5
  Coldest month   216, 220
  Hottest month   6, 7, 284
  Variability index   18–9, 31, 70, 74, 187, 196, 209, 214, 229, 231, 238, 246, 252–3, 282–3
  Variability, general   5, 18–20, 26, 30–2, 36, 40–5, 51–2, 58, 70, 74, 106–7, 110, 124–6, 129–33, 151, 155–9, 179, 186–7, 195–6, 209–10, 212, 214, 225–6, 229, 231–2, 238–40
  Of soil: *see* Soil temperature
  Base for vine growth   5, 263–4, 275, 282–3, 285
  Biological effectiveness for vines   6, 62–5, 71, 107, 158–9, 251
  Vine growth and phenological responses to   9, 11, 51, 62–5, 152, 184, 241
  19°C cut-off   62–3, 65, 70, 152, 183, 256
  Latitude, diurnal range and site adjustments to   40, 44, 69–71, 130–1, 186, 191, 233, 240, 263–4, 282–5
  Summations, general   5, 7, 18, 21–6, 40, 62–7, 70, 153, 184, 192, 246, 284–5
  Growing period   21–6, 51–2, 70, 131, 184, 195, 216, 246, 255–7, 264, 267–76, 279–81, 284–5
  Ripening period   8–11, 17–20, 31–2, 37, 51, 68, 71, 107, 110, 125–6, 129, 151–3, 156–7, 179, 184, 186–7, 192, 195, 209, 214, 225, 231–2, 240–1, 246, 276–7
  Effect on fruitfulness   11–3, 269

Adaptation of vine varieties to   12, 18

Interactions with light   51–2

Sea-level isotherms   73, 82, 105, 125, 152, 184–6, 271

Thermometer records   262–4, 266, 281–3

Forecast for 21st century   260, 267, 269, 271, 282

Thermal zone of slopes   40–1, 44, 81

Topography, general   40–5, 52–3, 82, 105–9, 129–32, 211, 246–54, 264, 272–6, 283

Trace elements, as nutrients   38, 53–4, 57–8, 277

Trellising: see Vine trellising and training

Urban warming   71, 106, 108, 124, 185, 245, 263–4, 282

Urethane: see Ethyl carbamate

Véraison   13–4, 19, 21, 28, 31, 52, 68

Vines:

Growth habit and dwarfing   11, 277, 279–80

Heading height   46–7

Rootstocks and grafting   60, 73, 265–6, 277, 279

Root systems   28, 34, 36–7, 53–4, 58, 60, 279–80

Spacing within and between rows   48–9

Orientation of rows   46–8

Trellising and training (see also Canopy management)   46–52, 54–5, 59–61, 75–6, 109

Vigour   12, 18, 48–51, 54–5, 60–1, 76, 196, 251, 269, 277, 279–80

Fruitfulness   12–3, 18, 20, 27, 46, 52, 183, 187, 268–9, 277, 279–80

Balance between vigour and fruitfulness   12, 18, 36, 48, 57, 60–1, 74, 108–9, 268–9, 275, 277

Phenology   6, 62–8, 71, 106, 155–6, 183–4, 256, 264–7

Winter dormancy   17, 77–8, 195, 277

Winter killing   216, 220, 253–4, 258–9

Water relations   7, 17, 27–30, 32–4, 38, 52–4, 60–1, 157, 275–7

Nutrition, general   37–40, 53–60, 268–9, 277

Age   60, 71

Varieties: see Grape varieties

Breeding   257, 264–5, 273, 277–8

As biological thermometer   71, 264–7

Vineyard management, general   2, 46–62, 109, 256, 265–7, 276–7

Vintage variability   17, 34, 106–7, 132, 255–60, 265–7, 273–4

Viridomology   261–4, 267, 269

Viticulture in future climates   269, 271

'Wahgunyah Doctor', Victoria   45

Waterlogging   28, 34, 38, 54

Weed control   59

Windbreaks   31, 59, 112, 277

Wind damage   30–1, 77–8, 127,183, 187

Windiness   30–1, 74, 77–83, 126–7, 183, 187, 238, 274–7

Wine:

Styles   1–2, 4–5, 8, 17, 21–6, 30–1, 33, 37–8, 61–2, 66–8, 71, 74–81, 104–8, 126–7, 129–32, 151–4, 156–7, 179, 186–7, 195–6, 209–20, 228, 239–40, 246, 252–3, 256–9, 269, 271, 273–6

Quality   1–2, 7, 9, 15, 17–21, 28–38, 40–5, 49–50, 52–3, 55–7, 60–1, 66–8, 110, 129, 151–3, 156, 158, 186–7, 195, 212, 238–9, 246, 251, 252–4, 258–9, 268–77

Temperatures for drinking   10, 68

In commerce and trade   1–2, 257–9

Winemaking technology   1–2, 56–7, 61, 151, 246, 256–8, 277

Winter dormancy: see Vines, winter dormancy

Winter killing see Vines, winter killing

Yield: see Grape yield

TP

TRIVINUM PRESS

www.ingramcontent.com/pod-product-compliance
Lightning Source LLC
Chambersburg PA
CBHW061104210326
41597CB00021B/3971